Multivariable Control

New Concepts and Tools

edited by

SPYROS G. TZAFESTAS
Electrical Engineering Department, University of Patras, Patras, Greece

D. Reidel Publishing Company

A MEMBER OF THE KLUWER ACADEMIC PUBLISHERS GROUP

Dordrecht / Boston / Lancaster

Library of Congress Cataloging in Publication Data

CIP

Main entry under title:

Multivariable control.

Papers from the Measurement and Control Symposium, held in Athens, Greece, Aug. 29–Sept. 2, 1983.

Includes bibliographies and index.

1. Control theory. I. Tzafestas, S. G., 1939- . II. Measurement and Control Symposium (1983 : Athens, Greece)

QA402.3.M847 1984 629.8′312 84-18166

ISBN 90–277–1829–6

Published by D. Reidel Publishing Company
P.O. Box 17, 3300 AA Dordrecht, Holland

Sold and distributed in the U.S.A. and Canada
by Kluwer Academic Publishers,
190 Old Derby Street, Hingham, MA 02043, U.S.A.

In all other countries, sold and distributed
by Kluwer Academic Publishers Group,
P.O. Box 322, 3300 AH Dordrecht, Holland

Printed in The Netherlands

CONTENTS

Chapter 6
DECOMPOSITIONS FOR GENERAL MULTILINEAR SYSTEMS
P.A.S.Veloso

Chapter 7
SIMPLIFICATION OF MODELS FOR STABILITY
ANALYSIS OF LARGE-SCALE SYSTEMS
I.Zambettakis, J.P.Richard and *F.Rotella*

PART II UNCERTAIN SYSTEMS AND ROBUST CONTROL

Chapter 8
REPRESENTATIONS OF UNCERTAINTY AND ROBUSTNESS
TESTS FOR MULTIVARIABLE FEEDBACK SYSTEMS
I.Postlethwaite and *Y.K.Foo*

PART III ALGEBRAIC AND OPTIMAL CONTROLLER DESIGN

Chapter 12
FREQUENCY ASSIGNMENT PROBLEMS IN LINEAR MULTIVARIABLE SYSTEMS:EXTERIOR ALGEBRA AND ALGEBRAIC GEOMETRY METHODS
N.Karcanias and *C.Giannakopoulos*

Chapter 13
ON THE STABLE EXACT MODEL MATCHING AND STABLE MINIMAL DESIGN PROBLEMS
A.I.G.Vardulakis and *N.Karcanias*

Chapter 14
POLE PLACEMENT IN DISCRETE MULTIVARIABLE SYSTEMS BY TWO AND THREE-TERM CONTROLLERS
H. Seraji

Chapter 15
LINEAR QUADRATIC REGULATORS WITH PRESCRIBED EIGENVALUES FOR A FAMILY OF LINEAR SYSTEMS
F. Heger and *P.M.Frank*

Chapter 16
SENSITIVITY REDUCTION OF THE LINEAR QUADRATIC OPTIMAL REGULATOR
C.Verde and *P.M.Frank*

Chapter 17
DESIGN OF LOW-ORDER DELAYED MEASUREMENT OBSERVERS FOR DISCRETE TIME LINEAR SYSTEMS
P.Stavroulakis and *S.G. Tzafestas*

Chapter 18
SINGULAR PERTURBATION METHOD AND RECIPROCAL TRANSFORMATION ON TWO-TIME SCALE SYSTEMS
G. Dauphin-Tanguy, O. Moreigne and *P. Borne*

Chapter 19
COORDINATED DECENTRALIZED CONTROL WITH MULTI-MODEL REPRESENTATION (CODECO)
Z. Binder, A.N. Hagras and *R. Perret*

Chapter 20
DESIGN OF TWO-LEVEL OPTIMAL REGULATORS WITH CONSTRAINED STRUCTURES
M. Mariton, M. Drouin, H. Aboukandil

PART IV MULTIDIMENSIONAL SYSTEMS

Chapter 21
A CANONICAL STATE-SPACE MODEL FOR m-DIMENSIONAL DISCRETE SYSTEMS
N.J.Theodorou and *S.G. Tzafestas*

Chapter 22
EIGENVALUE ASSIGNMENT OF 3-D SYSTEMS
T. Kaczorek

Chapter 23
FEEDBACK DEADBEAT CONTROL OF 2-DIMENSIONAL SYSTEMS
S.G.Tzafestas and *N.J.Theodorou*

Chapter 24
STATE OBSERVERS FOR 2-D AND 3-D SYSTEMS
S.Tzafestas, T.Pimenides and *P.Stavroulakis*

Chapter 25
EIGENVALUE-GENERALIZED EIGENVECTOR ASSIGNMENT USING PID CONTROLLER
A.I.A.Salama

CONTRIBUTORS

ABOUKANDIL, H., Laboratoire Signaux et Systemes, CNRS-ESE, France.

ANTSAKLIS,P., EE Dept.,Notre Dame University, U.S.A.

BINDER,Z., Laboratoire Automatique Grenoble, France.

BORNE, P., Laboratoire AII, IDN, Lille France.

*CHRISTODOULAKIS,N.,*Engineering Dept.,Cambridge University, U.K.

DANIEL, R., Engineering Sci.Dept.,Oxford University,U.K.

DAUPHIN-TANGUY,G., Laboratoire AII,IDN,Lille, France.

*DOYLE,J.,*Honeywell Inc.,Minneapolis, U.S.A.

*DROUIN,M.,*Laboratoire Signaux et Systemes, CNRS-ESE,France.

*FOO,Y.,*Engineering Sci. Dept.,Oxford University, U.K.

FRANK,P., Mess-und Regelungstechnik, Duisburg Univ.,FRG.

*GIANNAKOPOULOS,C.,*Control Eng.Centre,City University,U.K.

*HEGER,F.,*Mess-und Regelungstechnik,Duisburg Univ.,FRG.

*INCERTIS,F.,*IBM Scientific Centre, Madrid, Spain.

*KACZOREK,T.,*Electrical Eng.Faculty, Warsaw Tech.Univ.,Poland.

*KALOGEROPOULOS,G.,*Control Eng.Centre,City University,U.K.

*KARCANIAS,N.,*Control Eng.Centre,City University, U.K.

*KOUVARITAKIS,B.,*Engineering Sci.Dept.,Oxford University,U.K.

MARITON,M., Laboratoire Signaux et Systemes,CNRS-ESE,France.

MOREIGNE, O., Laboratoire AII, IDN, Lille, France.

PERRET, R., Laboratoire Automatique Grenoble, France.

PIMENIDES, T., Control Systems Lab., Patras University, Greece.

POSTLETHWAITE, I., Engineering Sci. Dept., Oxford University, U.K.

PUGH, A., Mathematics Dept., Loughborough Tech. Univ., U.K.

RICHARD, J., Lab. Systematique, USTL-UER d'IEEA, France.

ROTELLA, F., Lab. Systematique, USTL-UER d'IEEA, France.

SAFONOV, M., Eng. Systems Dept., Southern California Univ., U.S.A.

SAIN, M., EE Dept., Notre Dame University, U.S.A.

SERAJI, H., EE Dept., Tehran Tech. University, Iran.

SMITH, M., DFVLR, Oberpfaffenhofen, Wessling, FRG.

STAVROULAKIS, P., Mediterranean College, Athens, Greece.

THEODOROU, N., Airforce Research and Tech. Centre, P. Faliron, Greece.

TZAFESTAS, S., Control Systems Lab., Patras University, Greece.

VARDULAKIS, A., Mathematics Dept., Thessaloniki University, Greece.

VELOSO, A., Informatics Dept., Catholic Univ., RJ, Brazil.

VERDE, C., Mess-und Regelungstechnik, Duisburg Univ., FRG.

ZAMBETTAKIS, I., Lab. Systematique, USTL-UER d'IEEA, France.

PREFACE

The foundation of linear systems theory goes back to Newton and has been followed over the years by many improvements such as linear operator theory, Laplace Transformation etc. After the World War II, feedback control theory has shown a rapid development, and standard elegant analysis and synthesis techniques have been discovered by control system workers, such as root-locus (Evans) and frequency response methods (Nyquist, Bode). These permitted a fast and efficient analysis of simple-loop control systems, but in their original "paper-and-pencil" form were not appropriate for multiple-loop high-order systems. The advent of fast digital computers, together with the development of multivariable multi-loop system techniques, have eliminated these difficulties.

Multivariable control theory has followed two main avenues; the optimal control approach, and the algebraic and frequency-domain control approach. An important key concept in the whole multivariable system theory is "observability and controllability" which revealed the exact relationships between transfer functions and the state variable representations. This has given new insight into the phenomenon of "hidden oscillations" and to the transfer function modelling of dynamic systems.

The basic tool in optimal control theory is the celebrated matrix Riccati differential equation which provides the time-varying feedback gains in a linear-quadratic control system cell. Much theory presently exists for the characteristic properties and solution of this Riccati equation.

The basic concepts upon which multivariable frequency methodology is based are the "return ratio" and "return difference" quantities of Bode. These concepts, together with the "system matrix" concept of Rosenbrock, were utilized for developing the multivariable (vectorial) Nyquist and Bode theory as well as the algebraic modal (pole shifting) control theory.

The aim of the present book is to provide a number of recent new concept and tools in multivariable control theory through a set of selected contributions written by young and dinstinguished colleagues from several schools. An effort was made to have a well-balanced book. Out of the twenty four chapters, seven are from England, four from France, four from Greece, three from Germany, three from USA and one from Brazil, Iran and Poland. The contributors of the volume were participated at the Measurement and Control Symposium (MECO '83) in Athens (August 29 - September 2, 1983), and have been invited to write extended and enriched versions of the papers they have presented at this symposium.

I am really indeb ted to all of them for their enthusiastic response and their effort to provide excellent up-to-date contributions. Of course this volume is not intended to cover introductory aspects of multivariable control. Some background on linear system theory (transfer function and state-space approaches) is required for reading the book.

It is my hope that the book, which is much more than just a volume of Symposium Proceedings, will play a significant role in the progress and development of multivariable control theory, opening new perspectives of concepts and techniques for the reader.

MAY 1984 **SPYROS G. TZAFESTAS**

EDITORIAL

The book is divided in four parts; *Part I:* General Topics, *Part II:* Uncertain Systems and Robust Control, *Part III:* Algebraic and Optimal Controller Design, and *Part IV:* Multidimensional Systems.

Part I contains seven chapters dealing with general concepts and techniques of multivariable control theory. Chapter 1, by *Smith*, provides a unified approach to multivariable Nyquist and root-locus theory using the tools of algebraic functions and Riemann surfaces. In Chapter 2, *Giannakopoulos*, *Karcanias* and *Kalogeropoulos* give a survey of results concerning the theory of polynomial combinants $f(s,P,k)$, such as "coprimeness" and "almost coprimeness" of a polynomial set P, the concept of almost zero of P etc. *Pugh* in Chapter 3 examines the question of whether or not a closed-loop system under constant output feedback is proper and establishes the relevant conditions. *Incertis* in Chapter 4 provides some new representations of the continuous-time algebraic Riccati and Lyapunov equations, as matrix equations in the field of orthogonal matrices. The "algebraic Riccati generator" concept is formulated and used to derive new bounds for the solution of Riccati and Lyapunov equations. In Chapter 5, *Antsaklis* and *Sain*, derive some parameterizations of feedback controllers in a unifying manner using internal descriptions of polynomial matrices. The causality and hidden modes design issues are clarified using these parameterizations. Chapter 6, by *Veloso*, deals with composition and decomposition of time-varying multilinear systems. An important new way of obtaining multilinear systems from linear component systems is presented. The principal decomposition result is that any multilinear system can be decomposed into parallel linear systems interconnected by a memoryless tensor-product system followed by a linear system. Part I closes with a stability analysis , by *Zambettakis*, *Richard* and *Rotella*, that allows one to define a reduced-order model of a large-scale Lurie-Postnikov system. The stability conditions derived can be easily checked and can easily determine an admissible sector for variations of the nonlinear static gain.

Part II is devoted to the study of uncertain systems and the derivation of robust controllers. In Chapter 8, *Postlethwaite* and *Foo* examine various ways for representing uncertainty in control systems. Three kind of robustness test are given based on adequate representations of uncertainty. The first test is based on Nyquist stability criterion, the second on an inverse Nyquist criterion, and the third is a combination of the first two tests. In Chapter 9, *Daniel* and *Kouvaritakis* examine the characteristic locus stability method in the presence of uncertain parameters and extend the existing results to the case of multiplicative perturbations. The robustness properties of two alternative control schemes are tested on an open-loop unstable model. In Chapter 10, *Christodoulakis* develops a technique for designing a dynamic compensator giving "basically non-interactive" closed-loop performance for more than one model, representing the same system. The idea of this technique is to first construct a nominal diagonal model and then consider all other models as deviations around it. The final chapter of Part II is by *Safonov* and *Doyle* and deals with the problem of minimizing conservativeness of robustness singular values. It is shown that the function $\bar{\sigma}^2(DMD^{-1})$ is convex in D for any diagonal scaling matrix D=diag $(d_1,\ldots,d_n)$, which guarantees the convergence of algorithms for minimizing $\bar{\sigma}^2(DMD^{-1})$. The result finds application to stability margin investigation for multivariable control systems with structured uncertainty.

Part III involves chapters 12 to 20 dealing with design problems of optimal and algebraic (eigenvalue assignment, model matching etc), controllers. *Karcanias* and *Giannakopoulos*, in Chapter 12, examine the determinantal assignment problem (DAP) and reduce its multilinear structure to a linear problem of decomposing multivectors. They provide necessary conditions for the solvability of frequency assignment problems using the Plücker matrices which constitute a class of new system invariants. The resulting "reduced quadratic Plücker relations" are used for the study of linearizing compensators, and a new proof of the pole assignment theorem is given. *Vardulakis* and *Karcanias* in Chapter 13, present a survey of various results concerning the module structure of the set M* of all *proper* rational vectors having no poles inside a *forbidden* region Ω of the finite complex plane and contained in a given rational vector space T(s). In the light of these results, necessary and sufficient conditions for the solvability of the "stable exact model matching problem" are derived, and the minimal design and stable minimal design problems are studied. In Chapter 14, *Seraji* presents simple methods for the design of multivariable discrete-time PI,PD and PID pole assignment controllers. The PID and PI controllers ensure that in the steady-state the outputs of the system follow step reference inputs in the presence of inaccessible constant disturbances and parameter variations. In Chapter 15, *Heger* and *Frank* propose a new design procedure based on linear quadratic optimal and pole assignment control. Then this procedure is extended for designing robust control systems, and is applied to the control of an F4E aircraft. In Chapter 16, *Verde* and *Frank* show how sensitivity theory combined with linear optimal regulator theory can be used for designing robust fixed state feedback control for systems with non-infinitesimal parameter variations. It is also shown that a constant output feedback control for the F4E aircraft can be found which maintains the poles in prescribed regions under extreme flight conditions. In Chapter 17, *Stavroulakis* and *Tzafestas* provide a low-order delayed measurement observer for discrete-time systems, that implements the control directly (utilizing delayed measurements) without first estimating the state. It is shown that under certain conditions the dimension of this observer is much lower than the difference n-q of the system order n and the number q of linearly independent outputs. *Dauphin-Tanguy* and *Borne* in Chapter 18 propose a new transformation, called "reciprocal transformation" that inverses the dynamic and frequency behaviour of a system. The fast subsystem part which becomes the slow part of the reciprocal system is then decoupled by means of the singular perturbation techniques. An application to a singular optimal problem is considered, the solution of which is found in a quasi-optimal approximation form. Chapter 19, by *Binder*, *Hagras* and *Perret* consider the control of interconnected large-scale systems and propose a two-level control structure with multi-model representation at each level,which is well adapted for implementation on a distributed microcomputer network. The final chapter of *Part III* is by *Mariton*, *Drouin* and *Aboukandil*, and considers large scale systems that in addition to their multivariable nature possess a geographical distribution. Structure constraints are imposed, and a two level decomposition coordination method is developed. The first level consists of a feedback with constrained structures (typically decentralization). This level uses a local criterion to obtain an initial feedback structure which is then optimized by tuning a small number of parameters. The upper level of the hierarchy has the task of monitoring the static performance of the system and counterbacking unmeasured disturbances via slow feedback.

Finally, *Part IV* contains four chapters on multi-dimensional discrete systems. *Theodorou* and *Tzafestas* in Chapter 21 present a procedure for deriving a canonical state-space model for an (m+1)-D system described by its transfer function, which is based on the assumption that a canonical state-space representation of an m-D system in known. As state space models of 2-D and 3-D systems are available, this procedure permits the construction of state-space models of any dimension. In Chapter 22, *Kaczoreck* presents sufficient conditions for existence of a solution to the eigenvalue control problem of 3-D systems with separable closed-loop characteristic polynomials. He then gives three alternative methods, with corresponding implementation algorithms, for determining the appropriate state feedback gain matrix. *Tzafestas* and *Theodorou* in Chapter 23 consider a deadbeat controller design problem for 2-D systems in two particular cases. In the first, the transfer function of the system is given and a series compensator is found that leads to dead-beat closed-loop performance. In the second, the state-space model of the system is assumed, and the deadbeat controller is determined in static state feedback form. The book ends with a Chapter on 2-D and 3-D observers by *Tzafestas, Pimenides* and *Stavroulakis*. Starting with an examination of some basic 3-D structural characteristics (observability, controllability, transformation to canonical form), some 1-D results on adaptive state observers are extended to hold for the 2-D Roesser model. Then the state observer design problem for 3-D systems is considered, and a procedure is developed for constructing a 3-D observer of triangular form.

Multivariable control system theory has been recognized as a major engineering discipline and has attained considerable maturity and sophistication. From the above brief outline one can see that this book covers a wide range of questions and problems in multivariable control, and provides an important set of major tools and methods for their efficient treatment. The field is still expanding and so the book would hopefully be useful in answering open questions and formulating new ones.

PART I GENERAL TOPICS

CHAPTER 1

APPLICATIONS OF ALGEBRAIC FUNCTION THEORY IN MULTIVARIABLE CONTROL

M. C. Smith
DFVLR
Oberpfaffenhofen
8031 Wessling
Federal Republic of Germany

0. ABSTRACT

A unified approach to multivariable Nyquist and root-locus theory is given using the tools of algebraic functions and Riemann surfaces. The basic results of this area (and a proof of the generalised Nyquist stability criterion) are included together with some recent developments. The treatment is intended to highlight the close relationship between the different concepts and techniques.

1. INTRODUCTION

The aim of this chapter is to show how the theory of algebraic functions has a natural application in multivariable control. Our attention will be focused on the generalisations of the single-loop Nyquist and root-locus techniques to the matrix case. Two main goals in the approach will be (i) to achieve the greatest degree of unification, highlighting the connections between concepts and results which would otherwise appear unrelated; and (ii) to make the development as natural as possible.

A central theme of the chapter is that the Nyquist and root-locus diagrams are just different ways of looking at the same mapping. To move from the Nyquist approach - which studies gain as a function of frequency - to the root-locus approach - which studies (characteristic) frequency as a function of gain - involves an interchange in the role of dependent and independent variable in the characteristic gain-frequency equation. At least in the abstract sense, the existence of a common defining equation shows the equivalence of the two approaches; a fact which has only been appreciated as a result of the multivariable generalisations.

A proof of the generalised Nyquist stability criterion will be given using the tools of algebraic functions and Riemann surfaces. The approach has the advantage that the framework for the root-locus analysis is developed simultaneously, and that the existence of two distinct concepts of poles and zeros (those associated with the

S. G. Tzafestas (ed.), Multivariable Control, 3–26.

algebraic functions and those defined by the transfer-function) is brought into prominence. In the root-locus theory also, the two pole-zero concepts have a fundamental role to play. Here the question of pole-zero "structure" becomes important as a qualitative indicator of high and low gain behaviour. The connection between algebraic function expansions and the coefficients in the algebraic equation will be exploited to show how the two pole-zero structures are related in general and to indicate why they coincide generically.

The expansion of the solutions of an algebraic equation in the form of power series (Puiseaux expansions) is an important concept throughout the chapter. It is the leading exponents in the series which define the algebraic function poles and zeros. On the other hand the orientation of the departure and approach patterns of the root-loci are determined by the leading coefficients in the series and subsequent terms are of interest, for example, in the high gain asymptotic behaviour of the root-loci.

The analytic (meromorphic) character of the solutions of an algebraic equation, which is a consequence of the existence of the Puiseaux series, allows close links to be established between single-loop and multivariable root-locus diagrams. The concept of conformality and the closely related idea of "local multiplicity of mapping" will be used in a systematic comparison of root-locus behaviour in the scalar and matrix cases.

A general perspective on the Nyquist stability criterion will be presented which shows how the result is easily adapted to non-square plants and general one-parameter families of feedback gains. The formulation allows the inverse Nyquist criterion to be easily deduced.

Proofs have been largely omitted from the text unless essential in subsequent development. However, in order to make the chapter self-contained, an introductory discussion in the general theory of algebraic functions and Riemann surfaces is included.

2. PRELIMINARIES

We consider an m x m proper transfer-function matrix G(s) and a real scalar feedback parameter k. Our closed-loop system equations (in the Laplace transform domain) are

$$\begin{aligned} y &= G(s)e \\ e &= u - ky \end{aligned} \tag{2.1}$$

where y (output), e (error) and u (input) are m-dimensional vectors of Laplace transforms. We assume that k is chosen so that $\det[I+kG(s)]$ is not identically zero, otherwise the closed-loop system is not well-defined. Values of k for which this condition fails will be called non-admissible.

In the analysis to follow it will be convenient to represent the transfer-function in the form of a right-coprime matrix-fraction decomposition $G(s) = N(s)D(s)^{-1}$, where N(s) and D(s) are polynomial matrices and detD(s) is a monic polynomial (of degree n, say). The system (2.1) is (asymptotically) stable if and only if

$$\det\left[D(s) + kN(s)\right] = 0 \tag{2.2}$$

has no roots in the closed right-half plane.

The (open-loop) poles of the system are defined as the roots of detD(s) = 0 and the (closed-loop) poles are the roots of (2.2) (counting multiplicities) which agrees with the better known state-space definition. If G(s) (and thus N(s)) is non-singular the zeros are defined to be the roots of detN(s) = 0. The general case is given in section 10.

As the parameter k is increased continuously from 0 up to infinity the closed-loop poles trace out a set of paths in the complex plane. The resulting diagram we call the root-locus diagram (for positive k).

The Nyquist diagram is defined to be the set of loci traced out by the eigenvalues of G(s) as the contour in fig.1 is traversed. The imaginary axis indentations are made at the poles of G(s) and the parameter R is assumed "sufficiently large" - to be precise this means that all open and closed-loop poles in the closed right half plane (open left half plane) are included (excluded) by the contour.

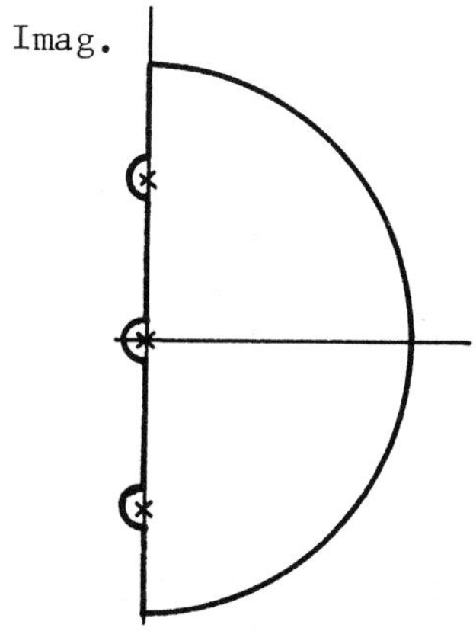

Fig.1 The classical Nyquist D_R-contour. The large semicircle has radius R and the small identations radius 1/R

3. CLASSICAL SINGLE-LOOP THEORY

Nyquist and root-locus diagrams for the case m = 1, are now regarded as part of "classical" control theory. Their use in analysis and design of control systems is widespread. Mathematically we can regard the diagrams as being defined by the single complex function G(s). The Nyquist diagram is, of course, the image of the D_R-contour under G(s). The root-locus diagram can be thought of as the inverse image of the negative real axis under G(s), since we are looking for solutions of the equation G(s) = -1/k. Thus the Nyquist diagram lives in the range of the function and the root-locus diagram in the domain. Fig.2 shows the qualitative situation for a transfer-function with three left half plane poles. The mapping G(s) is easily seen to be 3 → 1 and so the negative real axis has three inverse images - the three branches of the root-locus.

The Nyquist and root-locus diagrams each carry the same information with regard to stability. In the former case, the curves are parameterised by k and the ranges of gain for which all poles are in the open left half plane can be obtained from the diagram by inspection. For the latter case the stability criterion is

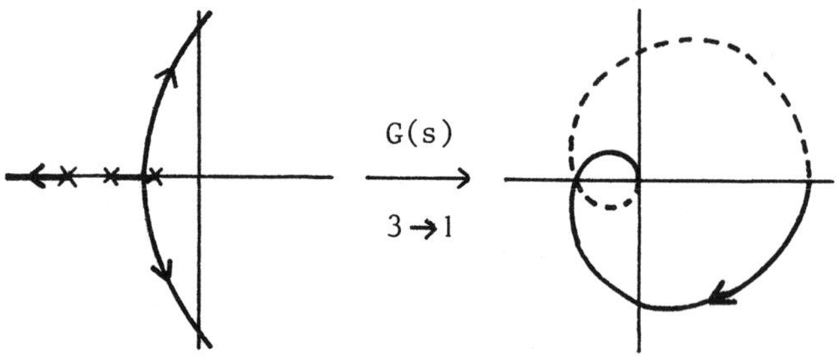

Fig.2 The Nyquist and root-locus diagrams for a 3-pole G(s)

Theorem 1 (Nyquist stability criterion m = 1)

Suppose G(s) has P poles in the closed right half plane. The system (2.1) is asymptotically stable if and only if the number of anticlockwise encirclements of $-1/k$ by the Nyquist diagram of G(s) is equal to P.

In the way we have approached the theory, the proof of theorem 1 follows straight away from a result of complex variable theory - the argument principle [12]. The result says: "given a region in the complex plane with a closed boundary curve orientated so that the region lies (locally) to the right of the boundary and given a complex function meromorphic in the region (with no poles or zeros on the boundary) then the number of anticlockwise encirclements about the origin of the complex plane by the image of the boundary curve equals the number of poles of the function in the region minus the number of zeros".

The region and boundary curve in our case have been given in section 2 (fig.1) and the relevant function is

$$1 + kG(s) = \frac{D(s) + kN(s)}{D(s)} \tag{3.1}$$

The poles and zeros of the argument principle are in this case the zeros of D(s) and D(s)+kN(s) respectively. Both for encirclement to be defined and for asymptotic stability D(s)+kN(s) must have no zeros on the imaginary axis. A direct application of the argument principle now gives

the number of anticlockwise encirclements of the Nyquist diagram of 1+kG(s) about the origin	=	the number of open-loop poles in the right half plane	-	the number of closed-loop poles in the right half plane

and theorem 1 follows immediately. (Note that with the indentations as in fig.1, imaginary axis poles of G(s) are counted as being inside the contour).

4. THE GENERALISED NYQUIST STABILITY CRITERION

A well-known generalisation of the Nyquist criterion to the case $m > 1$ is obtained from the so-called return-difference matrix

$$F(s) = I + kG(s) = [D(s) + kN(s)]D(s)^{-1} \tag{4.1}$$

whose determinant is

$$\det F(s) = \frac{\det[D(s) + kN(s)]}{\det D(s)} \tag{4.2}$$

For $m > 1$ we may have cancellations between the numerator and denominator in (4.2), but in any case an application of the argument principle gives

Theorem 2

Suppose G(s) has P poles in the closed right half plane. The system (2.1) is asymptotically stable if and only if the number of anticlockwise encirclements of the origin by the Nyquist diagram of detF(s) is equal to P.

A problem with this criterion for $m > 1$ is that, unlike the single-loop case, the Nyquist diagram of detF(s) is not linearly dependent on k, so to assess stability a new diagram has to be drawn for each value of the feedback parameter. This problem is alleviated in the generalised Nyquist stability. The criterion says that the Nyquist diagram of G(s) (the loci traced out by the eigenvalues of G(s) as s traverses the D_R-contour) comprises a set of closed curves such that the total number of anticlockwise encirclements about $-1/k$ is equal to P if and only if (2.1) is closed-loop stable. In summary this is

Theorem 3 (The generalised Nyquist stability criterion)

Theorem 1 is true for all m.

A partial proof of this criterion can be obtained from theorem 2 as follows. Let the eigenvalues of F(s) be denoted by $\{\rho_j(s) : j=1...m\}$ and the eigenvalues of G(s) by $\{\gamma_j(s) : j = 1...m\}$. Then it follows that $\rho_j(s) = 1 + k\gamma_j(s)$ (providing we correctly order), and we have

$$\det F(s) = \prod_{j=1}^{m} [1 + k\gamma_j(s)]$$

Regarding the number of encirclements as a measure of change in phase, it seems reasonable to say that the net sum of encirclements of the Nyquist diagrams of the $\gamma_j(s)$ about $-1/k$ is equal to the number of encirclements of the Nyquist diagram of $\det F(s)$ about the origin. The latter statement is not entirely correct though because it is possible that the Nyquist diagram of a single $\gamma_j(s)$ does not form a closed curve. It turns out that the eigenloci combine to form a set of closed curves and the normal concept of winding number applies. A proof of theorem 3 along these lines is given in [1]. We will pursue a different approach here developed in [2] which will make use of the theory of algebraic functions and Riemann surfaces. Using the same tools a complementary treatment of the root-locus will also emerge.

5. CHARACTERISTIC GAINS AND FIXED MODES

The characteristic gains (eigenloci, or generalised Nyquist loci) of $G(s)$ are determined by

$$\Delta(s,g) \equiv \det[gI_m - G(s)] = 0 \tag{5.1}$$

which is equal to the following expression on expanding the determinant

$$g^m - \mathrm{tr}(G(s))g^{m-1} + \sum(\text{all 2x2 pr. minors of } G(s))g^{m-2}$$
$$\ldots\ldots\ldots\ldots(-1)^m \det G(s) = 0 \tag{5.2}$$

The m-valued function "g(s)" determined by (5.1) is termed "algebraic" on account of the nature of its defining equation. In general (because the field of rational functions is not algebraically closed) the m values of "g(s)" will not be expressible as rational functions of s. However, the properties and behaviour of the more general algebraic functions is well understood, and we will present some of the basic theory in the next section. Note that when $m = 1$ $g(s) = G(s)$ is a single-valued rational function.

An initial step is usually performed on the algebraic equations to reduce them to being purely polynomial; in (5.2) we could multiply by the least common denominator of all the rational function coefficients of g^i for $i = 0\ldots.m-1$. However in order to keep track of all poles and zeros of $G(s)$ we choose to multiply throughout by the open-loop pole polynomial $\det D(s)$ to give

$$\Gamma(s,g) \equiv \det[gI_m - G(s)]\det D(s)$$
$$= \det[gD(s) - N(s)] \tag{5.3}$$

which is clearly a polynomial in s and g. In fact $\det D(s)$ is always a multiple (and possibly a strict multiple) of the least common denominator of the coefficients in (5.2). In the following $\Gamma(s,g)$ will be taken as the defining equation of the characteristic gains.

$\Gamma(s,g)$ also has a second interpretation as the closed-loop pole polynomial of the system (2.1) corresponding to a feedback of $g = -1/k$ (see (2.2)). For this reason we call $\Gamma(s,g)$ the <u>characteristic gain-frequency equation of $G(s)$</u>. Thus, solving $\Gamma(s,g) = 0$ in g for fixed $s = s_o$ gives the eigenvalues of $G(s)$ and solving as a polynomial in s for fixed $g = g_o$ $(= -1/k_o)$ gives the closed-loop of $G(s)$ under negative feedback $k_o I$. Generally speaking $\Gamma(s,g)$ may not be irreducible in the ring of polynomials in two variables and we can write

$$\Gamma(s,g) = e(s)\Gamma_1(s,g)\ \Gamma_2(s,g)\ldots\ldots\Gamma_t(s,g)\ c(g) \tag{5.4}$$

where $\{\Gamma_i(s,g) : i = 1\ldots\ldots t\}$ are irreducible polynomials and each $\Gamma_i(s,g)$ is dependent on both s and g. The term $e(s)$ is the product of all irreducible factors which are independent of g, and $c(g)$ is the product of all irreducible factors independent of s. $\Gamma_i(s,g)$ is a polynomial whose solutions (either "$g(s)$" or "$s(g)$") are non-constant and this leads to interesting interpretations for the factors $e(s)$ and $c(g)$. We call the roots of $e(s) = 0$ the <u>fixed modes</u>, since these are closed-loop poles for <u>any</u> feedback kI. The roots $c(g) = 0$ will be called the <u>fixed gains</u> since they represent eigenvalues of $G(s)$ which are independent of frequency. (For example, if $G(s)$ is identically singular then zero is a fixed gain, i.e. $c(0) = 0$). If g_o is a real, non-zero fixed gain then there is a further interpretation we can make. Putting $k_o = -1/g_o$ we see that

$$c(-1/k_o) = 0 \Leftrightarrow \det[D(s) + k_o N(s)] \equiv 0$$

$$\Leftrightarrow \det[I + k_o G(s)] \equiv 0$$

which is precisely the condition which makes k_o a "non-admissible" feedback gain (see section 2).

6. ALGEBRAIC FUNCTIONS AND RIEMANN SURFACES

We denote

$$\Gamma(s,g) = a_m(s)g^m + \ldots\ldots + a_o(s) \tag{6.1}$$

For $s = z$, $\Gamma(z,g)$ is a polynomial of degree m unless $a_m(z) = 0$, and we adopt the convention that $\Gamma(z,g)$ has $m\text{-}m^1$ "infinite roots" when $\Gamma(z,g)$ has degree $m^1 < m$. It can be shown [3] that $\Gamma(z,g)$ has <u>distinct</u> roots except when z satisfies the following (non-identically zero) polynomial in s - called the <u>discriminant</u> polynomial

$$0 = \det \begin{bmatrix} a_{m-1} & 2a_{m-2} & 3a_{m-3} & \cdots & ma_o & & \\ 0 & a_{m-1} & 2a_{m-2} & \cdots & (m-1)a_1 & ma_o & \\ & & & \vdots & & & \\ & & a_{m-1} & 2a_{m-2} & \cdots & (m-1)a_1 & ma_o \\ & & ma_m & (m-1)a_{m-1} & \cdots & 2a_2 & a_1 \\ & ma_m & (m-1)a_{m-1} & & & a_1 & 0 \\ & & & \vdots & & & \\ ma_m & (m-1)a_{m-1} & & \cdots & a_1 & & \end{bmatrix} \begin{matrix} \uparrow \\ m-1 \text{ rows} \\ \downarrow \\ \uparrow \\ m-1 \text{ rows} \\ \downarrow \end{matrix} \qquad (6.2)$$

For simplicity we assume for the moment that $\Gamma(s,g)$ is irreducible. A standard result of complex variable theory is the <u>implicit function theorem</u> [3] which states that the m roots of (6.1) $g_1(s) \dots g_m(s)$ can be expressed in the form of Taylor series expansions in a neighbourhood of a point z providing

(i) $a_m(z) \neq 0$

(ii) z is not a root of the discriminant polynomial (6.2)

We denote the points satisfying (i) and (ii) by $z_1 \dots z_v$. The m solutions of (6.1) are thus locally analytic functions except (possibly) at a finite number of points in the complex plane. Now let the $z_1 \dots z_v$ be joined together and to infinity by straight lines to give a simply connected region Ω as shown in fig.3. It follows

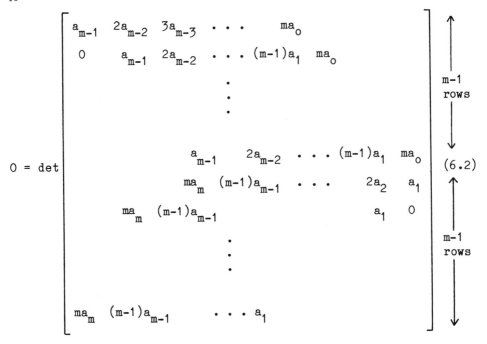

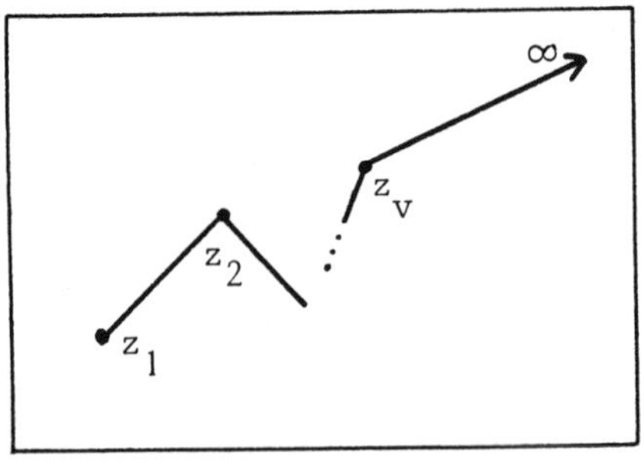

<u>Fig.3</u> The simply connected region Ω

from the <u>Monodromy theorem</u> [4] that in Ω we can define m separate functions $g_1(s) \dots g_m(s)$ satisfying $\Gamma(s,g) = 0$, <u>each of which is</u>

analytic in the whole of Ω. Although in this form the g_i's appear to be unrelated functions, this is far from being the case as we can see by observing the effect of moving across a cut. Consider a point c in Ω near to some to some z_j(= z, say) and a circle centre z passing through c (assume c is sufficiently close to z so that no other z_i with $i \neq j$ is contained in the circle). The functions $g_i(s)$ can now be continued analytically around the circle for a full revolution and back to c. In general such a continuation need not lead one function back to itself (as we can see by considering the "square root" function $g^2 - s = 0$). If after one revolution there is a g_i which has not continued to itself then we call z a branch point. The $g_i(s)$ are always arranged in a number of "cycles" in the neigbourhood of a branch point, for example we could have an r-cycle in which

$$\begin{array}{ccc} g_1(s) & \to & g_2(s) \\ g_2(s) & \to & g_3(s) \\ & \vdots & \\ g_r(s) & \to & g_1(s) \end{array}$$

after one revolution. A further consequence (we omit the details for which the reader is referred to standard texts e.g. [4]) is that the values of $g_1(s), \ldots, g_r(s)$ near z can be found by substituting the r choices of $(s-z)^{1/r}$ into a Laurent series expansion

$$\sum_{k=\alpha}^{\infty} a_k w^k \tag{6.3}$$

Alternatively we can write

$$s - z = w^r \qquad g = \sum_{k=\alpha}^{\infty} a_k w^k \tag{6.4}$$

$(a_\alpha \neq 0)$ which we stress is the general form for the solutions of an algebraic equation $\Gamma(s,g) = 0$. An expansion of the form (6.4) is sometimes called a Puiseaux series. Except at branch points we must have r = 1 in (6.4) and there must be m such series corresponding to a given z, one for each $g_i(s)$. The general form of (6.4) also holds for the point at infinity with the left hand side replaced by $s = w^{-r}$.

It should be noted that at this point we still have a multivalued function since corresponding to each s in the complex plane we have m function values. It is both possible, and very useful, to remove the multivaluedness, and we now indicate how this is done.

The idea is to extend the domain. We consider m copies of Ω which can be thought of as being placed vertically above one another in 3-dimensional space. To each copy of Ω we associate one of the branches $g_i(s)$ and we use the idea of analytic continuation to define a single surface. The association of the various sheets is done in the following way. If a branch $g_i(s)$ (with domain the ith sheet Ω) continues to a branch $g_j(s)$ (with domain the jth sheet Ω) across the cut between the points z_j and z_{j+1}, we "adjoin" the two corresponding edges of the cuts. If a branch continues to itself then the cut is simply removed. The procedure is summarised in fig.4.

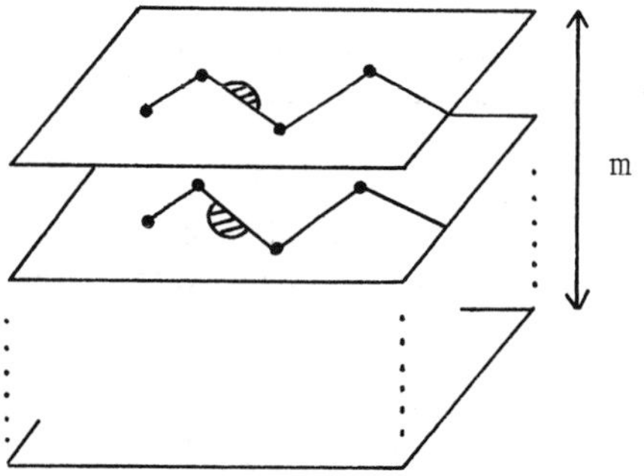

Fig.4 Construction of an m-sheeted Riemann surface

Such a surface (which is called a Riemann surface) cannot physically be constructed in 3-dimensional space because of the problem of self intersections. However, the topology of the surface can be precisely defined and it can be further shown that an irreducible $\Gamma(s,g)$ gives rise to a connected surface [5]. We have now arrived at a situation where a single-valued function $g(s)$ has been defined which represents the totality of the solutions of $\Gamma(s,g) = 0$.

Many results of complex variable theory generalise in a natural way to the algebraic function $g(s)$. One example is the argument principle [6] which we will use in the next section. In order to apply this result we need to know what the poles and zeros of $g(s)$ are. The definition is as follows. If in (6.4) $\alpha < 0$ $(\alpha > 0)$ $g(s)$ has a pole (zero) of multiplicity $|\alpha|$ at z. When r is not equal to 1 this means that the pole or zero is located at an r-cycle branch point i.e. it is shared between r sheets of the surface.

It is sometimes convenient to write the expansions about a point z in the form

$$\begin{aligned} g_1 &= k_1(s-z)^{\alpha_1} + \ldots\ldots \\ g_2 &= k_2(s-z)^{\alpha_2} + \ldots\ldots \\ &\;\;\vdots \\ g_m &= k_m(s-z)^{\alpha_m} + \ldots\ldots \end{aligned} \tag{6.5}$$

(or with (s-z) replaced by 1/s for the point at infinity) where the left hand side in (6.4) is substituted directly into the series expansion (so w is taken to be the principal r^{th} root of (s-z) multiplied by the r choices of the rth root of unity in turn). The α_i and the higher powers need not be integers and, for example, a pole of multiplicity α at an r-cycle now looks like r poles of multiplicity α/r.

7. THE RIEMANN SURFACE PROOF OF THE GENERALISED NYQUIST STABILITY CRITERION

For simplicity of exposition we continue to assume that $\Gamma(s,g)$ is irreducible. The general case has no extra difficulties, a detailed version of which can be found in [7].

The proof to be outlined here makes use of the argument principle on a Riemann surface [6]. This says "given a region on a Riemann surface whose boundary consists of a number closed curves (orientated so that the region lies locally to the right of the boundary), and given a function meromorphic on the surface (with no poles or zeros on the boundary), then the number of anticlockwise encirclements about the origin of the complex plane by the image of the boundary curve(s) equals the number of poles of function in the region minus the number of zeros."

The concept of a meromorphic function on a Riemann surface will not be defined here although it is true that the algebraic function g(s) always satisfies the conditions. Note however that the argument principle on the Riemann surface counts the poles and zeros of the algebraic function g(s) (see section 6), whereas to assess stability of the system (2.1) we need to know about the closed-loop poles of the transfer-function. The connection is given by the following result [7]

Lemma

Let $\Gamma(s,g) = a_m(s)g^m + \ldots + a_o(s)$ and let the expansions about $s = z$ be written in the form (6.5). Then

$$-\sum_{\alpha_i < o} \alpha_i = \text{no. of zeros of } a_m(s) \text{ at } z$$

$$\sum_{\alpha_i > o} \alpha_i = \text{no. of zeros of } a_o(s) \text{ at } z$$

The region of concern in the generalised Nyquist stability criterion is the projection of the region in fig.1 onto each sheet of the Riemann surface. We notice that if a point lying over some z on one particular sheet is contained in the region, then the points on all sheets over z are contained in the region. Thus it is enough that the lemma only gives the sum of the pole (zero) multiplicities over all expansions (i.e. over all sheets of the surface). It can also be seen that the boundary consists of a finite number of closed curves.

We apply the argument principle not to the algebraic function g(s), but to f(s) = 1 + kg(s), which is an algebraic function whose domain is the same Riemann surface as g(s). f(s) is the solution of the equation

$$\begin{aligned} k^m\Gamma(s,(f-1)/k) &\equiv \det[(f-1)D(s) - kN(s)] \\ &= \det[fD(s) - (D(s) + kN(s))] \\ &= b_m(s)f^m + \ldots\ldots + b_o(s) \end{aligned}$$

Notice that $b_m(s) = \det D(s)$ = open-loop pole polynomial, and $b_o(s) = (-1)^m \det[D(s) + kN(s)]$ = closed-loop pole polynomial. The image of our boundary curves under 1 + kg(s) is precisely the generalised Nyquist diagram of I + kG(s). The lemma together with the argument principle now show that "the number of anticlockwise encirclements of the origin by the generalised Nyquist diagram of I + kG(s) is equal to the number of open-loop poles minus the number of closed-loop poles in the right half plane" from which theorem 3 follows.

8. THE INVERSE NYQUIST STABILITY CRITERION AND ONE-PARAMETER FAMILIES OF FEEDBACK GAINS

It is very instructive to characterise theorem 3 in the following general way [8]. Consider an arbitrary polynominal in two variables s and g

$$\Gamma(s,g) \equiv a_m(s)g^m + \ldots\ldots + a_o(s) \tag{8.1}$$

A set of curves C(Γ) can be defined from Γ(s,g) by evaluating the roots of Γ(s,g) pointwise around the Nyquist D_R-contour, where the imaginary axis indentations are made at the roots of $a_m(s)$. Theorem 3 amounts to "the number of encirclements of C(Γ) about some point z is equal to the number of zeros of $a_m(s)$ in the right half plane minus the number of zeros of Γ(s,z) in the right half plane". In this form we can easily deduce a number of related results. For a non-singular G(s), the inverse Nyquist loci are determined by evaluating the roots of

$$\det[gI - G^{-1}(s)] = 0$$

pointwise around D_R (with the imaginary axis indentations at the zeros of G(s)). However, if we put Γ(s,g) = det[gN(s) - D(s)] then the inverse Nyquist diagram is precisely $C(\Gamma)$ as defined above and $\Gamma(s,-k)$ is the closed-loop pole polynomial of (2.1). We immediately get

Theorem 4 (The inverse Nyquist stability criterion)
Suppose $G(s)$ has Z zeros in the closed right half plane. The system (2.1) is asymptotically stable if and only if the number of anticlockwise encirclements of $-k$ by the inverse Nyquist diagram of $G(s)$ is equal to Z.

The interpretation given above for the generalised Nyquist criterion indicates the we really have a graphical stability test for a 1-parameter family of feedback gains. Consider a system

$$\begin{aligned} y &= G(s)e \\ e &= u - K(\lambda)y \end{aligned} \tag{8.2}$$

where $G(s)$ is not assumed "square" and $K(\lambda)$ is a rational matrix of appropriate dimension with $K(0) = 0$ (i.e. $\lambda = 0$ corresponds to "open-loop"). For λ not a "pole" of $K(\lambda)$ the closed-loop pole polynomial of the system is

$$\det[D(s) + K(\lambda)N(s)] = 0 \tag{8.3}$$

If we now form a left-coprime matrix-fraction decomposition

$$K(-g^{-1}) = Q^{-1}(g)P(g) \tag{8.4}$$

we can define

$$\Gamma(s,g) \equiv \det[Q(g)D(s) + P(g)N(s)] \tag{8.5}$$

Since $K(0) = 0$ then $K(-g^{-1})$ is strictly proper and it can be shown further that the coefficient of the highest power of g in (8.5) is a scalar multiple of $\det D(s)$. Thus

Theorem 5
Let $G(s)$ have P poles in the closed right half plane. The system (8.2) is asymptotically stable with feedback $K(k)$ if and only if the number of anticlockwise encirclements of $-1/k$ by $C(\Gamma)$ (as defined by (8.5)) is equal to P.

This result can easily be extended to the case where $K(0) \neq 0$.

9. THE MULTIVARIABLE ROOT-LOCUS

When $m = 1$ the principal features of the root-locus diagram can be encapsulated in a few simple rules, which enable the diagrams (for relatively simple systems) to be quickly and easily sketched. If the transfer-function takes the form

$$G(s) = \alpha \frac{\prod_{i=1}^{d} (s-z_i)}{\prod_{i=1}^{n} (s-p_i)} \qquad (d \leqslant n) \qquad (9.1)$$

and $\alpha > 0$ then the root-locus <u>for positive k</u> has the following properti-es

(i) the n branches start at the n open-loop poles

(ii) d branches tend to the zeros as $k \to \infty$ and (if $d < n$) n-d branches tend to infinity

(iii) poles cannot pass through the same point for different values of k

(iv) points on the real axis lie on the root-locus if and only if they lie to the left of an odd number of poles and zeros

(v) the breakaway points s_k (the points where two or more poles are coincident) are determined by

$$\frac{dG}{ds}(s_k) = 0 \qquad G(s_k) < 0$$

(vi) if we write $G(s) = c(s-z)^r + \ldots\ldots$ ($G(s) = c^{-1}(s-z)^{-r} + \ldots$) for z a zero (pole) of order r, then the r closed-loop poles which tend to z as $k \to \infty$ (depart from z as k is increased from zero) are asymptotic to r straight lines at angles satisfying

$$\theta = \text{phase } (-c^*)^{1/r}$$

where * denotes complex conjugation.

(vii) If $d < n$, n-d branches satisfy an equation of the form

$$s(k) \sim c_1 k^{1/(n-d)} + c_2 \qquad (9.2)$$

as $k \to \infty$, where c_1 and c_2 are real constants. In fact

$$(n-d)c_2 = \sum_{i=1}^{n} p_i - \sum_{i=1}^{d} z_i$$

In the root-locus for negative k, (iv) - (vi) only are modified. Note that the constant multiplier α in (9.1) does not affect the shape of the root-locus diagram, only its calibration in k.

When $m > 1$ (i) remains true as does (ii) if G(s) has full (normal) rank. If G(s) loses rank them (ii) can fail. (iii) is in general untrue as the following example shows

$$G(s) = \frac{1}{s+1}\begin{bmatrix} -2 & s+1 \\ -1 & 0 \end{bmatrix} \tag{9.3}$$

which has n=1 and d=0. The closed-loop poles are determined by

$$s + (1-k)^2 = 0$$

and the root-locus (for positive k) behaves as follows. As k is increased from 0 to 1 the single pole moves from s = -1 to 0. As k is further increased the pole moves off to infinity along the <u>negative</u> real axis. Thus the single branch of the root-locus "doubles back" on itself when k = 1, and any point between -1 and 0 is associated with two different values of feedback gain.

This behaviour can be seen in a different light though if we look at the Riemann surface. Once again we assume for the moment that $\Gamma(s,g)$ is irreducible and so there is a (connected) m-sheeted surface $\mathcal{R}$ making g(s) single-valued. Now we can see that the totality of points on $\mathcal{R}$ which are sent to a single complex number g_0 (say) are in fact the n solutions of the equation $\Gamma(s,g_0) = 0$. Thus for a feedback gain of $k = k_0 = -1/g_0$ we can find the n points on $\mathcal{R}$ which are the closed-loop poles corresponding to this feedback. We can evidently think of the root-locus as living on the Riemann surface and being the inverse image under g(s) of the negative real axis. Since g(s) is a single-valued function on the Riemann surface it is not possible for two portions of the root-locus to intersect unless they are mapped to the same complex number by g(s), that is, unless they correspond to the same value of feedback gain. In this way the "non-intersecting property" (iii) of the classical root-locus extends to the case $m > 1$.

In the above example the characteristic gain-frequency equation is

$$(s + 1)g^2 + 2g + 1 = 0$$

The Riemann surface is 2-sheeted and there are just two branch points at 0 and ∞ (see fig.5).

When $\Gamma(s,g)$ has the general factorisation (5.4) the non-intersecting property can still be restored in exactly the same way as above. The root-locus will live on m copies of the complex plane which we call the <u>extended domain</u>. The only difference from the irreducible case is that the extended domain will not be a single connected

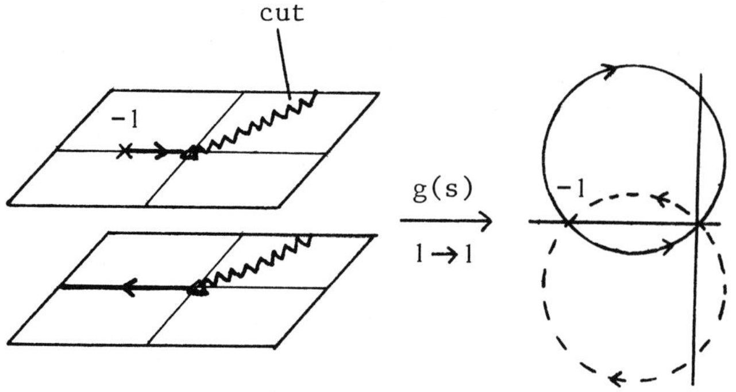

Fig.5 The Nyquist and root-locus diagrams of the system (9.3)

Riemann surface but several distinct surfaces, and in the extreme case when $t = m$, m 1-sheeted surfaces (i.e. complex planes).

Evidently a closed-loop pole may lie over a point z on the real axis for any of 0, 1, 2 ..... m different values of (positive) k, since this is precisely the number of distinct negative roots of $\Gamma(z,g)$. Rules can be derived [9] to compute this number although the complexity increases with m. For $m = 2$

$$\det[gI - G(s)] = g^2 - \mathrm{tr}G(s)g + \det G(s)$$

and the discriminant polynomial is $\mathrm{dsc}(s) = [\mathrm{tr}G(s)]^2 - 4\det G(s)$. The number of roots at a point on the real axis for positive k (pos.k) and negative k (neg.k) are determined as follows

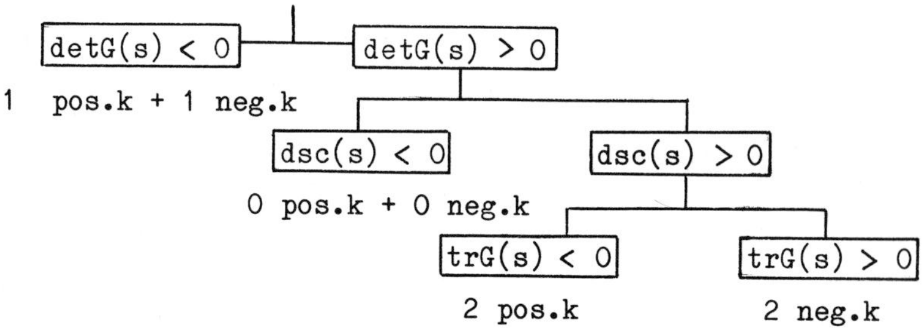

For the case $m = 1$, the breakaway points of the root-locus are the points in the complex plane where (for positive k) two or more poles are coincident. Equivalently, these are the places on the root-locus diagram where $G(s)$ is not locally one-to-one. $G(s)$ fails to be one-to-one precisely at the non-conformal points of $G(s)$ which are determined by $dG/ds = 0$ (see (v) above). When $m > 1$ it is not satisfactory to define the breakaways simply as points of coincidence of closed-loop poles. For example $G(s) = \mathrm{diag}\,\{2/(s + 1), 1/(s + 2)\}$ has both poles at -3 for $k = 1$ but clearly there is nothing to resemble a "breakaway". It appears to be most natural to define the breakaway points as those places in the extended domain where two poles come together. These are the non-conformal points of the algebraic functions $g_i(s)$ for i = 1 ..... t on their respective domains (see (5.4)). We will not define the concept of conformality on a Riemann surface, but suffice it to say that the concept is equivalent to (a) the mapping losing its local one-to-one property and (b) (providing we are not at a branch point) dg/ds = 0 in the sense of ordinary complex variable theory.

Denote the discriminant of $\Gamma(s,g)$ regarding Γ as a polynomial in g (resp. s) whose coefficients are functions of s (resp. g) by $D_g(s)$ (resp. $D_s(g)$) and assume that neither polynomial is identically zero - if the latter condition fails we could proceed as below with each irreducible factor in (5.4) in turn. Let $s_1 \ldots\ldots s_p$ be the roots of $D_g(s) = 0$ and $g_1 \ldots\ldots g_p$ the corresponding repeated g-values (an s-value may occur more than once with different g-values). Similarly $g^1 \ldots\ldots g^q$ are the roots of $D_s(g) = 0$ and $s^1 \ldots\ldots s^q$ the corresponding repeated s-values. We form the following list

$$\begin{array}{cc@{\qquad\qquad}cc} \multicolumn{2}{c}{D_g(s)} & \multicolumn{2}{c}{D_s(g)} \\ \hline s_1 & g_1 & s^1 & g^1 \\ \cdot & & \cdot & \\ \cdot & & \cdot & \\ s_p & g_p & s^q & g^q \end{array} \tag{9.4}$$

and we note that the pairs in the left (right) hand column are the solutions of $\Gamma(s,g) = 0$ which also satisfy $\partial\Gamma(s,g)/\partial g = 0$ ($\partial\Gamma(s,g)/\partial s = 0$). We have already seen that the branch points are a subset of $\{s_1 \ldots\ldots s_p\}$. It is also true that the non-conformal points are a subset of $\{s^1 \ldots\ldots s^q\}$. However, if we take a total derivative of $\Gamma(s,g) = 0$ with respect to s we get

$$\frac{\partial g}{\partial s} \cdot \frac{\partial}{\partial g}\Gamma(s,g) + \frac{\partial}{\partial s}\Gamma(s,g) = 0$$

It follows that if (s^i, g^i) is not an element of $\{(s_j, g_j)\ j = 1 \ldots p\}$ then s^i is a non-conformal point (i.e. breakaway point) in the extended domain.

10. POLES AND ZEROS

Rule (vi) in section 9 shows that the multiplicity r of the pole (or zero) in a single-loop transfer-function is enough to characterise the behaviour of the r closed-loop poles in the neighbourhood of the point for low (resp. high) k. To be precise, the closed-loop poles are asymptotic to a set of r equiangular straight lines - we will call this configuration a Butterworth pattern. It is easy to see that this characterisation extends to the point infinity if we define the local behaviour to be that at $w = 0$ after the transformation $s = 1/w$. How does this situation generalise when $m > 1$?

The rational matrix $G(s)$ can be reduced by pre- and post-multiplication with unimodular matrices to its (unique) Smith-McMillan form

$$\mathrm{diag}\left\{ \frac{\varepsilon_1(s)}{\psi_1(s)}, \frac{\varepsilon_2(s)}{\psi_2(s)} \;\ldots\; \frac{\varepsilon_r(s)}{\psi_r(s)}, 0 \ldots 0\right\} \tag{10.1}$$

where ε_i, ψ_i are monic coprime polynomials, $\varepsilon_1|\varepsilon_2 \ldots|\varepsilon_r$ and $\psi_r|\psi_{r-1} \ldots|\psi_1$. It can be shown that $\det D(s) = \psi_1(s)\,\psi_2(s) \ldots \psi_r(s)$ and, when $r = m$, $\det N(s) = \varepsilon_1(s)\,\varepsilon_2(s) \ldots \varepsilon_m(s) \times (\text{const})$. In general we define the zeros of $G(s)$ to be the roots of the polynomial $\varepsilon_1(s)\,\varepsilon_2(s) \ldots \varepsilon_r(s)$. For the rest of the section though we assume $r = m$.

It can be seen from the following simple examples: $G(s) = \mathrm{diag}\{1/s^3, 1/s\}$ and $G(s) = \mathrm{diag}\{1/s^2, 1/s^2\}$ that the multiplicity of the zeros of $\det D(s)$ is alone insufficient to characterise the root-locus behaviour when $m > 1$. We have respectively a third and a first order Butterworth pattern and two second order patterns. The form (10.1) allows a finer (algebraic) structure to be defined; for each complex number z we say

$$\mu_i(z) = \text{no. of zeros of } \varepsilon_i(s) \text{ at } z - \text{no. of zeros of } \psi_i(s) \text{ at } z \tag{10.2}$$

for $i = 1 \ldots m$, and we note that $\mu_1 \leq \mu_2 \ldots \leq \mu_m$. The integers μ_i will be called the Smith-McMillan indices at z, and the definition will be extended to infinity by defining $\mu_1(\infty) \ldots \mu_m(\infty)$ to be the indices of $G(1/w)$ at $w = 0$. Clearly the sum over all positive (negative) μ_i's at z is (minus) the total number of zeros (poles) at z.

In the above examples (which are already in Smith-McMillan form) we have $\mu_1(0) = -3$, $\mu_2(0) = -1$ and $\mu_1(0) = \mu_2(0) = -2$ respectively and the Smith-McMillan indices characterise the Butterworth patterns in the way we want. Unfortunately this is not always true as is shown by

$$G(s) = \begin{bmatrix} 0 & \frac{1}{s} \\ \frac{1}{s^2} & 0 \end{bmatrix}$$

Here $\mu_1(0) = -2$ and $\mu_2(0) = -1$ while we can check that the roots are arranged in a Butterworth pattern of order 3 (with a difference though since $s(k)^{3/2} \sim k$).

In (6.5) we assume without loss of generality that $\alpha_1(z) \leqslant \ldots \leqslant \alpha_m(z)$. It can be shown [10] that the α_i precisely characterise the root-locus Butterworth patterns, and we thus call these numbers the <u>root-locus indices</u>. In general, if we let $w = (s-z)$ - or $1/s$ for the point infinity - and precisely tq of the root-locus indices are equal to p/q (where p and q are coprime and t, q > 0) then there are tp closed-loop poles satisfying $w(k) \sim k^{-q/p}$ for large k if p > 0 (small k if p < 0). Moreover, they take the form of t $|p|^{th}$-order Butterworth patterns. We call p/q the <u>rate of approach</u> (departure) of the pattern according as p > 0 (p < 0), since this is the exponent governing the asymptotic behaviour. As an illustration suppose m = 5 and the root-locus indices for some point are -3/2, -3/2, -3/2, -3/2, 2 then (for low gain) we will have two third order departure patterns with rates 3/2 and (for high gain) a second order pattern with rate 2.

The root-locus indices can be calculated from $\Gamma(s,g)$ in specific cases using the <u>Newton diagram</u>. We now describe the procedure since it casts considerable light on the relationship between the Smith-McMillan and root-locus indices.

Select some complex number z. If the coefficient $a_i(s)$ of g^i in $\Gamma(s,g)$ is a non-zero polynomial we mark the following two points

(no. of zeros of $a_i(s)$ at z, i)
(deg $a_i(s)$, i)

on the diagram of fig.6 and the process is repeated for each i = 0 ... m. We now form the convex hull of the points occurring in diagram. The following is a standard result [3]: if a (complete) straight-line segment of the left-hand (right-hand) boundary of the

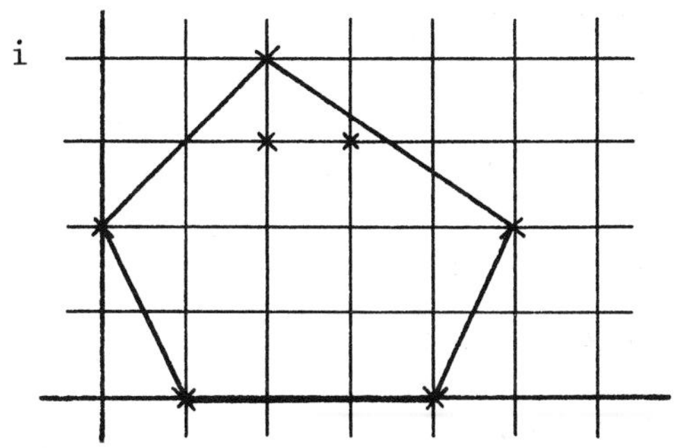

<u>Fig.6</u> Construction of a Newton diagram

Newton diagram has slope $-1/\alpha$ and the vertical displacement between endpoints is p then these are exactly p expansions of the form $g \sim (s-z)^{\alpha}$ ($g \sim s^{-\alpha}$) when written in the form (6.5).

It can be shown that the coefficients in $\Gamma(s,g)$ take the form

$$
\begin{aligned}
a_m(s) &= \psi_1(s)\ \psi_2(s)\ \psi_3(s)\ \dots\ \psi_m(s) \\
a_{m-1}(s) &= \varepsilon_1(s)\ \psi_2(s)\ \psi_3(s)\ \dots\ \psi_m(s)\ h_{m-1}(s) \\
&\ \ \vdots \\
a_1(s) &= \varepsilon_1(s)\ \varepsilon_2(s)\ \dots\ \varepsilon_{m-1}(s)\ \psi_m(s)\ h_1(s) \\
a_o(s) &= \varepsilon_1(s)\ \varepsilon_2(s)\ \dots\ \varepsilon_{m-1}(s)\ \varepsilon_m(s)\ h
\end{aligned}
$$

where $h_1(s)$, $h_2(s)$ ... $h_{m-1}(s)$ are arbitrary (possibly zero) and h is a non-zero constant. Inequality constraints can be established also on the degrees of the coefficients $a_i(s)$ [10]. It follows that $\alpha_i = \mu_i$ for all i at s = z (resp. infinity) if the polynomials $h_1(s) \dots h_{m-1}(s)$ have no zeros at z (resp. achieve their "maximal" degrees).

It is shown in [10] that coincidence of the two sets of indices occurs generically ("almost always") in the following sense. Let K be a constant m x m non-singular matrix. Then KG(s) has the same Smith-McMillan form independent of K. Furthermore, for almost any K the root-locus indices are <u>everywhere</u> the same as the Smith-McMillan indices.

Alternative approaches to the proof of "generic" coincidence of the Smith-McMillan and root-locus structure can be found in [13] and [14].

11. APPROACH AND DEPARTURE ANGLES AND MULTIVARIABLE PIVOTS

The angles of departure (approach) of the closed-loop poles from poles (to zeros) - i.e. the orientation of the Butterworth patterns - can be determined for m = 1 (rule (vii) section 9) by the leading <u>coefficients</u> of the appropriate power series expansion. The method generalises easily to the case $m > 1$.

To be precise, if there is an expansion of the form

$$
\begin{aligned}
g(s) &= c(s-z)^{p/q} + \dots\dots \\
&= c^{-1}(s-z)^{-p/q} + \dots\dots
\end{aligned}
$$

about some complex number z with p and q positive and coprime we can deduce: the straight line asymptotes of the associated q^{th}-order Butterworth pattern are orientated at angles satisfying

$$\theta = \text{phase}\ (-c^{-1})^{q/p} = \text{phase}\ (-c^*)^{q/p}$$

where * denotes complex conjugation. If there is an expansion about infinity of the form $g(s) = cs^{-p/q} \dots$ then the q poles tending to infinity are at asymptotic angles of $\theta = \text{phase}(-c)^{q/p}$.

If a group of poles tending to infinity satisfies an equation of the form (9.2) we call the number c_2 the pivot (in the single-loop case c_2 is called the asymptote intercept). The pivot as well as the constant c_1 in (9.2) may be complex when $m > 1$. It is also possible, as we shall see later in an example, that the poles tending to infinity may not be asymptotic to straight lines (i.e. pivots do not exist). Although pivots may fail to exist even when all the $\alpha_i(\infty)$ are integer, it turns out that existence is generic in the same sense as given at the end of section 10.

If G(s) is strictly proper the structure of the zeros at infinity is usually of especial importance. Let $G(s) = C(sI - A)^{-1}B$ be a minimal state space realisation. We can write

$$G(s) = \frac{CB}{s} + \frac{CAB}{s^2} + \frac{CA^2B}{s^3} + \cdots \tag{11.1}$$

The matrices CB, CAB ... are called the Markov parameters. If CB has rank $r < m$ it is easily shown that $\mu_1(\infty) = \ldots = \mu_r(\infty) = 1 < \mu_{r+1}(\infty)$. Let $s = 1/w$ and consider the equation

$$\det\left[gI - G(\tfrac{1}{w})\right] = \det\left[gI - CBw - CABw^2 \ldots\right] \tag{11.2}$$

near $w = 0$. With the substitution $g = w\lambda + h(w)$ (where $h(w)/w \to 0$ as $w \to 0$) we obtain $\det[\lambda I - CB] = 0$ in the limit as $w \to 0$. If the number of non-zero eigenvalues of CB is $v \leqslant r$ then it follows that $\alpha_1(\infty) = \ldots \alpha_v(\infty) = 1 < \alpha_{v+1}(\infty)$. As we have seen in the last section, it must be true generically that $v = r$ (which is called the simple null structure property because it means that all zero eigenvalues are associated with Jordan blocks of size 1). In any case though the v non-zero eigenvalues of CB give the asymptotic angles of the v closed-loop poles tending to infinity with approach rate 1. In particular if λ is a non-zero eigenvalue of CB, a closed-loop pole will have an asymptotic phase equal to $-\lambda$.

To obtain expressions for the pivots we can make use of the following lemma (which can be verified by direct substitution of one series into the other and balancing coefficients).

Lemma

If $g = d_1 s^{-p} + d_2 s^{-p-1}$ .... satisfies $\Gamma(s,g) = 0$ then there are p corresponding closed-loop poles satisfying $s(k) = c_1 k^{1/p} + c_2 + \ldots$ where $c_1 = (-d_1)^{1/p}$ and $c_2 = d_2/(d_1 p)$.

We proceed to find the next term in the expansion of g(w). Let λ be a non-zero eigenvalue of CB which is distinct from all other eigenvalues and suppose P is a matrix transforming CB to Jordan form

$$P(CB)P^{-1} = \left[\begin{array}{c|c} \lambda & 0 \\ \hline 0 & \diagup\!\!\diagup \end{array}\right] \qquad (11.3)$$

Denote the first row of P by $\underline{u}^t$ and the first column of P^{-1} by $\underline{v}$. Premultiplying the matrix in (11.2) by P and postmultiplying by P^{-1} leads to

$$\det \left[\begin{array}{c|c} \frac{h(w)}{w} - \underline{u}^t(CAB)\underline{v}\, w \ldots & wR_1 \ldots \\ \hline wR_2 \ldots & \frac{h(w)}{w} I + R_3 \ldots \end{array}\right] = 0 \qquad (11.4)$$

where $R_i : i = 1 \ldots 3$ are constant matrices with R_3 non-singular. Using Schur's formula for the evaluation of a partitioned determinant implies

$$\frac{h(w)}{w} - \underline{u}^t(CAB)\underline{v}\, w + \text{higher terms} = 0$$

from which we deduce $g = w\lambda + w^2\underline{u}^t(CAB)\underline{v} + \ldots$ By the lemma there must be a branch of the root-locus satisfying

$$s(k) = (-1/\lambda)k + \frac{\underline{u}^t(CAB)\underline{v}}{\lambda} + \ldots..$$

and the second term is the pivot. We had to assume λ was different from all other eigenvalues in order to be able to use Schur's formula. When the condition fails, pivots may not exist as is shown by

$$G(s) = \frac{1}{s}\begin{bmatrix} 1 & 0 \\ 2 & 1 \end{bmatrix} + \frac{1}{s^2}\begin{bmatrix} 1 & 2 \\ 0 & 1 \end{bmatrix}$$

whose asymptotic poles satisfy $s(k) = -k \pm 2ik^{1/2} + \ldots..$ (Note that the Smith-MacMillan and root-locus structure coincide at infinity).

For the second order and higher cases, expressions involving the Markov parameters for the number of Butterworth patterns, the coefficients governing the orientation and the pivots can be obtained by further applications of Schur's formula [11]. Alternative approaches using singular value decompositions can be found in [14] and [15].

12. ACKNOWLEDGEMENTS

Much of the work of this chapter was done while the author was with the University Engineering Department, Control and Management Systems Division, Cambridge. I would like to thank Prof. A.G.J. MacFarlane for his encouragement and all the members of the control group with whom I have had many fruitful discussions. The author now holds a European Science Exchange Programme Research Fellowship and would like to acknowledge the Royal Society for financial support. I am most grateful to Fr. A. Pittracher for typing the manuscript.

13. REFERENCES

[1] C.A. Desoer and Y. T. Wang, 1980, "On the generalised Nyquist stability criterion", IEEE Trans. on Autom. Contr., **25**, 187-196

[2] A.G.J. MacFarlane and I. Postlethwaite, 1977, "The generalised Nyquist stability criterion and multivariable root loci", Int. J. Contr., **25**, 81-127

[3] G.A. Bliss, 1933, Algebraic Functions, Amer. Math. Soc. Public.

[4] K. Knopp, 1945/7, Theory of Functions (part I/II), Dover

[5] B.F. Jones, 1971, Rudiments of Riemann Surfaces, Rice Univ. Lecture Notes in Mathematics

[6] G. Springer, 1957, Indroduction to Riemann Surfaces, Addison-Wesley

[7] M.C. Smith, 1981, "On the generalised Nyquist stability criterion", Int. J. Contr., **34**, 885-920

[8] P.K. Stevens, 1981, "A generalisation of the Nyquist stability criterion", IEEE Trans. on Autom. Contr., **26**, 664-669

[9] A.E. Yagle and B.C. Levy, 1982, "Multivariable root-loci on the real axis", Int. J. Contr., **35**, 491-507

[10] M.C. Smith, 1983, "Multivariable root-locus behaviour and the relationship to transfer-function pole-zero structure", Cambridge Univ. Eng. Dep. Internal Report TR 235

[11] B. Kouvaritakis and J. M. Edmunds, 1979, "Multivariable root loci: a unified approach to finite and infinite zeros", Int. J. Contr., **29**, 393-428

[12] A.F. Beardon, 1979, Complex Analysis: The Argument Principle in Analysis and Topology, New York: Wiley

[13] C.I. Byrnes and P.K. Stevens, 1982, "The McMillan and Newton polygons of a feedback system and the construction of root loci", Int. J. Contr., **35**, 29-53

[14] Y.S. Hung and A.G.J. MacFarlane, 1981, "On the relationships between the unbounded asymptote behaviour of multivariable root loci, impulse response and infinite zeros", Int. J. Contr., **34**, 31-69

[15] S.S. Sastry and C.A. Desoer, 1983, "Asymptotic Root Loci - Formulas and Computation", IEEE Trans. on Autom. Contr., **28**, 557-568

CHAPTER 2

THE THEORY OF POLYNOMIAL COMBINANTS IN THE CONTEXT OF LINEAR SYSTEMS

C. Giannakopoulos, N. Karcanias and G. Kalogelopoulos
Control Engineering Centre
School of Electrical Engineering and Applied Physics
The City University
London EC1V OHB, U.K.

ABSTRACT

The paper surveys a number of results of the theory of polynomial combinants $f(s,P,\underline{k})$, which are important for linear systems problems. The properties of coprimeness and "almost coprimeness" of a set P of polynomials, generating the combinants $f(s,P,\underline{k})$ are examined. The notion of the almost zero of P is discussed and its importance for the zero distribution properties of $f(s,P,\underline{k})$ is investigated. For the families of "strongly nonassignable" sets P, it is shown that the almost zeros act as "nearly" fixed zeros of all combinants, and disks trapping the zeros of $f(s,P,\underline{k})$, for all parameter vectors $\underline{k}$, are defined. Necessary and sufficient conditions for arbitrary assignment of the zeros of $f(s,P,\underline{k})$ are given; for nonassignable sets P, a sufficient condition for "approximate" zero assignment is discussed. Finally, the notion of a Hurwitzian set P is introduced and a necessary condition for P to be Hurwitzian is given. The surveyed results provide the means for a unifying treatment of frequency assignment and stabilization problems of linear systems, as well as for the interpretation of phenomena related to "almost" pole zero cancellations.

1. INTRODUCTION

Recent work for the unification of the pole, zero assignment problems, defined on a linear multivariable system, has led to the definition of the determinantal assignment problem (DAP), as the common formulation of all these problems [1,5]. The multilinear nature of DAP has motivated its reduction to a linear problem of zero assignment for polynomial combinants and a standard multilinear problem of decomposability of multivectors. Those properties of a set of polynomials, P, which are related to the zero distribution aspects of the corresponding polynomial combinants, are thus of crucial importance in the study of DAP. The aim of this paper is to survey various properties and results on sets of polynomials and associated polynomial combinants which are important for linear control problems.

For a set of polynomials $P = \{p_i(s): p_i(s) \in \mathbb{R}[s],\ i \in \underset{\sim}{m}\}$, the poly-

S. G. Tzafestas (ed.), Multivariable Control, 27–41.

nomial functions $f(s,P,\underline{k}) = \Sigma k_i p_i(s)$, where $k_i \in \mathbb{R}$, $i \in \underset{\sim}{m}$, have been defined as the k-polynomial combinants of P [2]. With a linear system, or linear system problems, we may always associate certain sets of polynomials and polynomial combinants generated by the corresponding sets. Concepts such as those of multivariable zeros [3,4] and decoupling zeros [4] are related to the greatest common divisor of certain sets P, associated with the system and they define fixed zeros of the associated combinants. The pole, zero assignment and stabilizability properties of linear systems are based on properties of corresponding combinants and thus on the structure of sets P, which generate these combinants. The examination of those properties of a set P which affect the assignability, stabilizability and "nearly fixed" zero phenomena of the corresponding combinants $f(s,P,\underline{k})$, is the subject of this survey.

The paper is structured as follows. The extension of the notion of a zero of P to that of an almost zero (AZ) [2] is considered first. The invariance properties under normal equivalence, features of the distribution in the complex plane and computational aspects of the set of AZs of a set P are examined. The role of AZs of P in the distribution properties of the associated combinants $f(s,P,\underline{k})$ is examined next. It is shown that the AZs act as "strong poles of attraction" for the zeros of $f(s,P,\underline{k})$. For the family of "strongly nonassignable" sets P, it is shown that there always exists a disk $\tilde{D}_m[z,\tilde{R}_m(z)]$ centered at an AZ, z, and with finite minimal radius $\tilde{R}_m(z)$, which contains at least one zero of all combinants of P; a criterion for finding upper bounds for $\tilde{R}_m(z)$ is given. It is this property that reveals the AZs of P as "nearly fixed" zeros of all combinants of P. The assignability properties of the zeros of $f(s,P,\underline{k})$ are discussed and necessary and sufficient conditions for arbitrary zero assignment are given. For nonassignable sets P, the problem of "approximate zero assignment" is examined; a sufficient condition for the assignment of the zeros of $f(s,P,\underline{k})$ in the neighbourhoods of a given symmetric set of points of the complex plane is given. Finally, the property of a set P to yield a Hurwitzian combinant $f(s,P,\underline{k})$ for at least one $\underline{k}$ is investigated; a necessary condition for P to be Hurwitzian is derived.

NOTATION: $\mathbb{R},\mathbb{C}$ will denote the fields of real, complex numbers respectively and $\mathbb{R}[s]$ the ring of polynomials with coefficients from $\mathbb{R}$. $\mathbb{R}^n[s]$ denotes the set of n-dimensional vectors with coefficients from $\mathbb{R}[s]$ and $\mathbb{R}^n$, $\mathbb{C}^n$ denote the n-dimensional real, complex vector spaces correspondingly. $\mathbb{R}^{m\times n}$, $\mathbb{C}^{m\times n}$ denote the sets of $m\times n$ matrices with elements from $\mathbb{R},\mathbb{C}$ respectively. Script capital letters denote vector spaces and Roman capital letters denote linear maps. $R(H)$, $N_r(H)$, $N_\ell(H)$ denote the range space, right null space, left null space respectively of a linear map H. Finally, if a property is said to be true for $i \in \underset{\sim}{n}$, this means that it is true for all $1 \le i \le n$.

2. SETS OF POLYNOMIALS: DEFINITIONS AND NORMAL EQUIVALENCE

Let $P = \{p_i(s)\colon p_i(s) \in \mathbb{R}[s],\ i \in \underset{\sim}{m},\ d_i = \deg p_i(s)\}$ be a set of polynomials and let $d = \max\{d_i,\ i \in \underset{\sim}{m}\}$. With P, we may always associate a polynomial vector $\underline{p}(s)$, $\underline{p}(s) \in \mathbb{R}^m[s]$, where

$$\underline{p}(s) = \begin{bmatrix} p_1(s) \\ \vdots \\ p_k(s) \\ \vdots \\ p_m(s) \end{bmatrix} = \begin{bmatrix} p_o^1 & p_1^1 & \cdots & p_{d_1}^1 & 0 & \cdots & 0 \\ \vdots & \vdots & & & & & \vdots \\ p_o^k & p_1^k & \cdots & \cdots & \cdots & \cdots & p_{d_k}^k \\ \vdots & \vdots & & & & & \\ p_o^m & p_1^m & \cdots & \cdots & p_{d_m}^m & 0 & \cdots\ 0 \end{bmatrix} \begin{bmatrix} 1 \\ s \\ s^2 \\ \vdots \\ s^d \end{bmatrix}$$

$$= [\underline{p}_o, \ldots, \underline{p}_d]\underline{e}_d(s) = P_d \underline{e}_d(s) \tag{2.1}$$

where $P_d \in \mathbb{R}^{m\times(d+1)}$ and $\underline{e}_d(s) \in \mathbb{R}^{d+1}[s]$. The polynomial vector $\underline{p}(s)$ is defined as a *vector representative* of P and $d = \deg \underline{p}(s)$ will be referred to as the *degree* of P. The matrix P_d characterises the properties of P and it is defined as a *basis matrix* of P. The set P will be called *reduced*, if the polynomials $p_i(s)$ are coprime; otherwise it will be called *nonreduced*. Finally, P will be called monic, if $\|\underline{p}_d\| = 1$ ($\|\cdot\|$ denotes the usual Euclidean norm).

The set P, or the vector representative $\underline{p}(s)$, may be represented by a Taylor expansion around $s = a$, $a \in \mathbb{C}$, as follows

$$\underline{p}(w) = \underline{b}_o + \underline{b}_1 w + \ldots \underline{b}_d w^d = [\underline{b}_o, \underline{b}_1, \ldots, \underline{b}_d] = B_d \underline{e}_d(s) \tag{2.2}$$

where $w = s-a$, $\underline{b}_i \in \mathbb{C}^m$, $i = 0,1,\ldots,d$, $B_d \in \mathbb{C}^{m\times(d+1)}$ and

$$\underline{b}_o = \underline{p}(a),\ \underline{b}_i = \frac{1}{i!}\left[\frac{d^i\{\underline{p}(s)\}}{ds^i}\right]_{s=\alpha} \quad i=1,\ldots,d-1 \text{ and } \underline{b}_d = \underline{p}_d \tag{2.3}$$

B_d is a basis matrix of P with respect to the point $\alpha \in \mathbb{C}$.

In the study of coprimeness and "almost coprimeness" properties of the polynomials of P, the following equivalence relation on sets of polynomials has been shown to be useful [2]:

<u>Definition (2.1)</u> Let P, P' be two sets of polynomials and let $\underline{p}(s) = P_d \underline{e}_d(s)$, $\underline{p}'(s) = P_d' \underline{e}_d(s)$ be their corresponding vector representatives, where $P_d \in \mathbb{R}^{m\times(d+1)}$ and $P_d' \in \mathbb{R}^{n\times(d+1)}$ are their corresponding basis matrices. The sets P, P' will be said to be *normally equivalent* (NE), and this will be denoted by PEP', if there exists an orthogonal square matrix Q such that

$$P_d = Q\begin{bmatrix} P_d' \\ \hline 0_{m-n} \end{bmatrix},\ Q \in \mathbb{R}^{m\times m}, \text{ if } m \ge n, \quad \underline{\text{OR}} \quad \begin{bmatrix} P_d \\ \hline 0_{n-m} \end{bmatrix} = QP_d', \tag{2.4}$$

$$Q \in \mathbb{R}^{n\times n}, \text{ if } n \ge m$$

It may be readily verified that the above relation satisfies all conditions of an equivalence relation; note that normal equivalence E is defined on sets of polynomials with the same degree, but not necessarily the same number of polynomials. The equivalence class, $E(P)$, of P under E is characterised by invariants and a canonical set P^*, as shown below [2].

Theorem (2.1) Let $P_d \in \mathbb{R}^{m\times(d+1)}$ be a basis matrix for the set of polynomials P, $\pi = \text{rank}\{P_d\} \leq \min\{m,d+1\}$ and let γ_i, $i \in \underset{\sim}{\pi}$, be the singular values of P_d. If $P_d = Y\Gamma U^t$ is the singular value decomposition of P_d, where $\Gamma = \text{diag}\{\gamma_i, i \in \underset{\sim}{\pi}\}$, $Y^tY = I_n = U^tU$, then:

(i) The part of the eigenstructure of $P_d^tP_d$ which is characterised by nonzero eigenvalues, is a complete invariant of $E(P)$.

(ii) The polynomial vector $\underline{p}^*(s) = \Gamma U^t \underline{e}_d(s)$ defines a canonical set P^*, which characterises uniquely the equivalence class $E(P)$. □

Corollary (2.1) The greatest common divisor of a set P is invariant under normal equivalence transformations. □

If we denote by $\underline{\gamma}, \bar{\gamma}$ the minimum and maximum singular value of P_d respectively, then $\theta = \bar{\gamma}/\underline{\gamma}$ is the condition number of P_d, and this is also a normal equivalence invariant of $E(P)$.

The coprimeness of a set of polynomials P may be investigated by using one of the standard resultant tests [11]. The notion of an "almost zero", or an "almost common divisor" of P [2] is discussed next.

3. ALMOST ZEROS: DEFINITION LOCATION AND COMPUTATION

When $s \in \mathbb{C}$, the vector representative $\underline{p}(s)$ of P defines a vector analytic function with domain $\mathbb{C}$ and codomain $\mathbb{C}^m$; we define the norm of $\underline{p}(s)$ (or the norm of P) as

$$\|\underline{p}(s)\| = \phi(\sigma,\omega) = \sqrt{\underline{p}(s^*)^t\underline{p}(s)} = \sqrt{\underline{e}_d(s^*)^tP_d^tP_d\underline{e}_d(s)} \qquad (3.1)$$

where s^* is the complex conjugate of s $(s = \sigma + j\omega)$. Note that if $q(s) = s+\alpha$ is a divisor of P, then $p(-\alpha) = \underline{0}$ and thus $\|\underline{p}(-\alpha)\| = 0$. This observation leads to the following definition.

Definition (3.1) Let P be a reduced set of polynomials. If $s = z$, $z \in \mathbb{C}$, is a local minimum of $\|\underline{p}(s)\|$, then z will be called an *almost zero* (AZ) of P and the value $\|\underline{p}(z)\| = \varepsilon$ will be referred to as the *order* of the AZ. If $s = \tilde{z}$ is the global minimum of $\|\underline{p}(s)\|$, then $\tilde{z}$ will be called the *prime almost zero* (PAZ) of the set P.

Clearly, if P is not reduced, then the set of AZs, which have order $\varepsilon = 0$, defines the zeros of P. Thus, the above definition unifies the notions of exact and "approximate" zeros, since both emerge as minima of a norm function of P. The order ε of an AZ indicates how well z may be considered as an "approximate" zero of P; we should note, however, that scaling of the polynomials of P by a $c \in \mathbb{R}$, $c \neq 0$, affects the order ε of an AZ, but not its location. In the following, $\mathcal{P}$ will be assumed monic ($\underline{p}(s)$ is assumed monic).

Proposition (3.1) The set of AZs $\{(z_i, \varepsilon_i)\}$ of $\mathcal{P}$ is invariant under normal equivalence transformations.

□

This result suggests that the canonical set $\mathcal{P}^*$ defined by Theorem (2.1) may be used for the computation and investigation of properties of AZs of $\mathcal{P}$. A number of results on the distribution of AZs in the complex plane are given next [2].

Proposition (3.2) Let $\underline{\gamma}, \bar{\gamma}$ be the min, max singular values of the basis matrix P_d of $\mathcal{P}$. Then,

$$\underline{\gamma} \| \underline{e}_d(s) \| \leq \phi(\sigma,\omega) \leq \bar{\gamma} \| \underline{e}_d(s) \| \tag{3.2}$$

□

This result establishes bounds for the function $\phi(\sigma,\omega)$ and yields the following results.

Theorem (3.1) Let d be the degree and θ the condition number of $\mathcal{P}$. The PAZ of $\mathcal{P}$ is always within the disk $\mathcal{D}[0,\tilde{\rho}]$, where $\tilde{\rho}$ is the unique positive real solution of the equation

$$1 + r^2 + r^4 + \ldots + r^{2d} = \theta^2 \tag{3.3}$$

□

The disk $\mathcal{D}[0,\tilde{\rho}]$ is called the *prime almost zero disk*. The radius $\tilde{\rho} = g(\theta,d)$ is characterised by the following properties:

Corollary (3.1.1) (i) $\tilde{\rho}$ is invariant under scaling of the polynomials of $\mathcal{P}$ by a $c \in \mathbb{R}$, $c \neq 0$.
(ii) $\tilde{\rho}$ is monotonically decreasing function of d and $1/\theta$.
(iii) $\tilde{\rho}$ is within the following intervals:

(a) if $d+1 > \theta^2$, then $0 < \tilde{\rho} < 1$.

(b) if $d+1 \leq \theta^2$, then $1 \leq \tilde{\rho} < \sqrt{\theta}$.

Remark (3.1) Well conditioned sets $\mathcal{P}$ ($\theta \cong 1$) have a very small $\tilde{\rho}$, even for very small values of d. Badly conditioned sets $\mathcal{P}$ ($\theta >> 1$) have very large $\tilde{\rho}$, even for large values of d.

The definition of AZs as minima of $\phi(\sigma,\omega)$ leads to the following criteria for their exact calculation [2].

Proposition (3.2) Necessary conditions for $z \in \mathbb{C}$ to be an AZ of the set $\mathcal{P}$, which is defined by the basis matrix P_d, are:

$$\underline{e}_d^t(z^*)\Delta^t P_d^t P_d \underline{e}_d(z) = 0 \quad \text{AND} \quad \underline{e}_d^t(z^*) P_d^t P_d \Delta \underline{e}_d(z) = 0 \tag{3.4}$$

where

$$\Delta = \begin{bmatrix} 0 & 0 & 0 & . & . & . & 0 & 0 \\ 1 & 0 & 0 & . & . & . & 0 & 0 \\ 0 & 2 & 0 & . & . & . & 0 & 0 \\ . & . & . & . & . & . & . & . \\ . & . & . & . & . & . & . & . \\ 0 & 0 & 0 & . & . & . & d & 0 \end{bmatrix} \in \mathbb{R}^{(d+1)\times(d+1)} \tag{3.5}$$

For those z for which conditions (3.4) are satisfied, a set of sufficient conditions for z to be an AZ of P are:

$$\underline{e}_d^t(z^*)\{\Delta^{t2}P_d^tP_d + 2\Delta^tP_d^tP_d\Delta + P_d^tP_d\Delta^2\}\underline{e}_d(z)\} > 0 \tag{3.6}$$

$$2\{\underline{e}_d^t(z^*)\Delta^tP_d^tP_d\Delta\underline{e}_d(z)\}^2 > \{\underline{e}_d^t(z^*)\Delta^{t2}P_d^tP_d\underline{e}_d(z)\}^2 + \\ + \{\underline{e}_d^t(z^*)P_d^tP_d\Delta^2\underline{e}_d(z)\}^2 \tag{3.7}$$

□

The above result may be used for the analytic computation of the AZs of P. In practice, such computations are tedious and a numerical technique, of the hill climbing type, has to be used for the computation of AZs. A program that plots the surfaces $\phi(\sigma,\omega) = c$, $c \in \mathbb{R}^+$, over a selected rectangle of the $\mathbb{C}$-plane, is used first to locate the regions where $\phi(\sigma,\omega)$ has a minimum. A standard minimization routine is then used to compute the AZ. For the case of the PAZ, Theorem (3.1) and its Corollary may be used for the estimation of the initial point of the numerical search.

Example (3.1) Let P be defined by

$$\underline{p}(s) = \begin{bmatrix} s+1.1 \\ s^2+s \end{bmatrix} = \begin{bmatrix} 1.1 & 1 & 0 \\ 0 & 1 & 1 \end{bmatrix} \begin{bmatrix} 1 \\ s \\ s^2 \end{bmatrix} = P_2\underline{e}_2(s)$$

For this set we have that $\underline{\gamma} = 1.048$, $\bar{\gamma} = 1.763$, $\tilde{\rho} = 0.970$. A plot of $\phi(\sigma,\omega)$ surfaces in the prime disk region is shown in Figure (3.1). The numerical search reveals that the PAZ is at $z = -1.046$ and that its order is $\varepsilon = 0.005$.

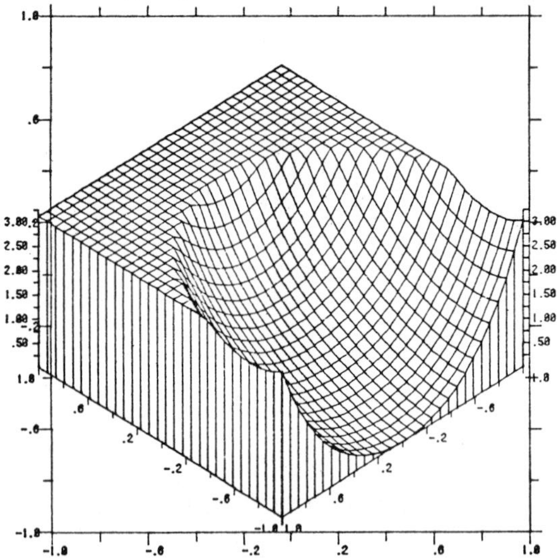

Figure (3.1): Plot of the surface $\|\underline{p}(s)\|$ in the prime region

The properties of the set of AZs of P are investigated next. It will be shown next that the similarities between exact and almost zeros extend beyond their common definition.

4. POLYNOMIAL COMBINANTS AND THE EXACT AND APPROXIMATE ZERO ASSIGNMENT PROBLEMS

Let $\underline{p}(s) = P_d \underline{e}_d(s)$ be a vector representative of the set of polynomials $P = \{p_i(s): p_i(s) \in \mathbb{R}[s], i \in \underset{\sim}{m}\}$ where $P_d \in \mathbb{R}^{m\times(d+1)}$ and let $\underline{k} \in \mathbb{R}^m$. The polynomial function of the parameter vector $\underline{k}$ defined by

$$f(s,P,\underline{k}) = \underline{k}^t P_d \underline{e}_d(s) = \sum_{i=1}^{m} k_i p_i(s) \tag{4.1}$$

is called a $\underline{k}$-*polynomial combinant* of P and shall be denoted in short by $f(s,\underline{k})$.

The problem of zero assignment of polynomal combinants has emerged as the linear subproblem of the determinantal assignment problem [1,5]. For a set of polynomials P represented by a basis matrix P_d, the zero assignment problem for polynomial combinants is defined as follows. Find $\underline{k} \in \mathbb{R}^m$, such that $f(s,\underline{k}) = \underline{k}^t P_d \underline{e}_d(s) = a(s)$, where $a(s) \in \mathbb{R}[s]$ arbitrary. It is clear that the maximum degree of $a(s)$ has to be equal to the degree d of P, and if $a(s) = a_0 + a_1 s + \ldots + a_d s^d = [a_0, a_1, \ldots, a_d]\underline{e}_d(s) = \underline{a}^t \underline{e}_d(s)$, then the problem is reduced to the solution of the equation

$$\tilde{P}_d \underline{k} = \underline{a}, \quad \tilde{P}_d = P_d^t \in \mathbb{R}^{(d+1)\times m}, \quad \underline{a} \in \mathbb{R}^{d+1} \tag{4.2}$$

A set P for which equation (4.1) has a solution for all $a \in \mathbb{R}^{d+1}$ will be called *completely assignable* (CA); otherwise, P will be referred to as *nonassignable* (NA). An important family of nonassignable sets are those for which there is no $\underline{k} \in \mathbb{R}^m$ such that $f(s,\underline{k}) = c$, $c \in \mathbb{R}$; such sets will be called *strongly nonassignable* (SNA) and they have the additional property that there is no combinant with all its zeros at $s = \infty$. Note that strong nonassignability implies nonassignability but the opposite is not always true.

Proposition (4.1) Let $P_d = [\underline{p}_0, \underline{p}_1, \ldots, \underline{p}_d] = [\underline{p}_0, \bar{P}_d] \in \mathbb{R}^{m\times(d+1)}$, with $\pi = \text{rank}\{P_d\}$ and $\bar{\pi} = \text{rank}\{\bar{P}_d\}$, where P_d is a basis matrix of a set P.

(i) P is completely assignable, if and only if $\pi = d+1$.

(ii) P is strongly nonassignable, if and only if $\bar{\pi} = m$. □

Proposition (4.2) Let P be a CA set, P_d be a basis matrix of P, and let $\tilde{P}_d = P_d^t$. If $\tilde{P}^{\perp}$ is a basis for $N_r(\tilde{P}_d)$ and $\tilde{P}^{+}$ a right inverse of $\tilde{P}_d$, then for every $\underline{a} \in \mathbb{R}^{d+1}$, the solution of equation (4.2) is given by

$$\underline{k} = \tilde{P}^{+}\underline{a} + \tilde{P}^{\perp}\underline{c}, \quad \text{where } \underline{c} \in \mathbb{R}^{m-d-1} \text{ arbitrary}$$

□

Part (i) of Proposition (4.1) states the rather obvious fact that the necessary and sufficient condition for completely assignability is that $R(\tilde{P}_d) = \mathbb{R}^{d+1}$. If $R(\tilde{P}_d) \subset \mathbb{R}^{d+1}$, then P is a NA set and assignment is possible only for those polynomials $a(s) \in \mathbb{R}[s]$ for which $\underline{a} \in R(\tilde{P}_d)$. For NA sets an extension of the "exact assignment" problem may be defined as follows: Find $\underline{k} \in \mathbb{R}^m$ such that $f(s,\underline{k}) = b(s)$ where $b(s)$ has its zeros in neighbourhoods of the zeros of a given polynomial $a(s)$. Such a problem will be called the *approximate zero assignment problem* (AZAP). The AZAP is meaningful for those $a(s)$ for which the vector coefficient $\underline{a} \notin R(\tilde{P}_d)$, otherwise an exact solution exists.

Lemma (4.1) [6] Let $f(s) = a_o + a_1 s + \ldots + a_n s^n = a_n \prod_{j=1}^{p} (s-z_j)^{m_j}$, $a_n \neq 0$ and $f'(s) = (a_o+\varepsilon_o) + (a_1+\varepsilon_1)s + \ldots + (a_{n-1}+\varepsilon_{n-1})s^{n-1} + a_n s^n$ be two polynomials of $\mathbb{R}[s]$ and let $0 < \tau_k < \min|z_k - z_j|$, $j = 1,2,\ldots,k-1,k+1,\ldots,p$. There exists $\varepsilon > 0$, $\varepsilon < \delta_k/M_k$, where

$$\delta_k = |a_n| \tau_k^{m_k} \prod_{j=1, j\neq k}^{p} (|z_k - z_j| - \tau_k)^{m_j}, \quad M_k = \sum_{j=0}^{n-1} (\tau_k + |z_k|)^j \tag{4.3}$$

such that if $|\varepsilon_i| \leq \varepsilon$, $i = 0,1,\ldots,n-1$, then $f'(s)$ has precisely m_k zeros in the disk $D_k[z_k, \tau_k]$. □

The above lemma readily yields the following result.

Theorem (4.1) Let P be a NA set, $P_d \in \mathbb{R}^{m\times(d+1)}$ be a basis matrix of P, and let $\tilde{P}_\ell^\perp$ be a basis for $N_\ell(\tilde{P}_d)$. Let $a(s) = \underline{a}^t \underline{e}_d(s)$, $\deg a(s) = d$, $\underline{a} \notin R(\tilde{P}_d)$ and let $\varepsilon > 0$, $\varepsilon < \min\{\delta_k/M_k\}$, where δ_k, M_k are defined by $a(s)$ as in Lemma 4.1; furthermore, let us denote by z_k, $k \in \underset{\sim}{p}$, the zeros of $a(s)$, by m_k the corresponding multiplicities, and by $\tau_k > 0$ the numbers defined by $a(s)$ as in Lemma (4.1). Then, for all $\underline{\varepsilon} = [\varepsilon_o, \varepsilon_1, \ldots, \varepsilon_{d-1}, 0]^t \in \mathbb{R}^{d+1}$, $|\varepsilon_i| \leq \varepsilon$, $i = 0.1.,,,.d-1$, such that

$$\tilde{P}_\ell^\perp(\underline{a}+\underline{\varepsilon}) = 0 \tag{4.4}$$

there always exists a $\underline{k} \in \mathbb{R}^m$ such that the polynomial combinant $f(s,\underline{k}) = \underline{k}^t P_d \underline{e}_d(s)$ has precisely m_k zeros, $k \in \underset{\sim}{p}$ in the disks $D_k[z_k, \tau_k]$. □

Example (4.1) Let P be defined by

$$\underline{p}(s) = \begin{bmatrix} s+1 \\ s^2+0.99s \end{bmatrix} = \begin{bmatrix} 1 & 1 & 0 \\ 0 & 0.99 & 1 \end{bmatrix} \begin{bmatrix} 1 \\ s \\ s^2 \end{bmatrix} = P_2 \underline{e}_2(s) \tag{4.5}$$

This set is clearly a NA set because $\text{rank}\{P_2\} = 2 < 3 = d+1$. Let $a(s) = s^2+3s+2 = [1\ 3\ 2]\underline{e}_2(s) = a^t \underline{e}_2(s) = (s+1)(s+2)$. Clearly $\underline{a} \notin R(\tilde{P}_2)$. So we have $z_1 = -1$, $z_2 = -2$, $0 < \tau_1, \tau_2 < 1$. We choose $\tau_1 = \tau_2 = 0.5$ and so

$$\delta_1 = 1\ \tau_1^1(|z_1 - z_2| - \tau_1)^1 = 0.25, \quad M_1 = 1 + \tau_1 + |z_1| = 2$$

$$\delta_2 = 1\ \tau_2^1(|z_2 - z_1| - \tau_2)^1 = 0.25, \quad M_2 = 1 + \tau_2 + |z_2| = 3.5$$

According to Theorem (4.1) $\varepsilon < \min\{\delta_1/M_1, \delta_2/M_2\} = 0.0714$. Let $[1\ -1\ 0.99]$ a basis for the $N_\ell(\tilde{P}_2)$. Then for all $\underline{\varepsilon} = [\varepsilon_o, \varepsilon_1, 0]^t \in \mathbb{R}^3$, $|\varepsilon_o|, |\varepsilon_1| < 0.07$ such that

$$[1\ -1\ 0.99]\begin{bmatrix}2+\varepsilon_o\\3+\varepsilon_1\\1\end{bmatrix} = 0 \quad \text{which means} \quad \varepsilon_o - \varepsilon_1 = 0.01 \tag{4.6}$$

there always exists a $\underline{k} \in \mathbb{R}^2$ such that the polynomial combinant $f(s,\underline{k}) = \underline{k}^t P_2 \underline{e}_2(s)$ has precisely one zero in the disk $D_1[-1,0.5]$ and one zero in the disk $D[-2,0.5]$.

□

Theorem (4.1) provides a sufficient condition for approximate zero assignment, but it may also be used as a test for investigating the sensitivity of the exact assignment problem due to uncertainties in the elements of the basis matrix P_d.

Thus, let $\tilde{P}'_d = \tilde{P}_d + E$ be the basis matrix of a CA set, where $\tilde{P}_d$ is its nominal value and E is a perturbation. If $\underline{k}$ is the nominal solution of equation (4.1) for a given $\underline{\alpha}$, then

$$\tilde{P}'_d\underline{k} = \tilde{P}_d\underline{k} + E\underline{k} = \underline{\alpha} + \underline{\varepsilon}(\underline{k}) \tag{4.7}$$

From equation (4.7) it is clear that Theorem (4.1) may also be used as a sensitivity test for the solutions of the exact assignment problem.

5. STRONGLY NONASSIGNABLE SETS AND ALMOST FIXED ZEROS

The case of SNA polynomial sets is considered next. It is shown that the property of an exact zero to be a zero of all combinants of P, also extends to the case of an AZ. The following results are based on the classical theory of polynomials in a complex variable [6,7]; proofs are given in [1].

Theorem (5.1) Let P be a SNA set, $\alpha \in \mathbb{C}$, and let $\underline{p}(w) = \underline{b}_o + \underline{b}_1 w + \dots + \underline{b}_d w^d$, $w = s-\alpha$, be the Taylor expansion of the vector representative of P at $s = \alpha$. For every $\underline{k} \in \mathbb{R}^m$, $f(s,\underline{k})$ has at least one zero in the finite, minimal radius disk $\mathcal{D}_m[\alpha, R_m(\alpha,\underline{k})] = \{s : |s-\alpha| \le R_m(\alpha,\underline{k})\}$, where $R_m(\alpha,\underline{k})$ is defined by

$$R_m(\alpha,\underline{k}) = \min\{[\tbinom{d}{i}|\underline{k}^t\underline{b}_o|/|\underline{k}^t\underline{b}_i|]^{1/i},\ i \in \underset{\sim}{d}\} \tag{5.1}$$

□

Corollary (5.1.1) Let P be a SNA set. For every $\alpha \in \mathbb{C}$ there exists a finite, minimal radius disk $\tilde{\mathcal{D}}_m[\alpha, \tilde{R}_m(\alpha)]$, where $\tilde{R}_m(\alpha) = \max\{R_m(\alpha,\underline{k})$, for all $\underline{k} \in \mathbb{R}^m\}$, which contains at least one zero of all combinants of P.

□

Some useful upper bounds for $R_m(\alpha,\underline{k})$ are defined below.

Corollary (5.1.2) An upper bound for $R_m(\alpha,k)$ is defined by

$$R_1(\alpha,\underline{k}) = \rho(\alpha,\underline{k})/\{2^{1/d}-1\} \quad (5.2)$$

where $\rho(\alpha,\underline{k})$ is the unique positive solution of

$$|\underline{k}^t\underline{b}_o| = |\underline{k}^t\underline{b}_1|\rho + |\underline{k}^t\underline{b}_2|\rho^2 + \dots + |\underline{k}^t\underline{b}_d|\rho^d \quad (5.3)$$

□

Corollary (5.1.3) Upper bounds $\hat{R}_i(\alpha,\underline{k})$, $i = d,d-1,\dots,j$ for $R_m(\alpha,\underline{k})$ are defined for families of vectors $\underline{k}^i \in \mathbb{R}^m$ as follows:

(i) If $\underline{k}^t\underline{p}_d \neq 0$, then

$$\hat{R}_d(\alpha,\underline{k}) = \{\|\underline{b}_o\|/|\underline{k}^t\underline{p}_d|\}^{1/d} \quad (5.4)$$

(ii) If $\underline{k}^t\underline{p}_d = \underline{k}^t\underline{p}_{d-1} = \dots = \underline{k}^t\underline{p}_{i+1} = 0$ and $\underline{k}^t\underline{p}_i \neq 0$, then

$$\hat{R}_i(\alpha,\underline{k}) = \{\tbinom{d}{i}\|\underline{b}_o\|/|\underline{k}^t\underline{p}_i|\}^{1/i} \quad (5.5)$$

where $\underline{p}_i$ are the vector coefficients of $\underline{p}(s)$. □

It is clear that the $\hat{R}_i(\alpha,\underline{k})$ bounds cover all vectors of $\mathbb{R}^m$. Note that Theorem (5.1) and its Corollaries make no distinction between a general point $\alpha \in \mathbb{C}$ and an AZ z of the polynomial set P. The feature that distinguishes an AZ of P from all other points of the complex plane is their "strong attractivity"; this is demonstrated by the following result.

Theorem (5.2) Let z be an AZ and z be the PAZ of P. Then,

(i) For all $k \in \mathbb{R}^m$, such that $\underline{k}^t\underline{p}_d \neq 0$

(a) $\hat{R}_d(z,k) < \hat{R}_d(\alpha,\underline{k})$ for all $\alpha \in \mathbb{C}$: $|\alpha - z| < \varepsilon$, $\varepsilon > 0$

(b) $\hat{R}_d(z,\underline{k}) < \hat{R}_d(\alpha,\underline{k})$ for all $\alpha \in \mathbb{C}$

(ii) For all $\underline{k} \in \mathbb{R}^m$ such that $\underline{k}^t\underline{p}_d = \dots = \underline{k}^t\underline{p}_i = 0$, $\underline{k}^t\underline{p}_i \neq 0$

(a) $\hat{R}_i(z,\underline{k}) < \hat{R}_i(\alpha,\underline{k})$ for all $\alpha \in \mathbb{C}$: $|\alpha - z| < \varepsilon$, $\varepsilon > 0$

(b) $\hat{R}_i(z,\underline{k}) < \hat{R}_i(\alpha,\underline{k})$ for all $\alpha \in \mathbb{C}$ □

The above result is a consequence of Corollary (5.1.3) and of the property that $\|\underline{b}_o(\alpha)\| = \|\underline{p}(\alpha)\|$ is locally minimized when α is an AZ, and it is globally minimized when α is the PAZ of P. The radii $\hat{R}_i(\alpha,\underline{k})$, or any other upper bound for $R_m(\alpha,\underline{k})$, provide a measure for the ability of α to attract the zeros of $f(s,\underline{k})$ for the given $\underline{k}$. Theorem (5.2) reveals the set of AZs as "strong poles of attraction" for the zeros of all combinants of P. It is this property that makes the AZs a special subset of the $\mathbb{C}$-plane.

In the following, attention is focused on the set of AZs. Corollary (5.1.1) establishes the existence of a minimal radius disk $\tilde{\mathcal{D}}_m[\alpha,\tilde{R}_m(\alpha)]$ which contains at least one zero of all combinants of P.

Clearly, such disks are also defined for the set of AZs; furthermore, Theorem (5.2) shows that the upper bound for this disk is minimized at an AZ. Computing $\tilde{R}_m(z)$ is extremely difficult and thus upper bounds for $\tilde{R}_m(z)$ are sought.

Theorem (5.3) Let $z \in \mathbb{C}$ be an AZ of a SNA set P, $w = s-z$, $\underline{p}(w) = \underline{b}_0+\underline{b}_1 w+ +\dots+\underline{b}_d w^d$, be the Taylor expansion of the polynomial vector representative of P, $\tilde{R}_m(z) = \max\{R_m(z,\underline{k})$, for all $\underline{k} \in \mathbb{R}^m\}$ and let $\chi \in \mathbb{R}^+$. A sufficient condition for $\tilde{R}_m(z) \leq R(z,\chi)$, where

$$R(z,\chi) = \chi/(2^{1/d}-1) \tag{5.6}$$

is that the matrix $Q(z,\chi)$ is positive semidefinite, where

$$Q(z,\chi) = \underline{b}_d\underline{b}_d^{*t}\chi^{2d} + \dots + \underline{b}_1\underline{b}_1^{*t}\chi^2 - \underline{b}_o\underline{b}_o^{*t} \tag{5.7}$$

□

This result can be used in two different ways:

(i) If $\chi \in \mathbb{R}^+$ is given, then the positive semidefiniteness of $Q(z,\chi)$ implies the existence of an upper bound of the type (5.6).

(ii) Find the minimum $\chi \in \mathbb{R}^+$ for which $Q(z,\chi)$ is positive semidefinite. In this case, the smallest of the (5.6) type bounds for $\tilde{R}_m(z)$ is defined.

The matrix $Q(z,\chi)$ is Hermitian and the following result may be used for testing its positive semidefiniteness.

Proposition (5.1) Let H be an n×n Hermitian matrix, $\rho = \text{rank}\{H\}$, and let $\phi(\lambda) = |\lambda I-H| = \lambda^n+(-1)^1 h_1\lambda^{n-1}+\dots+(-1)^m h_m\lambda^{n-m}+\dots+(-1)^n h_n$. H is positive semidefinite, if and only if for all $i \in \underset{\sim}{\rho}$, $h_i > 0$ and $h_i = 0$ for $i = \rho+1,\dots,n$.

□

These ideas are illustrated in the following example:

Example (5.1) For the polynomial set of Example (3.1) the set of combinants is defined by

$$f(s,\underline{k}) = [k \quad 1]\begin{bmatrix} s+1.1 \\ s^2+s \end{bmatrix} = k(s+1.1) + (s^2+s) \tag{5.8}$$

For this special case, the zero assignment is a standard root locus problem and thus the results of Theorem (5.3) may be compared with the standard root locus results. The root locus for equation (5.8) is shown in Figure (5.1). The polynomial set has an AZ at $z = -1.046$, and from Figure (5.1), the radius of the minimal disk, which is centered at $z = -1.046$, is found to be $\tilde{R}_m(z) = 0.43$.

The $Q(-1.046,\chi)$ may be shown to be

$$Q(-1.046,\chi) = \begin{bmatrix} \chi^2 - 0.01 & -\chi^2 \\ -\chi^2 & \chi^4 + \chi^2 \end{bmatrix}, \quad \chi \in \mathbb{R}^+$$

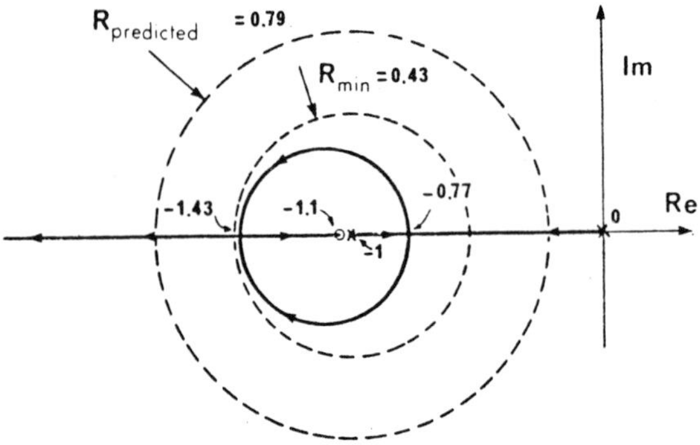

Figure (5.1):
Root Locus

By Proposition (5.1), the conditions for Q to be positive semi-definite are

$$\chi^4 + 2\chi^2 - 0.01 \geq 0 \quad \text{AND} \quad \chi^4 - 0.01\chi^2 - 0.01 \geq 0$$

By solving those two inequalities, it is found that $\chi \geq 0.3243$ and thus by equation (5.6) we have that the minimal predicted radius for $\chi = 0.3243$ is $R_{pr} = 0.79$; this clearly demonstrates that Theorem (5.3) provides a good estimate for $\tilde{R}_m(z)$.

6. STABILITY ASPECTS OF POLYNOMIAL COMBINANTS

In this section the problem of finding $\underline{k} \in \mathbb{R}^m$ such that the combinant $f(s,\underline{k}) = \underline{k}^t P_d \underline{e}_d(s) = a(s) = \underline{a}^t \underline{e}_d(s)$ is a Hurwitzian polynomial (its zeros have negative real parts), is considered. Such a problem will be referred to as the *polynomial combinants stabilization problem* (PCSP). Clearly, the PCSP is nontrivial for NA sets P.

Definition (6.1) A NA set P will be called *Hurwitzian*, if there exists at least one $\underline{k} \in \mathbb{R}^m$ such that $f(s,\underline{k})$ is a Hurwitzian polynomial; if there is no $\underline{k} \in \mathbb{R}^m$ for which $f(s,\underline{k})$ is Hurwitzian, then P will be called *non-Hurwitzian*.

For a NA set P defined by a basis matrix $P_d \in \mathbb{R}^{m\times(d+1)}$, the PCSP may be reduced to the following problem: Determine whether there exists a $\underline{k}$ such that the equation

$$\tilde{P}_d\underline{k} = \underline{a}, \quad \tilde{P}_d = P_a^t, \quad \underline{\alpha} = [\alpha_o, \alpha_1, \ldots, \alpha_d]^t \in \mathbb{R}^{d+1} \tag{6.1}$$

has a solution for some vector $\underline{\alpha}$ for which its coordinates α_i satisfy the Hurwitz inequalities (or equivalently the Lienard-Chipart inequalities) [8].

A necessary condition for a(s) to be Hurwitzian is that $a_o, a_1, \ldots, a_d > 0$; this necessary condition for P to be Hurwitzian is that the inequality system

$$\tilde{P}_d\underline{k} > 0 \tag{6.2}$$

is consistent. In the following we shall assume that $\mathrm{rank}\{P_d\} = \rho < d+1$ (the case of NA sets). The following result is a consequence of conditions (6.2).

Proposition (6.1) Necessary condition for P to be Hurwitzian is that P_d has no column which is identically zero. □

A number of results from the general theory of linear inequalities [9] may be used to derive necessary conditions for P to be Hurwitzian. Thus we have:

Lemma (6.1) [9] The system of inequalities

$$\underline{a}_i^t \underline{x} > 0 \ , \quad i \in \underset{\sim}{\tau} \tag{6.3}$$

is consistent, if and only if the zero vector is not in the convex hull of the vectors $\{\underline{a}_i, \ i \in \underset{\sim}{\tau}\}$.

Proposition (6.2) Let $P_d \in \mathbb{R}^{m\times(d+1)}$ be the matrix of a NA set, $\underline{b} = [0,0,\dots,0,1]^t \in \mathbb{R}^{m+1}$, $\underline{1}^t = [1,1,\dots,1]^t \in \mathbb{R}^{d+1}$ and S be the convex polyhedral cone generated by the columns of the matrix $Q = [P_d^t, \underline{1}]^t \in \mathbb{R}^{(m+1)\times(d+1)}$. Necessary condition for P to be Hurwitzian is that $\underline{b} \notin S$.

Proof: By Lemma (6.1), necessary and sufficient condition for the system $P_d^t \underline{k} > 0$ to be consistent is

$$P_d \underline{x} \neq \underline{0}, \quad \underline{x} \in \mathbb{R}^{d+1}, \quad \underline{x} \geq 0, \quad \sum_{i=1}^{d+1} x_i = 1 \tag{6.4}$$

or equivalently the $N_r(P_d)$ and the hyperplane $x_1+x_2+\dots+x_{d+1} = 1$ do not intersect in the positive orthant. This means

$$\begin{bmatrix} P_d \\ \underline{1}^t \end{bmatrix} \underline{x} \neq \begin{bmatrix} 0 \\ \underline{1} \end{bmatrix} = \underline{b}, \quad \underline{x} \geq 0 \tag{6.5}$$

or equivalently $\underline{b} \notin S$. □

The following result is an adaptation of a result on inconsistency of inequality systems of the type (6.2) [9].

Theorem (6.1) Let P_d be the basis matrix of a NA set P and let us assume that $P_d^t = \tilde{P}_d$ has no zero row. The set P is non-Hurwitzian, if certain r linearly dependent rows of $\tilde{P}_d$ contain an $r\times(r-1)$ submatrix T, such that

$$|T_i|\cdot|T_{i+1}| < 0, \quad \text{for} \quad 1 \leq i \leq r-1 \tag{6.6}$$

where $|T_i|$, $i \in \underset{\sim}{r}$, denotes the $(r-1)\times(r-1)$ minor of T, which is obtained by deleting the i-th row. □

This result implies an extremely useful algorithm for testing the non-Hurwitzian property of a set $\mathcal{P}$.

Corollary (6.1.1) Let $P_d = [\underline{p}_0, \underline{p}_1, \ldots, \underline{p}_d]$ be the basis matrix of a NA set $\mathcal{P}$. Necessary condition for $\mathcal{P}$ to be Hurwitzian is that the matrix

$$R = \underline{p}_{d-1}\underline{p}_{d-2}^t - \underline{p}_d\underline{p}_{d-3}^t \tag{6.7}$$

is positive semidefinite.

Proof: A necessary condition for the polynomial

$$f(s,\underline{k}) = \underline{k}^t\underline{p}_d s^d + \underline{k}^t\underline{p}_{d-1}s^{d-1} + \ldots + \underline{k}^t\underline{p}_o$$

to be Hurwitzian is:

$$\det\begin{bmatrix} \underline{k}^t\underline{p}_{d-1} & \underline{k}^t\underline{p}_{d-3} \\ \underline{k}^t\underline{p}_d & \underline{k}^t\underline{p}_{d-1} \end{bmatrix} > 0 \tag{6.8}$$

or equivalently $\underline{k}^t(\underline{p}_{d-1}\underline{p}_{d-2}^t - \underline{p}_d\underline{p}_{d-3}^t)\underline{k} > 0$, which clearly means that the matrix R has to be positive semidefinite. □

The results on the minimal radius disk associated with an AZ of $\mathcal{P}$, also provide sufficient conditions for $\mathcal{P}$ to be non-Hurwitzian. The result below follows from Theorem (5.3).

Proposition (6.3) Let $\mathcal{P}$ be a NA set, $z_i = \sigma_i + j\omega_i$ be the AZs of $\mathcal{P}$ for which $\sigma_i > 0$, and let us denote $x_i = \sigma_i/(2^{1/d}-1) \in \mathbb{R}^+$, where d is the degree of $\mathcal{P}$. A sufficient condition for $\mathcal{P}$ to be non-Hurwitzian is that for at least one z_i in the right half of the complex plane, the matrix $Q(z_i, x_i)$, defined by equation (5.7) is positive semidefinite. □

Note that if the above result holds true, then $\mathcal{P}$ is not only non-Hurwitzian, but also all combinants of $\mathcal{P}$ have at least one zero in the right half plane disks, centered at those AZs z_i for which $Q(z_i, x_i)$ is positive semidefinite.

7. CONCLUSIONS

The paper has surveyed a number of results on polynomial combinants. The emphasis here has been on those properties of polynomial combinants, which are of interest to control theory problems. Thus, the conditions for arbitrary assignment may be used to provide necessary conditions for arbitrary pole, zero assignability. The criteria for a set to be non-Hurwitzian may be used to check whether a system is not stabilizable by constant output feedback. The extension of the notion of an exact zero of $\mathcal{P}$ to that of an almost zero, provides the means for the extension of the system theoretic concepts of multivariable zeros and decoupling zeros to their "almost" versions. The trapping of the zeros of polynomial combinants by the almost zeros indicates that the similarities of exact and almost zeros extend beyond their common definition as minima of a norm function of P. The results on the stability properties of combinants are of a preliminary character; in fact, they are necessary conditions for P to be Hurwitzian, or sufficient conditions for P to be non-Hurwitzian. Sufficient conditions for a set to be Hurwitzian are under investigation. The theory of polynomial combinants forms the linear part of the frequency assignment problems of linear control theory; these results together with a number of results from exterior algebra and algebraic geometry [5] provide the means for the development of a unifying approach for frequency assignment and stabilization problems of linear systems.

REFERENCES

[1] Karcanias, N. and Giannakopoulos, C., 1983. "Grassman invariants, almost zeros and the determinantal zero, pole assignment problems of linear multivariable systems". The City Univ.Res.Rep., DSS/NK-CG/236.

[2] Karcanias, N., Giannakopoulos, C. and Hubbard, M., 1983. "Almost zeros of a set of polynomials". Int.J. Control, **38**, 1213-1238.

[3] MacFarlane, A.G.J. and Karcanias, N., 1976. "Poles and zeros of linear multivariable systems: A survey of Algebraic, Geometric and Complex variable Theory". Int.J. Control, **24**, 33-74.

[4] Rosenbrock, H.H., 1970. "*State space and Multivariable Theory*". Nelson, London.

[5] Karcanias, N. and Giannakopoulos, C., 1983. "Exterior Algebra and the problems of pole, zero assignment of linear systems". Proc. of MECO 83, Athens.

[6] Marden, M., 1949. "*The geometry of a polynomial in a complex variable*". American Mathematical Society Publication.

[7] Henrici, P., 1974. "*Applied and Computational Complex Analysis*". J. Wiley & Sons, New York.

[8] Gantmacher, F.R., 1959. "*The Theory of Matrices*", **2**, Chelsea, New York.

[9] Fan, K., 1956. "Systems of linear inequalities". In *Linear Inequalities and related systems*. Edit. by H.N. Kuhn and A.W. Tucker. Princeton University Press, 99-156.

[10] Marcus, M. and Minc, H., 1964. "*A survey of matrix theory and matrix inequalities*". Allyn and Bacon, Boston.

[11] Kailath, T., 1980. "*Linear systems*", Prentice Hall Inc., Englewood Cliffs, N.J., U.S.A.

CHAPTER 3

THE OCCURRENCE OF NON-PROPERNESS IN CLOSED-LOOP SYSTEMS AND SOME IMPLICATIONS

A.C. Pugh,
Department of Mathematics,
Loughborough University of Technology,
Loughborough, Leics.
England.

ABSTRACT:

The question of whether or not a closed-loop system under constant output feedback is proper is considered and conditions of two different types depending on the particular realisation available, are derived. The implications of these observations in the context of composite systems are then explored.

1. INTRODUCTION

This paper discusses the problem of closed-loop system nonproperness and derives various characterisations of those constant output feedback matrices F which produce non-proper closed-loop systems. The conditions presented are of two types depending on the realisation of G(s) that is employed.

If this transfer function matrix is written in terms of its strictly proper and polynomial parts which is essentially the case of state-space realisations then certain conditions will be shown to arise from the interrelation of F and the polynomial part of G(s). On the other hand if G(s) is written as a matrix fraction which is taken as minimal (in a sense to be defined) then certain conditions on F will be developed in relation to the high order coefficient matrix of this matrix fraction.

Subsequently the paper concerns itself with composite systems and makes specific comments on the series connection of two subsystems. Mostly work on composite systems deals with behaviour at finite frequencies and the paper goes on to suggest that equal consideration should be given to the point at infinity. In the case of the interconnection in series of two subsystems some detailed results will be presented.

S. G. Tzafestas (ed.), Multivariable Control, 43–63.

2. PRELIMINARY OBSERVATIONS

The effect of constant output feedback on the poles of an open-loop system represented by a transfer function matrix G(s) has long been of interest. A fundamental result, and one from which the ideas of this paper stem is

Lemma 1: The McMillan degree [1] of the open-loop transfer matrix G(s), denoted $\delta(G)$, is invariant under constant output feedback, however the least order of G(s) [ibid], denoted $\nu(G)$, may change.

Proof: The first part of this result was established in [2],[3], while the observation concerning the least order (together with the result as a whole) is illustrated by the following example.

Example 1: Consider

$$G(s) = \begin{pmatrix} \dfrac{1}{s^2+1} & \dfrac{1+s^2-s^3}{s(s^2+1)} \\ 0 & \dfrac{1-s}{s} \end{pmatrix} \tag{1}$$

Then G(s) may be written

$$G(s) = \begin{pmatrix} s^2+1 & -s^2-1 \\ 0 & s \end{pmatrix}^{-1} \begin{pmatrix} 1 & 1 \\ 0 & 1-s \end{pmatrix} \tag{2}$$

Now (2) constitutes a relatively left prime factorisation of the rational matrix G(s) and consequently $\nu(G)$ is simply the degree of the determinant of the denominator matrix. In the case of (2) it is readily apparent that

$$\nu(G) = 3 \tag{3}$$

Consider now the closed-loop system $G_1(s)$ obtained when output feedback as described by the matrix

$$F = \begin{pmatrix} 1 & 0 \\ 0 & 1 \end{pmatrix} \tag{4}$$

is applied. Then

$$G_1(s) \triangleq (I+G(s)F)^{-1} G(s)$$

$$= \begin{bmatrix} \dfrac{1}{s^2+2} & \dfrac{1+s^2-s^3}{s^2+2} \\ 0 & 1-s \end{bmatrix} \quad (5)$$

and a relatively left prime factorisation of this is

$$G_1(s) = \begin{bmatrix} s^2+2 & -s^2 \\ 0 & 1 \end{bmatrix}^{-1} \begin{bmatrix} 1 & 1 \\ 0 & 1-s \end{bmatrix} \quad (6)$$

It is immediately clear that

$$\nu(G_1) = 2 \quad (7)$$

so that the least order has not been preserved under this feedback control law.

For the case of the McMillan degree it is noted that if G(s) is written as

$$G(s) = G_S(s) + D(s) \quad (8)$$

where $G_S(s)$ is strictly proper and D(s) is polynomial then [1]

$$\delta(G) \triangleq \nu(G_S) + \nu(D(s^{-1})) \quad (9)$$

In respect of the matrices (1) and (5) it is seen that (8) gives

$$G(s) = \begin{bmatrix} \dfrac{1}{s^2+1} & \dfrac{s^2+s+1}{s(s^2+1)} \\ 0 & \dfrac{1}{s} \end{bmatrix} + \begin{bmatrix} 0 & -1 \\ 0 & -1 \end{bmatrix} \quad (10)$$

$$G_1(s) = \begin{bmatrix} \dfrac{1}{s^2+2} & \dfrac{2s-1}{s^2+2} \\ 0 & 0 \end{bmatrix} + \begin{bmatrix} 0 & 1-s \\ 0 & 1-s \end{bmatrix} \quad (11)$$

Using (9) it is then seen that

$$\delta(G) = 3 + 0 = 3$$
$$\delta(G_1) = 2 + 1 = 3 \qquad (12)$$

which verifies the invariance of the McMillan degree.

The quantities described in Lemma 1 have the following interpretations [1],[4].

Lemma 2: (i) The McMillan degree $\delta(G)$ represents the total number of poles (finite and infinite) of $G(s)$ counted according to their multiplicity and their degree.

(ii) The least order $\nu(G)$ represents the total number of finite poles of $G(s)$ counted according to their multiplicity and degree.

It is seen from these two results that the application of constant output feedback simply moves the poles of the open-loop system around the complex plane, and it is possible for certain feedbacks to place certain of the poles at the point at infinity. When this occurs there is a discrepancy between the least order and the McMillan degree which is directly attributable to the number of poles that are situated at the point at infinity. If there are no poles at infinity this discrepancy is zero for the particular closed-loop transfer function matrix $G_F(s)$ obtained by using a given feedback matrix F

$$\text{i.e} \quad \nu(G_F) = \delta(G_F) \qquad (13)$$

which is the precise requirement that $G_F(s)$ be proper [4],[5],[6].

Thus the problem of determining when a closed-loop system $G_F(s)$ possesses infinite poles is equivalent to that of determining when $G_F(s)$ is non-proper. A basic result concerning this issue indicates that non-properness is a non-generic property.

Lemma 3: For a finite constant feedback matrix F

(i) $G_F(s)$ is strictly proper if and only if $G(s)$ is strictly proper.

(ii) $G_F(s)$ is almost always proper if $G(s)$ is not strictly proper.

Proof: The result (i) was established in [6], while (ii) is noted in [7].

As a consequence of this result there remains just one interesting question and that is to characterise those feedback matrices F which produce a non-proper $G_F(s)$. It is this problem which is studied in the next section.

3. CONDITIONS FOR CLOSED-LOOP NONPROPERNESS

It is evident from lemma 3 that the question of whether or not a closed-loop system is proper only arises in case the open-loop system is not strictly proper. Accordingly it will be assumed that the open-loop transfer function matrix G(s) may be written in the form (8) where

$$D(s) \not\equiv 0 \qquad (14)$$

The conditions to be developed are of two types depending on the particular realisation of the open-loop G(s) that is available. In the first place assume that a least order state-space realisation of G(s) is known and let

$$P(s) = \begin{bmatrix} sI_\nu - A & B \\ -C & D(s) \end{bmatrix} \qquad (15)$$

be the system matrix corresponding to this realisation where $\nu = \nu(G)$,

$$G(s) = C(sI_\nu - A)^{-1} B + D(s) \qquad (16)$$

and D(s) in (16) is identical with D(s) in (8) and (14). The main result is

<u>Theorem 1:</u> If G(s) arises from the least order system matrix (15) with the polynomial matrix D(s) satisfying (14) then the closed-loop transfer function matrix $G_F(s)$ is proper if and only if

$$\delta(|I_m + D(s)F|) = \delta(D(s)) \qquad (17)$$

where $|\cdot|$ denotes the determinant of the indicated matrix.

<u>Proof:</u> Since P(s) in (15) has least order it follows from [1] and [6] that

$$P_F(s) = \left[\begin{array}{ccc|c} sI_\nu - A & B & 0 & 0 \\ -C & D(s) & I & 0 \\ 0 & -I & F & I \\ \hline 0 & 0 & -I & 0 \end{array}\right] \qquad (18)$$

is a least order realisation of the closed-loop transfer function matrix $G_F(s)$. By strict system equivalence [1], $P_F(s)$ can be reduced to the form

$$P'_F(s) = \left[\begin{array}{cc|c} sI_\nu - A & BF & B \\ -C & I+D(s)F & D(s) \\ \hline 0 & -I & 0 \end{array}\right] \qquad (19)$$

which also has least order. By the property of least order realisations

$$\nu(G_F) = \delta\left\{\begin{vmatrix} sI_\nu - A & BF \\ -C & I+D(s)F \end{vmatrix}\right\} \qquad (20)$$

Now the determinant

$$\begin{vmatrix} sI_\nu - A & BF \\ -C & I+D(s)F \end{vmatrix} \qquad (21)$$

can be expanded by the first ν rows using a Laplace expansion. Clearly the highest degree for determinants generated from the first ν rows is ν. Further the highest degree among minors of all orders of

$$(-C \qquad I+D(s)F) \qquad (22)$$

is its McMillan degree which since C is constant is $\delta(I+D(s)F)$. Now

$$\begin{aligned} \delta(I+D(s)F) &= \delta(D(s)F) \\ &\leqslant \delta(D(s)) \end{aligned} \qquad (23)$$

and so $\delta(D(s))$ is an upper bound for the degree of minors of all orders of (22). Consequently if

$$\delta(|I+D(s)F|) = \delta(D(s)) \qquad (24)$$

then the above Laplace expansion of (21) will contain a term of degree $\nu+\delta(D(s))$. Further from the form of the first ν rows of (21), it follows that this is the only term which can possess this degree, in fact all other terms in the Laplace expansion have degree strictly less than $\nu+\delta(D(s))$. Hence (21) has degree $\nu+\delta(D(s))$ if and only if (24) holds.

Suppose (21) has degree $\nu+\delta(D(s))$ then since $P'_F(s)$ has least order

$$\begin{aligned} \nu(G_F) &= \nu+\delta(D(s)) \\ &= \delta(G) \qquad \text{(by (9))} \\ &= \delta(G_F) \qquad \text{(by lemma 1)} \end{aligned}$$

i.e. $G_F(s)$ is proper.

Conversely we have

$$\delta(G_F) = \delta(G) = \nu+\delta(D(s))$$

Also, from (19),

$$\nu(G_F) = \delta\left(\begin{vmatrix} sI_\nu-A & BF \\ -C & I+D(s)F \end{vmatrix}\right)$$

Now if G(s) is proper then

$$\nu(G_F) = \delta(G_F)$$

$$\text{i.e. } \nu+\delta(D(s)) = \delta\left(\begin{vmatrix} sI_\nu-A & BF \\ -C & I+D(s)F \end{vmatrix}\right)$$

But as proved above this relation holds only if

$$\delta(|I+D(s)F|) = \delta(D(s)).$$

Conversely if $G_F(s)$ is proper then the above argument may be reversed to show that

$$\delta(|I+D(s)F|) = \delta(D(s))$$

which completes the proof.

There are various Corollaries of this theorem.

Corollary 1: The closed-loop system transfer function matrix $G_F(s)$ is non-proper if and only if

$$\delta(|I+DF|) < \delta(D(s)) \tag{25}$$

Proof: Obvious.

Corollary 2: If G(s) is proper so that in (16) $D(s) \equiv D$ a constant matrix then $G_F(s)$ is non-proper if and only if

$$|I+DF| = 0 \tag{26}$$

Proof: Since $D(s) \equiv D$ a constant $\delta(D) = 0$ and so from (25) $G_F(s)$ is nonproper if and only if

$$\delta(|I+DF|) < 0$$

Clearly this obtains if and only if $|I+DF| = 0$.

Example 2: To illustrate the above results reconsider Example 1. For G(s) of (1) the corresponding D is from (10)

$$D = \begin{bmatrix} 0 & -1 \\ 0 & -1 \end{bmatrix}$$

then for F of (4)

$$|I+DF| = \begin{vmatrix} 1 & -1 \\ 0 & 0 \end{vmatrix} = 0$$

This by Corollary 2 predicts that the resulting closed-loop $G_1(s)$ will be non-proper a fact which is verified by (11).

Note that if output feedback is applied around $G_1(s)$ according to the law described by

$$F_1 = \begin{bmatrix} -1 & 0 \\ 0 & -1 \end{bmatrix} \qquad (27)$$

then from (11)

$$D_1(s) = \begin{bmatrix} 0 & 1-s \\ 0 & 1-s \end{bmatrix}$$

and so

$$|I+D_1(s)F_1| = \begin{vmatrix} 1 & s-1 \\ 0 & s \end{vmatrix} = s$$

Hence

$$\delta(|I+D_1(s)F_1|) = 1 = \delta(D_1(s))$$

and so by theorem 1 the resulting closed-loop system is proper. This is readily verified since this feedback action recovers the original G(s) of (1) from $G_1(s)$.

The condition of theorem 1 provides a complete characterisation of those feedback matrices producing non-proper closed-loop systems. Although this result was proved by reference to a state-space realisation of G(s) it is not necessary to generate such a realisation, if only G(s) is available, in order to check the condition (17). All that is required is the direct feed through matrix D(s) of G(s) and this can be more simply generated via (8).

Despite this the condition (17) of theorem 1 may be rather difficult to check. If however one has a little more information concerning G(s), say a matrix fraction description, then a more readily verifiable condition may be generated as follows.

Definition 1: The left matrix fraction description

$$G(s) = T^{-1}(s)\ N(s) \tag{28}$$

is said to be MINIMAL in case the matrix

$$(T(s) \quad N(s)) \tag{29}$$

forms a minimal basis [8]. Similarly the right matrix fraction description

$$G(s) = N_1(s)\ T_1^{-1}(s) \tag{30}$$

is said to be minimal in case the matrix

$$\begin{pmatrix} T_1(s) \\ N_1(s) \end{pmatrix} \tag{31}$$

forms a minimal basis.

Now minimal factorisations possess an important structural property from a feedback point of view in that one is able to generate minimal factorisations of the closed-loop $G_F(s)$ from a minimal factorisation of the open-loop G(s) in a rather simple manner. The following result proved in [9] explains.

Lemma 4: If G(s) arises from the minimal factorisation (28) then

$$G_F(s) = (T(s) + N(s)\ F)^{-1}\ N(s) \tag{32}$$

is a minimal factorisation of $G_F(s)$ and further

$$(T(s)+N(s)F, \quad N(s)) \tag{33}$$

and (29) have identical row degrees. Analogously if G(s) is given by (30) then

$$G_F(s) = N_1(s)\ (T_1(s) + FN_1(s))^{-1} \tag{34}$$

is a minimal factorisation of G_F and further

$$\begin{pmatrix} T_1(s) + FN_1(s) \\ N_1(s) \end{pmatrix} \tag{35}$$

and (31) have identical column degrees.

The second type of condition for closed-loop non-properness is as follows.

Theorem 2: If the $m \times \ell$ open-loop transfer function matrix G(s) arises from the minimal factorisation (28) then the closed-loop transfer function matrix $G_F(s)$ is proper if and only if the constant $m \times m$ matrix

$$[T \quad N]_{hr} \begin{bmatrix} I_m \\ F \end{bmatrix} \tag{36}$$

is nonsingular, where $[\cdot]_{hr}$ denotes the high-order coefficient matrix for the rows of the indicated matrix.

Proof: By lemma 4 a minimal factorisation of $G_F(s)$ is

$$G_F(s) = (T(s) + N(s)F)^{-1}N(s) \tag{37}$$

Consider

$$(T(s)+N(s)F, \quad N(s)) \tag{38}$$

Now by [2] the sum of the row degrees of this matrix is $\delta(G_F)$ the McMillan degree of $G_F(s)$, which is an upper bound for the degrees of the minors of (38) of all orders. In particular then

$$\delta(|T(s) + N(s)F|) \leq \delta(G_F) \tag{39}$$

and it is required to characterise when equality obtains since

$$\delta(|T(s) + N(s)F|) \triangleq \nu(G_F) \tag{40}$$

Now from (38),

$$(T(s)+N(s)F, \quad N(s)) = (T(s) \quad N(s)) \begin{pmatrix} I_m & 0 \\ F & I_\ell \end{pmatrix} \tag{41}$$

and since (38) and (29) have the same row degrees it follows that

$$[T+NF, \quad N]_{hr} = [T, \quad N]_{hr} \begin{pmatrix} I_m & 0 \\ F & I_\ell \end{pmatrix} \tag{42}$$

Now only the nonsingular m×m minors of the left-hand matrix will correspond to m×m minors of (38) with degree equal to $\delta(G_F)$.

Consequently equality obtains in (39) if and only if the m×m minor formed from the first m columns of the left-hand matrix in (42) is nonsingular. Hence by (41) equality obtains in (39) if and only if (36) holds, which completes the proof.

Analogously it may be proved that

Theorem 3: If G(s) arises from the minimal factorisation (30) then $G_F(s)$ is proper if and only if the constant $\ell\times\ell$ matrix

$$[I_\ell \quad F]\begin{bmatrix} T_1 \\ N_1 \end{bmatrix}_{hc} \tag{43}$$

is nonsingular, where $[\cdot]_{hc}$ denotes the high-order coefficient matrix for the columns of the indicated matrix.

The conditions presented in theorems 2 and 3 are readily verifiable as the following example shows.

Example 3: Consider again Example 1. The factorisation (2) of G(s) is easily seen to be minimal. The factorisation (6) of $G_1(s)$ is also minimal and in fact has been obtained in the manner described by lemma 4. Consider the condition (36) then from (1) it is seen that

$$[T \quad N]_{hr} = \begin{pmatrix} 1 & -1 & \vdots & 0 & 0 \\ 0 & 1 & \vdots & 0 & -1 \end{pmatrix}$$

Hence (36) gives

$$\begin{pmatrix} 1 & -1 & \vdots & 0 & 0 \\ 0 & 1 & \vdots & 0 & -1 \end{pmatrix}\begin{pmatrix} 1 & 0 \\ 0 & 1 \\ \hline 1 & 0 \\ 0 & 1 \end{pmatrix} = \begin{pmatrix} 1 & -1 \\ 0 & 0 \end{pmatrix}$$

which is clearly singular. Hence $G_1(s)$ is non-proper by theorem 2 a fact which is easily verified from (11).

On the other hand suppose output feedback as described by the matrix F_1 of (27) is applied to $G_1(s)$ then the original G(s) results which is clearly proper from (10). In this case the condition (36) applied to the minimal factorisation (6) gives

$$\begin{bmatrix} 1 & -1 & \vdots & 0 & 0 \\ 0 & 0 & \vdots & 0 & -1 \end{bmatrix} \begin{bmatrix} 1 & 0 \\ 0 & 1 \\ \hline -1 & 0 \\ 0 & -1 \end{bmatrix} = \begin{bmatrix} 1 & -1 \\ 0 & 1 \end{bmatrix}$$

which is nonsingular and verifies theorem 2.

4. IMPLICATIONS FOR COMPOSITE SYSTEMS

For simplicity the case of two systems connected in series will be considered but it is possible to make analogous statements about other interconnections of two subsystems and about composite systems in general. Consider therefore the series connection of two transfer function matrices $G_1(s)$, $G_2(s)$ as shown in the figure 1.

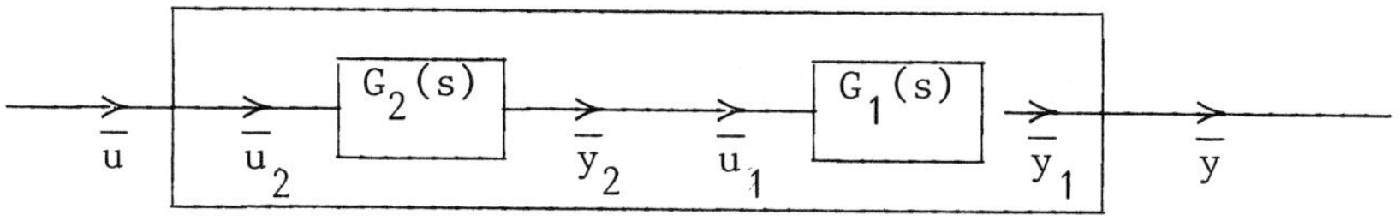

fig.1.

An interesting problem concerning this interconnection is that of providing conditions under which the composite system pole structure precisely reflects that of the component systems. Intuitively this occurs provided no pole of one subsystem cancels with a zero of the other when the product $G_1(s)\, G_2(s)$ is formed and indeed this phenomenon may be formally defined ([10]) as follows.

Denote the number of poles of the rational matrix G(s) occurring at a specific frequency $s_o \in C$ by $\delta_{s_o}(G(s))$, and the number of zeros of G(s) occurring at s_o by $\partial_{s_o}(G(s))$. The number of poles and zeros at the point at infinity of G(s) are similarly denoted by $\delta_\infty(G(s))$ and $\partial_\infty(G(s))$ respectively. It is noted that

$$\delta_\infty(G(s)) = \delta_o\left(G\left(\frac{1}{\omega}\right)\right) \quad ; \quad \partial_\infty(G(s)) = \partial_o\left(G\left(\frac{1}{\omega}\right)\right) \tag{44}$$

Following [10] we therefore say that

Definition 2: With $G_1(s)$ and $G_2(s)$ as above, the product $G_1(s)\ G_2(s)$ is said to contain NO POLE-ZERO CANCELLATION AT $s_o \in C \cup \{\infty\}$, in case

$$\delta_{s_o}(G_1(s)G_2(s)) = \delta_{s_o}(G_1(s)) + \delta_{s_o}(G_2(s)) \tag{45}$$

The following lemma is established in [10] and is stated here since it is required in the derivation of the new results.

Lemma 5: Let A(s), B(s), C(s), D(s), E(s), F(s) be polynomial matrices such that

$$G(s) = D^{-1}\ EF^{-1} = AB^{-1}C \tag{46}$$

for some rational matrix G(s) and let $s_o \in C$ be finite, then

(i) $$\delta_{s_o}(G(s)) = \partial_{s_o}(D(s)) + \partial_{s_o}(F(s)) \tag{47}$$

if and only if

$$(D(s_o)\ E(s_o)) \quad \text{and} \quad \begin{pmatrix} E(s_o) \\ F(s_o) \end{pmatrix} \tag{48}$$

have full rank.

(ii) $$\delta_{s_o}(G(s)) = \partial_{s_o}(B(s)) \tag{49}$$

if and only if

$$\begin{pmatrix} A(s_o) \\ B(s_o) \end{pmatrix} \quad \text{and} \quad (B(s_o) \quad C(s_o)) \tag{50}$$

have full rank.

Suppose now that the following relatively prime matrix decompositions of $G_1(s)$ and $G_2(s)$ of figure 1 are available

$$G_i(s) = T_{ri}^{-1}(s)\ N_{ri}(s) = N_{ci}(s)\ T_{ci}^{-1}(s) \qquad (i=1,2) \tag{51}$$

Then a fairly direct application of Lemma 5 yield the following theorem which forms one of the main results of [10]. This result is stated here for comparison purposes.

Lemma 6: There is no pole-zero concellation at the finite frequency $s_o \in C$ when forming the product $G_1(s)\ G_2(s)$ if and only if one of the following three equivalent sets of conditions hold:

$$\text{(i)} \quad \begin{bmatrix} N_{c1}(s_o) \\ T_{r2}(s_o)T_{c1}(s_o) \end{bmatrix} \quad \text{and} \quad (T_{r2}(s_o)T_{c1}(s_o) \quad N_{r2}(s_o)) \tag{52}$$

have full rank

$$\text{(ii)} \quad \begin{bmatrix} T_{c2}(s_o) \\ N_{r1}(s_o)N_{c2}(s_o) \end{bmatrix} \quad \text{and} \quad (T_{r1}(s_o) \quad N_{r1}(s_o)N_{c2}(s_o)) \tag{53}$$

have full rank.

$$\text{(iii)} \quad \begin{bmatrix} N_{r1}(s_o) \\ T_{r2}(s_o) \end{bmatrix} \quad \text{and} \quad (T_{c1}(s_o) \quad N_{c2}(s_o)) \tag{54}$$

have full rank.

In [10] a restriction to consideration of proper (i.e. causal) subsystems was made since the immediate application was to discrete systems. In that event neither subsystem transfer function possesses poles of infinity and so the possibility of pole-zero cancellations taking place at the point at infinity does not arise. Consequently there was no need in [10] to extend Lemma 6 to include the point at infinity.

If however the application of the results of Lemma 6 is in the context of continuous time systems then there are several reasons why this extension is essential. It may be of course that the subsystems themselves are non-proper since in the continuous time case this simply means that the system exhibits impulsive behaviour rather than that it is noncausal. On the other hand even if the subsystems themselves are proper and connected serially together it is possible, in view of what has been said above, that the implementation of local output feedback schemes around $G_1(s)$ and $G_2(s)$ may result in the serial connection of non-proper subsystems. Now it is clear that when non-proper subsystems (i.e. subsystems which possess poles at infinity) are considered, then pole-zero cancellations could well occur at the point at infinity when the product $G_1(s)\ G_2(s)$ is formed. Consequently from a continuous time point of view the extension of Lemma 6 to include the point at infinity is extremely desirable and it will be shown that a particularly simple and appealing extension is available.

To this end assume that the factorisations (51) of the subsystem transfer functions $G_1(s)$, $G_2(s)$ are minimal in the sense of definition 1.

Of course the factorisations (51) are already relatively prime and so minimal factorisations may be generated fairly readily by pre-or post-multiplication of the matrices from (51) corresponding to (29) or (31) by an appropriate unimodular matrix [11].

Now minimal factorisations have a further important structural property besides that described in Lemma 4.

Lemma 7: Let G(s) be an $m\times\ell$ rational matrix and

$$G(s) = T^{-1}(s)\, N(s) \tag{55}$$

a minimal factorisation with associated row degrees δ_i $(i=1,\ldots,m)$. Define the polynomial matrix

$$\Lambda_r(G) = \mathrm{diag}(\omega^{\delta_1},\ldots,\omega^{\delta_m}) \tag{56}$$

then

$$\left(\Lambda_r(G)\, T\left(\frac{1}{\omega}\right)\right)^{-1}\left(\Lambda_r(G)\, N\left(\frac{1}{\omega}\right)\right) \tag{57}$$

is a relatively left prime factorisation in ω of $G\left(\frac{1}{\omega}\right)$.

Proof: The result is a simple corollary of theorem 2 of reference [9].

Note: The matrix analogous to (56) for the factorisation (31) of G(s) will be denoted $\Lambda_c(G)$.

It is noted from the above result that a minimal factorisation of the rational matrix G(s) permits relatively prime factorisations of $G\left(\frac{1}{\omega}\right)$ to be readily constructed. In that the pole-zero structure of G(s) at the point at infinity is defined as that of $G\left(\frac{1}{\omega}\right)$ at $\omega = 0$ it is thus seen that a minimal factorisation provides a capacity to work with all frequencies, finite and infinite, simultaneously. It is the implication of this observation which permits the required extension of Lemma 6 to be obtained.

To state the extension of Lemma 6 for the point at infinity it will be assumed that the factorisations (51) are minimal and it will be necessary to refer to the individual blocks in the high order coefficient matrices corresponding to the factorisations (51). Thus write

$$\left.\begin{aligned}\begin{bmatrix} T_{ci}(s) \\ N_{ci}(s) \end{bmatrix}_{hc} &= \begin{pmatrix} (T_{ci})_{hc} \\ (N_{ci})_{hc} \end{pmatrix} \\ \begin{bmatrix} T_{ri}(s) & N_{ri}(s) \end{bmatrix}_{hr} &= \begin{pmatrix} (T_{ri})_{hr} & (N_{ri})_{hr} \end{pmatrix}\end{aligned}\right\} \quad (i=1,2) \tag{58}$$

then we have

<u>Theorem 4:</u> There is no pole-zero cancellation at the point at infinity when forming the product $G_1(s)\ G_2(s)$ if and only if one of the following three equivalent sets of conditions holds

$$\text{(i)} \quad \begin{pmatrix} (N_{c1})_{hc} \\ (T_{r2})_{hr}(T_{c1})_{hc} \end{pmatrix} \quad \text{and} \quad \begin{pmatrix} (T_{r2})_{hr}(T_{c1})_{hc}, & (N_{r2})_{hr} \end{pmatrix} \tag{59}$$

have full rank.

$$\text{(ii)} \quad \begin{pmatrix} (T_{c2})_{hc} \\ (N_{r1})_{hr}(N_{c2})_{hc} \end{pmatrix} \quad \text{and} \quad \begin{pmatrix} (T_{r1})_{hr}, & (N_{r1})_{hr}(N_{c2})_{hc} \end{pmatrix} \tag{60}$$

have full rank.

$$\text{(iii)} \quad \begin{pmatrix} (N_{r1})_{hr} \\ (T_{r2})_{hr} \end{pmatrix} \quad \text{and} \quad \begin{pmatrix} (T_{c1})_{hc}, & (N_{c2})_{hc} \end{pmatrix} \tag{61}$$

have full rank.

<u>Proof:</u> Note that

$$G_1\left(\frac{1}{\omega}\right)G_2\left(\frac{1}{\omega}\right) = N_{c1}\left(\frac{1}{\omega}\right)\Lambda_c(G_1)\left[\Lambda_r(G_2)T_{r2}\left(\frac{1}{\omega}\right)T_{c1}\left(\frac{1}{\omega}\right)\Lambda_c(G_1)\right]^{-1} \Lambda_r(G_2)N_{r2}\left(\frac{1}{\omega}\right) \tag{62}$$

is a polynomial factorisation in ω. To ensure no pole-zero cancellations at $\omega = 0$ we apply Lemma 5 which requires that

$$\begin{pmatrix} N_{c1}\left(\frac{1}{\omega}\right)\Lambda_c(G_1) \\ \\ \Lambda_r(G_2)T_{r2}\left(\frac{1}{\omega}\right)T_{c1}\left(\frac{1}{\omega}\right)\Lambda_c(G_1) \end{pmatrix} \tag{63}$$

and

$$\left(\Lambda_r(G_2)T_{r2}\left(\frac{1}{\omega}\right)T_{c1}\left(\frac{1}{\omega}\right)\Lambda_c(G_1) \qquad \Lambda_r(G_2)N_{r2}\left(\frac{1}{\omega}\right)\right) \tag{64}$$

both have full rank at $\omega = 0$.

In view of the fact that the factorisations (51) are minimal the matrix (63) when evaluated at $\omega = 0$ reduces by (58) to

$$\begin{pmatrix} (N_{c1})_{hc} \\ \\ (T_{r2})_{hr}(T_{c1})_{hc} \end{pmatrix} \tag{65}$$

Hence (63) has full rank at $\omega = 0$ if and only if the constant matrix (65) has full rank.

Similarly the condition that (64) has full rank at $\omega = 0$ reduces to the second of the conditions (59). This establishes the set of conditions (i).

For the sets of conditions (ii) and (iii) an analogous arguement concerning the polynomial factorisation

$$G_1\left(\frac{1}{\omega}\right)G_2\left(\frac{1}{\omega}\right)$$

$$= \left[\Lambda_r(G_1)T_{r1}\left(\frac{1}{\omega}\right)\right]^{-1}\Lambda_r(G_1)N_{r1}\left(\frac{1}{\omega}\right)N_{c2}\left(\frac{1}{\omega}\right)\Lambda_c(G_2)\left[T_{c2}\left(\frac{1}{\omega}\right)\Lambda_c(G_2)\right]^{-1} \tag{66}$$

yields (60), while consideration of the polynomial factorisations

$$G_1\left(\frac{1}{\omega}\right)G_2\left(\frac{1}{\omega}\right)$$

$$= \left[\Lambda_r(G_1)T_{r1}\left(\frac{1}{\omega}\right)\right]^{-1}\Lambda_r(G_1)N_{r1}\left(\frac{1}{\omega}\right)\left[\Lambda_r(G_2)T_{r2}\left(\frac{1}{\omega}\right)\right]^{-1}\Lambda_r(G_2)N_{r2}\left(\frac{1}{\omega}\right) \tag{67}$$

$$= N_{c1}\left(\frac{1}{\omega}\right)\Lambda_c(G_1)\left[T_{c1}\left(\frac{1}{\omega}\right)\Lambda_c(G_1)\right]^{-1}N_{c2}\left(\frac{1}{\omega}\right)\Lambda_c(G_2)\left[T_{c2}\left(\frac{1}{\omega}\right)\Lambda_c(G_2)\right]^{-1} \tag{68}$$

yields (61) which establishes the theorem.

Although at first sight the conditions of theorem 4 may appear complicated this is not the case and in fact these conditions are rather easy to construct and verify since they are merely constructed from high-order coefficient matrices of the relevant minimal factorisations. Thus the additional work in forcing the factorisations (51) to be minimal rather than relatively prime results in readily verifiable conditions for the absence of pole-zero cancellations at the point at infinity when the product $G_1(s)\ G_2(s)$ is formed. The closeness of the analogy between the conditions in theorem 4 and Lemma 6 is also noted.

Example 4: Suppose

$$G_1(s) = \begin{pmatrix} \frac{1}{s+2} & \frac{2+s-s^2}{(s+2)^2} \\ 0 & \frac{s^2+1}{s+2} \end{pmatrix} \tag{69}$$

$$G_2(s) = \begin{pmatrix} s & 0 \\ 0 & \frac{s}{s+4} \end{pmatrix} \tag{70}$$

Then

$$G_1(s) = \begin{pmatrix} 1 & 1 \\ 0 & s^2+1 \end{pmatrix} \begin{pmatrix} s+2 & s^2 \\ 0 & s+2 \end{pmatrix}^{-1} = N_{c1}(s)T_{c1}(s)^{-1} \tag{71}$$

$$G_2(s) = \begin{pmatrix} 1 & 0 \\ 0 & s+4 \end{pmatrix}^{-1} \begin{pmatrix} s & 0 \\ 0 & s \end{pmatrix} = T_{r2}^{-1}(s)N_{r2}(s) \tag{72}$$

are minimal factorisations with,

$$\Lambda_c(G_1) = \begin{pmatrix} s & 0 \\ 0 & s^2 \end{pmatrix} \quad ; \quad \Lambda_r(G_2) = \begin{pmatrix} s & 0 \\ 0 & s \end{pmatrix} \tag{73}$$

It may then be verified from (71) that $G_1(s)$ has zeros each of degree one at $s=-i$, $s=+i$ and $s=\infty$, and poles at $s=\infty$ of degree one and $s=-2$ of degree two.

Similarly it may be shown from (72) that $G_2(s)$ has two zeros each of degree one at s=0, and poles of degree one at s=-4 and s=∞.

Consider now

$$G_1(s)G_2(s) = \begin{bmatrix} \frac{s}{s+2} & \frac{(2+s-s^2)s}{(s+4)(s+2)^2} \\ 0 & \frac{(s^2+1)s}{(s+2)(s+4)} \end{bmatrix} \tag{74}$$

which has the minimal factorisation

$$G_1(s)G_2(s) = N_c(s)T_c^{-1}(s)$$

$$= \begin{bmatrix} s & s \\ 0 & (s^2+1)s \end{bmatrix} \begin{bmatrix} s+2 & s^2 \\ 0 & (s+2)(s+4) \end{bmatrix}^{-1} \tag{75}$$

with

$$\Lambda_c(G_1G_2) = \begin{bmatrix} s & 0 \\ 0 & s^3 \end{bmatrix} \tag{76}$$

It may then be verified that $G_1(s)\ G_2(s)$ has zeros of degree one at s=0, s=0, s=-i, s=i and poles at s=-2 of degree two and at s=-4, s=∞ of degree one. Notice then that

$$\delta_\infty(G_1G_2) = 1 \neq \delta_\infty(G_1) + \delta_\infty(G_2) = 1 + 1 = 2 \tag{77}$$

from which it is apparent that there has been a cancellation at the point at infinity when $G_1(s)\ G_2(s)$ was formed.

To illustrate theorem 4 notice that the form of the factorisations (71), (72) are such that the set of conditions (i) apply. Now

$$\begin{bmatrix} T_{c1} \\ N_{c1} \end{bmatrix}_{hc} = \begin{bmatrix} 1 & 1 \\ 0 & 0 \\ \hline 0 & 0 \\ 0 & 1 \end{bmatrix} = \begin{bmatrix} (T_{c1})_{hc} \\ (N_{c1})_{hc} \end{bmatrix} \tag{78}$$

$$\begin{bmatrix} T_{r2} & N_{r2} \end{bmatrix}_{hr} = \left(\begin{array}{cc|cc} 0 & 0 & 1 & 0 \\ 0 & 1 & 0 & 1 \end{array}\right) = \left((T_{r2})_{hr} \quad (N_{r2})_{hr} \right) \tag{79}$$

Thus,

$$(T_{r2})_{hr}(T_{c1})_{hc} = \begin{pmatrix} 0 & 0 \\ 0 & 1 \end{pmatrix} \begin{pmatrix} 1 & 1 \\ 0 & 0 \end{pmatrix} = \begin{pmatrix} 0 & 0 \\ 0 & 0 \end{pmatrix} \tag{80}$$

and so the matrices in (59) become

$$\begin{pmatrix} (N_{c1})_{hc} \\ (T_{r2})_{hr}(T_{c1})_{hc} \end{pmatrix} = \left(\begin{array}{cc} 0 & 0 \\ 0 & 1 \\ \hline 0 & 0 \\ 0 & 0 \end{array}\right) \tag{81}$$

$$\left((T_{r2})_{hr}(T_{c1})_{hc} \ , \quad (N_{r2})_{hr} \right) = \left(\begin{array}{cc|cc} 0 & 0 & 1 & 0 \\ 0 & 0 & 0 & 1 \end{array}\right) \tag{82}$$

While (82) has full rank it is apparent that (81) does not which therefore confirms the observation (77).

CONCLUSIONS:

Two types of conditions which ensure the properness of a closed-loop system under constant output feedback have been derived. It was seen in particular that if the open-loop transfer function matrix is written in terms of its strictly proper and polynomial parts then certain conditions arise from the interrelation of the polynomial part of $G(s)$ and the feedback matrix F. The condition for closed-loop properness generated may prove difficult to check since it relies on the computation of the degrees of certain determinants but one should realise that the knowledge required of $G(s)$ in this computation is not great, merely its polynomial part. If however one has a little more information concerning $G(s)$, say a minimal matrix fraction description then a more readily verifiable condition for closed-loop properness results. It is thus true to say that the more detailed the realisation of $G(s)$ the more simple the conditions for properness become.

The implications of these observations for composite systems have also been considered. Because of the possibility of local feedback loops occurring in such composite systems it is suggested by the above that non-proper subsystems should generally be considered. In that event the subsystems have significant infinite frequency behaviour and to what extent this is inherited by the composite system is not a question that is usually addressed. To illustrate these comments the paper has concentrated on the connection in series of two subsystems and has derived some simple and readily verifiable conditions for the properness of the composite system.

REFERENCES

[1] Rosenbrock, H.H., "State-space and Multivariable Theory", Nelson, 1970.

[2] Rosenbrock, H.H., and Hayton, G.E., "Dynamical Indices of a Transfer Function Matrix", Int. J. Control, 20, 1974, pp.177-189.

[3] Pugh, A.C., "On Composite Systems that Preserve the Degrees of their Subsystems", Int. J. Control, 21, 1975, pp.465-473.

[4] Pugh, A.C. and Ratcliffe, P.A., "On the Zeros and Poles of a Rational Matrix", Int. J. Control, 30, 1979, pp.235-243.

[5] Verghese, G.C., "Infinite Frequency Behaviour in Generalised Dynamical Systems", Ph.D. Dissertation, 1978, Stanford University, California.

[6] Rosenbrock, H.H. and Pugh, A.C., "Contributions to a Hierarchical Theory of Systems", Int. J. Control, 19, 1974, pp.845-867.

[7] Anderson, B.D.O. and Scott, R.W., "Comments on Conditions for a Feedback Transfer Function Matrix to be proper", I.E.E.E. Trans. Aut. Control, AC-21, 1976, pp.632-633.

[8] Forney, G.D., "Minimal Bases of Rational Vector Spaces", S.I.A.M. J. Control, 13, 1975, pp.493-520.

[9] Pugh, A.C. and Ratcliffe, P.A., "Infinite Frequency Interpretations of Minimal Bases", Int. J. Control, 32, 1980, pp.581-588.

[10] Anderson, B.D.O. and Gevers, M.R., "On Multivariable Pole-Zero Cancellations and the Stability of Feedback Systems", I.E.E.E. Trans. Circuits and Systems, CS-28, 1981, pp.830-833.

[11] Wolovich, W.A., "Linear Multivariable Systems", Springer-Verlag, 1974.

CHAPTER 4

SKEW-SYMMETRIC MATRIX EQUATIONS IN MULTIVARIABLE CONTROL THEORY

F. C. Incertis
IBM Scientific Center
P° de la Castellana, 4
Madrid 1
Spain

ABSTRACT. By introducing a set of analytical transformations new representations of the continuous-time algebraic Riccati and Liapunov equations are obtained as matrix equations in the field of real orthogonal matrices. Making use of different functional relationships, linking orthogonal and skew-symmetric real matrices, new canonical representations are derived on the field of real skew-symmetric matrices. As a main result of those reformulations a reduction in the number of coupled quadratic equations to be solved is achieved. Closed form analytical solutions are obtained when some special functional or structural conditions are fulfilled in the transformed spaces. As an immediate application of the transformation approach closed form solutions for general Riccati equations in $\mathbb{R}^{2x2}$ and near analytical solutions in $\mathbb{R}^{3x3}$ are given by the first time. The theory is then formalized to the 'Algebraic Riccati Generator' concept and new useful bounds for the solutions of Riccati and Liapunov equations are derived.

1. INTRODUCTION

In this paper we shall be concerned with the continuous-time algebraic Riccati equation (ARE)

$$C^T C + A_1^T K_1 + K_1 A_1 - K_1 B_1 R^{-1} B_1^T K_1 = 0 \qquad (1)$$

in the form it arises in the linear-quadratic-gaussian (LQG) optimal control problem [1, 2]. Here $A \in \mathbb{R}^{nxn}$, $B \in \mathbb{R}^{nxm}$, $C \in \mathbb{R}^{rxn}$, $R \in \mathbb{R}^{mxm}$, and $R = R^T > 0$. Also $m, r \leq n$. As a particularization of the theory we shall consider the continuous-time algebraic Liapunov equation (ALE)

$$C^T C + A_1^T K_1 + K_1 A_1 = 0 \qquad (2)$$

in the form it arises in stability analysis, parameter optimization, covariance analysis, Luenberger's observers, etc.

Kalman[3] first demonstrated that when the pair $< A_1, B_1 >$ is controllable and the pair $< A_1, C >$ is observable, (1) has a unique positive

S. G. Tzafestas (ed.), Multivariable Control, 65–84.

definite solution. Kucera[4] then relaxed this condition to that the stabilizability and detectability are the necessary and sufficient conditions for the existence of a unique nonnegative definite solution to (1).

An important research effort in multivariable control theory during the last two decades has been focussed towards algebraic (and differential) matrix Riccati and Liapunov equation problems. The solution of Riccati equations plays a crucial role in optimal estimation and filtering theories, in scattering theory and is basic to the design of continuous and sample-data control systems using the solution of the LQG optimal control problem[5-7]. Most algorithms for the solution of algebraic Riccati equations fall into two categories: direct methods, such as the Potter's[8] and Laub's[9] methods, or iterative methods, such as the Kleinman's[10] and Newton's methods or doubling[11]. However, a comparatively little attention has been paid to the development of transformation methods to reformulate and solve Riccati equations into more suitable matrix spaces for which the different geometrical[22], structural[12,13,25] or functional[14-20] properties of the ARE are made more explicit and useful for analytical or computational purposes.

The required solution of ARE (1) belongs to the field of the real symmetric matrices. The key idea of the transformation approach is to utilize a set of analytical matrix transformations to obtain new representations of ARE in some particular matrix fields for which the application of algorithmical tools is more efficient than in the original formulation. Hence, transformation methods can be classified as 'mixed' methods since they usually apply in two steps: analytical transformations and numerical methods.

Several special cases of Riccati equations have been investigated in the literature by restricting the system matrices A_1, B_1 and the functional matrices C^TC, R to fulfill particular structural properties or analytical relations on transformed spaces.

Mozhaev[13] investigates the use of symmetry in linear-quadratic optimal control problems. He shows that if the equations of motion and the functional are invariant relative to a compact group of linear transformations of the variables, then the control problem can be reduced, by group representation theory methods, to several independent ones, the total dimension of which equals the dimension of the original problem. Jones[14] considers the special case where matrix C is invertible and commutes with a symmetric system matrix $A_1 = A_1^T$. In Incertis and Martínez[15] the less restrictive set of conditions:

1) C nonsingular

2) C^TCA_1 is a symmetric matrix

are imposed and a closed analytical formula for the ARE positive definite solution is obtained in this particular case. In Incertis[16], by forcing matrix C to be nonsingular and by utilizing the similarity transformation $A = CA_1C^{-1}$, $B = CB_1$ and the symmetric and skew-symmetric components S and D, respectively, of matrix A, it is demonstrated that if $M = Q + AA^T - DD^T$ is a positive semidefinite matrix and $MD = DM$, then the unique positive definite solution of (1) is given by $K_1 = C^T(M^{1/2} - S)C$. In[17] the nonsingularity of C is relaxed and replaced by a less restrictive commutativity condition, extending the results of[14-16] to a wider class of Riccati problems.

Application of the results of[14-17] is restricted, by the imposed positive definiteness and commutativity conditions, to the solution of ARE problems in systems with only real roots. In[18] the nonsingularity of matrix C is preserved and the remainder restrictive conditions are cancelled. For this subclass of Riccati equations a canonical representation is obtained as a polynomial quadratic matrix equation. This equation is then shown to be equivalent to a linear matrix equation where the solution must fulfil an additional orthogonality condition. Making use of a functional relationship, linking orthogonal and skew-symmetric matrices, the original ARE problem becomes equivalent to the solution of a new quadratic equation, where the unknown belongs to the field of the skew-symmetric matrices. The reformulation of[18] reduces the system of $n(n+1)/2$ coupled quadratic equations, associated with the symmetric solutions of ARE in $\mathbb{R}^{nxn}$, to an equivalent system of $n(n-1)/2$ coupled quadratic equations, associated with the skew-symmetric solutions of the transformed equation. The skew-symmetric equation has been interpreted[18] as the 'dual' representation of the algebraic Riccati equation and the formulation has been applied in[19] to solve a subclass of differential Riccati equations with a positive definite state cost matrix C^TC. Other reformulations in the literature, such as[12,15] are very particular and restrictive.

The organization of this paper has been structured in two main parts. In the first part the nonsingularity condition of matrix C is cancelled and the transformation approach of[18] is extended to general continuous-time Riccati equations. Two versions of the method are presented, the first related with orthogonal representations and the second related with two different skew-symmetric representations, and derived from the orthogonal formulation. As in[18] the main theoretical result in this part is a reduction in the number of coupled quadratic equations to be solved in the transformed spaces. Moreover, several new analytical solutions of the ARE are obtained when some special, but not trivial, structural or functional conditions are fulfilled in the transformed spaces. Those conditions remain obscure in the original formulation but appear quite clear and are easily understood in the reformulated framework. As a particularization of this transformation approach a closed form solution in $\mathbb{R}^{2x2}$ and a near-analytical solution in $\mathbb{R}^{3x3}$ are given by the first time for general Riccati equations. Moreover, several new closed form analytical solutions are obtained in $\mathbb{R}^{nxn}$ when certain functional or structural conditions are fulfilled in the transformed spaces.

In the second part, by relaxing the symmetry condition of the algebraic Riccati equation (1) a new upper class of quadratic matrix equations is constructed and denoted as 'algebraic Riccati generators' (ARG)

$$C^T C + A_1^T K_1 + K_1^T A_1 - K_1^T B R^{-1} B^T K_1 = 0 \qquad (3)$$

The above equation gives rise to a natural generalization of the classical algebraic Riccati (1) and Liapunov (2) equations of optimal control where the solution set is constrained to belong to the field of real symmetric matrices $K_0 = K_0^T \in \mathbb{R}^{nxn}$. In this part the solutions of the ARG (3) are investigated from the analytical point of view. By introducing a set of matrix transformations a new coordinate space is constructed for which (3) reduces to the simple orthogonality equation $N^T N = I$. By reversing the transformations, for every orthogonal matrix $N = N^{-T} \in \mathbb{R}^{nxn}$ the corresponding solution $K_N = \Psi(N)$ of (3) is obtained, where $\Psi(\cdot)$ represents a function of operators in $\mathbb{R}^{nxn}$. Moreover, by restricting N to be the unique orthogonal matrix N_0 for which $\Psi(N_0) = \Psi^T(N_0) = K_1$ the unique positive definite solution of ARE (1) is formally represented in terms of orthogonal and skew-symmetric matrices. The formulation is then particularized to the ALE (2) leading to a 'dual' skew-symmetric representation of this equation. The theory of Riccati generators is illustrated with two interesting applications related with upper and lower bounds of ARE and ALE solutions.

Although the discussions in this paper are concentrated on the continuous-time control problem, they are directly applicable to matrix equations arising in estimation problems, by means of the well-known duality principle, and to sampled-time problems with a non-singular transition matrix.

The structure of the paper is as follows. Section 2 contains the development of the transformation method to obtain the orthogonal representation. Section 3 contains the development of skew-symmetric representations making use of different functions relating orthogonal and skew-symmetric matrices. Section 4 presents a closed form solution for ARE in $\mathbb{R}^{2x2}$ and a near-analytical solution in $\mathbb{R}^{3x3}$. Section 5 discusses some structural and functional considerations associated with the transformed equations and presents new analytical solutions in $\mathbb{R}^{nxn}$ for some rather interesting particular cases. Section 6 introduces the theory of 'algebraic Riccati generators' and its applications. Section 7 is the summary and conclusions.

NOTATION. Throughout the paper $M \in \mathbb{F}^{mxn}$ will denote an mxn matrix with elements in a field $\mathbb{F}$. The field will usually be the real numbers $\mathbb{R}$. For any symmetric or Hermitian matrix $M, M \geq 0$ and $M > 0$ imply that M is nonnegative definite and positive definite, M^T and M^{-T} indicate a transpose of M and inverse transpose of M, respectively. Also $[.,.]$ represents the "commutator product" defined as $[A,B] \triangleq AB - BA$ and $\{.,.\}$ denotes the "symmetric commutator product", defined here as $\{A,B\} \triangleq \{AB - B^T A^T\}$. Finally I will denote the identity matrix.

2. DERIVATION OF THE ORTHOGONAL REPRESENTATION

Let $C^T C \geq 0$ be represented into the canonical form

$$C^T C = D^T \Pi D \tag{4}$$

where Π is a nxn diagonal matrix $\Pi = \text{diag}(\Pi_1, \Pi_2, \ldots, \Pi_n)$; $\Pi_i \geq 0$; $i = 1,2,\ldots,n$ and D is an orthogonal matrix composed of the corresponding eigenvectors of $C^T C$. From (4) we can write the factorization

$$C^T C = D^T \Phi\ U\ \Phi\ D = F^T U F \tag{5}$$

where $F \triangleq \Phi D$ and Φ, U are nxn diagonal matrices such that

$$\Phi = \text{diag}(\Phi_1, \Phi_2, .., \Phi_n); \Phi_i = \pi_i^{1/2} \text{ if } \pi_i > 0, \Phi_i = 1 \text{ if } \pi_i = 0 \tag{6}$$

and $U = \text{diag}(u_1, u_2, ..., u_n)$; $u_i = 1$ if $\pi_i > 0$, $u_i = 0$ if $\pi_i = 0$, for all $i = 1,2,...,n$. Now since (5) and (6) F^{-1} exists and is given by $F^{-1} = [\Phi D]^{-1} = D^T \Phi^{-1}$. The pre-and postmultiplication of (1) by F^{-T} and F^{-1} respectively, then gives

$$U + A^T K + KA - KQK = 0 \quad , \quad K = K^T > 0 \tag{7}$$

with

$$A = FA_1 F^{-1}, \ Q = FB_1 R^{-1} B_1^T F^T \text{ and } K_1 = F^T K F \tag{8}$$

Now, because (8) is a similarity transformation the pair <A,B> is stabilizable and the pair <A,U> is completely detectable. Thus, the symmetric positive definite solution K of (7) exists and is unique; hence K^{-1} exists and is also unique and positive definite. After pre-and post-multiplication of (7) by K^{-1}, this equation becomes

$$K^{-1} U K^{-1} + K^{-1} A^T + AK^{-1} - Q = 0 \quad , \quad K^{-1} > 0 \tag{9}$$

Furthermore, let by definition

$$T = Q + A(2I - U)A^T = T^T > 0 \tag{10}$$

where I denotes the nxn identity matrix. From (10) T may always be diagonalized, so that

$$T = G \Lambda G^T \ ; \ \Lambda = \text{diag}(\lambda_1, \lambda_2, ..., \lambda_n) > 0 \qquad G^T = G^{-1} \tag{11}$$

Then the positive definite square root matrix $T^{1/2}$ is given by $T^{1/2} = G\Lambda^{1/2} G^T > 0$. Now, consider making the coordinate transformation

$$K^{-1} = HT^{1/2} - A^T \tag{12}$$

in (9). In this new matrix space the symmetry condition $K^{-1} = K^{-T}$ takes the form

$$K^{-1} = HT^{1/2} - A^T = T^{1/2} H^T - A = K^{-T} \tag{13}$$

Substituting (13) into (9) yields

$$(T^{1/2}H^T - A)U(HT^{1/2} - A^T) + (T^{1/2}H^T - A)A^T + A(HT^{1/2} - A^T) - Q = 0$$

which may be readily rewritten as

$$T^{1/2}H^{T}UHT^{1/2}+T^{1/2}H^{T}(I-U)A^{T}+A(I-U)HT^{1/2}-Q-A(2I-U)A^{T} = 0 \quad (14)$$

and by substituting T, as given by (10), into (14) yields

$$T^{1/2}H^{T}UHT^{1/2}+T^{1/2}H^{T}(I-U)A^{T}+A(I-U)HT^{1/2}-T = 0 \quad (15)$$

Hence, by introducing the notation

$$A_u = T^{1/2}A(I-U) \quad (16)$$

after pre-and postmultiplication of (15) by $T^{1/2}$, it is a routine matter to verify that

$$H^{T}UH + H^{T}A_u^{T} + A_uH-I = 0 \quad (17)$$

Thus, by collecting (13) and (17) the ARE problem (1) becomes equivalent to find a matrix H such that

$$T^{1/2}H^{T} - HT^{1/2}= A - A^{T} \quad (18a)$$

$$H^{T}UH + H^{T}A_u^{T} + A_uH - I = 0 \quad (18b)$$

$$HT^{1/2} - A^{T} > 0 \quad (18c)$$

where (18c) characterizes the positive-definiteness condition ($K > 0$) in the new coordinate space. Moreover, (18b) is equivalent to

$$U + A_u^{T}H^{-1} + H^{-T}A_u - H^{-T}H^{-1} = 0$$

which may be factored as

$$(H^{-T}-A_u^{T})(H^{-1}-A_u) = S_u \quad \text{with } S_u \triangleq U+A_u^{T}A_u = S_u^{T} > 0 \quad (19)$$

Furthermore, consider making the coordinate change

$$N = (H^{-1}-A_u)S_u^{-1/2} \quad (20)$$

in (18a-c). From (19) it is easily shown that

$$N^{T}N=S_u^{-1/2}(H^{-T}-A_u^{T})(H^{-1}-A_u)S_u^{-1/2}=S_u^{-1/2}S_uS_u^{-1/2}= I$$

from which the original ARE problem (1) can be cast in terms of finding a matrix N, such that

$$N^{T}N = NN^{T} = I \quad (21a)$$

$$T^{1/2}(S_u^{1/2}N^{T}+A_u^{T})^{-1}-(NS_u^{1/2}+A_u)^{-1}T^{1/2} = 2E \quad (21b)$$

$$(NS_u^{1/2}+A_u)^{-1}T^{1/2}-A^{T} > 0 \quad (21c)$$

where, by definition

$$E \triangleq 1/2 \quad (A - A^T) \tag{22}$$

In the above system the solution of matrix equation (21b) must simultaneously fulfil the orthogonality and positive-definiteness conditions (21a) and (21c), respectively. With this reformulation the ARE problem is of special interest: by utilizing a set of analytical transformations ARE (1), defined on the field of the real symmetric matrices, becomes equivalent to matrix equation (21b), defined on the field of the real orthogonal matrices. The derivation in this section leads to the following important theorem on the equivalence of both formulations.

Theorem 1: if $< A_1, B_1 >$ is stabilizable and $< A_1, C >$ is detectable then the unique symmetric positive definite solution of ARE (1) is given by

$$K_1 = F^T((NS_u^{1/2} + A_u)^{-1}T^{1/2} - A^T)^{-1}F \tag{23}$$

with matrices F, A, T, A_u, S_u as given by (5), (8), (10), (16) and (19); and where N is the unique orthogonal solution of (21b) that fulfils the positive-definiteness condition (21c).

Proof: Easily follows from the previous developments. Since all the utilized matrix transformations to get (21a-c) from (1) are invertible (regular) transformations, uniqueness of the solution of (21a-c) is a consequence of the positive definite solution uniqueness of (1). By successive substitution of transformation equations (8), (12) and (20) and taking into account the coefficient matrices definitions we can write

$$K_1 = F^T K F = F^T (HT^{1/2} - A^T)^{-1} F = F^T ((NS_u^{1/2} + A_u)^{-1} T^{1/2} - A^T)^{-1} F$$

which gives the required solution. This completes the proof. Theorem 1 has an important implication from the viewpoint of applications, since it leads to meaningful characterizations of the solutions and interesting particularizations in practice. The remainder of this section is devoted to this analysis. After some straightforward manipulations (21b) can be shown to be equivalent to

$$\begin{aligned} & S_u^{1/2}[T^{1/2} - 2EA_u^T]N - N^T[T^{1/2} + 2A_u E]S_u^{1/2} - 2S_u^{1/2} E S_u^{1/2} \\ & = N^T[2A_u E A_u^T + T^{1/2} A_u^T - A_u T^{1/2}]N \end{aligned} \tag{24}$$

The following corollary has been demonstrated in[18] in a less general theoretical framework and is reproduced here since it leads to a considerable simplification of (23) for a wide class of ARE problems.

Corollary 1.1: Assume that the pair $< A_1, B_1 >$ is stabilizable, the pair $<A_1, C>$ is detectable and C is nonsingular. Then the unique positive definite solution of (1) is given by

$$K_1 = C^T (T^{1/2} N - A)^{-1} C \tag{25}$$

where N is the unique orthogonal solution of linear matrix equation

$$T^{1/2}N - N^T T^{1/2} = 2E \tag{26}$$

that fulfils the positive-definiteness condition

$$T^{1/2}N - A > 0 \tag{27}$$

Proof: Since C is nonsingular then factorization (5) yields F=C and U=I. Thus, by direct substitution into definition equations (16) and (19), we have $A_u=0$ and $S_u=I$. Hence, by substituting those values into (24) and (23), and taking into account the orthogonality conditions $N^T=N^{-1}$, (26) and (27) are obtained. Moreover, condition (27) directly follows from the positive-definiteness of K_1, as given by (25).

A characterization of several ARE problems for which application of Theorem 1 leads to useful particularizations and closed form analytical solutions is given by the following corollaries.

Corollary 1.2: Assume that the stabilizability and detectability conditions of Theorem 1 are fulfilled and that A in (8) is a symmetric matrix. Then the unique positive definite solution of (1) is given by (23), where N is the unique orthogonal solution of linear matrix equation

$$NS_u^{1/2}T^{1/2}-T^{1/2}S_u^{1/2}N^T = T^{1/2}A_u^T-A_uT^{1/2} \tag{28}$$

that verifies positive-definiteness condition (21c).

Proof: Observe that if A=A , then from (22) E=0. By substituting this value into (24b) and after some routinary algebraic manipulations (28) is easily obtained.

Corollary 1.3: The conditions of Corollary 1.2 as well as the additional symmetry condition

$$T^{1/2}A_u^T = A_uT^{1/2} \tag{29}$$

are assumed. Then the required orthogonal positive-definite solution of (28) is given by

$$N = (T^{1/2}S_uT^{1/2})^{1/2}T^{-1/2}S_u^{-1/2} \tag{30}$$

Proof: If (29) is fulfilled then (28) takes the simple form

$$NS_u^{1/2}T^{1/2}-T^{1/2}S_u^{1/2}N^{-1} = 0 \quad ; \quad N^T = N^{-1}$$

The postmultiplication of this equation by $NS_u^{1/2}T^{1/2}$ yields

$$(NS_u^{1/2}T^{1/2})^2 = T^{1/2}S_uT^{1/2}$$

from which the positive definite solution (30) is computed.

Notice that in the above corollary a sufficient condition for (29) to be satisfied is that $A_u=0$. Thus, we can formulate without proof the following corollary.

Corollary 1.4: Assume that the conditions of Corollary 1.1 as well as the symmetry condition $A = A^T$ are fulfilled. Then, the unique positive definite solution of (1) is given by $K_1=C^T(T^{1/2}-A)^{-1}C$, with $T=Q+A^2$.

The above result, formerly demonstrated in [15], appears now as a particularization of the general orthogonal approach in this section.

3. SKEW-SYMMETRIC FORMULATIONS

In this section, by utilizing analytical relationships linking together orthogonal and skew-symmetric matrices, we derive two skew-symmetric formulations of the ARE problem from the orthogonal formulation (21a-c). Since a real orthogonal matrix $N \in \mathbb{R}^{n\times n}$ contains $n(n-1)/2$ degrees of freedom and orthogonality imposes $n(n+1)/2$ functional constraints, the basic idea is to represent orthogonal matrices in $\mathbb{R}^{n\times n}$ as functions of matrices with $n(n-1)/2$ structural degrees of freedom. The best suited domain for this purpose is the field of real skew-symmetric matrices. The main advantage of this approach is that system (21a,b) is reduced to a single matrix (quadratic) equation in the field of the real skew-symmetric matrices. In the framework of the restricted class of Riccati problems investigated in Incertis [18 - 20] this class of skew-symmetric matrix equations has been interpreted as a dual representation of the "primal" (symmetric matrix domain) problem. Before proceeding to deal with the new formulation, the following lemmas are reproduced from[20,21].

Lemma 1: Let N be a real orthogonal matrix. Then there exists a unique real skew-symmetric matrix $X = (N+I)^{-1}(N-I)$, $N^T = N^{-1}$, so that

$$N = (I+X)(I-X)^{-1}, \quad X^T = -X \quad \text{and} \quad X \le 1 \tag{31}$$

Lemma 2: Let N be a real orthogonal matrix. Then there exists a unique real skew-symmetric matrix $Y = (N-N^T)/2$, $N^T = N^{-1}$, so that

$$N = Y+(I+Y^2)^{1/2}, \quad Y^T = -Y \quad \text{and} \quad Y \le 1 \tag{32}$$

Moreover, $(I+Y^2)^{1/2} = (N+N^T)/2$ gives the symmetric component of N as a function of the skew-symmetric component Y.

Those lemmas play an important role in the subsequent derivation of skew-symmetric formulations. Now consider the equation (24), (equivalent to (21b)), written as

$$LN - N^TL^T - N^TWN - V = 0 \tag{33}$$

where, by definition

$$L = S_u^{1/2}[T^{1/2}-2EA_u^T] \tag{34a}$$

$$V = 2S_u^{1/2}ES_u^{1/2} = -V^T \tag{34b}$$

$$W = 2A_uEA_u^T + T^{1/2}A_u^T - A_uT^{1/2} = -W^T \tag{34c}$$

By utilizing the transformation (31), since Lemma 1 the orthogonality condition (21a) is satisfied for any skew-symmetric matrix X. After some straightforward manipulations on (33) and (21c) the ARE problem (1) becomes equivalent to find the unique skew-symmetric matrix X which solves the quadratic matrix equation

$$(V+W+L^T-L)-(L+L^T+V-W)X-X(L+L^T-V+W)-X(V+W-L^T+L)X = 0 \tag{35}$$

and satisfies the positive-definiteness condition

$$((I+X)(I-X)^{1/2}S_u^{1/2}+A_u)^{-1}T^{1/2}-A^T > 0 \tag{36}$$

A similar formulation follows from Lemma 2. By substituting N, as given by (32), into (33) and (21c) the ARE problem (1) is easily shown to be equivalent to determine the unique skew-symmetric matrix Y, such that solves

$$LY+YL^T=V+(I+Y^2)^{1/2}L^T-L(I+Y^2)^{1/2}+(-Y+(I+Y^2)^{1/2})W(Y+(I+Y^2)^{1/2} \tag{37}$$

and fulfils the positive-definiteness condition

$$((Y+(I+Y^2)^{1/2})S_u^{1/2}+A_u)^{-1}T^{1/2}-A^T > 0 \tag{38}$$

From the above reformulation the ARE problem is of special interest: By utilizing analytical formulas that produce orthogonal matrices in terms of skew-symmetric arguments the ARE problem (1), defined on the field of the real symmetric matrices, becomes equivalent, (via the orthogonal formulation(21a-c)), to the 'dual' equations (35) or (37), defined on the field of the real skew-symmetric matrices. Thus, the system of n(n+1)/2 coupled quadratic equations associated to (1) can be reduced to the equivalent systems of n(n-1)/2 coupled quadratic equations given by (35) or (37). From the foregoing developments we can present the following two alternative formulations for the fundamental Theorem 1 in Section 2.

Theorem 2: If $<A_1, B_1>$ is stabilizable and $<A_1, C>$ is detectable then the unique symmetric positive definite solution K_1 of ARE (1) is given by (23) with N as given by (31),((32)) and where X,(Y) is the unique skew-symmetric solution of (35),((37)) that fulfils the positive definiteness condition (36),((38)).

Proof: First notice that if X,(Y) is a solution of (35),((37)) then $-X^T$,$(-Y^T)$ is also a solution of this equation. Thus, the solution space of (35),((37)) belongs to the field of the skew-symmetric matrices. Moreover, the uniqueness of the solution of (35-36),((37-38)) is a consequence of the uniqueness of the solution of (21a-c) and that of X,(Y) in Lemma 1,(2). Finally, it is an easy matter to verify that the fulfilment of condition (36),((38)) implies the positive definiteness of K_1, as given by (23). This completes the proof.

The following sections are mainly devoted to the development of analytical techniques for the solution of the (dual) skew-symmetric equations (35) and (37). An efficient numerical iterative method for the solution of (35) when C is nonsingular has been presented in[18] with several computational examples and comparisons with other methods.

4. ANALYTICAL SOLUTIONS IN $\mathbb{R}^{2x2}$ and $\mathbb{R}^{3x3}$

4.1. The $\mathbb{R}^{2x2}$ Case

The application of Theorem 2 in a matrix space $\mathbb{R}^{2x2}$ reduces the solution of the ARE (1) to that of the solution of a simple scalar quadratic equation, thus leading to a closed form analytical formula of the ARE problem. A similar result has been obtained by the first time in[18] for the case in which matrix C is a nonsingular square matrix. We are now in conditions to extend this result to general Riccati equations.

Let us consider Riccati equation (1) in $\mathbb{R}^{2x2}$. Making use of (5), (8), (10), (16), (19), (22) and (46) we can compute matrices F, A, Q, T, A_u, S_u, E and then

$$L=\begin{bmatrix} l_1 & l_2 \\ l_3 & l_4 \end{bmatrix}, \quad V=\begin{bmatrix} 0 & v \\ -v & 0 \end{bmatrix}, \quad W=\begin{bmatrix} 0 & w \\ w & 0 \end{bmatrix} \tag{39}$$

in closed analytical form. Now, since Lemma 2 any real orthogonal matrix $N \in \mathbb{R}^{2x2}$ admits the representation (32) with

$$Y=\begin{bmatrix} 0 & y \\ -y & 0 \end{bmatrix}, \quad (I+Y^2)^{1/2}=(1-y^2)^{1/2}\begin{bmatrix} 1 & 0 \\ 0 & 1 \end{bmatrix}, \quad 0 \leq |Y| \leq 1 \tag{40}$$

Therefore, by substituting (39) and (40) into (37) and solving the resulting quadratic scalar equation for the element (1, 2), yields

$$y = [th-(t^2h^2-(t^2+s^2)(h^2-s^2))^{1/2}] / (t^2+s^2) \tag{41}$$

where, by definition

$$t = l_1+l_4, \quad s = l_3-l_2, \quad h = v+w \tag{42}$$

It is an easy matter to verify that y, as given by (41) is the unique solution that fulfils the norm bound $|y| \leq 1$ for all ARE problems and gives rise to a symmetric positive definite solution. Finally, since Theorem 2 the solution of (1) is given by (23) with N explicitely computed by means of (40) and (32).

Hence we conclude that the algebraic Riccati equation in $\mathbb{R}^{2x2}$ can be solved in closed analytical form. Interesting applications of this anlytic approach to the solution of the ARE problem can be the investigation of the geometry[22], the sensitivity and the structural and equilibrium analysis of the Riccati equation[23-25].

The following example illustrates the analytical procedure and shows the important advantages of the proposed approach for ARE computations in $\mathbb{R}^{2x2}$.

Example 1: The ARE (1) coefficients are given by

$$A_1=\begin{bmatrix} 0 & 1 \\ -1 & 0 \end{bmatrix}, \quad B_1=\begin{bmatrix} 0 \\ 1 \end{bmatrix}, \quad C=\begin{bmatrix} 1 & 0 \end{bmatrix}, \quad R=1$$

Following the main steps of the algorithm, we obtain

$$U=\begin{bmatrix} 1 & 0 \\ 0 & 0 \end{bmatrix}, \quad F=\begin{bmatrix} 1 & 0 \\ 0 & 1 \end{bmatrix}, \quad T^{1/2}=\begin{bmatrix} \sqrt{2} & 0 \\ 0 & \sqrt{2} \end{bmatrix}$$

and

$$E=\begin{bmatrix} 0 & 1 \\ -1 & 0 \end{bmatrix}, \quad A=\begin{bmatrix} 0 & \sqrt{2} \\ 0 & 0 \end{bmatrix}, \quad S_u=\begin{bmatrix} 1 & 0 \\ 0 & 2 \end{bmatrix}$$

Also

$$L=\begin{bmatrix} 0 & 0 \\ 0 & 1 \end{bmatrix}, \quad V=\begin{bmatrix} 0 & \sqrt{2} \\ -\sqrt{2} & 0 \end{bmatrix}, \quad W=\begin{bmatrix} 0 & -1 \\ 1 & 0 \end{bmatrix}$$

from which, by means of (42), (41) and (40)

$$y = \sqrt{2} - 1 \quad , \quad N = \begin{bmatrix} \sqrt{2(\sqrt{2}-1)} & \sqrt{2}-1 \\ 1-\sqrt{2} & \sqrt{2(\sqrt{2}-1)} \end{bmatrix}$$

Finally, by substituting N in (23) the positive definite solution of (1) is found to be

$$K_1 = \begin{bmatrix} 2\sqrt{\sqrt{2}-1} & \sqrt{2}-1 \\ \sqrt{2}-1 & \sqrt{2}\sqrt{\sqrt{2}-1} \end{bmatrix}$$

An alternative formulation in $\mathbb{R}^{2x2}$ can be performed from Lemma 1 and Theorem 2. We omit here the details of this alternative skew-symmetric formulation.

4.2. The $\mathbb{R}^{3x3}$ Case

Let us consider the ARE (1) in $\mathbb{R}^{3x3}$. To simplify the algebra, let us further consider the particular case of Corollary 1.1 in which matrix C is nonsingular. The orthogonal representation is given by (26) and by introducing in this equation the similarity transformation $M=G^T NG=M^{-T}$ from (11) and the skew-symmetric representation $M=(I+X)(I-X)^{-1}$, from Lemma 1, after some straightforward manipulations, the 'dual' equation (35) takes the form

$$\Lambda^{1/2}(I-X)^{-1}-(I+X)^{-1}\Lambda^{1/2}= D_0 \quad , \quad D_0 = G^T EG \tag{43}$$

Now, when n=3, $\Lambda^{1/2}$, D_0, X admit the representation

$$\Lambda^{1/2}=\begin{bmatrix}\lambda_1^{1/2} & & 0\\ & \lambda_2^{1/2} & \\ 0 & & \lambda_3^{1/2}\end{bmatrix}, \quad D_0=\begin{bmatrix}0 & d_1 & d_2\\ -d_1 & 0 & d_3\\ -d_2 & -d_3 & 0\end{bmatrix}, \quad X=\begin{bmatrix}0 & x_1 & x_2\\ -x_1 & 0 & x_3\\ -x_2 & -x_3 & 0\end{bmatrix}$$

and by introducing the coefficients

$$r_1 = (\lambda_1^{1/2}- \lambda_2^{1/2})/(\lambda_1^{1/2}+ \lambda_2^{1/2}) \quad , \quad f_1 = d_1/(\lambda_1^{1/2}+ \lambda_2^{1/2})$$
$$r_2 = (\lambda_3^{1/2}- \lambda_1^{1/2})/(\lambda_1^{1/2}+ \lambda_3^{1/2}) \quad , \quad f_2 = d_2/(\lambda_1^{1/2}+ \lambda_3^{1/2})$$
$$r_3 = (\lambda_2^{1/2}- \lambda_3^{1/2})/(\lambda_2^{1/2}+ \lambda_3^{1/2}) \quad , \quad f_3 = d_3/(\lambda_2^{1/2}+ \lambda_3^{1/2})$$

and the determinantal intermediate variable

$$\Delta \triangleq | I+X | = 1 + x_1^2 + x_2^2 + x_3^2$$

The problem (43) becomes equivalent to the quadratic system

$$x_1 = r_1x_2x_3 + f_1 \Delta$$
$$x_2 = r_2x_1x_3 + f_2 \Delta$$
$$x_3 = r_3x_1x_2 + f_3 \Delta \tag{44}$$

where $| x_i | \leq 1$, i=1,2,3. A procedure for solving (44) is given by the following Algorithm: Define the intermediate variables

$$\alpha_1 = f_2+f_1r_2x_3 \quad , \quad \alpha_2 = f_1+f_2r_1x_3$$

which are linearly dependent on x_3. Then, solve the polynomial

$$P(x_3) = [x_3(\alpha_1^2+\alpha_2^2)+\alpha_1\alpha_2r_3(1+x_3^2)]^2-$$
$$-(1-r_1r_2x_3^2)^2 [x_3-f_3(1+x_3^2)] [r_3\alpha_1\alpha_2+f_3(\alpha_1^2+\alpha_2^2)]$$

for a real root in the interval $x_3\epsilon[-1,1]$. By utilizing this root, compute

$$\Delta=[x_3(\alpha_1^2+\alpha_2^2)+\alpha_1\alpha_2r_3(1+x_3^2)]/[r_3\alpha_1\alpha_2+f_3(\alpha_1^2+\alpha_2^2)]$$

and then

$$x_1=[f_1+f_2r_1x_3]\Delta/(1-r_1r_2x_3^2)\ ,\ x_2=[f_2+f_1r_2x_3]\Delta/(1-r_1r_2x_3^2)$$

from which the required skew-symmetric matrix X is obtained.

Finally, the solution of ARE (1) is computed by means of (25), where $N = GMG^T$ and $M = (I+X)(I-X)^{-1}$.

Thus, we conclude that the algebraic Riccati equation problem in $\mathbb{R}^{3x3}$ is equivalent to find a real root of an scalar polynomial in the interval [-1,1]. This problem is far less complex than the solution of the coupled system of six quadratic equations in the original 'primal' formulation and the algorithm gives a near-analytical approach for Riccati equations in $\mathbb{R}^{3x3}$.

Because the solution of the skew-symmetric (dual) equations (35) or (37) determines the solution of the ARE (1) it is highly interesting to investigate analytical closed form solutions of (35) and (37) in some rather important particular cases.

5. ANALYTICAL SOLUTIONS IN $\mathbb{R}^{nxn}$

Let us now consider the skew-symmetric equation (35), rewritten here as

$$XE_0X + X(L_0-E_1) + (L_0+E_1)X - E_2 = 0 \tag{45}$$

where $E_0 \triangleq V+W+L-L^T$, $E_1 \triangleq V-W$, $E_2 \triangleq V+W-L+L^T$, $L_0 \triangleq L+L^T$. Now, assume that there exists a matrix Ψ such that

$$[(L_0 + E_1),E_0] = 2\Psi E_0 \tag{46}$$

where $[A,B] \triangleq AB-BA$ denotes the "commutator product" of A and B. Introducing the notation

$$\Omega = E_1-\Psi \tag{47a}$$

$$M = \Psi^2+L_0^2+E_0E_2-2\Psi L_0+[E_1,(L_0-\Psi)] \tag{47b}$$

and following parallel arguments to that of Incertis[17], we state here, without proof, the following

Theorem 3: If a solution Ψ of (46) exists for which M, as given by (47b) is positive definite and if $[M^{1/2},\Omega] = 0$, then the unique skew-symmetric solution X of (35) that fulfils (36) coincides with the unique solution of the linear Liapunov matrix equation

$$A_\alpha^TX+XA_\alpha=E_2,\ A_\alpha=1/2(M^{1/2}+\Psi+L_0)-E_1 \tag{48}$$

Observe that if $[(L_0+E_1),E_0] = 0$ then from (46) it is readily seen that $\Psi = 0$. Hence $\Omega = E_1$ and $M = L_0^2+E_0E_2+[E_1,L_0]$. We can now state without proof the following corollary.

Corollary 3.1: If $[(L_0+E_1),E_0] = 0$, and if $M = L_0^2+E_0E_2+[E_1,L_0]$ is such that $M \geq 0$ and $[M^{1/2}, E_1] = 0$ then the unique skew-symmetric solution X of (35), (36) is given by the unique solution of linear matrix equation (48), with $\psi = 0$.

When C is nonsingular, from (5), (6), (10), (16) and (19) one obtains the following values $U=I$, $F=C$, $T=Q+AA^T$, $A_u=0$, $S_u=I$, from which, and by means of (34) one gets $L=L^T=T^{1/2}$, $V=2E$, $W=0$. This implies that (35) can be alternatively expressed by

$$XEX + (T^{1/2}+E)X + X(T^{1/2}-E)-E = 0$$

Particularizing Corollary 3.1 to this case leads to the following result, formerly demonstrated in[16] and[18].

Corollary 3.2: If $[T^{1/2},E] = 0$ and if $M = T + E^2$ is positive definite, then the unique solution of the dual skew-symmetric problem (35), (36) is given by

$$X = E(T^{1/2}+ M^{1/2})^{-1}$$

Therefore, the results in this section generalize [18] to the case in which matrix C is nonsquare or singular. Also, the main result in [15] appears now as a particular case of Corollary 3.2 where $E = 0$ and then $X = 0$.

6. THE ALGEBRAIC RICCATI GENERATOR

The algebraic Riccati generator equation (3) has been introduced in Incertis[26] as an upper class of the ARE (1) by relaxing the symmetry condition of the unknown matrix K_1. Now, by following a complementary argumentation to that of Section 2, let $B_1R^{-1}B_1^T \geq 0$ be represented into the canonical form $B_1R^{-1}B_1^T = G_0^TVG_0^T$, where $V = \text{diag}(v_1,v_2,\ldots,v_n)$ is such that $v_i=1$ when the corresponding eigenvalue of $BR^{-1}B^T$ is positive and $v_i=0$ otherwise. After pre-and postmultiplication of (3) by G_0 and G_0^T respectively, one gets the normalized equation

$$P+F_0^TK_0+K_0^TF_0-K_0^TVK_0 = 0 \tag{49}$$

where $F_0=G_0^{-T}A_1G_0^T$, $P=G_0C^TCG_0^T$ and $K_0=G_0K_1G_0^T$. Furthermore, let by definition

$$T_0 = P + F_0^T(2I-V)F_0 = T_0^T > 0 \tag{50}$$

from which the positive definite square-root matrix $T_0^{1/2}$ can be computed. Now, consider making the coordinate transformation $K_0=H_0T_0^{1/2}+F_0$ into (49). After some straightforward manipulations this equation becomes

$$I+T_0^{-1/2}F_0^T(I-V)H_0+H_0^T(I-V)F_0T_0^{-1/2}-H_0^TVH_0 = 0 \tag{51}$$

Hence, by introducing the notation

$$F_V = (I-V)F_0T_0^{-1/2} \tag{52}$$

after pre-and postmultiplication of (51) by H_0^{-T} and H_0^{-1} respectively, it is an easy matter to verify that

$$(H_0H_0^T)^{-1} + H_0^{-T}F_V^T + F_VH_0^{-1} - V = 0$$

which may be factored as

$$(H_0^{-T}+F_V)(H_0^{-1}+F_V^T)=S_V \ , \ S_V \triangleq V+F_VF_V^T > 0 \tag{53}$$

Thus, by computing $S_V^{1/2}$ and by introducing the coordinate transformation $N = (H_0^{-1}+F_V^T)S_V^{-1/2}$ into (53), this equation finally becomes equivalent to the simple orthogonality condition $N^TN = NN^T = I$.

The above derivation leads to the following important theorem on the analytical solution of the ARG problem:

Theorem 4: Let F_0, T_0, F_V, S_V as defined above and let $N \in \mathbb{R}^{nxn}$, then

$$K_1 = G_0^{-1}[(N\,S_V^{1/2} - F_V^T)^{-1}T_0^{1/2} + F_0]G_0^{-T} \tag{54}$$

is a real solution of the ARG (3) iff N is orthogonal.

Proof: The necessity part of the theorem has been demonstrated in the previous developments: If K_0 is a real solution of (3) then N must be an orthogonal matrix. To demonstrate the sufficiency part, let us start from (54), rewritten here as

$$K_0=(N\,S_V^{1/2} - F_V^T)^{-1}T_0^{1/2} + F_0 \ , \ K_0 = G_0K_1G_0^T \tag{55}$$

from which matrix N, as a function of K_0, is given by

$$N = [T_0^{1/2}(K_0-F_0)^{-1} + F_V^T]S_V^{-1/2} \tag{56}$$

Now, by computing N^TN from (56) and taking into account the definitions of F_V and T_0, as given by (52) and (50), after some routinary manipulations, N_0 and K_0 are found to satisfy the equation

$$P+F_0^TK_0+K_0^TF_0-K_0^TVK_0=(K_0^T-F_0^T)S_V^{1/2}(NN^T-I)S_V^{1/2}(K_0-F_0) \tag{57}$$

Therefore, if N is orthogonal then $K_1=G_0^{-1}K_0G_0^{-T}$, with K_0 as given by (55) is a real solution of the ARG (3). This completes the sufficiency and ends the theorem's proof.

In the sequel we shall denote by K_N the solution of (3) associated to the orthogonal matrix N. From Theorem 4 this solution is obtained by means of (54), compactly written here as $K_N = \Psi(N)$. Now, since ARE (1) is a particular case of ARG (3) where the solution set is constrained to belong to the field of the real symmetric matrices $K_1=K_1^T$, we can formulate the following corollary.

Corollary 4.1: If $\langle A_1, B_1 \rangle$ is stabilizable and $\langle A_1, C \rangle$ is detectable then the unique symmetric positive definite solution of ARE (1) is given by

$$K_1 = \Psi(N_1) \tag{58}$$

where N_1 is the unique orthogonal matrix that fulfils the symmetry and positive-definiteness conditions

$$\Psi(N_1) = \Psi^T(N_1) \quad \text{and} \quad \Psi(N_1) > 0 \tag{59}$$

Proof: Immediately follows from Theorem 4 and the Kucera's[4] existence and uniqueness conditions for the positive definite solution of the algebraic Riccati equation.

6.1. Particularization to the Algebraic Liapunov Equation

The above formulation can be immediately applied to the solution of the linear Liapunov matrix equation (2) as a degenerate case of Riccati equation (1) where $B_1R^{-1}B_1^T = 0$ and thus $V = 0$. Following the main course of mathematical developments, we get $V=0$, $G_0=I$, $F_0=A_1$, $P=C^TC$, $T_0=C^TC + 2A_1A_1^T$, and thus $F_v \triangleq J = A_1T_0^{-1/2}$ and $S_u = JJ^T$. Moreover, since Theorem 4, for every orthogonal matrix $N=N^{-T}$ the corresponding solution of the associated 'Algebraic Liapunov Generator' (ALG)

$$C^TC + A_1^TK_1 + K_1^TA_1 = 0 \tag{60}$$

is given by $K_N = J^{-T}[(N(JJ^T)^{1/2}J^{-T}-I)^{-1} + J]T_0^{1/2}$. Now, let by definition

$$M \triangleq N(JJ^T)^{1/2}J^{-T} = M^{-T} \tag{61}$$

In terms of this new orthogonal variable the corresponding solution of the ALG (60) is expressed by

$$K_M = (J^{-T}(M-I)^{-1} + J)T_0^{1/2} \tag{62}$$

Furthermore, for every $M=M^{-T}$ there exists a unique skew-symmetric matrix $Z=-Z^T$, such that $M=-(I+Z)(I-Z)^{-1}$. Moreover $(M-I)^{-1}=-\frac{1}{2}(I-V)$ and by utilizing this relationship (62) becomes

$$K_Z=[-1/2\, J^{-T}(I-Z)+J]T_0^{1/2} \quad , \quad Z=-Z^T, \quad ||Z|| \leq 1 \tag{63}$$

from which, after some straightforward manipulations $K_Z^TK_Z$ is easily found to be given by

$$K_Z^TK_Z=T_0^{1/2}[J^TJ-I+1/4(I+Z)(J^TJ)^{-1}(I-Z)]T_0^{1/2} \tag{64}$$

Hencefor , for every skew-symmetric matrix $Z=-Z^T$, $||Z||\leq 1$ we get the following bounds

$$[K_Z^TK_Z]_{min}= T_0^{1/2}[J^TJ-I+1/4(J^TJ)^{-1}]T_0^{1/2} \tag{65a}$$

$$[K_Z^T K_Z]_{max}=T_0^{1/2}[J^TJ-I+1/2(J^TJ)^{-1}]T_0^{1/2} \quad (65b)$$

In terms of the ALE (2) coefficients the above bounds are given by

$$[K_Z^T K_Z]_{min}=1/4\ C^TC(A_1^TA_1)^{-1}C^TC \quad (66a)$$

$$[K_Z^T K_Z]_{max}=C^TC+A_1^TA_1+1/2\ C^TC(A_1^TA_1)^{-1}C^TC \quad (66b)$$

and the solution of this equation satisfies the inequality

$$[K_Z^T K_Z]_{min}^{1/2} \le K_1 \le [K_Z^T K_Z]_{max}^{1/2}$$

Finally, by introducing the symmetry condition $K_Z=K_Z^T$ in (63) the skew-symmetric Liapunov equation takes the form

$$J^TJ^{-1}A_1Z+ZA_1^TJ^{-T}J=J^TJ^{-1}A_1(2J^TJ-I)-(2J^TJ-I)A_1^TJ^{-T}J \quad (67)$$

From the foregoing developments we see that the skew-symmetric matrix equations play a 'dual' role in some of the most important matrix problems in control systems theory. The main advantage of this 'dual' approach is a reduction of the number of linear or quadratic coupled equations to be solved in the skew-symmetric matrix domain. The numerical solution of skew-symmetric equations of the form of (35) or (37) is the main object of Incertis[18] and comparisons with other widely used numerical methods are discussed.

6.2. Algebraic Riccati Equation Solution Bounds

Let us consider the ARG (3) in the particular case: Rank $[B_1R^{-1}B_1^T]=n$. From (55) we can write

$$K_1=G_0^{-1}K_0G_0^{-T}\ ,\ K_0=(M+J)T_0^{1/2}\ ,\ I\ge J^TJ \quad (68)$$

where $M\triangleq N^T=M^{-T}$, $J=F_0T_0^{-1/2}$, $T_0=P+F_0^TF_0$. Therefore

$$K_0^TK_0=T_0^{1/2}(M^T+J^T)(M+J)T_0^{1/2}=T_0^{1/2}(I+J^TJ+J^TM+M^TJ)T_0^{1/2}$$

Now, for every J there exist two orthogonal matrices M_0 and M_1, such that $J^TM_0=-(J^TJ)^{1/2}$ and $J^TM_1=(J^TJ)^{1/2}$. By utilizing those matrices, we can write the bounds

$$[K_0^TK_0]_{min}=T_0^{1/2}[I-(J^TJ)^{1/2}]^2T_0^{1/2} \quad (69a)$$

$$[K_0^TK_0]_{max}=T_0^{1/2}[I+(J^TJ)^{1/2}]^2T_0^{1/2} \quad (69b)$$

Hence, we conclude that

$$T_0^{1/2}[I-(J^TJ)^{1/2}]^2T_0^{1/2} \le K_1^2 \le T_0^{1/2}[I+(J^TJ)^{1/2}]^2T_0^{1/2} \quad (70)$$

which gives useful upper and lower bounds for the ARE (1) in this nonsingularity case. Finally, let $M=(I+Z)^{-1}(I-Z)$, $Z^T=-Z$, $||Z||\leq 1$ and let, by definition $E_0 \triangleq 1/2(JT_0^{1/2}-T_0^{1/2}J^T)$. By imposing the symmetry condition $K_0=K_0^T$ in (68) the skew-symmetric 'dual' Riccati equation takes the form

$$ZE_0Z+(E_0+T_0^{1/2})Z+Z(T_0^{1/2}-E_0)-E_0=0,\ Z^T=-Z,\ ||Z||\leq 1 \tag{71}$$

7. SUMMARY AND CONCLUSIONS

We have presented two new formulations of the algebraic Riccati and Liapunov equations in orthogonal and skew-symmetric real matrix spaces. The main contribution of this paper is the unification and extension of previous transformation methods [11-20] for particular cases to the general algebraic Riccati equation problem.

REFERENCES

1 Athans, M., 'The role and use of the stochastic linear-quadratic Gaussian problem in control system design', IEEE *Trans. Automat. Contr.*, **16**, 529-552 (1971).

2 Jacobson, D.H. *Extensions of Linear-Quadratic Control, Optimization and Matrix Theory*, Academic, New York, 1977.

3 Kalman, R.E., 'New methods and results in linear prediction and estimation theory', RIAS *Rep.* **61**-1, Baltimore, MD (1961).

4 Kucera, V., 'A contribution to matrix quadratic equations', IEEE *Trans. Automat. Contr.*, **17**, 344-347 (1972).

5 Anderson, B.D.O. and J.B. Moore, *Optimal Filtering*, Englewood Cliffs, N.J. Prentice-Hall, 1979.

6 Casti, J.L. and E.T. Tse, 'Optimal linear filtering theory and radiative transfer: Comparisons and interconnections', *J. Math. Anal, Appl.*, **40**, 45-54 (1972).

7 Wonham, W.M., *Linear Multivariable Control - A Geometric Approach*, Springer-Verlag, New York, 1974.

8 Potter, J.E., 'Matrix quadratic solutions', SIAM *J. Applied Math.*, **14**, 496-501 (1966).

9 Laub, A.J., 'A Schur method for solving algebraic Riccati equations', IEEE *Trans. Automat. Contr.*, **24**, 913-921, (1979).

10 Kleinman, D.L., 'On an iterative technique for Riccati equations computations', IEEE *Trans. Automat. Contr.*, **13**, 114-115 (1968).

11 Lainiotis, D.G., 'Partitional Riccati solutions and integration-free

doubling algorithms', IEEE *Trans. Automat. Contr.*, **21**, 677-689 (1976).

12 Udilov, V.V., 'On the optimization of linear control systems with a symmetric main part', *J. Inst. Kibernetik Akad. Nauk*, Ukrain, Kiev URSS, 63-71 (1968).

13 Mozhaev, G.V., 'Use of symmetry in linear optimal control problems with a quadratic performance index', I and II, Dnepropetrovsk. Translated from *Automatika i Telemekhanika*, **6**, 22-30 (1975).

14 Jones, E.L., 'A reformulation of the algebraic Riccati equation problem', IEEE *Trans. Automat. Contr.*, **21**, 113-114 (1976).

15 Incertis, F. and J.M. Martínez, 'An extension on a reformulation of the algebraic Riccati equation problem', IEEE *Trans. Automat. Contr.*, **22**, 128-129 (1977).

16 Incertis, F., 'A new formulation of the algebraic Riccati equation problem', IEEE *Trans. Automat. Contr.*, **26**, 768-770 (1981).

17 Incertis, F., 'An extension on a new formulation of the algebraic Riccati equation problem', IEEE *Trans. Automat. Contr.*, **28**, 235-238 (1983).

18 Incertis, F., 'A dual formulation of the algebraic Riccati equation problem', *Optimal Control: Applications and Methods* (in press) (1983).

19 Incertis, F., 'On closed form solutions for the differential matrix Riccati equation problem', IEEE *Trans. Automat. Contr.*, **28**, 845-848 (1983).

20 Incertis, F., 'A skew-symmetric formulation of the algebraic Riccati equation problem', IEEE *Trans. Automat. Contr.*, **29**, (in press) (1984).

21 C.C. Mac Duffee, *The Theory of Matrices*, Chelsea, New York, 1946.

22 Rodríguez Cabanal, J., 'Geometry of the Riccati equation', *Stochastics*, **1**, 129-149 (1973).

23 Willems, J.C., 'Least squares stationary optimal control and the algebraic Riccati equation', IEEE *Trans. Automat. Contr.*, **16**, 621-634 (1971).

24 Thom, R., 'Topological models in biology', *Topology*, **8**, 313-335 (1969)

25 Bucy, R.S., 'Structural stability for the Riccati equation', SIAM, *J. Control*, **13**, 749-753 (1975)

26 Incertis, F., 'Skew-symmetric matrix equations: A new mathematical approach in control systems theory', 6th IASTED International Symposium on Measurement and Control, (in press), Athens, August 29-31, (1983).

CHAPTER 5

FEEDBACK CONTROLLER PARAMETERIZATIONS: FINITE HIDDEN MODES AND CAUSALITY

P.J. Antsaklis and M.K. Sain
Department of Electrical Engineering
University of Notre Dame
Notre Dame, Indiana 46556
USA

ABSTRACT

Parameterizations of feedback controllers are derived in a unifying way, using polynomial matrix internal descriptions, and the important design issues of causality and hidden modes are clarified.

1. INTRODUCTION

The design of multivariable control systems can often be simplified if an appropriate parameterization of the feedback controller is used to incorporate important design objectives such as internal stability. Youla et al. [1] were the first to introduce a parameterization of all stabilizing controllers for linear multivariable systems. Since then, knowledge of this parameterization has been increased [2,16] and has been extended to more general classes of systems [3,4]; alternative controller parameterizations have also been introduced [5-8,17].

If feedback controller parameterizations are to be used effectively to control a system, the issues of causality and hidden modes must be clarified. In particular, note that certain parameterizations, the ones more closely related to the internal description of the plant (eg. Youla's), offer good control of the closed loop eigenvalues but might lead to a nonproper controller; other parameterizations involving rational matrices can easily solve the properness problem but they have less direct control over the closed loop eigenvalues and so can result in hidden modes and high order compensators (eg. using Zames' parameter [5,9] and proper, stable matrices or generalized polynomials [3,4,10].)

In this paper, controller parameterizations are derived in a unifying way, using polynomial matrix descriptions. In Proposition 1, all stabilizing controllers are characterized using parameters K [1] and (D_k,N_k) [2]. Alternative internal stability conditions are derived in Proposition 2; some of these conditions can be used as alternative definitions for internal stability [3,5]. Theorem 3 is fundamental in deriving known as well as novel parameterizations involving rational matrices as parameters; note that the use of internal de-

S. G. Tzafestas (ed.), Multivariable Control, 85–104.

scriptions in the analysis allows the detailed study of the internal structure of the feedback system when rational parameters are used to characterize all stabilizing controllers. The relation of stabilizing controllers to observers of the state is established in Proposition 4. Proposition 5 characterizes all stabilizing controllers using rational parameters, and Corollaries 5.1 and 5.2 study the special cases of stable and nonsingular plants; Corollary 5.3 introduces an additional test for internal stability. Causality of the controller is then discussed and methods to obtain proper controllers when using parameterizations are introduced; proper controllers are obtained in Proposition 6 working over stable and proper matrices [3,4]. The hidden modes of the feedback system are fully characterized and they are identified as the uncontrollable and/or unobservable modes in the case of the single degree of freedom feedback configurations {G,I;P} and {I,H;P}; this is done in terms of the transfer matrices of the plant and the controller, and also in terms of the parameters characterizing the stabilizing controllers. The discussion on hidden modes makes possible the characterization of the closed loop eigenvalues in terms of poles of loop quantities, which in turn leads to additional tests for internal stability. Finally in Proposition 7 a parameterization of all stabilizing controllers is employed to achieved desired command/output-response and command/control-response when {G,I;P} or {I,H;P} feedback configurations are used to compensate the plant.

2. MAIN RESULTS

Consider

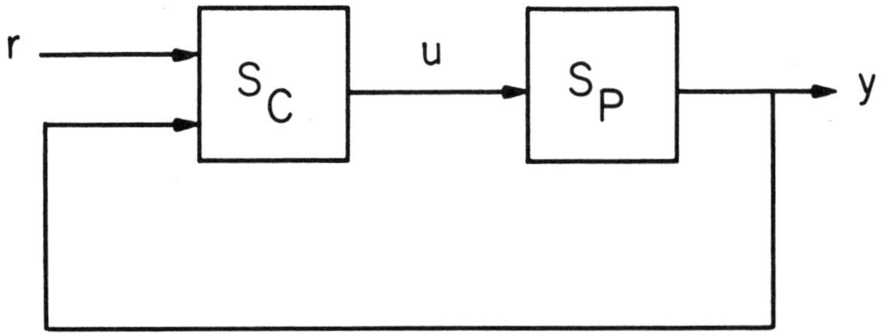

where S_P is the given plant and S_C the controller. Assume S_P, S_C controllable and observable, and let their transfer matrices be given by

$$y = Pu \quad , \quad u = [-C_y,\ C_r]\begin{bmatrix} y \\ r \end{bmatrix} \tag{1}$$

where y is the output, u the control input and r the command input. If the feedback loop is well defined, that is if $|I + C_yP| \neq 0$, the closed loop transfer matrix T between y and r is given by

$$T = PM \tag{2}$$

where

$$M \overset{\Delta}{=} (I + C_yP)^{-1}C_r \ . \tag{3}$$

Notice that T characterizes the command/output-response $y = Tr$, while the command/control-response $u = Mr$ is characterized by M.

We are interested in causal controllers which internally stabilize the system. The input r does not affect internal stability; and for such studies it will be taken to be zero. In addition, we are interested in the hidden modes of the system. These are affected by C_r. We shall study the hidden modes of the following single degree of freedom feedback systems:

$\{G,I;P\}$ where $u = Ge$, $e = r - y$ ($C_y = G$, $C_r = G$). This is the error or unity feedback configuration with compensator G in the feedforward path.

$\{I,H;P\}$ where $u = -Hy + r$ ($C_y = H$, $C_r = I$) with compensator H in the feedback path.

2.1 <u>Internal Stability</u>. Let $u = -Cy$ ($r = 0$, $C = C_y$ for notational convenience) and consider the controllable and observable internal operator polynomial matrix descriptions [11]:

$$S_P : \underline{D}z = \underline{N}u, \ y = \underline{z} \qquad ; \ (\underline{D},\underline{N}) \ \ell p \tag{4}$$

$$S_C : \underline{D}_c\underline{z}_c = -\underline{N}_c y, \ u = \underline{z}_c \ ; \ (\underline{D}_c,\underline{N}_c) \ \ell p \tag{5}$$

(ℓp or rp will be used for left or right prime polynomial matrices). Then

$$\underline{A}\begin{bmatrix} \underline{z}_c \\ \underline{z} \end{bmatrix} = 0 \ , \ y = [0 \quad I]\begin{bmatrix} \underline{z}_c \\ \underline{z} \end{bmatrix} \quad ; \quad \underline{A} \overset{\Delta}{=} \begin{bmatrix} \underline{D}_c & \underline{N}_c \\ -\underline{N} & \underline{D} \end{bmatrix} \tag{6}$$

is the closed loop internal description. If the dual descriptions

$$S_P : Dz = u \ , \ y = Nz \qquad ; \quad (D,N) \ rp \tag{7}$$

$$S_C : D_c z_c = -y \ , \ u = N_c z_c \quad ; \quad (D_c,N_c) \ rp \tag{8}$$

are used, then

$$A\begin{bmatrix} z \\ z_c \end{bmatrix} = 0 \ , \ y = [N \quad 0]\begin{bmatrix} z \\ z_c \end{bmatrix} \quad ; \quad A \overset{\Delta}{=} \begin{bmatrix} D & -N_c \\ N & D_c \end{bmatrix}. \tag{9}$$

Note that, in transform terms, which we shall use hereafter unless otherwise noted,

$$P = \underline{D}^{-1}\underline{N} = ND^{-1} \quad , \quad C = \underline{D}_c^{-1}\underline{N}_c = N_cD_c^{-1} \tag{10}$$

are prime factorizations of P and C corresponding to the above internal descriptions. Furthermore,

$$\underline{A}\, A = \begin{bmatrix} \underline{D}_k & 0 \\ 0 & D_k \end{bmatrix} \tag{11}$$

where

$$\underline{D}_k \overset{\Delta}{=} \underline{D}_c D + \underline{N}_c N \quad , \quad D_k \overset{\Delta}{=} \underline{D} D_c + \underline{N} N_c \ . \tag{12}$$

Notice that other operator systems, such as

$$\underline{D}_k z = 0 \ , \ y = Nz \quad ; \quad D_k z_c = 0 \quad , \quad y = -D_c z_c \tag{13}$$

are also closed loop internal descriptions, equivalent to (6) and (9). $|A|$, $|\underline{A}|$, $|D_k|$ and $|\underline{D}_k|$ are therefore alternative expressions for the closed loop characteristic polynomial; and they are equal within a multiplicative constant. Note that similar results can be derived using matrix identities to evaluate $|\underline{A}|$ and $|A|$ from (6) and (9).

<u>Definition</u>. The feedback loop is <u>well defined</u> if the closed loop internal description is well defined, that is if $|\underline{A}| \neq 0$.

Notice that

$$(I + CP) = \underline{D}_c^{-1}\, \underline{D}_k D^{-1} \tag{14}$$

which, in view of the assumption that $|\underline{D}_c|$, $|D| \neq 0$ and the fact that $|\underline{A}| = k|\underline{D}_k|$, directly implies the known result, namely:

The feedback loop is well defined if and only if $|I + CP| \neq 0$.

<u>Definition</u>. The closed loop system is <u>internally stable</u> if $\underline{A}^{-1}$ exists and is stable.

Since the roots of $|\underline{A}|$ are the closed loop eigenvalues, the system will be internally stable when all eigenvalues lie in the open left half of the s-plane (continuous system) or inside the open unit disc in the z-plane (discrete system).

We shall now <u>parametrically characterize all stabilizing controllers</u> C. Here we shall follow the development in [2]:

Consider a unimodular matrix U so that $U \begin{bmatrix} D \\ N \end{bmatrix} = \begin{bmatrix} I \\ 0 \end{bmatrix}$ and of the form:

$$U = \begin{bmatrix} x_1 & x_2 \\ -\underline{N} & \underline{D} \end{bmatrix}, \quad U^{-1} = \begin{bmatrix} D & -\underline{x}_2 \\ N & \underline{x}_1 \end{bmatrix} . \tag{15}$$

It is known [2,11] that such a matrix U does exist. Postmultiply $\underline{A}$ of (6) by U^{-1} and premultiply A of (9) by U to obtain:

$$\underline{A}\ U^{-1} = \begin{bmatrix} \underline{D}_k & \underline{N}_k \\ 0 & I \end{bmatrix}, \quad U\ A = \begin{bmatrix} I & -N_k \\ 0 & D_k \end{bmatrix} \tag{16}$$

where $\underline{N}_k$, N_k are polynomial matrices. In view of the definitions of $\underline{A}$ and A, (16) directly implies that

$$[\underline{D}_c\ \underline{N}_c] = [\underline{D}_k\ \underline{N}_k]\ U \tag{17}$$

and

$$\begin{bmatrix} -N_c \\ D_c \end{bmatrix} = U^{-1} \begin{bmatrix} -N_k \\ D_k \end{bmatrix} . \tag{18}$$

Since U is unimodular, it is clear that $(\underline{D}_c, \underline{N}_c)$ ℓp $((N_c, D_c)$ rp) if and only if $(\underline{D}_k, \underline{N}_k)$ ℓp $((N_k, D_k)$ rp). Furthermore, if the product $(\underline{A}U^{-1})(UA)$ is determined using (16), then in view of (11),

$$\underline{D}_k N_k = \underline{N}_k D_k . \tag{19}$$

As it has been shown in [2,12], (17) and (18) characterize any and all solutions of equations (12) where $(D, N, \underline{D}_k)$, $(\underline{D}, \underline{N}, D_k)$ are given and $\underline{N}_k, N_k$ are arbitrary polynomial matrices of appropriate dimensions. Furthermore, in the system context, $\underline{D}_k$ and $\underline{N}_k$ must satisfy $|\underline{D}_k| \neq 0$, $|\underline{D}_k x_1 - \underline{N}_k \underline{N}| = |\underline{D}_c| \neq 0$ for the loop and controller to be well defined; and $\underline{D}_k^{-1}$ must be stable for internal stability. In view of (10), it follows that [2]:

<u>Proposition 1.</u> Any and all stabilizing controllers are given by

$$\begin{aligned} C &= (\underline{D}_k x_1 - \underline{N}_k \underline{N})^{-1}(\underline{D}_k x_2 + \underline{N}_k \underline{D}) = (x_1 - K\underline{N})^{-1}(x_2 + K\underline{D}) \\ &= (\underline{x}_2 D_k + D N_k)(\underline{x}_1 D_k - N N_k)^{-1} = (\underline{x}_2 + DK)(\underline{x}_1 - NK)^{-1} \end{aligned} \tag{20}$$

where $(\underline{D}_k, \underline{N}_k)$ $((N_k, D_k))$ are any polynomial matrices with appropriate dimensions such that $\underline{D}_k^{-1}$ (D_k^{-1}) is stable and $|\underline{D}_k x_1 - \underline{N}_k \underline{N}| \neq 0$ $(|\underline{x}_1 D_k - N N_k| \neq 0)$ or, alternatively, K is any stable rational matrix such that $|x_1 - K\underline{N}| \neq 0$ $(|\underline{x}_1 - NK| \neq 0)$.

Note that

$$K = \underline{D}_k^{-1} \underline{N}_k = N_k D_k^{-1} \tag{21}$$

which shows that the poles of K are the desired closed loop eigenvalues. It should be noted that the parameter K and the expression $C = (\underline{x}_2 + DK)(\underline{x}_1 - NK)^{-1}$ were first introduced in [1] using an alternative method.

Proposition 1 parametrically characterizes all stabilizing feedback controllers. The parameters are either the polynomial matrices D_k, N_k ($\underline{D}_k, \underline{N}_k$) or the rational matrix K. Note that this parameterization requires the knowledge of prime polynomial matrix factorizations of the plant transfer matrix P. It is clear that using (20), the designer has control over the closed loop internal descriptions; furthermore, complete and arbitrary closed loop eigenvalue assignment is easy to achieve by appropriately choosing D_k ($\underline{D}_k$) or K. The problem of causality is addressed in a later section of this paper.

Internal stability in the feedback loop can also be determined directly from the transfer matrices of the plant P and the controller C without using internal descriptions. In particular, let α_P, α_C denote the characteristic polynomials of P, C respectively and

$$S_1 \stackrel{\Delta}{=} (I + PC)^{-1}, \quad S_2 \stackrel{\Delta}{=} (I + CP)^{-1}, \qquad Q \stackrel{\Delta}{=} C\, S_1 = S_2\, C \,. \tag{22}$$

Note that $\underline{D}(I + PC)y = \underline{D}S_1^{-1}y = 0$ and $\underline{D}_c(I + CP)u = \underline{D}_c S_2^{-1}u = 0$, when interpreted in an operator sense.

Proposition 2. The following statements are equivalent.

(a) The closed loop system is internally stable.

(b) $\alpha_P \alpha_C\, |S_1^{-1}| = \alpha_P \alpha_C\, |S_2^{-1}|$ is Hurwitz.

(c) The zero polynomial of $\begin{bmatrix} I & C \\ -P & I \end{bmatrix}$ is Hurwitz.

(d) $\begin{bmatrix} S_2 & -Q \\ PS_2 & S_1 \end{bmatrix}$ is stable.

Proof: (b) $S_2^{-1} = I + CP = \underline{D}_c^{-1}\underline{D}_k D^{-1}$ and $\alpha_P = |D|$, $\alpha_C = |\underline{D}_c|$. Then $\alpha_P \alpha_C\, |S_2^{-1}| = |\underline{D}_k| = k\,|A|$ which is Hurwitz by definition. Also $|S_2^{-1}| = |I + CP| = |I + PC| = |S_1^{-1}|$.

To show (c), note that $\begin{bmatrix} I & C \\ -P & I \end{bmatrix} = \begin{bmatrix} \underline{D}_c & 0 \\ 0 & \underline{D} \end{bmatrix}^{-1} \underline{A}$, which is a ℓp factorization. The zero polynomial is $|\underline{A}|$ and therefore Hurwitz. (d) The zero polynomial in (c) is the characteristic polynomial of the inverse system $\begin{bmatrix} I & C \\ -P & I \end{bmatrix}^{-1} = \begin{bmatrix} (I+CP)^{-1} & -C(I+PC)^{-1} \\ P(I+CP)^{-1} & (I+PC)^{-1} \end{bmatrix}$, which is the matrix in (d). ΔΔΔ

Note that (b) is a well known stability test, while the matrix in (d) has been used by Zames and Desoer in [3,5] and earlier papers to define stable feedback systems. Here internal stability is defined using internal descriptions from which the results of Proposition 2 are easily derived as alternative tests for internal stability.

The feedback controller parameterizations of Proposition 1 are closely related to internal system descriptions; and it is clear that they can be used to solve directly the design problems of eigenvalue assignment and stabilization. However, if the main design objective is to obtain desired input/output maps between certain signals in the loop, then it may be more convenient to use alternative controller parameterizations directly related to those maps. The following basic feedback compensation theorem introduces such parameterizations and establishes their relation to the internal system descriptions.

Theorem 3. Given a plant $y = Pu$ with $P = ND^{-1}(=\underline{D}^{-1}\underline{N})$ prime polynomial matrix factorizations, and controller $u = -Cy$, the closed loop system is internally stable if and only if

$$C = \underline{L}_2^{-1}\underline{L}_1 \quad (C = L_1L_2^{-1}) \tag{23}$$

where $\underline{L}_1$, $\underline{L}_2$ (L_1, L_2) are stable rational matrices with $|\underline{L}_2|$ ($|L_2|$) $\neq 0$ which satisfy

$$\underline{L}_2D + \underline{L}_1N = I \quad (\underline{D}L_2 + \underline{N}L_1 = I) \ . \tag{24}$$

Furthermore, if

$$[\underline{L}_2\ \underline{L}_1] = \underline{D}_k^{-1}\ [\underline{D}_c\ \underline{N}_c] \ \left(\begin{bmatrix} L_2 \\ L_1 \end{bmatrix} = \begin{bmatrix} D_c \\ N_c \end{bmatrix} D_k^{-1}\right) \tag{25}$$

are prime polynomial matrix factorizations, the closed loop internal description is given in operator terms by

$$\underline{D}_kz = 0,\ y = Nz \ (D_kz_c = 0,\ y = -D_cz_c). \tag{13}$$

Proof: The part in parentheses will not be shown as it follows in a similar way. Let $[\underline{L}_2\ \underline{L}_1]$ stable satisfy (23), (24) and write a ℓp factorization as in (25). Then $\underline{D}_cD + \underline{N}_cN = \underline{D}_k$ where $(\underline{D}_c\ \underline{N}_c)$ ℓp, $\underline{D}_c^{-1}$ and $\underline{D}_k^{-1}$ exist with $\underline{D}_k^{-1}$ stable. Therefore the feedback loop with controller $C = \underline{L}_2^{-1}\underline{L}_1 = \underline{D}_c^{-1}\underline{N}_c$ is well defined, and it is internally stable with internal description (13). Assume now that the closed loop system is internally stable, that is $\underline{D}_k^{-1}$ in (13) exists and is stable. (17) implies that $\underline{D}_k^{-1}\ [\underline{D}_c\ \underline{N}_c] = [I\ K]\ U$; note that $(\underline{D}_k, [\underline{D}_c\ \underline{N}_c])$ is ℓp since $(\underline{D}_c, \underline{N}_c)$ is ℓp. Let $\underline{L}_2 = \underline{D}_k^{-1}\underline{D}_c$, $\underline{L}_1 = \underline{D}_k^{-1}\underline{N}_c$; the rest easily follows. ΔΔΔ

The stable rational matrices $\underline{L}_1$, $\underline{L}_2$, which satisfy (24) and characterize all stabilizing output controllers, also characterize all state observers.

Proposition 4. $[\underline{L}_2\ \underline{L}_1]\begin{bmatrix} u \\ y \end{bmatrix}$, where $\underline{L}_1$, $\underline{L}_2$ are stable, is a partial state observer if and only if $\underline{L}_2D + \underline{L}_1N = I$.

Proof: In view of the plant description (7) $Dz=u$ $y=Nz$, $\underline{L}_2u + \underline{L}_1y = (\underline{L}_2D + \underline{L}_1N)z$ which is equal to the partial state z if and only if $\underline{L}_2D + \underline{L}_1N = I$. ΔΔΔ

Observers of linear functionals of the state of the form Fz can also be easily derived as follows:

Let D be column proper (reduced) and let F be such that the column degrees of D, $\partial_{ci}D > \partial_{ci}F$; F is a desired state feedback matrix ($u = Fz$) [11]. Let $D_F = D-F$; then in view of (24)

$$(I - D_F\underline{L}_2)D + (-D_F\underline{L}_1)N = F$$

which implies

$$(I - D_F\underline{L}_2)u + (-D_F\underline{L}_1)y = Fz.$$

This means that $[I - D_F\underline{L}_2, -D_F\underline{L}_1]$ is an observer with output Fz. Furthermore, if

$$[I - D_F\underline{L}_2, -D_F\underline{L}_1] = L^{-1}\ [K_1,K_2]$$

is a left prime factorization, then

$$K_1D + K_2N = LF$$

which is precisely the relation used in [11] to derive linear state feedback realizations via an observer compensation scheme.

These results establish the exact relation between the factors of stabilizing compensators C and observers of the state of the plant. Also note that if $\underline{L}_1$ is chosen as discussed in a later section to guarantee C proper, then the corresponding observer of Fz derived above will also be proper.

If the output feedback configuration {G,H;P}, where $u = G(r - HY)$, is used to realize linear state feedback then the appropriate choices for G and H are [13]:

$$G = (L - K_1)^{-1}L = (D_F\underline{L}_2)^{-1} \quad , \quad H = -L^{-1}K_2 = D_F\underline{L}_1.$$

It is clear that the feedback path compensator H is stable; however, the feedforward path compensator G is stable if and only if $\underline{L}_2^{-1}$ is stable (D_F^{-1} is chosen to be stable) that is, if and only if C, in Theorem 3, is stable. Therefore, stabilizing the plant P via a stable controller C is a problem in precisely the same spirit as that of realizing a state feedback control law via {G,H;P} compensation with G and H stable. Furthermore note that in view of Theorem 3, the above choice for G and H stabilize the plant; this is an alternative proof of the known result, namely that any coprimely represented P can be stabilized with the aid of an observer.

It is of interest to notice that any stabilizing controller C of the plant P can be written, in view of Theorem 3, as a product of two factors; one of the factors is stable ($\underline{L}_1$) while the other ($\underline{L}_2^{-1}$) has stable transmission zeros [18]. This observation also implies that PL_2^{-1} (or $P(D_F\underline{L}_2)^{-1}$) can be stabilized by a stable controller $\underline{L}_1$ ($D_F L_1$).

In view of Theorem 3, a number of controller parameterizations can now be readily derived. In particular, we have the following.

Proposition 5. Any and all stabilizing controllers are given by:

(a) $C = \underline{L}_2^{-1}\underline{L}_1 \; (= L_1 L_2^{-1})$ (23)

where $\underline{L}_2, \underline{L}_1$ (L_2, L_1) satisfy (24);

(b) $C = S_2^{-1}Q \; (= QS_1^{-1})$ (26)

where $D^{-1}[S_2 \; Q]$ $(S_1\underline{D}^{-1}, Q\underline{D}^{-1})$ stable with $|S_2|$ $(|S_1|) \neq 0$ satisfying $S_2 + QP = I$ $(S_1 + PQ = I)$; (27)

(c) $C = [(I - \underline{L}_1 N)D^{-1}]^{-1}\underline{L}_1$ (28)

where $(I - \underline{L}_1 N)D^{-1}$, $\underline{L}_1$ stable with $|I - \underline{L}_1 N| \neq 0$;

(d) $C = Q(I - PQ)^{-1}$ (29)

where $(I - PQ)\underline{D}^{-1}$, $Q\underline{D}^{-1}$ stable with $|I - PQ| \neq 0$.

Proof: (a) is clear in view of Theorem 3. $S_2 + QP = I$ can be written as $(D^{-1}S_2)D + (D^{-1}Q)N = I$ which in view of (a) directly implies (b). If $\underline{L}_2$, S_2 are expressed in terms of $\underline{L}_1$, Q respectively (c) and (d) are derived; note that the dual of (c) and (d) are also true. ΔΔΔ

It should be noted that these parameterizations are related to internal descriptions of the closed loop system via (25). In this way, the effect of the particular choice for the parameter on the closed loop eigenvalues and, in general, on the closed loop internal description can be determined.

When C is expressed in terms of Q, S_2, S_1 as above, (22) are satisfied. These parameters are important design maps related to feedback and response properties. For example, $S_1 = (I + PC)^{-1}$ is the well known comparison sensitivity matrix which provides a measure of the effect of the parameter variations in P on the output y. Expressing the internal stability criteria directly in terms of these maps can provide significant insight in design. Note that the parameter Q is the parameter introduced by Zames in [5] using an alternative method; furthermore $\underline{L}_1$ is in the case of error feedback configuration the design parameter X discussed by Sain, et. al. [6-8]. Notice that there is a one-to-one correspondence between $\underline{L}_1$ or Q and the stabilizing compensators C given by (28) or (29).

In Propositions 1 and 5 all stabilizing controllers have been parametrically characterized. The parameters involved are of course related and their exact relation is easily derived to be:

$$\underline{L}_2 = D^{-1}S_2 = x_1 - \underline{KN} \; (L_2 = S_1\underline{D}^{-1} = \underline{x}_1 - NK),$$
$$\underline{L}_1 = D^{-1}Q = x_2 + \underline{KD} \; (L_1 = Q\underline{D}^{-1} = \underline{x}_2 + DK). \tag{30}$$

In view of these relations, additional parameterizations involving combinations of these parameters (e.g. S_1 and K) can also be obtained.

When the plant P is unstable it is clear that the conditions (c) and (d) impose restrictions on the structure of the parameters $\underline{L}_1$ and Q. When the plant is stable, the conditions on the parameters in Proposition 5 are simplified [5,8]:

Corollary 5.1. If P is stable, any and all stabilizing controllers are given by:

(a) $C = [(I - \underline{L}_1N)D^{-1}]^{-1}\underline{L}_1$

where $\underline{L}_1$ is stable with $|I - \underline{L}_1N| \neq 0$; (31)

(b) $C = Q(I - PQ)^{-1}$

where Q is stable with $|I - PQ| \neq 0$. (32)

Proof: When P is stable, D^{-1} is stable. Note that the dual of (a) and (b) are also true. ΔΔΔ

When P is stable, these parameterizations are simple to use because, for internal stability, the only restriction imposed on the parameter is that it must be stable; furthermore, as it will be shown in the section on causality, if P is proper and Q or $D\underline{L}_1$ are chosen to be strictly proper, then C will be proper. Q and (32) are used in [5,9] where it is assumed that the plant P is proper and stable; Q is then chosen to satisfy additional design requirements.

When P is square and nonsingular, that is $|P| \neq 0$, additional parameterizations can be derived:

Corollary 5.2. If $|P| \neq 0$, any and all stabilizing controllers are given by:

(a) $C = \underline{L}_2^{-1}(I - \underline{L}_2D)N^{-1}$

where $\underline{L}_2$, $(I - \underline{L}_2D)N^{-1}$ are stable with $|\underline{L}_2| \neq 0$. (33)

(b) $C = (S_1P)^{-1}(I - S_1)$

where $N^{-1}S_1P$, $N^{-1}(I - S_1)$ are stable with $|S_1| \neq 0$. (34)

Proof: The proof of (a) is similar to (c) of Proposition 5. (b) can be shown directly: notice that $[N^{-1}S_1P]D + [N^{-1}(I - S_1)]N = I$ and $I + PC = I + P(S_1P)^{-1}(I - S_1) = S_1^{-1}$ which in view of Theorem 3 implies the result. The dual of (a) and (b) are also true. ΔΔΔ

If P has no zeros in the closed right half s-plane (or outside the closed unit disc) then N^{-1} is stable and stabilizing controllers can be derived by choosing any stable $\underline{L}_2$ or S_1, with S_1P stable, in (a) and (b) above; this is true for stable or unstable plants P. However, using these parameterizations, proper C is more difficult to obtain. Nevertheless, Corollary 5.2 points to the fact that if P has stable zeros then it can be easily stabilized via a, not necessarily proper, controller C.

Corollaries 5.1 and 5.2 are examples of cases where special properties of P are used to derive alternative parameterizations. It is clear that for a given plant, additional parameterizations could be derived depending on the particular properties of P.

In the following corollary an internal stability test is presented:

Corollary 5.3. The closed loop system is internally stable if and only if $|I + CP| = |I + PC| \neq 0$ and

$$D^{-1}\ [(I + CP)^{-1},\ (I + CP)^{-1}C]\ \left(\begin{bmatrix}(I + PC)^{-1}\\ C(I + PC)^{-1}\end{bmatrix}\underline{D}^{-1}\right)\ \text{stable} \qquad (35)$$

Proof: See (d) of Proposition 5 and (22).

Notice that in (d) of Proposition 2, four rational matrices must be stable for internal stability. In Corollary 5.3, by using the denominator D ($\underline{D}$) of the plant P, internal stability depends on the stability of only two rational matrices.

2.2 Causality. We are interested in proper controllers C. In view of

$$C = Q(I - PQ)^{-1} \qquad (29)$$

if Q is proper and (I -PQ) at (∞) is finite and nonsingular (that is (I -PQ) and its inverse are proper, or (I -PQ) is biproper) then C is proper. If in addition P is proper then any strictly proper Q satisfies the requirements and results, by (29), in a strictly proper controller C.

It is therefore clear that proper stabilizing controllers C are derived from (d) of Proposition 5 if the additional requirement of Q proper and (I -PQ) biproper is added. Using the dual of (d) the equivalent conditions Q proper and (I -QP) biproper are derived. These conditions on Q are easily translated into conditions on $\underline{L}_1$ (L_1) using (30) to give $D\underline{L}_1$ proper and $(I - N\underline{L}_1)$ biproper; in this way (c) of Proposition 5 can also be used to obtain proper stabilizing controllers C.

In view of (30)

$$K = (\underline{L}_1 - x_2)\underline{D}^{-1} = (D^{-1}Q - x_2)\underline{D}^{-1}. \tag{36}$$

For any Q or $\underline{L}_1$ which satisfies the causality conditions or, with a proper P for simplicity, for any Q strictly proper, the corresponding K, if used in (20) of Proposition 1 will give a proper C. For stability, K must also be stable or Q must satisfy the conditions in Proposition 5 unless P is stable, in which case Q stable suffices (Corollary 5.1). It is apparent that it is more difficult to choose K (or $\underline{D}_k,\underline{N}_k$), than Q or $\underline{L}_1$, to guarantee causality of the controller. In general, for C proper, the order of K (or $\partial|\underline{D}_k|$ in (12)) must be higher than the order of P by an amount which depends on the structure of N and D [8]. This can be seen from the known result [14] namely that all closed loop eigenvalues can be arbitrarily assigned with a controller C of order min $(\mu-1,\nu-1)$ where μ and ν are the controllability and observability indices of P; this result implies that for $\partial|\underline{D}_k|$ in (12) large enough proper C does exist. Note that for particular (N,D) it might be possible to derive lower order proper C, that is proper solution $C = \underline{D}_c^{-1}\underline{N}_c$ of (12) might exist for lower $\partial|\underline{D}_k|$; this is the case when for example eigenvalue assignment via constant output feedback is possible.

Proper C can also be achieved if one works over the proper and stable matrices [3,4]. This is shown here to be a direct consequence of Theorem 3.

Proposition 6. Let P be strictly proper. Any and all proper stabilizing controllers are given by

$$C = L_2'^{-1}\, L_1' \tag{37}$$

where $L_2'\, D' + L_1'\, N' = I$, $|L_2'| \neq 0$, $P = N'\, D'^{-1}$ with all (') matrices proper and stable.

Proof: If $L_2'D' + L_1'N' = I$, then $L_2' + L_1'\, P = D'^{-1}$ which implies D'^{-1} proper, that is D' biproper. Also, if C is proper then $D' + CN' = L_2'^{-1}$ which implies that $L_2'^{-1}$ is proper, that is L_2' biproper. Let $\begin{bmatrix} D' \\ N' \end{bmatrix} = \begin{bmatrix} D \\ N \end{bmatrix} D_1^{-1}$ [6-7] and let $D_1^{-1}\,[L_2'\; L_1'] = [\underline{L}_2,\underline{L}_1]$; then $\underline{L}_2D + \underline{L}_1N = I$ which in view of Theorem 3 implies that the proper $C = L_2'^{-1}\, L_1' = \underline{L}_2^{-1}\underline{L}_1$ is a stabilizing controller. Assume that the proper $C = \underline{L}_2^{-1}\underline{L}_1$ satisfies the conditions of Theorem 3. Let D_1 be a polynomial matrix such that D_1^{-1} is stable, $D_1\underline{L}_2$ is biproper and $D_1\underline{L}_1$ is proper. Note that such a matrix exists because P strictly proper and C proper imply $D\underline{L}_1$ proper, $D\underline{L}_2 = I - D\underline{L}_1P$ biproper (see causality arguments).

Now $\underline{L}_2 D + \underline{L}_1 N = I$ implies $(D_1\underline{L}_2)(DD_1^{-1}) + (D_1\underline{L}_1)(ND_1^{-1}) = I$ from which $(D_1\underline{L}_2) + (D_1\underline{L}_1) P = (DD_1^{-1})^{-1}$ which implies that (DD_1^{-1}) is biproper; in view of P proper, (ND_1^{-1}) is also proper. If $[L_2', L_1'] = D_1[\underline{L}_2\ \underline{L}_1]$ and $D' = DD_1^{-1}$, $N' = ND_1^{-1}$ the result is obtained. ΔΔΔ

All solutions of $L_2'D' + L_1'N' = I$ over the proper and stable matrices are given by

$$[L_2' L_1'] = [x_1' - K'\underline{N}' \quad x_2' + K'\underline{D}'] \tag{38}$$

with K' any proper and stable matrix corresponding to parameter K in (20) [3,4,12]. Proper stabilizing controllers C are easily obtained by appropriately choosing K'. Note that in view of the above proof and (25), the closed loop eigenvalues are the zeros of $|\underline{D}_k|$ where

$$D_1^{-1}\ [x_1' - K'\underline{N}' \quad x_2' + K'\underline{D}'] = \underline{D}_k^{-1}\ [\underline{D}c\ \underline{N}c] \tag{39}$$

a left prime polynomial factorization with $\begin{bmatrix} D' \\ N' \end{bmatrix} = \begin{bmatrix} D \\ N \end{bmatrix} D_1^{-1}$ [6-7].

This shows that the closed loop eigenvalues (zeros of $|\underline{D}_k|$) depend on x_1', x_2' in addition to K'. Clearly in this case a desired set of closed loop eigenvalues cannot be achieved easily via K'. Note that similar difficulty is encountered when the parameter Q [5,9] or the λ-generalized polynomials [10] are used. In the latter method, a transformation $\lambda = 1/(s+a)$ is introduced to transform the proper transfer matrices into polynomial matrices in λ; in this way causality is easily achieved but multiple closed loop eigenvalues at $-a$ tend to appear. The discussion on hidden modes sheds more light upon this issue.

2.3. <u>Hidden Modes</u>. Hidden modes are those modes of the closed loop system which do not appear as poles of the closed loop transfer matrix. They correspond to closed loop eigenvalues which are uncontrollable and/or unobservable. If the internal stability conditions are satisfied, it is clear that unstable hidden modes do not exist in the loop. It is possible, however, to have stable hidden modes which might cause undesirable signal behavior in the loop; furthermore these hidden modes unnecessarily increase the order of the controller [8].

The <u>unobservable</u> eigenvalues can be determined using the internal operator <u>description</u> (13) since an output y has already been specified. They are the zeros of the determinant of a greatest right divisor (grd) of $(\underline{D}_k, N)$ [11]. This grd is equal to a grd of $(\underline{D}_c D, N)$ since

$$\begin{bmatrix} \underline{D}_k \\ N \end{bmatrix} = \begin{bmatrix} I & \underline{N}_c \\ 0 & I \end{bmatrix} \begin{bmatrix} \underline{D}_c D \\ N \end{bmatrix} \tag{40}$$

which implies, in view of $N(\underline{D}_c D)^{-1} = P\underline{D}_c^{-1}$, that <u>the unobservable</u>

eigenvalues are exactly those poles of C which cancel in the product PC.

If parameterizations involving $\underline{D}_k$ or K and (20) are used, one can easily control the number of unobservable hidden modes. However, when parameterizations involving $\underline{L}_1$, Q and so forth, as in Proposition 5, are used, it is not as clear how this can be achieved. This is studied next. The unobservable eigenvalues are exactly those poles of $\underline{L}_1$ which cancel in $N\underline{L}_1$. This can be shown using (25) as follows: The unobservable eigenvalues are those zeros of $|\underline{D}_k|$ which cancel in $N\underline{D}_k^{-1}$ or in $N\underline{D}_k^{-1}[\underline{D}_c\ \underline{N}_c] = N[(I - \underline{L}_1N)D^{-1}\ \underline{L}_1]$, since $(\underline{D}_k, (\underline{D}_c, \underline{N}_c))$ are prime and $\underline{D}_k^{-1}$ contains all poles of $\underline{L}_1$. The only poles which cancel in $N(I - \underline{L}_1N)D^{-1} = (I - N\underline{L}_1)P$ with N, are the poles of $\underline{L}_1$ which cancel $N\underline{L}_1$. Furthermore note that in view of (23) all the unobservable eigenvalues appear as poles of C.

To discuss uncontrollable eigenvalues we must specify an input r. Consider first unity or error feedback:

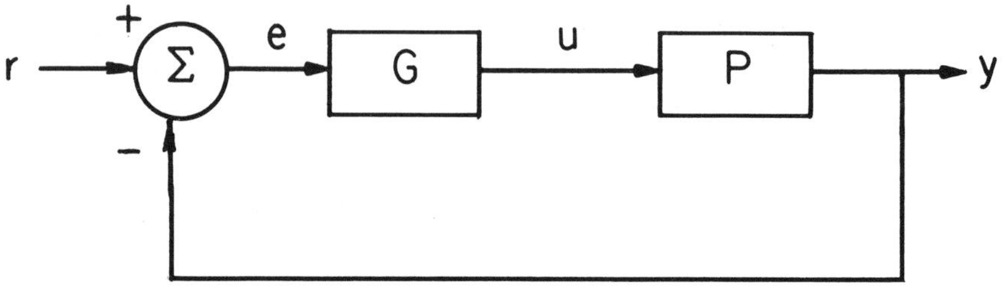

$$\{G,I;P\}: \quad u = Ge, \ e = r-y.$$

In view of (1-3), $C_y = C = C_r = G$ and

$$T = PM_G \quad , \quad M_G = (I + GP)^{-1}G \tag{41}$$

where T (y=Tr) is the closed loop transfer matrix and M_G ($u=M_Gr$) characterizes the control action u. Using the internal operator description $Dz = u$, $y = Nz$ and $\underline{D}_c\underline{z}_c = \underline{N}_ce$, $u = \underline{z}_c$ for the plant P and the controller G respectively with $e = r-y$, the closed loop internal description is

$$\underline{D}_kz = \underline{N}_cr \quad , \quad y = Nz \tag{42}$$

with $\underline{D}_k = \underline{D}_cD + \underline{N}_cN$ as in (12).

The uncontrollable eigenvalues are the zeros of the determinant of a greatest common left divisor ($g\ell d$) of $(\underline{D}_k, \underline{N}_c)$. This $g\ell d$ is equal to a $g\ell d$ of $(\underline{D}_cD, \underline{N}_c)$ since

$$[\underline{D}_k\ \underline{N}_c] = [\underline{D}_cD\ \underline{N}_c]\begin{bmatrix} I & 0 \\ N & I \end{bmatrix} \tag{43}$$

which implies, in view of $(\underline{D}_c D)^{-1}\underline{N}_c = D^{-1}G$, that in the {G,I;P} feedback configuration the uncontrollable eigenvalues are exactly those poles of P which cancel in the product PG. Note that, as it was shown above, the unobservable eigenvalues are exactly those poles of G which cancel in the product PG.

When (20) and parameters D_k or K are used then one can easily control the number of uncontrollable hidden modes. However, if parameters $\underline{L}_1$, Q and so forth are used, the issue becomes quite complicated and it is treated below. First note that, as it was shown above, the unobservable eigenvalues are those poles of $\underline{L}_1$ which cancel in $N\underline{L}_1$. It is now shown that the uncontrollable eigenvalues are exactly those poles of P which do not cancel in $(I - \underline{L}_1 N)D^{-1}$:

In view of (25), $[(I - \underline{L}_1 N)D^{-1}\ \underline{L}_1] = D_k^{-1}\ [\underline{D}_c, \underline{N}_c]$ is ℓp. Let $\underline{L}_1 = D_1^{-1}N_1$ be a left prime factorization. Then $[\underline{D}_c\ \underline{N}_c] = D_k\ D_1^{-1}$ $[(D_1 - N_1 N)D^{-1}, N_1]$. Since $\underline{N}_c$ is a polynomial matrix, $D_k = D_2 D_1$ and $\underline{N}_c = D_2 N_1$; also $\underline{D}_c = D_2(D_1 - N_1 N)D^{-1}$. Clearly the poles of D^{-1} (of P) which do not cancel in $(D_1 - N_1 N)$ must cancel D_2 completely since $\underline{D}_c$ is a polynomial matrix and $(\underline{D}_c, \underline{N}_c)$ are ℓp. This shows the result because D_2 is a gℓd of $(D_k, \underline{N}_c)$ and it contains the uncontrollable eigenvalues. Note that the uncontrollable eigenvalues appear in the numerator $\underline{N}_c$ of the controller G.

It is clear that in view of $\underline{L}_1 = D^{-1}Q$, the parameter Q can also be used to characterize the hidden modes. If (25) is expressed in terms of Q then

$$D^{-1}[I - QP \quad Q] = D_k^{-1}\ [\underline{D}_c\ \underline{N}_c]\ \ell p. \tag{44}$$

It follows that if Q is chosen to be proper and stable as in the case of stable P [5,9] then at least all of the poles of P will tend to appear as uncontrollable hidden modes in the loop thus increasing the complexity of the controller. In order to eliminate the possibly undesirable stable hidden modes (poles of P) and simplify the controller, the designer must use the above results on hidden modes, which impose additional structural restrictions on Q (or $\underline{L}_1$) similar to the ones imposed when P is unstable. This of course reduces the ease of implementation of this parameterization.

Notice that if $G = \underline{L}_2^{-1}\ \underline{L}_1$ in (41), then $M_G = D\underline{L}_1$ (=Q) and $T = N\underline{L}_1$; that is $\underline{L}_1$ is the design matrix X (z=Xr) introduced in [6,7] and used in [8] to show similar results.

We consider next another of a variety of possible feedback configurations which have come to be known as single degree of freedom types. The previous case, of course, which is output error feedback, is probably the most frequently studied example in the class. However, the case following is also of considerable interest, and has been commonly studied in the classical literature. In addition, it plays an important role in the applications, due to the use of sensors in the feedback path. Of course, the two cases under study here do not exhaust the set of possibilities; but they are representative of the techniques by which such problems may be approached. Consider, then, the following diagram:

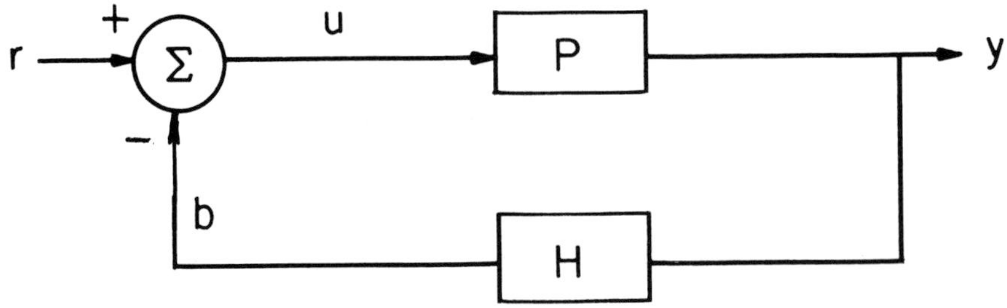

$$\{I,H;P\} : u = -Hy + r.$$

In view of (1-3), $C_y = C = H$, $C_r = I$ and

$$T = PM_H \quad , \quad M_H = (I + HP)^{-1} . \tag{45}$$

Using the internal operator descriptions $Dz = u$, $y = Nz$ and $\underline{D}_c \underline{z}_c = \underline{N}_c y$, $b = \underline{z}_c$ for the plant P and the controller H respectively with $u = r-b$, the closed loop internal description is

$$\underline{D}_k z = \underline{D}_c r \quad , \quad y = Nz \tag{46}$$

with $\underline{D}_k$ as in (12). Proceeding similarly, the uncontrollable eigenvalues are the zeros of the determinant of a gℓd of $(\underline{D}_k,\underline{D}_c)$ or of $(\underline{N}_c N,\underline{D}_c)$ which implies, in view of $\underline{D}_c^{-1}\underline{N}_c N = HN$ that they are exactly those poles of H which cancel in the product HP. Note that, as it was shown above, the unobservable eigenvalues are those poles of H which cancel in the product PH.

If parameters $(\underline{L}_1,\underline{L}_2)$ are used, it was shown that the unobservable eigenvalues are those poles of $\underline{L}_1$ which cancel in $N\underline{L}_1$. It is now shown that, in the $\{I,H;P\}$ feedback configuration, the uncontrollable eigenvalues are those poles of $\underline{L}_1$ which cancel in $\underline{L}_1 N$:

Let $\underline{L}_2 = D_1^{-1}N_1$ be a left prime factorization. Then in view of (25), $\underline{D}_c = \underline{D}_k\underline{L}_2 = \underline{D}_k D_1^{-1}N_1$ which implies that $\underline{D}_k = D_2D_1$ and $\underline{D}_c = D_2N_1$. D_2 is a gld of $(\underline{D}_k,\underline{D}_c)$ and it contains all the uncontrollable eigenvalues. Note that $\underline{N}_c = \underline{D}_k\underline{L}_1 = D_2D_1\underline{L}_1$; since $\underline{N}_c$ is a polynomial matrix ℓp to $\underline{D}_c$, the poles of $\underline{L}_1$ must cancel all zeros of $|D_2|$. Now the gℓd of $(\underline{N}_c N,\underline{D}_c)$ contains all the zeros of $|D_2|$. Therefore all the poles of $\underline{L}_1$ which correspond to the uncontrollable eigenvalues must cancel with N in $\underline{N}_c N = D_2D_1\underline{L}_1 N$. Note that the uncontrollable eigenvalues appear in the denominator $\underline{D}_c$ of the controller H.

It is clear that, in view of $\underline{L}_1 = D^{-1}Q$, the parameter Q can also be used to characterize the hidden modes. Notice that if $H = \underline{L}_2^{-1}\underline{L}_1$ in (45) then $M_H = D\underline{L}_2$ and $T = N\underline{L}_2$, that is $\underline{L}_2$ is the design matrix X [6-8]; furthermore $\underline{L}_1 = XH$ and $Q = M_H H = D\underline{L}_1$. For no hidden modes, $\underline{L}_1$ must be chosen so that no cancellations take place in $\underline{L}_1 N$ or in $N\underline{L}_1$.

In view of the above discussion on hidden modes, the closed loop eigenvalues can be easily described in terms of poles of certain transfer matrices in the loop. This provides additional insight and leads to additional tests for internal stability. The results presented below can be shown either by using the hidden modes results derived above for the single degree of freedom feedback configurations or by using internal descriptions and derivations similar to the ones used in the above proofs:

Consider the {G,I;P} feedback configuration. The closed loop eigenvalues are:

(i) The poles of $PG(I + PG)^{-1} = T$ and the poles of P and G which cancel in PG (uncontrollable and unobservable eigenvalues).

(ii) The poles of $(I + PG)^{-1} = S_1$ and the poles of P and G which cancel in PG. Notice that $S_1 = I - T$.

(iii) The poles of $(I + GP)^{-1} = S_2$ and the poles of P and G which cancel in GP.

(iv) The poles of $G(I + PG)^{-1} = M_G$ and the poles of P which cancel in PG and GP.

(v) The poles of $P(I + GP)^{-1}$ and the poles of G which cancel in PG and GP.

Notice that (ii) implies that the closed loop system is internally stable if and only if $(I + PG)^{-1}$ is stable and no unstable cancellations take place in PG; this is the main result of [15].

(vi) The poles of X and the poles of P which cancel in PG (uncontrollable eigenvalues). Note that the poles of X which cancel in NX are the unobservable eigenvalues; the poles of P which cancel in GP are the poles of X which cancel in DX.

As a direct consequence of Theorem 3 and the relation between X and $\underline{L}_1$, $\underline{L}_2$ the following important result is presented, an alternative proof of which was given in [8].

Proposition 7

$\begin{bmatrix} T \\ M \end{bmatrix} = \begin{bmatrix} N \\ D \end{bmatrix} X$ can be realized with internal stability via:

(a) {G,I;P} compensation if and only if

$$X, \ (I - XN)D^{-1} \text{ stable and } |I - XN| \neq 0; \tag{47}$$

then $G = [(I - XN)D^{-1}]^{-1}X = M(I - PM)^{-1}$;

(b) {I,H;P} compensation if and only if
X stable, $|X| \neq 0$ and there exists stable $\underline{X}$ such that

$$XD + \underline{X}N = I; \tag{48}$$

then $H = X^{-1}\underline{X}$

Proof In (a) $X = \underline{L}_1$ and in (b) $X = \underline{L}_2$, $\underline{X} = \underline{L}_1 = XH$ of Theorem 3. This proves the result. If (25) is used the corresponding internal description can be derived. ΔΔΔ

In (b), if P^{-1} is stable then an appropriate H (not necessarily proper) always exists; also note that H is stable if and only if X^{-1} is stable. Furthermore, note that in view of the causality discussion above, if P is proper then for DX strictly proper G is proper and for $D\underline{X}$ strictly proper (DX biproper) H is proper.

It should be noted that the comparison sensitivity matrix is $S_1 = S = (I + PG)^{-1} = I - T = I - NX$. Notice that the poles of S are the poles of T. The zeros of S are the poles of PG, that is: all of the poles of P except the uncontrollable hidden modes and all of the poles of G except the unobservable hidden modes [8].

Proposition 7 is an example of expressing the conditions for internal stability directly in terms of the design parameter of interest; here X characterizes the command/output-reponse y = Tr and the command/control-response u = Mr. The internal stability conditions can also be expressed in terms of other design parameters, such as Q or the comparison sensitivity matrix S_1, as it was shown above, and the designer must choose the parameterization which best fits his or her data and design objectives. In general, when designing using controller parameterizations, constraints must be imposed on the parameters so that other design objectives in addition to internal stability are attained. (eg. [1,5,8,9,13]).

3. CONCLUSIONS

In this paper, a number of stabilizing controller parameterizations were presented. Certain parameterizations are closely related to polynomial matrix internal descriptions and they allow complete control of the closed loop eigenvalues, but they might lead to nonproper controllers. Other parameterizations involving rational matrices can easily solve the properness problem but they have less direct control over the closed loop eigenvalues and so can result in (stable) hidden modes and high order compensators. Internal polynomial matrix descriptions were used in the analysis and the relation of all the parameterizations to the internal structure of the feedback system was established. Tests for internal stability were also presented throughout the paper.

The theory of all stabilizing controllers as presented here can be used to derive other parameterizations as well as additional tests for internal stability. Having established their strengths and weaknesses the designer can choose the parameterization which best suits the given data and the design objectives.

4. ACKNOWLEDGEMENT

This work has been carried out with the support of the National Science Foundation under Grant ECS-81-02891.

5. REFERENCES

[1] D.C. Youla, J.J. Bongiorno and H.A. Jabr, "Modern Wiener-Hopf Design of Optimal Controllers-Part II", IEEE Trans. Autom. Contr., Vol. AC-21, pp. 319-338, June 1976.

[2] P.J. Antsaklis, "Some Relations Satisfied by Prime Polynomial Matrices and Their Role in Linear Multivariable System Theory", Theory", IEEE Trans. Autom. Contr., Vol. AC-24, pp. 611-616, August 1979.

[3] C.A. Desoer, R.W. Liu, J. Murray and R. Saeks, "Feedback Systems Design: The Fractional Representation Approach to Analysis and Synthesis", IEEE Trans. Autom. Contr., Vol. AC-25, No. 3, pp. 399-412, June 1980.

[4] M. Vidyasagar, H. Schneider and B.A. Francis, "Algebraic and Topological Aspects of Feedback Stabilization", IEEE Trans. Autom. Contr., Vol. AC-27, pp. 880-894, August 1982.

[5] G. Zames, "Feedback and Optimal Sensitivity: Model Reference Transformations, Multiplicative Seminorms, and Approximate Inverses", IEEE Trans. Autom. Contr., Vol. AC-26, No. 2, pp. 301-320, April 1981.

[6] M.K. Sain, B.F. Wyman, R.R. Gejji, P.J. Antsaklis and J.L. Peczkowski, "The Total Synthesis Problem of Linear Multivariable Control, Part I: Nominal Design", Proc. 20th Joint Autom. Contr. Conf., Paper WP-4A, June 1981.

[7] M.K. Sain, P.J. Antsaklis, B.F. Wyman, R.R. Gejji and J.L. Peczkowski, "The Total Synthesis Problem of Linear Multivariable Control, Part II: Unity Feedback and the Design Morphism", Proc. 20th IEEE Conf. Decision and Contr., pp. 875-884, December 1981.

[8] P.J. Antsaklis and M.K. Sain, "Unity Feedback Compensation of Unstable Plants", Proc. 20th IEEE Conf. Decision and Contr., pp. 305-308, December 1981.

[9] C.A. Desoer and M.J. Chen, "Design of Multivariable Feedback Systems with Stable Plant", IEEE Trans. Autom. Contr., Vol. AC-26, No. 2, pp. 408-415, April 1981.

[10] L. Pernebo, "An Algebraic Theory for the Design of Controllers for Linear Multivariable Systems-Parts I and II", IEEE Trans. Autom. Contr., Vol. AC-26, No. 1, pp. 171-194, February 1981.

[11] W.A. Wolovich, Linear Multivariable Systems. New York: Springer-Verlag, 1974.

[12] M.K. Sain, "The Growing Algebraic Presence in Systems Engineering: An Introduction", IEEE Proceedings Vol. 64, No. 1, pp. 96-111, January 1976.

[13] P.J. Antsaklis and M.K. Sain, "Feedback Synthesis with Two Degrees of Freedom: {G,H;P} Controller", Proc. of the 9th World Congress of the International Federation of Automatic Control, Budapest, Hungary, July 2-6, 1984.

[14] F.M. Brasch and J.B. Pearson, "Pole Placement Using Dynamic Compensators", IEEE Trans. on Automatic Control, Vol. AC-15, pp. 34-43, February 1970.

[15] B.D.O. Anderson and M.R. Gevers, "On Multivariable Pole-Zero Cancellations and the Stability of Feedback Systems", IEEE Trans. on Circuits and Systems, Vol. CAS-28, pp. 830-833, August 1981.

[16] V. Kucera, Discrete Linear Control, New York: Wiley, 1979.

[17] R. Liu and C.H. Sung, "On Well-Posed Feedback Systems-In an Algebraic Setting", Proc. 19th IEEE Conference Decision and Control, pp. 269-271, December 1980.

[18] B.F. Wyman and M.K. Sain, "The Zero Module and Essential Inverse Systems", IEEE Trans. on Circuits and Systems, Vol. 27, No. 2, pp. 112-126, February 1981.

CHAPTER 6

DECOMPOSITIONS FOR GENERAL MULTILINEAR SYSTEMS

Paulo A.S. Veloso
Catholic University
Dept. of Informatics
22453 Rio de Janeiro RJ
Brazil

ABSTRACT. Research reported herein deals with composition and mainly decompositions of (time-varying) multilinear systems, which are defined as an extension of the concept of a linear system as a module of input-output functions. Realizations of multilinear systems have both practical and theoretical interest, which motivates the study of decomposition as well as composition of multilinear systems. Composition of linear and multilinear systems is shown to produce multilinear systems. A natural and important way to obtain multilinear systems from linear component systems and one tensor-product system is also presented. The main decomposition result is that this natural way is also a general one: any multilinear system of a quite large class can be decomposed into parallel linear systems interconnected by a memoryless tensor-product system followed by a linear system. Some finer decompositions are also obtained to clarify the general structure and alternative interconnection patterns for the realization of systems in this class.

1. INTRODUCTION

Multilinear systems form an interesting subclass of the class of non-linear systems. They can be viewed as a generalization of the familiar linear systems in that its input-output relation is multilinear rather than linear. So, a multilinear system is not linear, in general.However, its nonlinearity is a very special one. Namely, the action of its inputs on the output is not linear, but each input acts linearly on the output, provided that the other inputs are held constant. Due to these reasons it might the expected that multilinear systems have a structure similar to that of linear systems. Our decompositions justify this conjecture by showing that the non-linearity of a class of multilinear systems can be localized in a memoryless tensor-product system, all memory components being linear. This implies that in addition to the usual linear components only one extra building block is needed to realize all multilinear systems in this class.

Differential bilinear systems in control (input) and state have a wide application in nuclear reactor control and in the study of some

S. G. Tzafestas (ed.), Multivariable Control, 105–123.

biological systems. The controllability of a class of bilinear stationary differential systems has been studied in [9]. Also, an interesting decomposition for a class of discrete-time stationary multilinear systems has been obtained in [1,4] by using state-space construction techniques.

The major features of the present decompositions are that a larger class of multilinear systems is encompassed and that these decomposition are to a large extent independent of any state assignment for the component systems. Moreover, the decompositions described herein are quite general, holding not only for time-invariant systems, but also for time-varying systems, regardless of any hipotheses about discreteness or continuity of time.

A version [10] of a General Time-Systems Theory developped by Windeknecht [14,16], is used to place the problem in a general setting. Section 2 outlines the basic concepts and results to be used. The third section is concerned with the concepts of linearity and multilinearity and with composition of linear and multilinear systems. Then, algebraic techniques for multilinear functions [3,5,6], are used to obtain the main results, decompositions for multilinear systems. They are presented in section 4. The first and the fifth sections contain introductory and concluding remarks, respectively.

So, composition and decompositions of static, causal and non-anticipatory multilinear systems defined over an arbitrary commutative ring with unit are the central theme of this paper. In a non-anticipatory system the output up to an instant is not influenced by inputs after this instant, whereas the input up to a given instant is sufficient to uniquely determine the value of the corresponding output at this instant if the system is causal. A static system is simply a pointwise map in that the input at each instant determines the corresponding output at the same instant. These concepts are made precise in the following section.

Understanding of what interconnection patterns of linear systems are sufficiently general to include realizations for all multilinear systems is a theoretical reason for this research. But, this is also a practical motivation. For one would like to know what basic systems are necessary to realize any multilinear system. For instance, what modules are needed on an analog computer to simulate a multilinear system. It might appear at first sight that the universality of the tensor product, [5,6], would provide a complete solution for this problem. This is indeed true for static systems. However, for other kinds of multilinear systems it is not so. All we know about the tensor-product system - then with memory, probably - is its existence and universality (apart from its uniqueness up to isomorphism). Not much insight into its structure is given. It is not necessarily a static system, and static tensor-product systems are the ones easily simulated on an analog computer by means of multiplier modules (of course, with finite-dimensional input vector spaces). Thus, decompositions of multilinear systems into linear systems and static non-linear systems would be of interest. Here we provide a conceptual framework for the study of these problems in a quite general setting as well as some (partial) solutions at an abstract level.

2. TIME-SYSTEMS

This section presents some basic definitions, notations and results from a General Time-Systems Theory [10] that will be nedeed in the subsequent sections. Its objective is mainly to establish notation and terminology by defining the following concepts: time, system causality, and system composition.

The definition of time to be used is sufficiently general to include discrete (usually modelled by the naturals) and continuous time (usually modelled by some subset of the reals) [17,18].

2.1 Definition. Let the totally ordered non void set T denote the time set. The element t of T will be referred to as the instant t. The symbols '≤', '>', '≥', etc., have the usual meaning concerning the ordering of the time set T. For a given instant t in T we define the following subsets of T

$$T^t \mathrel{\hat{=}} \{t' \in T \ / \ t' < t\} \qquad T_t \mathrel{\hat{=}} \{t' \in T \ / \ t' \geq t\}$$

Given two instants $t_1 \leq t_2$ in T, let

$$[t_1, t_2) \mathrel{\hat{=}} \{t \in T \ / \ t_1 \leq t < t_2\} = T^{t_1} \cap T_{t_2}$$

We now give notations related to time functions. In the following two definitions W denotes any non-empty set and W^T denotes the set of all functions from T into W.

2.2 Definition. Let $U : T \to W$ be an element of W^T. Given an instant t in T, the t-segment of u (denoted u^t) is the restriction of u to the set T^t and the t-section of u (denoted u_t) is the restriction of u to the set T_t. Given two instants $t_1 \leq t_2$ in T, let $u_{t_2}^{t_1}$ denote the restriction of u to the set $[t_1, t_2)$.

2.3 Definition. Let V be a non-void subset of W^T. For a given instant t in T we define the following sets

$$V^t \mathrel{\hat{=}} \{u^t \ / \ u \in V\}, \quad V_t \mathrel{\hat{=}} \{u_t \ / \ u \in V\}, \quad V(t) \mathrel{\hat{=}} \{u(t) \ / \ u \in V\}$$

We will also need the following (surjective) function

$$val(t) : V^{t_0} \to V(t) \quad ; \quad u^{t_0} \mapsto u(t)$$

for a fixed instant t_0 and $t < t_0$, in T. Given two instants $t_1 \leq t_2$ in T, let

$$V_{t_1}^{t_2} \mathrel{\hat{=}} \{u_{t_1}^{t_2} \ / \ u \in V\}$$

Following Zadeh [17,18], Windeknecht [14,15,16] and others, a system is defined essentially as a relation between input and output time-functions.

2.4 Definition. A T-system with input alphabet I, output alphabet Z, input space U and output space Y is a five-tuple $\mathcal{S}=\langle I,Z,U,Y,S\rangle$, such that U and Y are non-void subsets of, resp., I^T and Z^T and $\mathcal{S}$ is a non-void relation contained in $U \times Y$, with domain U.

Thus, U may be regarded as the space of admissible inputs.

This definition appears to be sufficiently general to include most usual systems. It is, however, much too general; we must classify systems according to a causality criterion for it to become mathematically powerful and to relate such general systems to the physical ones. In doing this, we follow quite closely the ideas, if not exactly the terminology, of Windeknecht [14,16]. There is, however, an apparently important restriction we chose to lift, namely the time set is not necessarily bounded from below. We may regard the notion of causality to be embodied in assertions such as "Knowledge of such and such part of the input signal (and, possible, also a certain part of the prior output signal) is sufficient to determine uniquely some specified part of the output of the system". That is, the present behavior of the system is in some sense determined by its past behavior, future inputs having no influence on present outputs. We formalize these ideas by means of convenient relations and functions. The restriction on the interpretation that might be caused by the use of functions instead of relations is weakened by not demanding a lower bound for the time set. Clearly we gain in mathematical power and simplicity.

2.5 Definition. Let $\mathcal{S} = \langle I,Z,U,Y,S\rangle$ be a T-system and t an instant in the time set T. Consider the following relations

$$S^t \mathrel{\hat=} \{(u^t,y^t) \in U^t \times Y^t \; / \; uSy\}$$

$$S^+ \mathrel{\hat=} \{(u^{t'},y^{t'}) \; / \; uSy \;\&\; t' \in T\} \subseteq U^+ \times Y^+$$

$$St \mathrel{\hat=} \{(u^t,y(t)) \in U^t \times Y(t) \; / \; uSy\}$$

$$\hat{S} \mathrel{\hat=} \{(u^{t'},y(t')) \; / \; uSy \;\&\; t' \in T\} \subseteq U^+ \times Z$$

$$S(t) \mathrel{\hat=} \{(u(t),y(t)) \in U(t) \times Y(t) \; / \; uSy\}$$

$$\dot{S} \mathrel{\hat=} \{(u(t_1),y(t_2)) \; / \; uSy \;\&\; t_1,t_2 \in T\} \subseteq I \times Z$$

where $U^+ \mathrel{\hat=} \bigcup_{t\in T} U^t$ and $Y^+ \mathrel{\hat=} \bigcup_{t\in T} Y^t$

Then

$\mathcal{S}$ is t-non-anticipatory if and only if $S^t : U^t \to Y^t$

$\mathcal{S}$ is uniformly non-anticipatory if $S^+ : U^+ \to Y^+$

$\mathcal{S}$ is t-causal if and only if $St : U^t \to Y(t)$

$\mathcal{S}$ is uniformly causal iff $\hat{S} : U^+ \to Z$

S is t-static if and only if $S(t) : U(t) \to Y(t)$

S is uniformly static iff $\dot{S} : I \to Z$

Examples of static systems are purely combinational switching circuits (for the discrete-time case) and electrical networks containing only ideal resistors and diodes, say, as continuous-time systems. A sequential machine with a fixed initial state is an example of a non-anticipatory or causal $\mathbb{N}$-system, depending whether the Mealy or the Moore model, respectively, is used, [2].

Some of the causality concepts introduced in the previous definition are interrelated in the following two results.

Corollary 2.6 clarifies the meaning to be attached to the word 'uniformly' in definition 2.5, whereas proposition 2.7 shows that 'non-anticipation' is a property shared by some classes of systems.

2.6 Corollary. Let S be a T-system. S is uniformly non-anticipatory (respectively uniformly causal) if and only if S is t-non-anticipatory (respectively t-causal) for every t in T. If S is uniformly static then S is t-static for every t in T.

Proof. Straightforward from the definitions. QED

2.7 Proposition. Let S be a T-system and t_0 an instant in T. If S is t-static, for every $t < t_0$, or t-causal, for every $t < t_0$, then S is t_0-non-anticipatory.

Proof. $S^{t_0} : U^{t_0} \to Y^{t_0}$ is well defined by the assignment

$u^{t_0} \mapsto \{(t, S(t)(u(t))) / t < t_0\}$, in the first case, or

$u^{t_0} \mapsto \{(t, St(u)) / t < t_0\}$, in the second one.

Thus, S is t_0-non-anticipatory in either case. QED.

Another important class of systems is introduced in the next definition. It differs from the previous ones in that the value of the output is also relevant to completely specify the behavior of the system.

2.8 Definition. Let $S = \langle I,Z,U,Y,S \rangle$ be a T-system and t_1 and t_2 be two instants in T, such that $t_1 \leq t_2$. Consider the set

$$_{t_1}S^{t_2} \hat{=} \{((y(t_1),u_{t_1}^{t_2}) ,y(t_2)) \in (Y(t_1) \times U_{t_1}^{t_2}) \times Y(t_2) / uSy\}$$

System S is t_1-t_2-transitional if and only if

$$_{t_1}S^{t_2} : Y(t_1) \times U_{t_1}^{t_2} \to Y(t_2)$$

Thus, a transitional system can be viewed as a system presenting a sort of feedback of the output, which also serves as its state. An example of such a system is the linear differential system characterized by the equations

$$\dot{x}(t) = F(t)x(t) + G(t)u(t)$$

$$y(t) = H(t)x(t)$$

provided that H(t) is a non-singular square matrix. It can be proved [10] that transitional systems form, as might be expected, the very core memory of non-anticipatory and causal systems. But we will need this result only for the special case of linear systems.

Before closing this preliminary section we still must present some basic facts about composition of time-systems.

2.9 Definition. Given n T-systems

$$\mathcal{S}_j = \langle I_j, Z_j, U_j, Y_j, S_j \rangle \quad , j = 1,2,\ldots,n \quad ,$$

their cartesian product is defined to be the T-system

$$\mathop{\mathsf{X}}_{j=1}^{n} \mathcal{S}_j \mathrel{\hat{=}} \langle \mathop{\mathsf{X}}_{j=1}^{n} I_j \ , \ \mathop{\mathsf{X}}_{j=1}^{n} Z_j \ , \ \mathop{\mathsf{X}}_{j=1}^{n} U_j \ , \ \mathop{\mathsf{X}}_{j=1}^{n} Y_j \ , \ \mathop{\mathsf{X}}_{j=1}^{n} S_j \rangle$$

where (with $\underline{n} \mathrel{\hat{=}} \{1,2,\ldots,n\}$)

$$\mathop{\mathsf{X}}_{j=1}^{n} S_j \mathrel{\hat{=}} \{(\mathop{\mathsf{X}}_{j=1}^{n} u_j \ , \ \mathop{\mathsf{X}}_{j=1}^{n} y_j) \ / \ \forall j \in \underline{n} : u_j S_j y_j\}$$

Of course, $\mathop{\mathsf{X}}_{j=1}^{n} u_j$ and $\mathop{\mathsf{X}}_{j=1}^{n} y_j$ denote the (cartesian) product of the u_j's and of the y_j's, respectively [5,6].

Clearly, the cartesian-product system $\mathop{\mathsf{X}}_{j=1}^{n} \mathcal{S}_j$ is t-non-anticipatory, t-causal, t-static if and only if every $\mathcal{S}j$ is t-non-anticipatory, t-causal, t-static, respectively.

2.10 Definition. Given two T-systems $\mathcal{S} = \langle I,J,U,V,S \rangle$ and $\mathcal{R} = \langle J,Z,V,Y,R \rangle$, their (series) composition $\mathcal{S}$ followed by $\mathcal{R}$ is the T-system $\mathcal{R}.\mathcal{S} \mathrel{\hat{=}} \langle I,Z,U,Y,R,S \rangle$, where $R.S \mathrel{\hat{=}} \{(u,y)\ UxY \ / \ \exists v \in V \ ; \ uSv \ \& \ vTy\}$.

The component systems are causally related to their series system in the next result.

2.11 Theorem. Let $\mathcal{R}$ and $\mathcal{S}$ be T-systems as above and t be an instant in T. Let $\mathcal{R}'$ denote the T-system obtained by the restriction of $\mathcal{R}$ to the range of $\mathcal{S}$. If $\mathcal{S}$ is t-non-anticipatory then $\mathcal{R}.\mathcal{S}$ is t-non-anticipatory (resp. t-causal) if and only if $\mathcal{R}'$ is t-non-anticipatory (resp. t-causal). If $\mathcal{S}$ is t-static then $\mathcal{R}.\mathcal{S}$ is t-static if and only if $\mathcal{R}'$ is t-static.

Proof. Composition of functions("if" part) and factorization of functions through their images ("only if" part). QED.

A simple, and probably expected, special case that will be needed concludes this section.

2.12 Proposition. Let $\mathcal{R}$, $\mathcal{S}$ and t be as above. If $\mathcal{S}$ is t-causal and $\mathcal{R}$ is t-static then $\mathcal{R}.\mathcal{S}$ is also t-causal.

Proof. According to definitions 2.5 and 2.10, $(R.S)t : U^t \to Y^t$ is well defined by the assignment $u^t \mapsto R(t)[St(u^t)]$. So, $R.S$ is t-causal, QED

This preliminary section establishes the context in which the concepts of linearity and multilinearity will be examined in the next section. A more detailed presentation of this or similar approaches to a General Time-Systems Theory can be found in [4,7,10,16]. But the contents of this section will be sufficient for our work in sections 3 and 4.

3. LINEAR AND MULTILINEAR SYSTEMS

In this section linear and multilinear systems are defined and compared and their composition examined. It is shown that composition of convenient multilinear systems yields multilinear systems. Also, linear systems are shown to be a special case of multilinear systems.

Henceforth let K denote a commutative ring with unit. Linear and multilinear systems over K will be treated in this and the next section.

3.1 Definition. A T-system $L = \langle I,Z,U,Y.L\rangle$ is linear (over K) if and only if I and Z are K-modules, U and Y are K-submodules of I^T and Z^T, respectively, and L is a K-submodule of UxY.

In much the same way that a linear system is defined as a linear relation between the input and output modules [4] or vector spaces [17, 18], a multilinear system will be defined as a multilinear relation [4, 10]. This is a generalization of the concept of a multilinear function [3,5,6], introduced in the next definition.

3.2 Definition. Given (n+1) K-modules $U_1,U_2,\dots,U_n$, Y and a relation

$$M \subseteq (\mathop{X}_{k=1}^{n} U_k)xY$$

fix an index i in $\underline{n} \mathrel{\hat{=}} \{1,2,\dots,n\}$. By choosing (n-1) elements u_j in the U_j's, for j in $\underline{n}^i \mathrel{\hat{=}} \underline{n}-\{i\}$, a binary relation on $U_i xY$ is induced as follows

$$M^i(u_1,\dots,u_{i-1},u_{i+1},\dots,u_n) \mathrel{\hat{=}} \{(u_i,y)\in U_i xY \;/\; (u_1,\dots,u_i,\dots,u_n)My\}$$

The relation

$$M \subseteq (\mathop{X}_{k=1}^{n} U_k)xY$$

is an n-linear relation (multilinear relation of order n) if and only if for every index i in $\underline{n}$ and every (n-1)-list $(u_1,\dots,u_{i-1},u_{i+1},\dots,u_n)$, the induced relation $M^i(u_1,\dots,u_{i-1},u_{i+1},\dots,u_n)$ is a K-submodule of $U_i xY$.

3.3 Definition. A T-system

$$M = \langle \mathop{X}_{j=1}^{n} I_j \;,\; Z,\; \mathop{X}_{j=1}^{n} U_j,\; Y,\; M\rangle$$

is n-linear (over K) if and only if all I_j (for j in n) and Z are K-modules, for each j in n U_j is a K-submodule of I_j^T, Y is a K-submodule of Z^T, and M is an n-linear relation (over K) on

$$(\underset{j=1}{\overset{n}{X}} U_j)xY.$$

According to this definition multilinear systems might be viewed as a generalization of linear systems. The following proposition justifies this, by showing that linear systems are exactly the 1-linear systems[4].

3.4 Proposition. Let $M = \langle I,Z,U,Y,M\rangle$ be a T-system such that I, Z, U and Y are K-modules. So, M is linear if and only if M is 1-linear (over K).

Proof. For a 1-linear system all the induced relations of definition 3.2 reduce to only one, namely M itself. Thus, M is 1-linear if and only if M is linear, for in either case M is a K-module. QED.

Linearity and multilinearity are related to the causality concepts introduced in the previous section in the next result.

3.5 Corollary. Let M be an n-linear T-system (over K) and t an instant in T. So, M is t-non-anticipatory (resp. uniformly non-anticipatory, t-causal, uniformly causal, t-static, uniformly static) iff the relation M^t (resp. M^+, Mt, $\hat{M}$, $M(t)$, $\dot{M}$) is an n-linear function (over K).

Proof. Straightforward from definitions 2.5 and 3.3, for $I_1,\ldots,I_n,Z$ induce module structures on $I_1^T,\ldots,I_n^T,Z^T$. QED.

Clearly, a similar result for linear T-systems is obtained by setting n=1 in the previous proposition, due to proposition 3.4.

Composition of linear and multilinear systems is now shown to produce multilinear systems. We first examine the general case with all the systems multilinear and then particularize by letting some of them be linear.

3.5 Theorem. Let the T-system

$$M = \langle \underset{i=1}{\overset{n}{X}} J^i,\ Z,\ \underset{i=1}{\overset{n}{X}} W^i,\ Y,\ M\rangle$$

be n-linear and let also the n T-systems

$$M^i = \langle \underset{j=1}{\overset{p_i}{X}} I_j^i,\ J^i,\ \underset{j=1}{\overset{p_i}{X}} U_j^i,\ W^i,\ M^i\rangle$$

be p_i-linear for each i = 1,2,...,n. Let N be the T-system

$$N \hat{=} M.(\underset{i=1}{\overset{n}{X}} M^i) = \langle I,\ Z,\ U,\ Y,\ N\rangle,\text{where}$$

$$U \hat{=} \underset{i=1}{\overset{n}{X}} (\underset{j=1}{\overset{p_i}{X}} U_j^i) \quad , \quad N \hat{=} M.(\underset{i=1}{\overset{n}{X}} M^i).$$

Thus, the T-system N is p-linear, where $p \hat{=} \sum_{i=1}^{n} p_i$.

Proof. The idea of the proof can be outlined as follows. Consider the system N, fix all its inputs except one, the kth.input, say. This input will act linearly on the output of its component system. Since all other inputs to M are fixed its output will vary linearly with the non-fixed input to N. We now make this argument precise.

Let, for k in $\underline{n}$ and $j \leq p_k$,

$$(u_1^1, \ldots, u_j^k, \ldots, u_{p_n}^n) Ny \quad \text{and} \quad (u_1^1, \ldots, v_j^k, \ldots, u_{p_n}^n) Nz.$$

Thus, by definition 2.10 there exists

$$(w^1, \ldots, w^k, \ldots, w^n) \in \mathop{X}_{i=1}^{n} W^i \quad ,$$

such that, for every i in $\underline{n}$,

$$(u_1^i, \ldots, u_{p_i}^i) \; M^i \; w^i.$$

Moreover $(w^1, \ldots, w^k, \ldots, w^n) My$. There also exists x^k in W^k such that

$$(u_j^k, \ldots, v_j^k, \ldots, u_{p_k}^u) \; M^k \; x^k \quad \text{and}$$

$$(w^1, \ldots, x^k, \ldots, w^n) Mz.$$

Therefore, for every r and s in the ring K, definitions 3.2 and 3,3 yield both

$$(w^1, \ldots, r.w^k + s.x^k, \ldots, w^n) \; M(r.y + s.z)$$

$$(u_1^k, \ldots, r.u_j^k + s.v_j^k, \ldots, u_{p_k}^k) \; M^k(r.w^k + s.x^k).$$

Hence, by definition 2.10,

$$(u_1^1, \ldots, r.u_j^k + s.v_j^k, \ldots, u_{p_n}^n) \; N(r.y + s.z)$$

Thus N is n-linear over K. QED

The two corollaries to follow, due to proposition 3.4, give the result of compositions of linear and multilinear systems.

3.6 Corollary. If the T-systems M is n-linear and L is linear, then the T-system $L.M$ is n-linear.

3.7 Corollary. If the T-system M is n-linear and the n T-systems L_k are linear, for k=1,2,...,n, then the T-system

$$M.(\mathop{X}_{k=1}^{n} L_k) \text{ is n-linear.}$$

It is well known that any multilinear function can be written as the composition of a special multilinear function and a linear. We will need this fact later for our decompositions of multilinear systems. So, the universality of the tensor product is recorded, for completeness, in the next result.

3.8 Lemma. Given n K-modules $A_1, A_2, \ldots, A_n$, there exist a K-module D (the tensor-product module) and an n-linear function

$$p : \mathop{\times}_{i=1}^{n} A_i \to D$$

(the tensor-product map) that are universal in the sense that for every K-module B and every n-linear function

$$m : \mathop{\times}_{i=1}^{n} A_i \to B$$

there exists a (unique) K-linear function $L : D \to B$, such that $m = L.p$. Moreover, the tensor-product module of $A_1, \ldots, A_n$ is unique, up to an isomorphism, and will often be denoted by

$$\bigotimes_{i=1}^{n} A_i$$

Proof. See, for instance, [3]p. 5-27, [5]p. 408-410, or [6]p. 319-322.

Besides showing that every multilinear function can be factored into a linear function following a universal multilinear function, this theorem coupled with corollaries 3.6 and 3.7 suggests a natural way to obtain multilinear systems. Let us first define a special system, which those familiar with analog computation will recognize as a generalized multiplier when modules are restricted to finite-dimensional vector spaces.

3.9 Definition. Given n K-modules $V_1, V_2, \ldots, V_n$, let

$$\bigotimes_{i=1}^{n} V_i$$

denote their tensor-product module (over K), with tensor-product map

$$p : \mathop{\times}_{i=1}^{n} V_i \to \bigotimes_{i=1}^{n} V_i$$

The T-system

$$P = \left(\mathop{\times}_{i=1}^{n} V_i \,,\ \bigotimes_{i=1}^{n} V_i \,,\ \mathop{\times}_{i=1}^{n} V_i^T,\ \left(\bigotimes_{i=1}^{n} V_i\right)^T,\ p\right), \quad \text{where}$$

$$P : \mathop{\times}_{i=1}^{n} V_i^T \to \left(\bigotimes_{i=1}^{n} V_i\right)^T$$

is defined by the assignment $(u_1, \ldots, u_n) \mapsto \{(t, p[u_1(t), \ldots, u_n(t)]) \,/\, t \in T\}$, will be called the static tensor-product system with input alphabet

$$\mathop{\times}_{i=1}^{n} V_i \,.$$

Now, a bilinear system could be realized on an analog computer by feeding the scalar outputs, say, of two linear systems into a multiplier and, possibly, connecting its output to the input of another linear system. A generalization of this construction is essentialy the content of the next result.

3.10 Proposition. Let $L_k = \langle I_k, Z_k, U_k, Z_k^T, L_k \rangle$, for k=0,1,2,...,n, be (n+1) linear T-systems (over K) and let P denote the static tensor-product system with input alphabet

$$\underset{k=1}{\overset{n}{\times}} Z_k.$$

Then the system

$$L_0.P.(\underset{k=1}{\overset{n}{\times}} L_k) \qquad \text{is n-linear (over K).}$$

Proof. By corollary 3.7, the system

$$P.(\underset{k=1}{\overset{n}{\times}} L_k)$$

is n-linear. Hence, the result follows from corollary 3.6. QED

In the next section we show a converse, in a way, of this proposition. Namely, any multilinear system can be obtained by such a construction, provided that it is uniformly causal or non-anticipatory. And, of course, a similar statement holds for static multilinear systems.

4. DECOMPOSITIONS FOR MULTILINEAR SYSTEMS

This section contains the main results of this paper. Decompositions for static multilinear systems and a general decomposition theorem are presented first. Then causal and non-anticipatory multilinear systems are decomposed into linear component systems and a static non-linear system. Finally, finer decompositions are shown.

The first decomposition for static systems is due to the universality of the tensor product.

4.1 Proposition. Any t-static n-linear T-system M can be decomposed, at the instant t, into a t-static tensor-product system P followed by a t-static linear system L. Moreover, if M is uniformly static then both P and L are also uniformly static.

Proof. Since M is t-static and n-linear, by corollary 3.5,

$$M(t) : \underset{j=1}{\overset{n}{\times}} U_j(t) \to Y(t) \quad \text{is an n-linear function.}$$

So, it can be factored, by lemma 3.8, as M(t) = L(t).P(t), where,

$$P(t) : \underset{j=1}{\overset{n}{\times}} U_j(t) \to \underset{j=1}{\overset{n}{\otimes}} (U_j(t))$$

defines the t-static tensor-product system P with input alphabet

$$\underset{j=1}{\overset{n}{\times}} U_j(t)$$

as in definition 3.9. The linear function $L(t) : \underset{j=1}{\overset{n}{\otimes}} (U_j(t)) \to Y(t)$ characterizes the linear t-static system L, according to corollary 3.5 and proposition 3.4. If M is uniformly static then

$$\dot{M} = \underset{j=1}{\overset{n}{X}} I_j \rightarrow Z$$

can be written as $\dot{M} = \dot{L} . \dot{P}$, where

$$\dot{P} : \underset{j=1}{\overset{n}{X}} I_j \rightarrow \underset{j=1}{\overset{n}{\otimes}} I_j$$

gives a uniformly static tensor-product system P and

$$\dot{L} : \underset{j=1}{\overset{n}{\otimes}} I_j \rightarrow Z$$

defines a uniformly static linear system L. QED

The following basic result about multilinear functions contains the concept of multilinear kernels as a generalization of the familiar null-spaces for linear functions. Since a similar argument will be used to decompose multilinear systems, it is stated and proved here to avoid repetitions.

4.2 Lemma. Let $A_1, A_2, \ldots, A_n, B$ be (n+1) K-modules and the function

$$m : \underset{i=1}{\overset{n}{X}} A_j \rightarrow B$$

be n-linear (over K). Define for each index i in $\underline{n}$ the set:

$$N_i \mathrel{\hat{=}} \{x_i \in A_i \,/\, \forall j \in \underline{n}^i : m(x_1, \ldots, x_j, \ldots, x_n) = 0\}$$

Then, for each i in $\underline{n}$, N_i is a K-submodule of A_i, giving rise to the quotient K-module $\tilde{A}_i \mathrel{\hat{=}} A_i/N_i$ with canonical projection $L_i : A_i \rightarrow \tilde{A}_i$; and there exists an n-linear function

$$\tilde{m} : \underset{i=1}{\overset{n}{X}} A_i \rightarrow B \quad , \text{ such that } \quad m = \tilde{m}.(\underset{i=1}{\overset{n}{X}} L_i)$$

Proof. Clearly the zero of A_i is in N_i; which is a suset of A_i. To show that N_i is a K-submodule of A_i we prove that if u_i and v_i are in N_i then, for any r and s in K, the linear combination $(r.u_i+s.v_i)$ is in N_i. Indeed, for any $x_1, \ldots, x_{i-1}, x_{i+1}, \ldots, x_n$ in the respective A_i's, due to the multilinearity of m, we have $m(x_1, \ldots, r.u_i + s.v_i, \ldots, x_n) =$ $= r.m(x_1, \ldots, u_i, \ldots, x_n) + s.m(x_1, \ldots, v_i, \ldots, x_n)$. Now, since both u_i and v_i are in N_i, we also have $m(x_1, \ldots, u_i, \ldots, x_n) = 0 = m(x_1, \ldots, v_i, \ldots, x_n)$. Thus, $m(x_1, \ldots, r.u_i+s.v_i, \ldots, x_n) = 0$ and $(r.u_i+s.v_i)$ belongs to N_i, as claimed. Now, consider the canonical projections, for $i = 1,2,\ldots,n$ $L_i : A_i \rightarrow \tilde{A}_i$ such that $x_i \mapsto \tilde{x}_i \mathrel{\hat{=}} x_i + N_i$. The function

$$\tilde{m} : \underset{i=1}{\overset{n}{X}} \tilde{A}_i \rightarrow B$$

is defined by the assignment $(\tilde{x}_1, \ldots, \tilde{x}_i, \ldots, \tilde{x}_n) \mapsto m(x_1, \ldots, x_i, \ldots, x_n)$. To show that $\tilde{m}$ is well defined by the above assignment, we prove that its value independs of the particular representative chosen for a $\tilde{x}_i$. Indeed, suppose $L_i(x_i) = \tilde{x}_i = L_i(u_i)$, for all i in $\underline{n}$. Then there exist w_i in N_i such that $u_i = x_i+w_i$, for $i = 1,2,\ldots,n$. Due to the multilinearity of m, $m(u_1,u_2,\ldots,u_n) = m(x_1,u_2,\ldots,u_n) + m(w_1,u_2,\ldots,u_n)$. Since

w_1 is in N_1, $m(w_1,u_2,\ldots,u_n)=0$, so $m(u_1,u_2,\ldots,u_n)=m(x_1,u_2,\ldots,u_n)$. The same argument can be applied to $u_2,u_3,\ldots,u_n$. We obtain $m(u_1,u_2,\ldots,u_n)=m(x_1,x_2+w_2,\ldots,u_n)=m(x_1,x_2,\ldots,u_n)+m(x_1,w_2,\ldots,u_n)=$ $=m(x_1,x_2,\ldots,u_n)=\ldots\ldots=m(x_1,x_2,\ldots,x_n)+m(x_1,x_2,\ldots,w_n)$. Thus, $m(u_1,u_2,\ldots,u_n)=m(x_1,x_2,\ldots,x_n)$, showing that $\tilde{m}$ is well defined. It remains to show that $\tilde{m}$ is n-linear. Indeed, since each L_i is linear, $r.\tilde{u}_i+s.\tilde{v}_i=r.L_i(u_i)+s.L_i(v_i)=L_i(r.u_i+s.v_i)=(r.u_i+s.v_i)+N_i$. So, we can write, since $\tilde{m}$ is well defined $\tilde{m}(\tilde{x}_1,\ldots,r.\tilde{u}_i+s.\tilde{v}_i,\ldots,\tilde{x}_n)=$ $=m(x_1,\ldots,r.u_i+s.v_i,\ldots,x_n)=r.m(x_1,\ldots,u_i,\ldots,x_n)+$ $+s.m(x_1,\ldots,v_i,\ldots,x_n)=r.\tilde{m}(\tilde{x}_1,\ldots,\tilde{u}_i,\ldots,\tilde{x}_n)+s.\tilde{m}(\tilde{x}_1,\ldots,\tilde{v}_i,\ldots,\tilde{x}_n)$. Thus, we can finally conclude that $\tilde{m}$ is a well defined n-linear function such that

$$m=\tilde{m}\,.\,(\mathop{X}_{i=1}^{n} L_i),$$

by its very definition. QED.

This result, coupled with proposition 4.1, immediately gives a finer decomposition for multilinear static systems.

4.3 Corollary. Any t-static n-linear T-system M can be decomposed, at the instant t, into n t-static linear systems $L_1,\ldots,L_n$ in parallel, interconnected by a t-static tensor system P, followed by a linear t-static system L_0, so that

$$M(t)=L_0(t).P(t).\,[\mathop{X}_{i=1}^{n} L_i(t)]\,.$$

Moreover, if M is uniformly static then P and $L_0,L_1,\ldots,L_n$ are uniformly static, as well.

We now turn to the main results of this article. Both lemmas 3.8 and 4.2 are used to obtain internal decompositions for multilinear systems.

4.4 Theorem. Any uniformly non-anticipatory n-linear system M can be decomposed as

$$M=L\,.\,\tilde{R}\,.\,[\mathop{X}_{j=1}^{n}(L_j\,.\,\tilde{L}_j)]$$

into 2(n+1) systems, where the (2n+1) systems $\tilde{L}_1,\ldots,\tilde{L}_n,L,L_1,\ldots,L_n$ are linear and uniformly non-anticipatory and the system $\tilde{R}$ is n-linear and t-static for every instant t in T.

Proof. For every instant t in T,

$$M^t : \mathop{X}_{i=1}^{n} U_i^t \to Y^t$$

is an n-linear function, by proposition 2.6 and corollary 3.5. For each i=1,2,...,n , define

$$N_i^t \mathrel{\hat{=}} \{u_i^t\in U_i^t \,/\, \forall j\in \underline{n}^i : \forall u_j^t\in U_j^t : M^t(u_1^t,\ldots,u_i^t,\ldots,u_n^t)=0^t\}.$$

By lemma 4.2, N_i^t is a submodule of U_i^t with canonical projection

$$\tilde{L}_i^t : U_i^t \to \tilde{U}_i^t \triangleq U_i^t / N_i^t \quad ; \quad u_i^t \mapsto \tilde{u}_i^t$$

Thus, the induced function

$$\tilde{M} : \mathop{X}_{i=1}^{n} \tilde{U}_i^t \to Y^t$$

is n-linear and such that

$$M^t = \tilde{M}^t . (\mathop{X}_{i=1}^{n} \tilde{L}_i^t).$$

Since $\tilde{M}^t$ is n-linear, by lemma 3.8 , there exist a module W^t and functions

$$R^t : \mathop{X}_{i=1}^{n} \tilde{U}_i^t \to W^t ,$$

n-linear, and $L^t : W^t \to Y^t$, linear, such that $\tilde{M}^t = L^t.R^t$.
Fix an instant t_0 in T and for each i=1,2,...,n and $t < t_0$, let

$$\tilde{N}_i(t) \triangleq \{\tilde{u}_i^{t_0} \in \tilde{U}_i^{t_0} / \forall j \in \underline{n}^i : \forall \tilde{u}_j^{t_0} \in \tilde{U}_j^{t_0} : [R^{t_0}(\tilde{u}_1^{t_0}, \ldots, \tilde{u}_n^{t_0})](t) = 0\}.$$

Since $val(t).R^{t_0}$ is n-linear, again by lemma 4.2, $\tilde{N}_i(t)$ is a submodule of $\tilde{U}_i^{t_0}$ with canonical projection

$$L_i^{t_0}(t) : \tilde{U}_i^{t_0} \to V_i(t) \triangleq \tilde{U}_i^{t_0} / \tilde{N}_i(t)$$

The induced n-linear function

$$\tilde{R}(t) : \mathop{X}_{i=1}^{n} V_i(t) \to W(t)$$

is such that

$$val(t) . R^{t_0} = \tilde{R}(t) . [\mathop{X}_{i=1}^{n} L_i^{t_0}(t)]$$

and characterizes a t-static n-linear system $\tilde{\mathcal{R}}$.
Now, for each i in $\underline{n}$, the assignment

$$\tilde{u}_i^{t_0} \mapsto \{(t , L_i^{t_0}(t)(\tilde{u}_i^{t_0})) / t < t_0\}$$

defines a linear function $L_i^{t_0} : \tilde{U}_i^{t_0} \to V_i^{t_0}$, where

$$V_i \triangleq \bigcup_{t<t_0} [\{t\} x V_i(t)].$$

We can now write, for $t < t_0$,

$$val(t).R^{t_0} = \tilde{R}(t) . val(t) . (\mathop{X}_{i=1}^{n} L_i^{t_0})$$

Associated to each linear function $\tilde{L}_i^t$ above, there is t-non-anticipatory linear system $\tilde{\mathcal{L}}_i$. Also, a t-non-anticipatory linear system $\mathcal{L}_i$ corresponds to L_i^t . By proposition 2.7, $\tilde{\mathcal{R}}$ is t-non-anticipatory. And so is $\mathcal{R}$, by theorem 2.1.
Now, we have $\mathcal{M} = \mathcal{L}.\mathcal{Q}$, where the system

$$Q \hat{=} R\,[\mathop{\times}_{i=1}^{n} (L_i . \tilde{L}_i)]$$

is t-non-anticipatory. By theorem 2.11, thus, the restriction of L to the range of Q is also t-non-anticipatory. So, L can be made t-non-anticipatory without loss of generality. To conclude the proof, it suffices to note now that the linear systems $L_1,\ldots,L_n$, L, $\tilde{L}_1,\ldots,\tilde{L}_n$ are uniformly non-anticipatory, because of proposition 2.6. QED

From this theorem and corollary 4.3, we are immediately led to the following refinement.

4.5 Corollary. The uniformly non-anticipatory n-linear system M can also be decomposed as

$$M = L \,.\, L_0 \,.\, P \,.\, [\mathop{\times}_{i=1}^{n} (L_i \,.\, \tilde{L}_i)]$$

where $L_1,\ldots,L_n$, L, $\tilde{L}_1,\ldots,\tilde{L}_n$ are as in the theorem 4.4, P is a tensor-product system and L_0 is a linear system, both t-static for every instant t in T.

Similar results might be expected for uniformly causal multilinear systems. The following theorem and corollary show how.

4.6 Theorem. Let M be a uniformly causal n-linear T-system. Then M can be decomposed as

$$M = L \,.\, \tilde{R} \,.[\mathop{\times}_{i=1}^{n} (L_i \,.\, \tilde{L}_i)]$$

into 2(n+1) systems. The component systems are classified as follows: the n-linear system $\tilde{R}$ is t-static for every t in T, the linear system L is uniformly causal, the 2n linear systems $L_1,\ldots,L_n$, $\tilde{L}_1,\ldots,\tilde{L}_n$ are uniformly non-anticipatory.

Proof. Due to propositions 2.6 and 2.7, M is also uniformly non-anticipatory . So, the same argument of theorem 4.4 holds, yielding the desired decomposition. It only remains to see that the linear system L is indeed uniformly causal, besides being uniformly non-anticipatory. But, we have $M = L.Q$, with the system

$$Q \hat{=} R\,[\mathop{\times}_{i=1}^{n} (L_i \,.\, \tilde{L}_i)]$$

t-non-anticipatory, as in the end of the proof of theorem 4.4. Since M is t-causal, theorem 2.11 assures that L restricted to the range of Q is t-causal, as well. Thus, L can be made t-causal, and, by proposition 2.6, uniformly causal. QED.

4.7 Corollary. The uniformly causal n-linear system can also be decomposed as

$$M = L \,.\, L_0 \,.\, P \,.\, [\mathop{\times}_{i=1}^{n} (L_i \,.\, \tilde{L}_i)]$$

where $L_1,\ldots,L_n$, L, $\tilde{L}_i,\ldots,\tilde{L}_n$ are as in the theorem 4.6, P is a tensor-product system and L_0 is a linear system, both t-static for every instant t in T.

Now it is clear that the natural way to obtain multilinear systems suggested in proposition 3.10 is also a general way to realize such systems. This is recorded in the next corollary.

4.8 Corollary. If the n-linear T-system $\mathcal{M}$ is (i) uniformly causal, or (ii) uniformly non-anticipatory, then it can be decomposed into the parallel interconnection of n linear systems $(\mathcal{L}_1,\ldots,\mathcal{L}_n)$ by a tensor-product static system $\mathcal{P}$, followed by a linear system $\mathcal{L}_0$. Then n linear systems in parallel, $\mathcal{L}_1,\ldots,\mathcal{L}_n$, are uniformly non-anticipatory and the linear system $\mathcal{L}_0$ is (i) uniformly causal, or (ii) uniformly non-anticipatory.

Proof. Immediate, from theorems 2.11 and either 4.6 or 4.4. QED

In order to present a finer decomposition that lays bare the internal structure of a realization for causal and non-anticipatory multilinear systems, we will need two basic facts about linear systems. They are simple decompositions for causal and non-anticipatory linear systems. These results can be proved [10], from the definition of linear system as a linear relation between input and output modules together with the concepts of uniform causality and non-anticipation, but they are quite intuitive. As a matter of fact, they are often used nearly as a starting point for the theory of linear systems [2,4,18].

4.9 Fact. Any uniformly causal linear T-system $\mathcal{L} = \langle I,Z,U,Y,L\rangle$ can be decomposed into the series composition of two linear T-systems: $\mathcal{E}$ followed by $\mathcal{H}$, where $\mathcal{H}$ is t-static for every t in T and $\mathcal{E}$ is uniformly causal and t_1-t_2-transitional for all $t_1 \le t_2$ in T. Moreover, $\mathcal{E}$ can be characterized by two linear functions

$$F(t_1,t_2) : X(t_1) \rightarrow X(t_2) \quad , \quad G(t_1,t_2) : U_{t_1}^{t_2} \rightarrow X(t_2) \quad ,$$

so that

$$_{t_1}E^{t_2}\,[x(t_1),u_{t_1}^{t_2}] = F(t_1,t_2)[x(t_1)] + G(t_1,t_2)[u_{t_1}^{t_2}]$$

for all $t_1 \le t_2$ in T.

An example of this fact is provided by linear discrete-time (possibly time-varying) systems, characterized by equations such as

$$x(t+1) = F(t)x(t) + G(t)u(t) \quad , \quad y(t) = H(t)x(t)$$

where F(t), G(t) and H(t) are matrices of compatible dimensions (and over the same field), for every t=0,1,2,..., say.

4.10 Fact. Any uniformly non-anticipatory linear T-system $\mathcal{L} = \langle I,Z,U,Y,L\rangle$ can be decomposed into the parallel interconnection of two linear systems interconnected by a uniformly static adder. One of the systems in parallel $\mathcal{D}$ is t-static for every t in T. The other, being uniformly causal, can be written as $\mathcal{C}.\mathcal{E}$, as in the fact 4.9. So, a uniformly non-anticipatory linear T-system $\mathcal{L}$ can be decomposed as $\mathcal{L} = \pm\,.\,[\mathcal{D}\times(\mathcal{C}.\mathcal{E})]$, where $\pm$

denotes the adder. System E can be characterized by two linear functions usually denoted by A and B.

Continuous-time linear differential systems described by equations like the following two

$$\dot{x}(t) = A(t)x(t) + B(t)u(t)$$

$$y(t) = C(t)x(t) + D(t)u(t) \quad ,$$

where A and B characterize the causal and transitional system E, are examples for this fact.

We can now turn to the finer decompositions.

4.11 Proposition. Any uniformly causal n-linear T-system can be decomposed into 4(2n+1) systems, (8n+3) of which are linear and one (the tensor-product system $\otimes$) is n-linear and t-static for every t in T, so that

$$M = H.E.L_0.\otimes.\ \underset{j=1}{\overset{n}{\times}}\{\pm.[\mathcal{D}_j \times (C_j.E_j)].\pm.[\tilde{\mathcal{D}}_j \times (\tilde{C}_j.\tilde{E}_j)]\}$$

The linear systems are classified as follows: (2n+1) systems ($E_1,\ldots,E_n$, $E,\tilde{E}_1,\ldots,\tilde{E}_n$) are uniformly causal and t_1-t_2-transitional for all $t_1 \le t_2$ in T, and the remaining 2(3n+1) are all static, [being 2n uniformly static adders, and 2(2n+1) systems ($L_0,C_1,\ldots,C_n,\mathcal{D}_1,\ldots,\mathcal{D}_n,\tilde{C}_1,\ldots,\tilde{C}_n$, $\tilde{\mathcal{D}}_1,\ldots,\tilde{\mathcal{D}}_n,H$) t-static for every t in T].

Proof. From corollary 4.7, by decomposing the linear causal system L as in fact 4.9 and the 2n linear non-anticipatory systems as in fact 4.10. QED

The corresponding decomposition for non-anticipatory multilinear systems is presented in the next result.

4.12 Proposition. Any uniformly non-anticipatory n-linear T-system can be decomposed into 2(4n+3) systems, of which (8n+5) are linear and one (the tensor-product system $\otimes$) is n-linear and t-static for every t in T, so that

$$M = \pm.[\mathcal{D}\times(C.E)].L_0.\otimes.\ \underset{j=1}{\overset{n}{\times}}\{\pm.[\mathcal{D}_j \times (C_j.E_j)].\pm.[\tilde{\mathcal{D}}_j \times (\tilde{C}_j.\tilde{E}_j)]\}$$

The linear systems are classified as follows: (2n+1) systems ($E_1,\ldots,E_n$, $E,\tilde{E}_1,\ldots,\tilde{E}_n$) are uniformly causal and t_1-t_2-transitional for all $t_1 \le t_2$ in T, and the remaining 2(3n+2) systems are all static [being (2n+1) uniformly static adders, and (4n+3)systems ($L_0,C_1,\ldots,C_n,\mathcal{D}_1,\ldots,\mathcal{D}_n,C,\mathcal{D}$, $\tilde{C}_1,\ldots,\tilde{C}_n,\tilde{\mathcal{D}}_1,\ldots,\tilde{\mathcal{D}}_n$) t-static for every t in T].

Proof. From corollary 4.5, by decomposing the (2n+1) linear non-anticipatory systems $L_1,\ldots,L_n,L,\tilde{L}_1,\ldots,\tilde{L}_n$ as in fact 4.10. QED

5. CONCLUSIONS

First of all, let us briefly outline the main conceptual thread followed. A general definition of time-system is introduced and conveniently particularized to several classes of dynamical systems by means of a classification according to their causality properties. Within this framework linear and multilinear systems are defined over an arbitrary commutative ring with unit. Then, composition and especially decomposition of multilinear time-systems are investigated from an algebraic as well as system-theoretic viewpoint.

The main fact concerning composition is that the composition of linear and multilinear systems yields multilinear systems. The decompositions obtained for multilinear systems have some important consequences. Namely, they provide insight into the structure of multilinear systems by explicity showing its general patterns. Furthermore, they indicate how to realize multilinear systems by composing linear systems and one memoryless tensor-product system.

These decompositions have three major features due mainly to the general algebraic methods employed. Firstly, they include both discrete-time and continuous-time systems. Secondly, these results are valid not only for time-invariant systems but also - and more remarkably - for time-varying systems. And thirdly, except for the finer decompositions, all the others are independent of any state-assignment either to the multilinear system or to the component linear systems. Decompositions based on state assignment appear in [11,13].

Thus, many - though not all - problems concerning multilinear systems can be reduced to studying the effect of a memoryless tensor-product interconnection on the solution of the corresponding problems for the component linear systems. For instance, any discrete-time linear system can be constructed by using only three primitive blocks, which are adders, scalers and unit-delays [2]. A consequence of these decompositions for the case of discrete-time multilinear systems is that only one new building block - the tensor product - must be included to permit the construction of multilinear systems, as well.

There is also a methodological - so to speak - conclusion to be drawn from this work. It does illustrate the adequacy of such a General Time-Systems formalism to handle problems of a general or abstract nature concerning dynamical systems. Once the problems are so formulated, simple but powerful algebraic methods can be readily and effectively applied to solve them.

REFERENCES

1. M.A.Arbib, 'A Characterization of Multilinear Systems', IEEE Trans. on Automatic Control, AC-11 (1969), p.699-704.

2. T.L Booth, Sequential Machines and Automata Theory, Wiley, New York, 1967.

3. W.H.Greub, Multilinear Algebra, Springer-Verlag, Berlin, 1967.

4. L.H.Kerschberg, An Algebraic Approach to Linear and Multilinear Systems Theory, Case Western Reserve Univ., Cleveland, Ohio, 1969; (Ph.D. Dissertation).

5. S.Lang, Algebra. Addison-Wesley, Reading, Mass., 1967.

6. S.MacLane and G.Birkhoff, Algebra, MacMillan, New York, 1965.

7. M.D.Mesarovic, 'Foundations for a General Systems Theory", in M.D. Mesarovic (ed.) Views on General Systems Theory, Wiley, New York, 1964.

8. M.D.Mesarovic, 'New Directions in General Theory of Systems', in J.F. Hart and S. Takasu (eds.) Systems and Computer Science, Toronto Univ. Press, Toronto, 1967.

9. R.E.Rink and R.R.Mohler, 'Completely Controllable Bilinear Systems', SIAM J. on Control, 6 (1968), p.477-486.

10. P.A.S.Veloso, Algebraic Decompositions for Additive and Multilinear Causal Systems, COPPE, Federal Univ. of Rio de Janeiro, Brasil, 1970; (M.Sc. thesis; in Portuguese).

11. P.A.S.Veloso, 'A Decomposition for Causal Multilinear Systems' (in Portuguese), Proc. 1st. Congress of the Brazilian Society for Automatics, São Paulo, Brasil, 1976, p.18.1-18.8.

12. P.A.S.Veloso, 'On the Structure of Multilinear Systems', in D.G. Lainiotis and N.S. Tzannes (eds.) Advances in Control, Reidel, Dordrecht, 1980, p.87-93.

13. P.A.S.Veloso and L.Kerschberg, 'A Decomposition for a Class of Multilinear Systems', Proc. IEEE International Conference on Systems, Networks and Computers, Mexico, 1971, p.154-158.

14. T.G.Windeknecht, 'Mathematical Systems Theory: Causality', Math. Systems Theory, 1 (1967), p.279-289.

15. T.G.Windeknecht, 'Concerning an Algebraic Theory of Systems', in J.F.Hart and S.Takasu (eds.) Systems and Computer Science, Toronto Univ. Press, Toronto, 1967.

16. T.G.Windeknecht, General Dynamical Processes: a Mathematical Introduction, Academic Press, New York, 1971.

17. L.A.Zadeh, 'The Concepts of System, Aggregate and State in System Theory', in L.A. Zadeh and E.Polak (eds.), System Theory, McGraw-Hill, New York, 1969, p.3-42.

18. L.A.Zadeh and C.A.Desoer, Linear System Theory: the State Space Approach, McGraw-Hill, New York, 1963.

CHAPTER 7

SIMPLIFICATION OF MODELS FOR STABILITY ANALYSIS OF LARGE-SCALE SYSTEMS

I. Zambettakis[1], J.P. Richard[1,2], F. Rotella[1]
(1) Laboratoire de Systématique - U.S.T.L. - U.E.R. d'I.E.E.A.
59655 Villeneuve d'Ascq Cédex - France
(2) Laboratoire d'Automatique et d'Informatique Industrielle
Institut Industriel du Nord - B.P. 48
59651 Villeneuve d'Ascq Cédex - France

ABSTRACT

The method of stability analysis presented in this study allows a definition to be made of a reduced order model of a high-order, Lurie-Postnikov system. This reduced order system has two main properties for our purpose : linear conjecture can be applied to it, and the stability of its equilibrium involves the same property for the equilibrium of the original system. Moreover these stability conditions may be easily checked and they directly determine an admissible sector for variations of the non-linear static gain. A common algebraic condition appears in the different stated theorems. A section deals with solving this condition, and a method based on the use of a particular algebraic array is proposed. In order to sum up and apply these results, a methodology for systems analysis is implemented, which both computes the expression of the reduced order system and the associated stability conditions. An example of a fourth-order system is then presented to illustrate the different steps of the study.

INTRODUCTION

The stability analysis of high-order Lurie-Postnikov systems involves the use of comparison techniques (Lyapunov, 1949) (Grujić, Gentina, Borne, 1976) or frequency domain criteria (Popov, 1973) (Narendra,Taylor, 1973). In some cases of poles-zeros configuration, linear stability criteria are available (Benrejeb, 1980). This latter work introduced a new state-space modelling : the "generalized arrow form" matrix.

Recent work (Rotella, 1983) proposed such a type of modelling for the particular case of single input - single output Lurie-Postnikov systems. All the coefficients of the evolution matrix are to be found only in the last line, the last column and in canonical blocks in the main diagonal, and such representations thus simplify the application of two different stability criteria (Gentina, Borne, 1972) (Zambettakis, 1983) for these systems (Rotella, Richard, 1983).

We now propose the simultaneous use of these criteria on the studied systems. This point of view leads to a definition of reduced order

S. G. Tzafestas (ed.), Multivariable Control, 125–148.

models the stability of which ensures the same property for the original system.

1. SYSTEM DESCRIPTION AND STATE-SPACE REPRESENTATION

In this study we consider continuous, single input - single output Lurie-Postnikov systems. They are composed of a linear part of transfer function $N(s) / D(s)$, controlled by a non-linear gain $f^*(\varepsilon)$, which depends on input-output difference (see Figure 1). Throughout the study, the linear part is assumed non-degenerate (Popov, 1973) ; this means $N(s)$ and $D(s)$ have no common zero. To simplify the presentation, let us denote $f^*(\varepsilon)$ by f^*, the set $\{i, i+1, \ldots, j-1, j\}$ where i and j are integers by I_i^j, the set of real numbers by $\mathcal{R}$ and the set of complex numbers by $\mathcal{C}$.

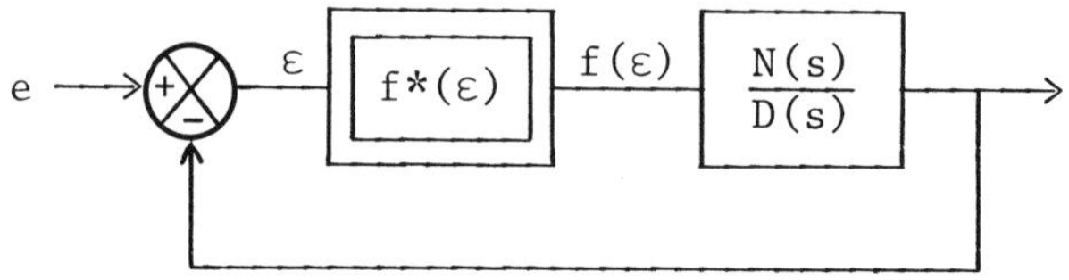

Figure 1.

with :

$$\begin{cases} N(s) = \sum_{i=0}^{q-1} b_i \, s^i \\ D(s) = s^q + \sum_{i=0}^{q-1} a_i \, s^i \end{cases}$$

$$\forall i \in I_0^{q-1}, \ (a_i, b_i) \in \mathcal{R}^2 \ ; \ b_{q-1} \geq 0$$

$$\forall \varepsilon \in \mathcal{R} - \{0\} \quad f(\varepsilon) = f^*(\varepsilon).\varepsilon \ ; \ f^*(0) = \left[\frac{df(\varepsilon)}{d\varepsilon}\right]_{\varepsilon=0}$$

$$\nexists \lambda \in \mathcal{C}, \ N(\lambda) = D(\lambda) = 0 \qquad (1)$$

A "symbolic polynomial" $p(\lambda,\varepsilon)$ can be associated to this structure (Richard, 1981 a), defined by :

$$\forall \lambda \in \mathcal{C}, \forall \varepsilon \in \mathcal{R}, \ p(\lambda,\varepsilon) = D(\lambda) + f^* N(\lambda) \qquad (2)$$

Under the above condition (1), this polynomial is an invariant of representation for the system ; i.e., considering an arbitrary state-space modelling as below :

$$\mathring{x}(t) = A(\varepsilon) \, x(t) + B(\varepsilon) \, e(t)$$

with :

$$\begin{cases} \forall t \in T = [0\ ,\ +\infty[,\ x(t) \in R^q,\ \varepsilon \in R,\ e(t) \in R \\ \forall \varepsilon \in R,\ B(\varepsilon) \in R^q,\ A(\varepsilon) \in R^{q \times q} \end{cases}$$

the matrix $A(\varepsilon)$ is suitable iff :

$$\forall \lambda \in C,\ \forall \varepsilon \in R,\ \det\ (\lambda I_q - A(\varepsilon)) = p\ (\lambda,\varepsilon)$$

Note that, if the last component of $x(t)$ is the error signal ε, the vector $B(\varepsilon)$ is the opposite of the terms appearing with f^* in the last column of $A(\varepsilon)$ (Rotella, 1983).

Several previous works have introduced an interesting class of evolution matrices, known as "arrow form" matrices (Benrejeb, 1980). As a specific case, we consider here the following representation, defined by (q-1) arbitrary (but different) parameters λ_i, $i \in I_1^{q-1}$:

$$A(\varepsilon) = \begin{bmatrix} \lambda_1 & & 0 & & k_1^* \\ & \ddots & & & \vdots \\ & 0 & & \lambda_{q-1} & k_{q-1}^* \\ 1 & \cdots & \cdots & 1 \quad -a_{q-1} - b_{q-1} & f^* - \sum_{i=1}^{q-1} \lambda_i \end{bmatrix} \qquad (3)$$

$$B^T(\varepsilon) = [h_1^*,\ \ldots,\ h_{q-1}^*,\ b_{q-1} f^*]$$

with : $\forall i \in I_1^{q-1}$

$*\ \lambda_i \in R$

$*\ \forall j \in I_1^{q-1},\ \lambda_i \neq \lambda_j \quad \text{iff} \quad i \neq j$

$*\ \forall \varepsilon \in R$

$$k_i^* = - \frac{p(\lambda_i,\varepsilon)}{\prod_{\substack{j=1 \\ j\neq i}}^{q-1} (\lambda_i - \lambda_j)}\ ;\ h_i^* = \frac{N(\lambda_i)}{\prod_{\substack{j=1 \\ j\neq i}}^{q-1} (\lambda_i - \lambda_j)}\ f^*$$

Previous works have pointed out the importance of the choice of the parameters λ_i in the stability analysis. Indeed, it is interesting to obtain the proportionnality of the first (q-1) terms of the last column of $A(\varepsilon)$. Different cases have thus been considered and lead to the following models.

a) $\underline{\lambda_i}$ roots of N(s) (Benrejeb, 1976) (Benrejeb, Borne, Laurent, 1982)

When N(s) presents (q-1) real separate roots z_i, $i \in I_1^{q-1}$: $\forall i \in I_1^{q-1}$, $N(z_i) = 0$, $\forall j \in I_1^{q-1}$, $z_i \neq z_j$ iff $i \neq j$, it is possible to choose in (3) $\lambda_i = z_i$, $i \in I_1^{q-1}$.

Thus giving :

$$A(\varepsilon) = \begin{bmatrix} z_1 & & & & & \vdots \\ & \ddots & & 0 & & \vdots \\ & & z_i & & & -\left[\frac{D(s)}{N(s)}(s-z_i)\right]_{s=z_i} \\ & 0 & & \ddots & & \vdots \\ & & & & z_{q-1} & \vdots \\ 1 & \text{---} & & \text{---} & 1 & -a_{q-1}+\frac{b_{q-2}}{b_{q-1}}-b_{q-1}\,f^* \end{bmatrix} \quad (4)$$

The first (q-1) terms of the last column are constant.
Remark : in this case, it is necessary to have $b_{q-1} \neq 0$.

b) $\underline{\lambda_i}$ roots of D(s) (Richard, 1981 b)

When D(s) has (q-1) real separate roots p_i, $i \in I_1^{q-1}$, and another one p_o : $\forall i \in I_1^{q-1}$, $D(p_i) = 0$, $\forall j \in I_1^{q-1}$, $p_i \neq p_j$ iff $i \neq j$, $D(p_o) = 0$, it is possible to choose in (3) $\lambda_i = p_i$, $i \in I_1^{q-1}$.

Thus giving :

$$A(\varepsilon) = \begin{bmatrix} p_1 & & & & & \vdots \\ & \ddots & & 0 & & \vdots \\ & & p_i & & & -\left[\frac{N(s)(s-p_o)(s-p_i)}{D(s)}\right]_{s=p_i}.f^* \\ & 0 & & \ddots & & \vdots \\ & & & & p_{q-1} & \vdots \\ 1 & \text{---} & & \text{---} & 1 & p_o - b_{q-1}\,f^* \end{bmatrix} \quad (5)$$

The first (q-1) terms of the last column are proportionnal to f*.
Remark : p_o is not necessarily distinct from a p_i, $i \in I_1^{q-1}$; if p_o is equal to zero, all the terms of the last column are proportionnal to f*.

c) λ_i roots of $D(s) + a\,N(s)$, $a \in \mathcal{R}$ (Rotella, Zambettakis, Richard, 1982)

If there exists a real a, such that the polynomial P(s) defined by (6) has (q-1) real separate roots μ_i, $i \in I_1^{q-1}$, and another one μ_o, not necessarily distinct from the μ_i, $i \in I_1^{q-1}$:

$$P(s) = D(s) + a\,N(s) \tag{6}$$

$\forall i \in I_1^{q-1}$, $P(\mu_i) = 0$, $\forall j \in I_1^{q-1}$, $\mu_i \neq \mu_j$ iff $i \neq j$, $P(\mu_o) = 0$, it is possible to choose in (3) $\lambda_i = \mu_i$, $i \in I_1^{q-1}$. In this case this gives :

$$A(\varepsilon) = \begin{bmatrix} \mu_1 & & & & \vdots \\ & \ddots & 0 & & \vdots \\ & & \mu_i & & -\left[\dfrac{N(s)(s-\mu_o)(s-\mu_i)}{P(s)}\right]_{s=\mu_i} . g^* \\ & 0 & & \ddots & \vdots \\ & & & \mu_{q-1} & \vdots \\ 1 & \cdots & \cdots & 1 & \mu_o - b_{q-1}\,g^* \end{bmatrix} \tag{7}$$

where $g^* = f^* - a$.
Remark : This case can be considered as a generalization of the previous one and corresponds to an assignement of the poles to the transfer function. In the same way, the case a) can be generalized by considering a polynomial $Q(s) = b\,D(s) + N(s)$ and choosing its roots as λ_i.

The modellings presented in a), b), c) are the particular ones we will use. The studied systems must therefore verify the following hypothesis (H) :

(H) $\exists$ $(c,d) \in \mathcal{R}^2$, $c\,D(s) + d\,N(s)$ has (q-1) real separate zeros

2. STABILITY ANALYSIS

Considering the equilibrium 0 of the free motion, the evolution matrix A(ε) is taken as read, concerning the stability analysis.

In this part we will use the practical criterion of Borne and Gentina (Gentina, Borne, 1972) and the direct method of Lyapunov (Lyapunov, 1949), proposing a quadratic plus integral type, candidate Lyapunov function (Lasalle, Lefshetz, 1961), also called Lurie function. In some cases, when the first (q - 1) coefficients of the last column of A(ε) are all positive (linear conjecture) (Benrejeb, 1976) or all negative (Zambettakis, Richard, Laurent, 1983) these two methods are particularly suitable. However, the more general case needs to introduce a comparison system (Gentina, 1976) or majoration techniques (Rotella, Richard, Zambettakis, 1983).

We propose to conjugate the two methods and to study models a), b) or c) when signs of the coefficients of the last column are different.

In order to unify the notations, the system considered in this part will be described by the following state-space representation in free motion :

$$\overset{\circ}{x} = C(\varepsilon)\, x \tag{8}$$

with :

$$C(\varepsilon) = \begin{bmatrix} \lambda_1 & & & & & & -k_1^+ \alpha^* \\ & \ddots & & 0 & & & \vdots \\ & & \lambda_\ell & & & & -k_\ell^+ \alpha^* \\ & & & \lambda_{\ell+1} & & & -k_{\ell+1}^- \alpha^* \\ & 0 & & & \ddots & & \vdots \\ & & & & & \lambda_{q-1} & -k_{q-1}^- \alpha^* \\ 1 & \text{———} & \text{———} & \text{———} & \text{———} & 1 & -\beta^* \end{bmatrix}$$

where : $x^T = (x_+^T, x_-^T, \varepsilon)$

$x_+^T = (x_1, \ldots, x_\ell)$; $x_-^T = (x_{\ell+1}, \ldots, x_{q-1})$

$\forall\ i \in I_1^\ell,\ k_1^+ > 0$

$\forall\ i \in I_{\ell+1}^{q-1},\ k_i^- < 0$

$\forall\ (i,j) \in (I_1^{q-1})^2,\ i \neq j \quad \text{iff} \quad \lambda_i \neq \lambda_j$

$\forall\ \varepsilon \in R,\ \alpha^* . \varepsilon = \alpha(\varepsilon),\ \beta^* . \varepsilon = \beta(\varepsilon),\ (\alpha(\varepsilon), \beta(\varepsilon)) \in R^2$

Moreover, it should be noted that α^* and β^* can be dependant on ε or not, as the case may be.

Throughout this section, we consider that the sign of α^* is constant in a neighbourhood of $\varepsilon = 0$, and assume without any restriction that it is positive :

$$\exists\, h > 0,\ \forall\ \varepsilon \in R,\ |\varepsilon| < h \Rightarrow \alpha^* > 0$$

An interpretation of previous works (Hahn, 1963) (Zambettakis, 1983) on those types of systems leads us to propose the following candidate Lyapunov function :

$$V(x) = \frac{1}{2}\left[x_+^T D_+ x_+ + x_-^T D_- x_- \right] + \int_0^{\varepsilon} \alpha(z)\, dz \tag{9}$$

where

$$\begin{cases} D_+ = \text{diag}\left\{\frac{1}{k_1^+}, \frac{1}{k_2^+}, \dots, \frac{1}{k_\ell^+}\right\}, \quad D_+ \in \mathcal{R}^{\ell\times\ell} \\ \\ D_- = \text{diag}\left\{-\frac{1}{k_{\ell+1}^-}, -\frac{1}{k_{\ell+2}^-}, \dots, -\frac{1}{k_{q-1}^-}\right\}, \quad D_- \in \mathcal{R}^{(q-\ell-1)\times(q-\ell-1)} \end{cases} \tag{10}$$

This function is positive, defined in a neighbourhood of the origin, and we assume :

$$\begin{cases} \Lambda_+ = \text{diag}\,\{\lambda_1, \lambda_2, \dots, \lambda_\ell\}, \quad \Lambda_+ \in \mathcal{R}^{\ell\times\ell} \\ \\ \Lambda_- = \text{diag}\,\{\lambda_{\ell+1}, \lambda_{\ell+2}, \dots, \lambda_{q-1}\}, \quad \Lambda_- \in \mathcal{R}^{(q-\ell-1)\times(q-\ell-1)} \end{cases} \tag{11}$$

The eulerian derivative of V(x) is :

$$\overset{\circ}{V}(x) = x_+^T D_+ \Lambda_+ x_+ + \left[x_-^T, \alpha(\varepsilon)\right] \left[\begin{array}{c|c} D_- \Lambda_- & \begin{matrix}1\\ | \\ 1\end{matrix} \\ \hline 1 \text{---} 1 & -\frac{\beta^*}{\alpha^*} \end{array}\right] \left[\begin{array}{c} x_- \\ \hline \alpha(\varepsilon) \end{array}\right] \tag{12}$$

Based on this last expression, we propose the following theorem :

Theorem 2.1 :

Let two disconnected systems be defined by :

$$(S_+) : \overset{\circ}{z}_+ = \Lambda_+ z_+$$

$$(S_-) : \overset{\circ}{z}_- = S_-(\varepsilon)\, z_-$$

with :

$$S_-(\varepsilon) = \left[\begin{array}{c|c} \Lambda_- & \begin{matrix} -k_{\ell+1}^- \cdot \alpha^* \\ \vdots \\ -k_{q-1}^- \cdot \alpha^* \end{matrix} \\ \hline 1 \text{------} 1 & -\beta^* \end{array}\right]$$

where Λ_- and Λ_+ are defined in (11) and k_i^-, $i \in I_{\ell+1}^{q-1}$, α^* and β^* are defined by (8).

If (S_+) is asymptotically stable and (S_-) satisfies the dollowing conditions :

$$\begin{cases} \Lambda_- \text{ is negative defined} \\ \exists\, h > 0,\ \forall\, \varepsilon \in R,\ |\varepsilon| < h,\ (-1)^{q-\ell} \det S_-(\varepsilon) > 0 \end{cases}$$

(equilibrium 0 of (S_-) is asymptotically stable), then the equilibrium 0 of the system (8) is asymptotically stable. □

Proof :

a) The stability condition of (S_-) comes from the application of linear-systems stability conditions of Koteliansky (Gentina, Borne, 1972) : the linear conjecture holds for (S_-) because off-diagonal terms are positive and non-constant elements are located in the last row.
b) Considering the quadratic form occuring in $\mathring{V}(x)$ (12), we have :

$$\det \left[\begin{array}{c|c} D_- \Lambda_- & \begin{matrix} 1 \\ | \\ 1 \end{matrix} \\ \hline 1 \text{———} 1 & -\dfrac{\beta^*}{\alpha^*} \end{array} \right] = \frac{1}{\alpha^*} \det (D_-)^{-1} . \det S_-(\varepsilon)$$

$\det (D_-)$ is positive, therefore the Koteliansky lemma (Gantmacher, 1966) holds for this quadratic form and ensures its negativity.

This theorem ensures that the stability study of the whole system (8) can be reduced to the analysis of an obvious system (S_+) and a reduced order one (S_-), for which the linear conjecture is valid.

3. APPLICATION TO THE DEFINITION OF THE REDUCED ORDER SYSTEM

The previous part has shown that, if ℓ is the number of negative coefficients $(-k_i^+)$ arising in representation (8) for a choice of the (λ_i), the stability study can be made on a $(q - \ell)$ order system, denoted (S_-). This order therefore depends on the (λ_i) choice.

In the case of the system Figure1, supposing the hypothesis (H) satisfied, the modelling (8) holds. The results of section 2 can lead to sufficient stability conditions in every case of proposed representations a), b) or c). The representation (8) then satisfies :

$$\exists\, (a, b, c, d) \in R^4 \qquad \beta^* = a + b f^* \qquad \alpha^* = c + d f^*$$

From (S_-) and its symbolic polynomial (Richard, 1981 a); we can construct a reduced order process with the structure defined Figure 2, where

$N_r(s)$ and $D_r(s)$ are defined by (13).

$$\forall\, \lambda \in \mathcal{C},\ \forall\, \varepsilon \in \mathcal{R},\ \det\,(\lambda I_{q-\ell} - S_-(\varepsilon)) = D_r(\lambda) + f^* N_r(\lambda) \tag{13}$$

where $I_{q-\ell}$ is the identity matrix of order $q-\ell$.

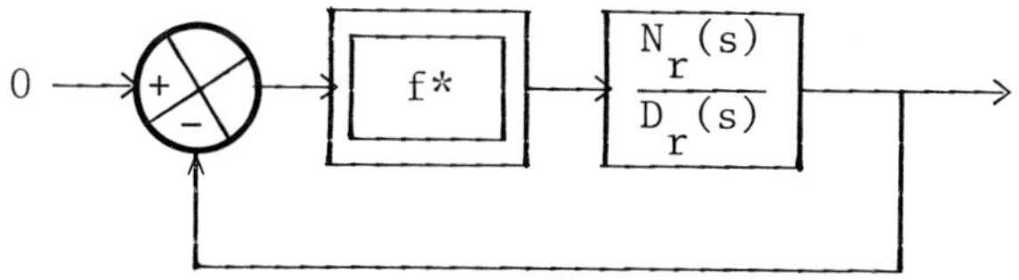

Figure 2.

The last condition appearing in Theorem 2.1 is then equivalent to :

$$\forall\, \varepsilon \in \mathcal{R},\ |\varepsilon| < h \quad D_r(0) + f^* N_r(0) > 0 \quad \text{and} \quad \alpha^* > 0 \tag{14}$$

We intend to develop this condition for different choices of (λ_i).

3.1. $N(s)$ has $(q-1)$ separate real roots, z_i

Under this condition the modelling (4) holds. Let us now define the two complementary sets $\mathcal{Z}^+$ and $\mathcal{Z}^-$ with :

$$\forall\, i \in \mathcal{I}_1^{q-1},\ k_i = \left[\frac{D(s)\ (s-z_i)}{N(s)}\right]_{s=z_i}$$

$$\begin{cases} \mathcal{Z}^+ = \{z_i,\ i \in \mathcal{I} \subset \mathcal{I}_1^{q-1},\ N(z_i) = 0,\ k_i > 0\} \\ \\ \mathcal{Z}^- = \{z_i,\ i \in \mathcal{J} \subset \mathcal{I}_1^{q-1},\ N(z_i) = 0,\ k_i < 0\} \end{cases} \tag{15}$$

and :

$$\mathcal{I} \cap \mathcal{J} = \emptyset,\ \mathcal{I} \cup \mathcal{J} = \mathcal{I}_1^{q-1}$$

From (13) the following definitions of $N_r(s)$ and $D_r(s)$ emerge :

$$\begin{cases} N_r(s) = b_{q-1}\left[\prod\limits_{i\in\mathcal{J}} (s-z_i)\right] \\ D_r(s) = \left[\prod\limits_{i\in\mathcal{J}} (s-z_i)\right](s + a_{q-1} - \dfrac{b_{q-2}}{b_{q-1}}) + \sum\limits_{i\in\mathcal{J}} k_i \prod\limits_{\substack{j\in\mathcal{J}\\ j\neq i}} (s-z_j) \end{cases}$$

So, from (14), we can state the following result :

Theorem 3.1 :

If Z^+ and Z^- are strictly negative sets, a sufficient condition of asymptotic stability of the equilibrium 0 of the system Figure 1 is :

$$\forall\ \varepsilon \in R,\ f^*(\varepsilon) > -\frac{1}{b_{q-1}}\left[a_{q-1} - \frac{b_{q-2}}{b_{q-1}} - \sum_{i\in J}\frac{k_i}{z_i}\right] \qquad \square$$

3.2. P(s) has (q-1) separate real roots, μ_i, and another, μ_o

As in the previous part, we define the two complementary sets P^+ and P^- from (7) with :

$$\forall\ i \in I_1^{q-1},\ k_i = \left[\frac{N(s)\ (s-\mu_i)\ (s-\mu_o)}{P(s)}\right]_{s=\mu_i}$$

$$\begin{cases} P^+ = \{\mu_i,\ i \in I \subset I_1^{q-1},\ P(\mu_i) = 0,\ k_i > 0\} \\ P^- = \{\mu_i,\ i \in J \subset I_1^{q-1},\ P(\mu_i) = 0,\ k_i < 0\} \end{cases} \tag{16}$$

and :

$$I \cap J = \emptyset\ ;\ I \cup J = I_1^{q-1}$$

So, bearing this in mind, we have in this case :

$$\begin{cases} N_r(s) = b_{q-1} \prod_{i\in J}(s-\mu_i) + \sum_{i\in J} k_i \prod_{\substack{j\in J \\ j\neq i}} (s-\mu_j) \\ \\ D_r(s) = (s-\mu_o) \prod_{i\in J}(s-\mu_j) - a\,N_r(s) \end{cases}$$

where a is defined by :

$$P(s) = D(s) + a\,N(s) = (s-\mu_o) \prod_{i\in I_1^{q-1}} (s-\mu_i)$$

The interpretation of (14) leads to the result :

Theorem 3.2 :

If P^+ and P^- are strictly negative sets, and $N_r(0) > 0$, a sufficient condition of asymptotic stability of the equilibrium 0 for the system Figure 1 is :

$$\forall\, \varepsilon \in R,\ f^*(\varepsilon) > -\frac{D(\mu_o)}{N(\mu_o)} + \mathrm{Max}\left(0,\ \frac{\mu_o \prod_{i\in J}(-\mu_i)}{N_r(0)}\right) \quad \square$$

It should be noted that, if we are in the particular case where $b_{q-1} \neq 0$ and $\mu_o = 0$, the condition obtained in this theorem is thus :

$$\forall\, \varepsilon \in R \quad f^*(\varepsilon) > -\frac{D(0)}{N(0)}$$

So, we generalize this classical condition (Benrejeb, 1980) to a very large class of processes by this corollary :

Corollary :

If the system described Figure 1 is such that $\left[D(s) - \frac{D(0)}{N(0)} N(s)\right]$ has (q-1) real negative different roots with $b_{q-1} > 0$ and $N_r(0) > 0$, then a sufficient condition of asymptotic stability of the origin is :

$$\forall\, \varepsilon \in R,\ f^*(\varepsilon) > -\frac{D(0)}{N(0)} \quad \square$$

4. ALGEBRAIC CONDITION FOR THE APPLICATION OF THE RESULTS

At the end of part 1., it was noted that the systems concerned in the study must verify hypothesis (H). Furthermore, after stability analysis (part 2.) we can say that the roots of $c\,D(s) + d\,N(s)$ must be strictly negative ; so, hypothesis (H) must finally be replaced by the following condition :

(H') $\exists\, (c,d) \in R^2$, $c\,D(s) + d\,N(s)$ has (q-1) real separate strictly negative zeros

If this condition is satisfied, (H) is verified and P^+ and P^- (or Z^+ and Z^-) are obviously strictly negative sets. Theorems of part 3. can be immediately applied to the system. This underlines the importance of having a criterion for deciding if hypothesis (H') is verified or not.

4.1. Geometrical criterion

Let us now consider the linear system obtained from the original non-linear one described Figure 1, replacing the non-constant gain f* by a constant gain k, $k \in R$. The Evans locus (Gille, Decaulne, Pelegrin, 1971) of this linear system is the evolution in the complex plane of the roots of $D(s) + k\,N(s)$, k varying from $-\infty$ to $+\infty$. So, the existence of a couple of real numbers (c , d) such that (H') is satisfied, can be deduced theoretically from the Evans locus of the linear system (Rotella, Zambettakis, Richard, 1982). This method is a geometrical one and enables the application of a classical method of linear systems to the analysis of non-linear processes ; but, in practice, for the fixed purpose, it presents two main drawbacks.

In the first place, the departure points from the real axis in the Evans locus are of importance, and these points must be known in function of k. This is not very easy to do in every case. So we cannot know the exact boundaries in k between real and complex roots ; particular cases exist where it is not possible to decide rapidly if all roots are real or not.

A second disavantage is that, at the beginning, the Evans locus needs, calculation of the roots of the polynomials N(s) and D(s) without anything being known about their roots (number of real or complex zeros, orders of multiplicity), and so, in the more general case, this geometrical method is difficult to apply and to use.

When hypothesis (H') is verified, we must know the values of the roots for a chosen value of (c , d), in order to apply the theorems about stability. A knowledge of Evans locus of the linear system is thus useful in a neighbourhood of k ($k = d/c$) to have an approximate value of the zeros, and then to initialize a classical algorithm for the calculation of roots. The Evans locus can be obtained by computation (Paillet, 1981), and thus be easily displayed.

In the next part we propose an algebraic method to obtain exact limits on k in order to have all real negative separate roots in the Evans locus.

4.2. Algebraic criterion

Let us consider a real parametric polynomial P(s,k) defined by :

$$P(s,k) = \sum_{i=0}^{n} f_i(k)\, s^i, \quad s \in \mathcal{C},\ k \in \mathcal{R}$$

with :

$$\forall\, k \in \mathcal{R},\ \forall\, i \in I_o^n,\ f_i(k) \in \mathcal{R},\ f_n(k) \neq 0$$

That is to say that P(s,k) is a real polynomial with its coefficients depending on a real parameter k. Bearing this in mind, the problem put by hypothesis (H') can be formulated in two points as follows :

> What are the conditions on k to have (n-1) roots of P(s,k) with properties :
> a) all real and distinct ?
> b) all negative ?

To answer the point a), a method from Hermite's works can be employed (Hermite, 1856) which uses Newton's formulas (Rosenbrock, 1970) (Deif, 1982). The signature of a quadratic form gives the number of real distinct roots of a polynomial (Gantmacher, 1966). The coefficients of the quadratic form are deduced from the coefficients of the studied polynomial after a matrix inversion. In the case of a parametric polynomial this method cannot be easily used.

We propose, to answer simultaneously the points a) and b), a method using Sturm's results about Cauchy's index of rational functions (Gantmacher, 1966).

4.2.1. Construction of a Modified Routh Array. Considering two real parametric polynomials A(s,k) and B(s,k) defined by :

$$A(s,k) = \sum_{i=0}^{n} a_i(k)\, s^i \quad ; \quad B(s,k) = \sum_{i=0}^{n-1} b_i(k)\, s^i \qquad s \in \mathcal{C},\ k \in \mathcal{R}$$

with :

$$\forall\, k \in \mathcal{R}$$

$$\forall\, i \in I_0^n \qquad a_i(k) \in \mathcal{R},\ a_n(k) \neq 0$$

$$\forall\, i \in I_0^{n-1} \qquad b_i(k) \in \mathcal{R},\ b_{n-1}(k) \neq 0$$

To simplify notations, we will now note, $a_i(k)$ and $b_i(k)$ by a_i and b_i. We propose to deduce another parametric polynomial C(s,k) from A(s,k) and B(s,k) by the following rules :

1) From A(s,k) and B(s,k) is deduced an intermediate polynomial $C_I(s,k)$ defined by :

$$C_I(s,k) = \sum_{i=0}^{n-1} c'_i\, s^i$$

with :

$$\forall\, i \in I_1^{n-1} \qquad c'_i = \frac{b_{n-1}\, a_i - a_n\, b_{i-1}}{b_{n-1}} \qquad c'_o = a_o$$

2) From B(s,k) and $C_I(s,k)$ is deduced the parametric polynomial C(s,k) defined by :

$$C(s,k) = \sum_{i=0}^{n-2} c_i\, s^i$$

with :

$$\forall\, i \in I_0^{n-2} \qquad c_i = \frac{c'_{n-1}\, b_i - b_{n-1}\, c'_i}{b_{n-1}}$$

These rules are summed up by the following diagrams :

1) $A(s,k) \longrightarrow a_n \; . \; . \; . \; . \; . \; . \; a_i \quad a_{i-1} \; . \; . \; . \; a_1 \quad a_0$

$B(s,k) \longrightarrow b_{n-1} \; . \; . \; . \; . \; . \; b_{i-1} \quad b_{i-2} \; . \; . \; . \; b_0$

$\Downarrow$

c'_i

2) $B(s,k) \longrightarrow b_{n-1} \; . \; . \; . \; . \; b_i \quad b_{i-1} \; . \; . \; . \; b_0$

$C_I(s,k) \longrightarrow c'_{n-1} \; . \; . \; . \; . \; . \; c'_i \quad c'_{i-1} \; . \; . \; . \; c_0$

$\Downarrow$

c_i

By analogy with the construction of a Routh Array, we propose to call Modified Routh Array from A(s,k) and B(s,k) the array formed as follows :

1st line : coefficients of A(s,k) : $a_n, a_{n-1}, \ldots, a_1, a_0$

2nd line : coefficients of B(s,k) : $b_{n-1}, b_{n-2}, \ldots, b_1, b_0$

3rd line : coefficients of C(s,k) : $c_{n-2}, c_{n-1}, \ldots, c_0$ where C(s,k) is deduced from A(s,k) and B(s,k) by the above rules

4th line : coefficients of D(s,k) : $d_{n-3}, \ldots, d_0$ where D(s,k) is deduced from B(s,k) and C(s,k) by the above rules

and so, every new line of this array is deduced from the two previous ones by the rules 1) and 2) ; the last line of this array occures when the next deduced polynomial is null. The Modified Routh Array is constituted by m successive lines and has a "triangular" form :

$$\begin{array}{ccccccc} \bullet & \bullet & \bullet & \bullet & \bullet & \bullet & \bullet \\ \bullet & \bullet & \bullet & \bullet & \bullet & \bullet & \\ \bullet & \bullet & \bullet & \bullet & \bullet & & \\ \bullet & \bullet & \bullet & \bullet & & & \end{array}$$

In the next part we will call the first column and the last diagonal of the Modified Routh Array from A(s,k) and B(s,k) respectively the series $(a_n, b_{n-1}, c_{n-1}, \ldots, y_{n-m+1})$ and $(a_0, b_0, c_0, \ldots, y_0)$. These series are formed by upper and lower degree coefficients of the deduced polynomials. In the next part we will indicate a property of this type of array.

4.2.2. Property of the Modified Routh Array. First, we have to define the real derivate polynomial dP(s,k) / ds from P(s,k) by :

$$\text{if} \quad P(s,k) = \sum_{i=0}^{n} f_i(k)\, s^i, \quad \text{thus} \quad \frac{dP(s,k)}{ds} = \sum_{i=0}^{n-1} g_i(k)\, s^i$$

$$\text{with } \forall\, i \in I_0^{n-1}, \quad g_i(k) = (i+1)\, f_{i+1}(k)$$

We can now state the following theorem :

Theorem 4.1 :

The number F of sign changes in the first column and the number L of sign changes in the last diagonal of the Modified Routh Array from P(s,k) and dP(s,k) / ds give the number N of negative real separate roots of P(s,k) by :

$$N = m - F - L - 1$$

where m is the number of lines of the Modified Routh Array. □

Before proving this property, we must state some classical theorems about Cauchy's Index (Gantmacher, 1966).

Definition 1 :

The Cauchy's Index of a real rationnal function R(x) between real limits a and b ($a < b$) is the number of gaps of R(x) from $-\infty$ to $+\infty$ minus the number of gaps of R(x) from $+\infty$ to $-\infty$ when x describes $]a\,,b[$ from a to b. □

This function is noted $I_a^b\, R(x)$.

Definition 2 :

A series of real polynomials $f_1(x)$, $f_2(x)$, ..., $f_m(x)$ is called Sturm's series in a real interval $]a\,,b[$ if the following conditions are verified :

$$\text{a)}\ \forall\, x \in\,]a\,,b[\qquad f_m(x) \neq 0$$

$$\text{b)}\ \forall\, x \in\,]a\,,b[,\ \forall\, k \in I_2^{m-1} \qquad f_k(x) = 0 \implies f_{k-1}(x)\, f_{k+1}(x) < 0 \qquad \square$$

With these notions we can state the following theorems.

Sturm's Theorem :

If $f_1(x)$, $f_2(x)$, ..., $f_m(x)$ is a Sturm's series for $]a\,,b[$ and if V(x) is the number of sign changes in the series, then :

$$I_a^b \frac{f_2(x)}{f_1(x)} = V(a) - V(b) \qquad \square$$

Gantmacher's Theorem :

The number of real separate roots of a polynomial f(x) in $]a, b[$ is equal to :

$$I_a^b \frac{f'(x)}{f(x)}$$

where f'(x) is the derivate polynomial from f(x). $\square$

To solve our problem by this method we must construct a Sturm's series from $f(x) = f_1(x)$ and $f'(x) = f_2(x)$. We can now state the following property.

Property :

The construction of a Modified Routh Array from two real polynomials $f_1(x)$ and $f_2(x)$, such as $d°f_2(x) = d°f_1(x) - 1$, gives, line after line the coefficients of successive real polynomials which constitute a Sturm's series. $\square$

This property can be proved with the help of the following remarks. The polynomial c(x) deduced from two others a(x) and b(x), $d°b(x) = d°a(x) - 1$, with the rules 1) and 2) of part 4.1., is the opposite of the remainder in the Euclidian division of a(x) by b(x). According to the construction of the Modified Routh Array, every line is constituted by the coefficients of a polynomial which is the opposite of the remainder in the Euclidian division of the two polynomials given by the two previous lines. Thus, from (Gantmacher, 1966) all these polynomials are a Sturm's series.

Let us now suppose we have, with the Modified Routh Array, a Sturm's series $f(x) = f_1(x)$, $f'(x) = f_2(x)$, $f_3(x)$, ..., $f_m(x)$. Using Sturm's and Gantmacher's theorems we can state that the number N of real negative distincts roots of f(x) is given by :

$$N = V(-\infty) - V(0)$$

But, if $f_m(x)$ is the last remainder non null, we have also :

$$V(-\infty) = m - 1 - V(+\infty)$$

If we now remark that $V(+\infty)$ and V(0) are respectively the number of sign changes in the first column and the last diagonal of the Modified Routh Array from f(x) and f'(x) the theorem 4.1 emerges.

On the other hand we can state another theorem useful in studying real polynomials and which answers to hypothesis (H) :

Theorem 4.2 :

The number F of sign changes in the first column of the Modified Routh Array from $P(s,k)$ and $dP(s,k)/ds$ gives the number R of real separate roots of $P(s,k)$ by :

$$R = m - 1 - 2F$$

where m is the number of lines of the Modified Routh Array. □

To prove this latter theorem, it is sufficient in the previous proof to replace the relation $N = V(-\infty) - V(0)$ by $R = V(-\infty) - V(+\infty)$.

If we compare the two theorems, for a same parametric real polynomial, it is necessary to have $N \leqq R$, and therefore, in a Modified Routh Array, to have $L \geqq F$. This final remark can be used to verify a Modified Routh Array.

4.3. Application on hypothesis (H')

To solve the problem set by hypothesis (H'), we can apply the theorem 4.1. to the real parametric polynomial $P(s,k) = D(s) + k\,N(s)$ and to deduce conditions on k, in order to have N equal to q or (q-1), if q is the order of the system.

But the quantities F and L are not negative, so it is necessary to have only three possible solutions in the Modified Routh Array from $P(s,k)$ and $dP(s,k)/ds$. These three possible configurations are :

$$(m\,,F\,,L) \in \{(q+1\,,0\,,0)\ ,\ (q+1\,,0\,,1)\ ,\ (q\,,0\,,0)\}$$

therefore there is no sign changes in the first column and, at most, one sign changes in the last diagonal ; and, in the case where $m = q$, this means that $D(s) + k\,N(s)$ has a double root.

By this method, we do not have to distinguish the case where $c = 0$, because, when k tends to an infinite value, the roots of $P(s,k)$ are the roots of $N(s)$.

5. METHODOLOGY AND IMPLEMENTATION

5.1. Methodology

In order to summarize the results of the previous parts, we propose a methodology for the analysis of systems described Figure 1, relating to the use of the modellings of type (3). The main question lies in the choice of parameters λ_i and of the stability criterium.

We propose to choose as λ_i the zeros of a polynomial $c\,D(s) + d\,N(s)$, where c and d have to verify the hypothesis (H') (i.e. $c\,D(s) + d\,N(s)$ has (q-1) real separate negative roots). The choice of (q-1) parameters λ_i is then reduced to the (c,d) one. The existence of (c,d) can be interpreted directly on the Evans locus of the linear part $N(s)/D(s)$ of the system but must be proved by the calculation of a Modified Routh Array.

The Modified Routh Array use gives the exact limits for c/d to verify hypothesis (H'). For a fixed value of c/d, the values of the roots of c D(s) + d N(s) must then be calculated. With the aid of graphic methods for linear systems (Paillet, 1981) to display the Evans locus, an algorithm to calculate roots of a real polynomial can be easily initialized.

The previous works, about stability analysis, considered the two following cases, resulting from the choice of (c,d) :

* $Z^+ = \emptyset$ or $P^+ = \emptyset$; all the off-diagonal terms in (8) are positive. Application of Borne and Gentina's criterion lead to the validation of linear conjecture.

* $Z^- = \emptyset$ or $P^- = \emptyset$; all the off-diagonal terms in (8) are negative. The quadratic plus integral candidate Lyapunov function enables results to be checked with great ease. In the case $d/c = -D(0)/N(0)$, the sufficient conditions of stability are the same as if the system was linear.

The proposed study enables the intermediate cases to be taken into, when Z^+ or P^+, and Z^- or P^- are not empty. This is a more general case, and the results obtained are may be better compared to a separate application of Borne and Gentina's criterion, or Lurie-type candidate Lyapunov function.

5.2. Algorithm

A computer implementation is possible, since all coefficients can be calculated in the above mentioned modellings.

In the first steps, the existence and the choice of (c,d) are tested by analytic and graphic means. The terms k_i (8) are calculated, leading to the sets Z^- and Z^+ (or P^- and P^+). The elements of matrix $C(\varepsilon)$ (8), which correspond to the Z^+ (or P^+) set, can be suppressed, and the remaining elements constitute the reduced order matrix. A sufficient stability condition is then tested on the sign of the reduced order matrix determinant. If necessary, another choice of (c,d) can be considered.

These steps are summarized in the Figure 3. The second step ((c,d) choice) is in fact the choice of the only quotient d/c, supposing c different from zero. A simple variation on this parameter is then possible (step 7) in order to obtain a larger condition on f* variations.

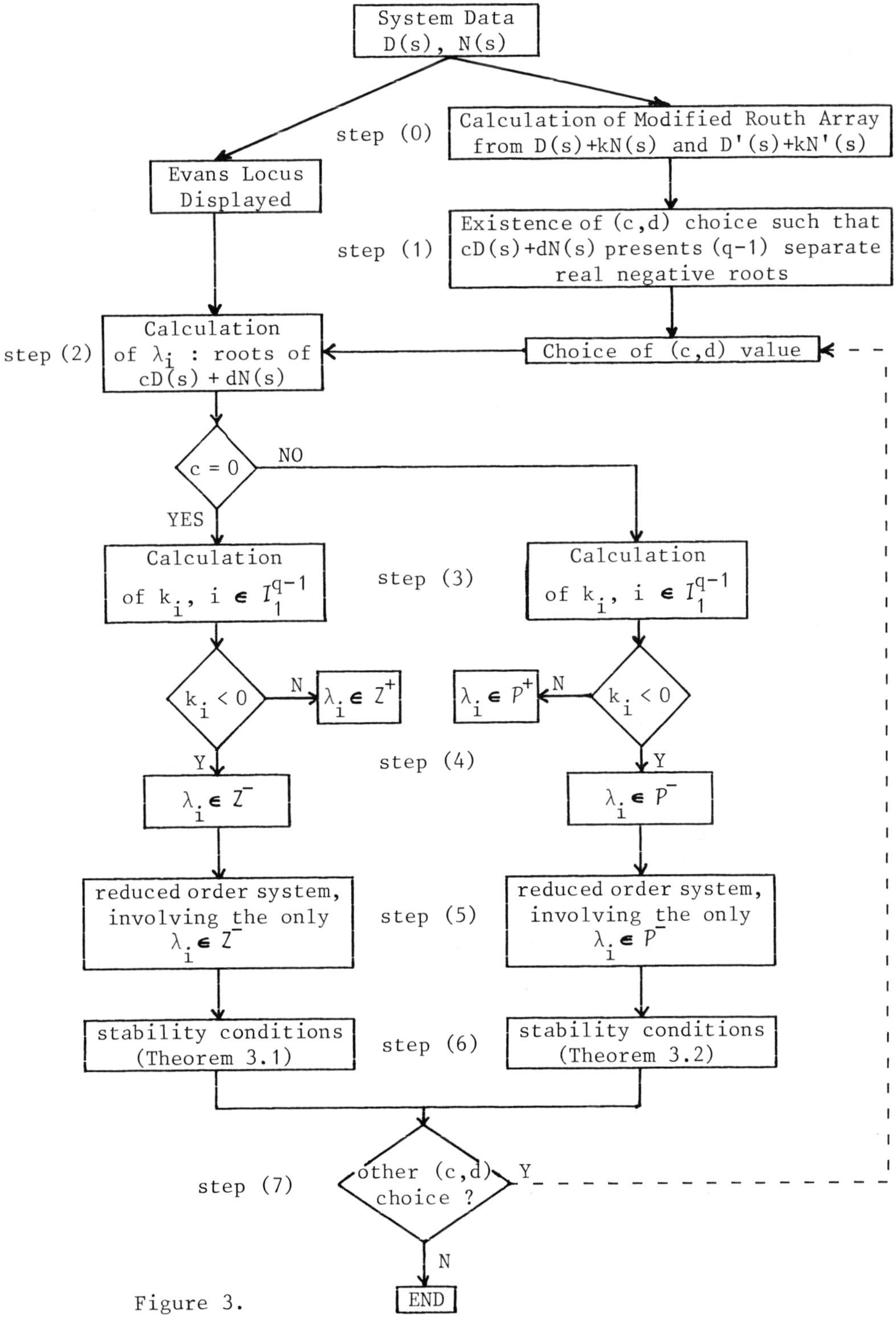
System Data
D(s), N(s)
step (0)
Calculation of Modified Routh Array from D(s)+kN(s) and D'(s)+kN'(s)
Evans Locus Displayed
step (1)
Existence of (c,d) choice such that cD(s)+dN(s) presents (q-1) separate real negative roots
step (2)
Calculation of λi : roots of cD(s) + dN(s)
Choice of (c,d) value
c = 0
NO
YES
Calculation of ki, i ∈ I1q-1
step (3)
Calculation of ki, i ∈ I1q-1
ki < 0
N
λi ∈ Z+
λi ∈ P+
N
ki < 0
Y
step (4)
Y
λi ∈ Z−
λi ∈ P−
reduced order system, involving the only λi ∈ Z−
step (5)
reduced order system, involving the only λi ∈ P−
stability conditions (Theorem 3.1)
step (6)
stability conditions (Theorem 3.2)
step (7)
other (c,d) choice ?
Y
N
END

Figure 3.

5.3. Example

Let us consider the following system described by :

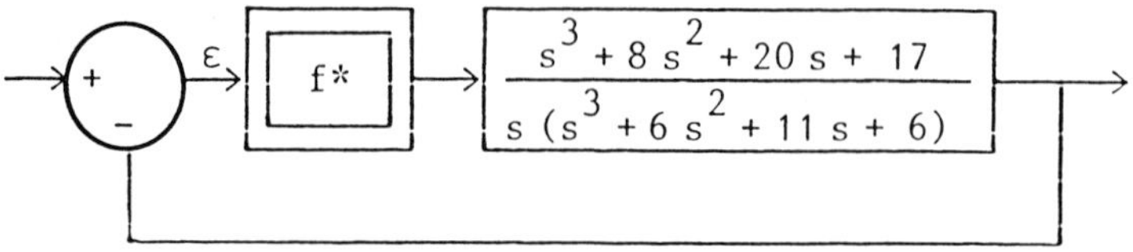

Step (0) : a) Modified Routh Array from N(s) and N'(s)

$N(s) = s^3 + 8 s^2 + 20 s + 17$

$\Longrightarrow$

1	8	20	17
3	16	20	
8/9	7/9		
$-\frac{531}{64}$			

From theorems 4.1 and 4.2 it emerges that N(p) has only one real negative root.

b) Modified Routh Array from D(s) and D'(s)

$D(s) = s\,(s^3 + 6 s^2 + 11 s + 6)$

D(s) has a zero root, thus we consider the polynomial $s^3 + 6 s^2 + 11 s + 6$ to construct a Modified Routh Array :

$\Longrightarrow$

1	6	11	6
3	12	11	
2/3	4/3		
1			

Applying theorem 4.1 we can see that D(s) / s has 3 real negative roots ; therefore the choice of (c,d) = (1,0) is available (steps (1 and 2)). The λ_i are thus : -1, -2, -3.

Step (3) : The coefficients k_i are numerically calculated in the following array :

λ_i	k_i
-1	2
-2	-1
-3	1

Step (4) : This leads to define P^+ and P^- by :

$$P^+ = \{-1\ ,\ -3\}\ ;\ P^- = \{-2\}$$

Step (5) : These considerations give the expression of the reduced order matrix :

$$\begin{pmatrix} -2 & f^* \\ 1 & -f^* \end{pmatrix}$$

it is associated with the reduced transfer function :

$$\frac{N_r(s)}{D_r(s)} = \frac{s+1}{s\ (s+2)}$$

Step (6) : We have $N_r(0) > 0$, and so can apply the corollary of the theorem, from which the following stability condition emerges :

$$\forall\ \varepsilon \in R,\ f^*(\varepsilon) > 0$$

By using the method described above, the linear conjecture will hold for the reduced order system. The sufficient conditions are the same as if this system was linear, and therefore the step (7) is useless.

CONCLUSION

The main interest of this study is to propose a way of applying two classical methods at the same time : the application of a Lurie-type candidate Lyapunov function leads to a reduced order system, which then can be studied with the aid of the Borne-Gentina practical criterion. It is to be noted that this last system verifies the linear conjecture and its sufficient stability conditions can be easily checked. The conditions thus worked out are less restrictive than those obtained by a separate application of the criteria. A computer implementation is possible. The only unexplained point of the proposed method is the optimisation of the (c,d) choice to ensure the largest permissible variation of the non-constant gain $f^*(\varepsilon)$. This point of importance will be studied in our future works.

REFERENCES

BENRENJEB M.
"Sur la synchronisation des systèmes continus non linéaires en régime forcé"
Thèse Docteur Ingénieur, Lille, n° 186, Juin 1976.

BENREJEB M.
"Sur l'analyse et la synthèse de processus complexes hiérarchisés, Application aux systèmes singulièrement perturbés"
Thèse de Doctorat ès Sciences Physiques, Lille, n° 479, 1980.

BENREJEB M., BORNE P., LAURENT F.
"Sur une application de la représentation en flèche à l'analyse des processus"
R. A. I. R. O. Automatique, **16**, n° 2, p. 133 - 146, 1982.

GANTMACHER F.R.
"Théorie des matrices"
Collection Universitaire de Mathématiques, Dunod, 1966.

GENTINA J.C.
"Contribution à l'analyse et à la synthèse des systèmes continus non linéaires de grande dimension"
Thèse de Doctorat ès Sciences Physiques, Lille, n° 347, 1976.

GENTINA J.C., BORNE P.
"Sur une condition d'application du critère de stabilité linéaire à certaines classes de systèmes continus non linéaires"
C. R. A. S., Paris, t. 275, 1972.

GILLE J.C., DECAULNE P., PELEGRIN M.
"Théorie et calcul des asservissements linéaires"
Dunod, 1971.

GRUJIĆ Lj.T., GENTINA J.C., BORNE P.
"General aggregation of large-scale systems by vector Lyapunov functions and vector norms"
Int. J. Control, **24**, n° 4, p. 529 - 550, 1976.

HAHN W.
"Theory and application of Lyapunov's direct method"
Prentice Hall, Englewood Cliffs, 1963.

HERMITE C.
"Sur le nombre de racines d'une équation algébrique comprises entre des limites données"
J. Reine Angew. Math., **52**, p. 39 - 51, 1856.

LASALLE J.P., LEFSHETZ S.
"Stability by Lyapunov's direct method with applications"
Academic Press, 1961.

LYAPUNOV A.M.
"Problème général de la stabilité du mouvement"
University Press, Princeton, 1949.

NARENDRA K.S., TAYLOR J.H.
"Frequency domain criteria for absolute stability"
Academic Press, New-York and London, 1973.

PAILLET D.
"P A A S, Programme d'aide à l'analyse des systèmes"
1st International Conference, Applied Modelling and Simulation, Lyon, **1**, p. 231 - 233, 1981.

POPOV V.M.
"L'hyperstabilité des systèmes automatiques"
Dunod, 1973.

RICHARD J.P.
"Sur la mise en équation d'état des systèmes continus non linéaires par une méthode de calcul symbolique : définition d'un invariant de représentation"
Thèse de Doctorat d'Ingénieur, Lille, n° 259, 1981 a.

RICHARD J.P.
"Sur les limitations de la conjecture linéaire dans l'étude de la stabilité"
1st International Conference, Applied Modelling and Simulation, Lyon, **1**, p. 91 - 94, 1981 b.

ROSENBROCK H.H.
"State-space and multivariable theory"
Nelson, 1970.

ROTELLA F.
"Détermination de nouvelles représentations d'état adaptées à l'analyse et à la synthèse des systèmes continus non linéaires"
Thèse de Docteur Ingénieur, Lille, n° 327, 1983.

ROTELLA F., RICHARD J.P.
"Modélisation et synthèse de systèmes monovariables de type Lurie Postnikov"
1st IASTED Symposium, Applied Informatics, **3**, p. 173 - 176, Lille, 1983.

ROTELLA F., RICHARD J.P., ZAMBETTAKIS I.
"Stability of non linear continuous systems"
Congrès IASTED Robotics and Automation, Lugano, Switzerland, 1983.

ZAMBETTAKIS I.
"Contribution à l'étude de systèmes à non-linéarités multiples. Application aux systèmes électromécaniques"
Thèse de Docteur Ingénieur, Lille, n° 326, 1983.

ZAMBETTAKIS I., RICHARD J.P., LAURENT F.
"Etude des systèmes électromécaniques non linéaires par la méthode directe de Lyapunov"
1st IASTED Symposium, Applied Informatics, **3**, p. 169 - 172, Lille, 1983.

DEIF A.S.
"Advanced matrix theory for scientists and Engineers"
Abacus Press, 1982.

PART II UNCERTAIN SYSTEMS AND ROBUST CONTROL

CHAPTER 8

REPRESENTATIONS OF UNCERTAINTY AND ROBUSTNESS TESTS FOR MULTIVARIABLE FEEDBACK SYSTEMS

Ian Postlethwaite and Yung Kuan Foo
Department of Engineering Science
Oxford University
Parks Road, Oxford OX1 3PJ

Abstract. In this essay we examine several different ways in which uncertainty can be represented in a control system. In general, not all of the commonly used representations can adequately represent all possible perturbations in the plant model. Given adequate representations of uncertainty we present three kinds of robustness test. The first kind is derived from Nyquist's stability criterion, and can be used providing the plant and perturbed plant models have the same number of unstable poles. The second is derived from an inverse Nyquist stability criterion, and can be used providing the plant and perturbed plant models have the same number of non-minimum phase zeros. The third is a combination of the other two tests, but does not require the plant and perturbed plant models to have either the same number of poles or the same number of zeros in the right-half complex plane.

1. INTRODUCTION

The usefulness of a nominal plant model in control system design is greatly enhanced if it is accompanied by some specification of the uncertainties involved. At present the most common way of doing this is to use singular values to define a set of perturbed plant models in the neighbourhood of a nominal model in the belief that the true plant model lies in this set. This approach has been adopted by research workers interested in the robustness properties of multivariable control systems, and several representations of uncertainty can be found in [1] - [11].

In this paper, we begin by examining the adequacy of these representations, that is, their ability to represent all possible perturbations of the plant. Then given adequate representations of uncertainty we describe three kinds of test which can be used to assess the robustness of the closed-loop stability property in a multivariable control system.

The detailed layout of the paper is as follows. In section 2, we consider the validity of characterizing uncertainty by additive or multiplicative perturbations. In section 3, the robustness tests are

S. G. Tzafestas (ed.), Multivariable Control, 151–160.

presented, and our conclusions are summarised in section 4.

The article draws heavily on work reported in [8] and [12] where further details and proofs can be found.

2. REPRESENTATIONS OF UNCERTAINTY

In this section we examine the validity of characterizing uncertainty in a nominal transfer function matrix G(s) by the following forms, [1], [3] - [6], [8]:

$$\{G_p(s)\}_E \triangleq \{G(s) + E(s) \mid \bar{\sigma}[E(j\omega)] \leq e(\omega)\} \tag{2.1}$$

$$\{G_p(s)\}_{\Delta_o} \triangleq \{(I + \Delta_o(s))G(s) \mid \bar{\sigma}[\Delta_o(j\omega)] \leq \delta_o(\omega)\} \tag{2.2}$$

$$\{G_p(s)\}_{\Delta_i} \triangleq \{G(s)(I + \Delta_i(s)) \mid \bar{\sigma}[\Delta_i(j\omega)] \leq \delta_i(\omega)\} \tag{2.3}$$

$$\{G_p(s)\}_{\tilde{E}} \triangleq \{G(s)^{\dagger} + \tilde{E}(s))^{\dagger} \mid \bar{\sigma}[\tilde{E}(j\omega)] \leq \tilde{e}(\omega)\} \tag{2.4}$$

$$\{G_p(s)\}_{\tilde{\Delta}_o} \triangleq \{(I + \tilde{\Delta}_o(s))^{-1}G(s) \mid \bar{\sigma}[\tilde{\Delta}_o(j\omega)] \leq \tilde{\delta}_o(\omega)\} \tag{2.5}$$

$$\{G_p(s)\}_{\tilde{\Delta}_i} \triangleq \{G(s)(I + \tilde{\Delta}_i(s))^{-1} \mid \bar{\sigma}[\tilde{\Delta}_i(j\omega)] \leq \tilde{\delta}_i(\omega)\} \tag{2.6}$$

where $\bar{\sigma}[A]$ denotes the maximum singular value of A, and $A^{\dagger}$ is the left or right inverse of A, whichever exists, such that if A has more rows than columns, then $A^{\dagger}A = I$. Physical interpretations of (2.4) - (2.6) have been given in [8] when both G(s) and $G_p(s)$ are invertible. Note that although the representations of uncertainty in [2] and [7] are in a slightly different form from those above the arguments that follow apply to these as well.

In each of the representations (2.1) - (2.6) a bound on a maximum singular value is used to define a set of perturbed plant models. The question that arises is whether the set so defined includes all possible physical perturbations in G(s). If it does not, then the set is clearly inadequate for representing uncertainty. If the result of a robustness test is to be valid, the representation of uncertainty used must be an adequate representation.

Obviously, if the representation and the actual perturbation happen to be at the same point in the loop, then the representation will be adequate, provided its maximum singular value bound is at least as large as that of the actual perturbation. However, in a real system, uncertainty occurs at several points in the loop. In such cases, it has been suggested that it may be possible to reflect all uncertainty to one point and combine the individual contributions in a single perturbation as in the forms (2.1) to (2.6). It is then important to know whether the chosen representation of uncertainty can adequately represent perturbations from all possible sources in the system.

It is trivially true that given any perturbed plant model G_p and a nominal plant model G, we can always find an uncertainty model E of the same field, such that $G_p = G + E$. Hence the additive perturbation E can be an adequate representation of all the different forms of perturbation under consideration. We may therefore adopt the viewpoint that all perturbations are additive in nature. In other words, no matter what the uncertainties actually are they will show up as an additive perturbation and robustness against this perturbation implies, and is implied by, robustness against the actual perturbations.

When using the multiplicative forms of perturbation, more care needs to be exercised. To give an example of the difficulties involved, the reader may verify that there is no input multiplicative perturbation which can cause

$$G(s) = [\frac{1}{s+1} \quad \frac{1}{s+2}]^T \text{ to become}$$

$$G_p(s) = [\frac{1}{s+3} \quad \frac{1}{s+4}]^T; \text{ where } A^T \text{ denotes the transpose of A.}$$

The problem of knowing when G(s) can be transformed to $G_p(s)$ will now be considered in more detail. The results are not thought to be new and are therefore stated without proof. The same problem has received considerable attention in the areas of model matching and system inversion; see [13], [14] and [15] for example.

Definition. Given two ℓxm matrices A and B, we say that A is linearly transformable to B if there exists a linear operator T such that TA = B.

It follows that:

(i) If G is linearly transformable to G_p, i.e. there exists an L_o such that $L_o G \triangleq (I + \Delta_o)G = G_p$, then G and the set $\{\Delta_o\}$ can combine to give an adequate representation of the true plant G_p.

(ii) If G^T is linearly transformable to G_p^T, i.e. there exists an L_i such that $L_i^T G^T \triangleq (I + \Delta_i)^T G^T = G_p^T$, then G and the set $\{\Delta_i\}$ can combine to give an adequate representation of the true plant G_p.

Necessary and sufficient conditions for G to be linearly transformable to G_p are given in theorem 2.1 below.

Theorem 2.1. Let G and G_p be two ℓ x m matrices of the same scalar field, then the following statements are equivalent:

(i) G is linearly transformable to G_p,

(ii) the row-span of G_p is contained in the row-span of G,

(iii) rank G = rank F
where F is the partitioned matrix $\begin{bmatrix} G_p \\ G \end{bmatrix}$.

Note that theorem 2.1 implies that rank $(G_p) \leq$ rank (G) is a necessary condition for G to be linearly transformable to G_p. Therefore because the nominal model of most physical plants has full column rank we give the following corollary.

Corollary 2.1. Let G and G_p be two ℓ x m matrices of the same scalar field with $\ell \geq m$ and rank G = m, then G is linearly transformable to G_p.

In simple terms, corollary 2.1 implies:

(i) If G has full rank and at least as many outputs as inputs, then the output multiplicative perturbation can be an adequate representation of uncertainty.

(ii) If G has full rank and at least as many inputs as outputs, then the input multiplicative perturbation can be an adequate representation of uncertainty.

If we choose to represent uncertainty by one of the inverse forms (i.e. $\tilde{E}$, $\tilde{\Delta}_i$ or $\tilde{\Delta}_o$), then more restrictions have to be observed. In particular:

(i) For the case of $\tilde{E}$, both G and G_p have to be invertible or pseudo-invertible (in the sense that they are left or right invertible).

(ii) For $\tilde{\Delta}_o$, $(I + \tilde{\Delta}_o)$ has to be invertible (i.e. $\det(I + \tilde{\Delta}_o) \neq 0$).

(iii) For $\tilde{\Delta}_i$, $(I + \tilde{\Delta}_i)$ has to be invertible (i.e. $\det(I + \tilde{\Delta}_i) \neq 0$).

The invertibility of $(I + \tilde{\Delta}_o)$, or $(I + \tilde{\Delta}_i)$, was implicitly assumed in [5] without discussion, and was ensured in [8] by explicitly requiring the existence of inverses or pseudo-inverses of G and G_p. In this paper we drop this latter requirement on G and G_p thus extending our discussion to non-invertible plants.

<u>Theorem 2.2.</u> There exists an $\ell \times \ell$ matrix L_o which is non-singular and satisfies $L_oG = G_p$ if, and only if, the two $\ell \times m$ matrices G and G_p have equal rank and G is linearly transformable to G_p.

<u>Corollary 2.2.</u> Let two $\ell \times m$ matrices G and G_p have equal rank, and assume that there exists an $\ell \times \ell$ singular matrix L' such that $L'G = G_p$. Then there exists another $\ell \times \ell$ matrix L which is non-singular and satisfies $L'G = G_p = LG$.

We end this section by making a few remarks concerning the possibility of modelling complete actuator or sensor failures with multiplicative perturbations which are invertible. If G is full rank and has more outputs than inputs, and if there is an actuator failure in one or more of the loops, then G_p will have lower rank than G, and it follows that no invertible L_i or L_o exists such that $GL_i = G_p$ or $L_oG = G_p$. However, if there is a sensor failure in the loops, then we may still be able to find an invertible L_o such that $L_oG = G_p$. Hence the tests described in [5] and [8] may indeed be capable of handling sensor failures. For example let

$$G = \begin{bmatrix} \dfrac{1}{s+1} \\ \\ \dfrac{1}{s+2} \end{bmatrix} \qquad \text{and } G_p = \begin{bmatrix} \dfrac{1}{s+1} \\ \\ 0 \end{bmatrix}$$

due to a sensor failure in output channel 2. A natural choice of L_o would be

$$L_o = \begin{bmatrix} 1 & 0 \\ 0 & 0 \end{bmatrix}$$

which is non-invertible. However, an invertible L_o is given by

$$L_o = \begin{bmatrix} 1 & 0 \\ \frac{1}{s+2} & \frac{-1}{s+1} \end{bmatrix}$$

3. ROBUSTNESS TESTS

This section is concerned with methods for assessing the robustness of the stability property of a finite dimensional linear time-invariant design given an adequate representation of uncertainty. If follows from section 2, that if a plant has more outputs than inputs, as is usual, then the perturbed plant models (2.3) and (2.6) are not adequate representations of uncertainty. For this reason representations (2.3) and (2.6) will not be considered further, other than to say that if the plant is square, then a robustness test in terms of $\Delta_i(s)$ (or $\tilde{\Delta}_i(s)$) will be available analgous to the test in terms of $\Delta_o(s)$ (or $\tilde{\Delta}_o(s)$) presented later. For simplicity, the tests will be presented only for square plants.

We will consider the feedback configuration shown in figure 1, where G(s) is a nominal model used for design purposes and K(s) is a controller such that the nominal feedback system is stable. It follows that there can be no pole-zero cancellations in the right-half plane in G(s)K(s). For practical reasons G(s)K(s) and $G_p(s)K(s)$ are assumed to be strictly proper transfer functions.

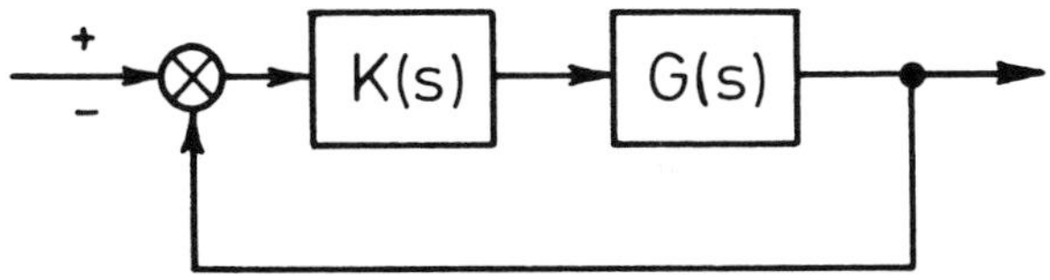

Fig. 1. Feedback Configuration

3.1 Nyquist based tests

The robustness tests of this section are based on the following well known generalization of Nyquist's stability criterion:

Theorem 3.1 The feedback system of figure 1 is stable if, and only if, the image of the Nyquist D-contour D_D under $\det[I + G(s)K(s)]$ encircles the origin P times anticlockwise, where P is the number of open-loop poles in D_D.

From this theorem we see that if $G_p(s)$ is a perturbed plant model with the same number of poles in D_D as $G(s)$, then the corresponding perturbed feedback system will be stable if the image of D_D under $\det[I + G_p(s)K(s)]$ encircles the origin the same number of times as the image of D_D under $\det[I + G(s)K(s)]$. This argument is central to the development of the robustness tests conveyed in the following two theorems, one for each of the representations of uncertainty given in (2.1) and (2.2). Note that the tests are not new and stem from the early work of Doyle and Stein [1].

Theorem 3.2 The feedback system of figure 1 will remain stable when the nominal plant model is replaced by any perturbed plant model taken from the set $\{G_p(s)\}_E$ if:

(i) $G_p(s)$ and $G(s)$ share the same number of poles in D_D, and

(ii) $\underline{\sigma}[(I + (G(s)K(s))^{-1})G(s)] > e(s)$, $\forall s \varepsilon D_D$, where $\underline{\sigma}$ denotes the minimum singular value.

Theorem 3.3 The feedback system of figure 1 will remain stable when the nominal plant model is replaced by any perturbed model taken from the set $\{G_p(s)\}_{\Delta_o}$ if:

(i) $G_p(s)$ and $G(s)$ share the same number of poles in D_D, and

(ii) $\underline{\sigma}[I + (G(s)K(s))^{-1}] > \delta_o(s)$, $\forall s \varepsilon D_D$.

The major disadvantage of theorems (3.2) and (3.3) is that, because $G_p(s)$ and $G(s)$ must have the same number of poles in D_D, they cannot be used when $G(s)$ has any nominal poles whose regions of uncertainty extend across the imaginary axis.

3.2 Inverse Nyquist based tests

In this section we use the representations of uncertainty in (2.4) and (2.5) and present two robustness tests based on the generalized inverse Nyquist stability criterion given below.

Theorem 3.4 The feedback system of figure 1 is stable if, and only if, the image of the inverse Nyquist D-contour D_I under $\det[I + (G(s)K(s))^{-1}]$ encircles the origin Z times anticlockwise, where Z is the total number of zeros of $G(s)$ and $K(s)$ in D_I.

Based on this stability criterion the following robustness tests can be derived.

Theorem 3.5 The feedback system of figure 1 will remain stable when the nominal plant model is replaced by any perturbed plant model taken from the set $\{G_p(s)\}_{\tilde{E}}$ if:

(i) $G_p(s)$ and $G(s)$ share the same number of zeros in D_I, and

(ii) $\underline{\sigma}[G(s)^{-1}(I + G(s)K(s))] > \tilde{e}(s)$, $\forall s \varepsilon D_I$.

Theorem 3.6 The feedback system of figure 1 will remain stable when the nominal plant model is replaced by any perturbed plant model taken from the set $\{G_p(s)\}_{\tilde{\Delta}_o}$ if:

(i) $G_p(s)$ and $G(s)$ share the same number of zeros in D_I, and

(ii) $\underline{\sigma}[I + G(s)K(s)] > \tilde{\delta}_o(s)$, $\forall s \varepsilon D_I$

An interesting feature of the robustness tests of this section is that when the multiplicative-type perturbation $I + \tilde{\Delta}_o(s)$ is used to characterize uncertainty the tests are in terms of the minimum singular value of the return difference matrix $(I + G(s)K(s))$, or equivalently the maximum singular value of the sensitivity function $(I + G(s)K(s))^{-1}$. That is, the function normally used to assess the performance of a system with respect to disturbance rejection and sensitivity is also shown here to be useful for assessing the robustness of the stability property. It follows that if there are no uncertain zeros which may cross the imaginary axis then good sensitivity and robust stability are compatible objectives.

3.3 Nyquist and inverse Nyquist combined

In this section we present a robustness test which is a combination of a Nyquist based test from section 3.1, and an inverse Nyquist based test from section 3.2. The virtue of this combination is that $G_p(s)$ need not have the same number of poles or zeros in the closed right-half complex plane as the nominal model $G(s)$. The improvement is achieved at the expense of a more refined, yet no less practical, definition of the set $\{G_p(s)\}$.

The set $\{G_p(s)\}$ is assumed to be arcwise connected in the topology of unstable plants defined in [16]. A brief description of the topology and a definition of arcwise connectedness is given in [8]. In simple terms the definition of $\{G_p(s)\}$ means that for any $G_p(s)$ in the set, $G(s)$ can be perturbed along a path to $G_p(s)$ so that at any point on the path an arbitrarily small neighbourhood can be defined within which all points on the path have arbitrarily close pole-zero structures in the Nyquist D-contour, D, which has indentations to the left of imaginary axis poles and zeros. We believe this definition to be a practical one with wide application. For example, when the frequency response of each of a plant's elements is known to lie within given bounds then the set of possible plant models will usually be arcwise connected. Note also that the topology we use is the weakest one for which closed-loop stability is a robust property [16].

Theorem 3.7 Let $\{G_p(s)\}$ be an arcwise connected set of transfer function matrices in the topology of unstable plants. Assume that for every member $G_p(s)$ in $\{G_p(s)\}$ there exists either $\Delta_o(j\omega)$, or $\tilde{\Delta}_o(j\omega)$, or both, such that

$$G_p(j\omega) = (I + \Delta_o(j\omega))G(j\omega) = (I + \tilde{\Delta}_o(j\omega))^{-1}G(j\omega) \ \forall\omega$$

where $G(s)$ is a member of $\{G_p(s)\}$, known as the nominal model, such that the closed loop system $(I + G(s)K(s))^{-1}G(s)K(s)$ is stable.

Define

$$\delta_o(\omega) = \sup_{\forall G_p(s)\varepsilon\{G_p(s)\}} \{\bar{\sigma}[\Delta_o(j\omega)]\}$$

$$\tilde{\delta}_o(\omega) = \sup_{\forall G_p(s)\varepsilon\{G_p(s)\}} \{\bar{\sigma}[\tilde{\Delta}_o(j\omega)]\}.$$

Then under these conditions, the perturbed closed-loop system $(I + G_p(s)K(s))^{-1}G_p(s)K(s)$ is stable for any member $G_p(s)$ from $\{G_p(s)\}$ if, at each $\omega\varepsilon[0,\infty]$, either

(i) $\underline{\sigma}[I + G(j\omega)K(j\omega)] > \tilde{\delta}_o(\omega)$, or

(ii) $\underline{\sigma}[I + (G(j\omega)K(j\omega))^{-1}] > \delta_o(\omega)$.

It will be clear to the reader that we may use more than two representations of uncertainty in order to define an even more refined set $\{G_p(s)\}$. For example if the perturbations $\Delta_o(s)$, $\tilde{\Delta}_o(s)$, $\Delta_i(s)$, $\tilde{\Delta}_i(s)$, $E(s)$ and $\tilde{E}(s)$ each represent an adequate representation of uncertainty then theorem 3.7 may be extended to include six corresponding inequality tests only one of which need be satisfied for each $\omega\varepsilon[0,\infty]$ for stability to be maintained. It follows that theorem 3.7 and its extensions may be used to reduce conservatism, even in cases where there is no likelihood of pole or zero movement across the imaginary axis.

4. CONCLUSIONS

In this paper we have discussed several representations of uncertainty and examined the adequacy of each. For example, it was shown that an input multiplicative perturbation cannot be used to represent an arbitrary perturbation in $G(s)$ if the plant has more outputs than inputs. We emphasize that a robustness test can only guarantee robustness against uncertainty if the uncertainty is adequately characterized by its representation.

We described three kinds of robustness test:

(i) a direct Nyquist based test,

(ii) an inverse Nyquist based test, and

(iii) a combined direct and inverse Nyquist based test.

The direct test is applicable when the plant and perturbed plant models (representing uncertainty) have the same number of poles in the direct Nyquist D-contour. It can therefore cope with zero movement across the imaginary axis as a result of parameter variations, and the sudden appearance or disappearance of zeros (anywhere in the complex plane) arising from large errors in the nominal model.

The inverse Nyquist based test is applicable when the plant and perturbed plant models have the same number of zeros in the inverse Nyquist D-contour, and can therefore cope with the sudden appearance or disappearance of poles as well as their movement across the imaginary axis.

The problem of simultaneous pole and zero crossings of the imaginary axis is solved by combining a Nyquist based test from section 3.1 with an inverse Nyquist based test from section 3.2. The resulting test is applicable to a large class of transfer function matrices $G_p(s)$ whose frequency responses are known to lie within given bounds. In addition, if any two elements in the class have arbitrarily close frequency responses then they must also have arbitrarily close poles and zeros in the Nyquist D-contour, D.

ACKNOWLEDGEMENT

I. Postlethwaite would like to thank the U.K. Science and Engineering Research Council for financial support.

REFERENCES

1. J. C. Doyle and G. Stein. "Multivariable feedback design: concepts for a classical modern synthesis", IEEE Trans., 26, pp 4-16, 1981.

2. I. Postlethwaite, J. M. Edmunds, and A. G. J. MacFarlane. "Principal gains and principal phases in the analysis of linear multivariable feedback systems", IEEE Trans., 26, pp 32-46, 1981.

3. M. G. Safonov, A. J. Laub, and G. L. Hartmann. "Feedback properties of multivariable systems: the role and use of the return difference matrix", IEEE Trans., 26, pp 47-65, 1981.

4. J. B. Cruz, J. S. Freudenberg, and D. P. Looze, "A relationship between sensitivity and stability of multivariable feedback systems", IEEE Trans., 26, pp 66-74, 1981.

5. N. A. Lehtomaki, N. R. Sandell, and M. Athans. "Robustness results in linear quadratic Gaussian based multivariable control designs", IEEE Trans., 26, pp 75-92, 1981.

6. J. S. Freudenberg, D. P. Looze, and J. B. Cruz. "Robustness

analysis using singular value sensitivities", Int. J. Control, pp 95-116, 1982.

7. B. Kouvaritakis and I. Postlethwaite. "Principal gains and phases: insensitive robustness measures for assessing the closed-loop stability property", Proc. IEE, Pt-D, 129, pp 233-341, 1982.

8. I. Postlethwaite and Y. K. Foo. "Robustness with simultaneous pole and zero movement across the jω-axis", (i) Oxford University Engineering Laboratory (OUEL) report no. 1465/83 (ii) revised, OUEL report no. 1505/83 (iii) presented at IFAC World Congress, Budapest, 1984, and submitted to Automatica.

9. J. C. Doyle. "Analysis of feedback systems with structured uncertainties", Proc. IEE, Pt-D, 129, pp 242-251, 1982.

10. M. G. Safonov and M. Athans. "Gain and phase Margin for Multiloop LQG Regulators", IEEE Trans., 22, pp 173-179, 1977.

11. K. M. Sobel, J. C. Chung, and E. Y. Shapiro. "Application of MIMO phase and gain margins to the evaluation of a flight control system". Proc. ACC., pp 1286-1287, 1983.

12. Y. K. Foo and I. Postlethwaite. "Adequate/inadequate representations of uncertainty and their implications for robustness analysis", Oxford University Engineering Laboratory (OUEL) report no. 1498/83, submitted for publication.

13. S. H. Wang and E. J. Davison. "A Minimization Algorithm for the Design of Linear Multivariable Systems", IEEE Trans, 18, pp 220-225, 1973.

14. G. D. Forney, Jr., "Minimal Bases of Rational Vector Spaces, with Applications to Multivariable Linear Systems", SIAM J. Control, 13, pp 493-520, 1975.

15. T. Kailath, Linear Systems, Prentice-Hall, 1980.

16. M. Vidyasagar, H. Schneider and B. A. Francis. "Algebraic and topological aspects of feedback stabilization", IEEE Trans., AC-27, pp 880 - 894, 1982.

CHAPTER 9

ADDITIVE, MULTIPLICATIVE PERTURBATIONS AND THE APPLICATION OF THE CHARACTERISTIC LOCUS METHOD

R.W. Daniel and B. Kouvaritakis
University of Oxford
Oxford, England

ABSTRACT The Characteristic Locus method gives necessary and sufficient stability conditions for systems with an exact description. In the presence of model uncertainty however, due to eigenvalue sensitivity problems, the Characteristic Loci of the nominal description may lead to an unreliable assessment of stability. Eigenvalue inclusion results developed recently can be used to construct bands which contain the Characteristic Loci of additively perturbed models and thus lead to an extension of the Generalised Nyquist criterion. The present paper gives a brief description of the relevant methods, extends these to the case of multiplicative perturbations, and applies them to an open-loop unstable model in order to appraise the robustness properties of two alternative control schemes.

INTRODUCTION

Nyquist diagrams combine gain and phase characteristics in one single plot and thus provide a convenient graphical means of conveying vital information concerning feedback systems. The popularity of the associated stability criterion for the case of single-input single-output systems is hardly surprising. Recent research work on the Inverse Nyquist Array(1) furnished an extension of the Nyquist approach to the multivariable case, whereas the Characteristic Locus method(2,3) led to the generalisation of the Nyquist criterion to systems with many inputs and many outputs. Despite their successful application to a variety of industrial problems, both these methods require an exact model description. Given however that one of the main reasons for the use of feedback is the presence of uncertainty in plant models, more often than not, it is not sensible to assume that plants are described exactly by a nominal model, such as, say, a nominal transfer function matrix $G_o(s)$. A far more realistic representation arises out of the assumption that $G_p(s)$, the plant transfer function matrix, is generated by $G_o(s)$ according to the additive perturbation equation $G_p=G_o+\Delta$ or the multiplicative perturbation equation $G_p=(I+D)G_o$, where $\Delta(s)$, $D(s)$ are largely unknown but are bounded such that $\bar{\sigma}(\Delta(s))\leq\delta(s)$ or $\bar{\sigma}(D(s))\leq d(s)$, where $\bar{\sigma}(M)$

S. G. Tzafestas (ed.), Multivariable Control, 161–178.

denotes the maximim singular value of M.

Recent work[4], based on singular values, has given consideration to the feedback properties of systems in the presence of uncertainty and has culminated in appropriate stability criteria. These however are developed around a single gain plot (against frequency) and as such convey little information about the properties of multivariable plants. Furthermore, they do not have a Nyquist interpretation and thus do not provide the natural medium for the extension of the classical frequency response design techniques.

In an attempt to overcome these difficulties, a new approach[5,6] considers the problem of constructing useful inclusion regions for the eigenvalues of additively perturbed matrices. Such regions applied to the frequency response of $G_o(s)$ result in a stability criterion which, unlike the singular values results, has a Nyquist interpretation and which unlike the Characteristic Locus method does not suffer from the eigenvalue sensitivity problem. Furthermore, the new approach enables the simultaneous assessment of tolerance to additive uncertainty as well as the gain/phase margins, leads to a systematic design procedure and facilitates the use of information on the structure of additive perturbations.

The purpose of the present paper is to give a brief account of the eigenvalue inclusion method, to extend this to the case of multiplicative perturbations and to demonstrate its application to the model of an open-loop unstable chemical reactor.

EIGENVALUE INCLUSION REGIONS

Consider first the case of "known" plants where $G_p(s)=G_o(s)$. Then it is possible to plot the eigenvalues of $G_o(s)$ (and hence of G_p) for $s=jw$, $0\leqslant w<\infty$ to form the Characteristic Loci[2] of G_o, denoted by $\lambda_i(G_o)$ and to subsequently postulate necessary and sufficient conditions for stability under unity feed-back; namely that the net sum of counter-clockwise encirlcements of the critical point must be equal to the number of open-loop unstable poles.

In the case of "uncertain" plants, G_p is no longer equal to $G_o(s)$ and we require the Characteristic Loci $\lambda_i(G_p)$of $G_p(s)$, not of G_o, to give the correct number of critical point encirclements; yet because of the uncertainty $\Delta(s)$ the $\lambda_i(G_p)$ are unknown. Furthermore, the $\lambda_i(G_o)$ may differ substantially from the $\lambda_i(G_p)$ and as such they cannot be used for the analysis of stability, even when $\Delta(s)$ is small. To illustrate this for the case of additive perturbations consider G_p to be given as

$$G_p(s) = G_o(s) + \Delta(s) \tag{1}$$

take the nominal model

$$G_o(s) = \frac{1}{s+1}\begin{bmatrix} 1 & 100 \\ 0 & 3 \end{bmatrix} \tag{2}$$

and let $\Delta(s)$ be a small perturbation such that $\bar{\sigma}(\Delta) \leqslant 0.09/|s+1|$. The characteristic Loci of G_o, $\lambda_i(G_o)$, are $1/(s+1)$, $3/(s+1)$ and predict infinite gain margins; yet a $\Delta(s)$

$$\Delta(s) = \frac{1}{s+1} \begin{bmatrix} 0 & 0 \\ 0.09 & 0 \end{bmatrix} \tag{3}$$

results in a G_p with Characteristic Loci $-1.16/(s+1)$, $5.16/(s+1)$ which of course predict instability for a feedback gain as small as 1.

This gross descrepancy between $\lambda_i(G_p)$ and $\lambda_i(G_o)$ can only occur when $G_o(s)$ is not normal, namely when $G_o^*(s)G_o(s) \neq G_o(s)G_o^*(s)$ where $(.)^*$ denotes transposition and complex conjugation. If $G_o(s)$ however is normal, then it can be shown that[7]

$$|\lambda_i(G_p(s)) - \lambda_i(G_o(s))| \leqslant \delta(s) \tag{4}$$

Thus if G_o is not normal the $\lambda_i(G_o)$ are poor indicators of stability and it is advantageous to replace $G_o(jw)$ by it nearest normal approximation $G_N(jw)$ according to the equation

$$G_o(jw) = G_N(jw) + E(jw) \tag{5}$$

where $E(jw)$ is the error of approximation with $\bar{\sigma}(E) = \varepsilon$. This combines with eqn. (1) to give

$$G_p(jw) = G_N(jw) + [E(jw) + \Delta(jw)] \tag{6}$$

It is then possible to substitute $G_o(jw)$ in eqn. (4) by $G_N(jw)$, so long as $\delta(jw)$ is increased to $\delta(jw)+\varepsilon(jw)$ in order to take into account the error of approximation. Thus we have

$$|\lambda_i(G_p) - \lambda_i(G_N)| < \delta(jw) + \varepsilon(jw) \tag{7}$$

or in words[6]:

Result 1. The eigenvalues of $G_p(jw)$ lie inside the union of discs $C[\lambda_i(G_N), \delta+\varepsilon]$ centred at the eigenvalues of $G_N(jw)$ and of radius $\delta(jw)+\varepsilon(jw)$.

The eigenvalues of $G_p(jw)$ however are also given by

$$\lambda_i(G_p) = \frac{w_i^* G_p w_i}{w_i^* w_i} = \frac{w_i^* G_o w_i}{w_i^* w_i} + \frac{w_i^* \Delta w_i}{w_i^* w_i} = z + \delta z \tag{8}$$

for w_i an eigenvector of G_p, where $z = w_i^* G_o w_i / w_i^* w_i$ is a point in the Numerical Range of G_o, $\mathcal{N}(G_o)$, and δz is always less than δ in modulus. Thus[6]:

Result 2. The eigenvalues of $G_p(jw)$ lie inside $\mathcal{N}(G_o, \delta)$, the envelope of circles centred at all the boundary points of $\mathcal{N}(G_o)$ and of radius $\delta(jw)$.

Clearly Results 1 and 2 combine to give a convenient eigenvalue inclusion result[6]:

Result 3. The eigenvalues of $G_p(jw)$ lie in the intersection of the union of the discs $C[\lambda_i(G_N),\delta+\varepsilon]$ of Result 2 and the regions $\mathcal{N}(G_o,\delta)$ of Result 3.

Fig. 1 gives an illustration of the type of regions one gets for a 3x3 example with

$$G_o(jw) = \begin{bmatrix} 1-j3 & 3+j6 & 1+j \\ 4-j8 & -1-j4 & -0.5-j0.5 \\ 1-j2 & 1+j0.5 & 0.5-j \end{bmatrix} \tag{9}$$

The circles of Result 1 are shown in solid lines for $\delta=0$ and in dotted lines for $\delta=0.5$. The remaining solid and dotted closed curves shown in the same diagram correspond to $\mathcal{N}(G_o)$ and $\mathcal{N}(G_o,\delta)$. The shaded regions of the Figure are the eigenvalue inclusion regions of Result 3.

In order to make the regions of Result 3 as tight as possible, it is necessary to choose G_N to be the nearest normal approximation of G_o in the sense that it minimises $\bar{\sigma}(E)=\varepsilon$. It is easy to show that

$$\varepsilon \geq \left|\frac{x^*Ex}{x^*x}\right| = \left|\frac{x^*G_o x}{x^*x} - \frac{x^*G_N x}{x^*x}\right| = d(x) \tag{10}$$

where $x^*G_ox/x^*x = z_1(x)$, $x^*G_Nx/x^*x = z_2(x)$ are two points on $\mathcal{N}(G_o)$ and $\mathcal{N}(G_N)$, respectively. The problem thus becomes how to choose G_N and $\mathcal{N}(G_N)$ so as to minimise the supremum, over x, of the distance d(x) between the points $z_1(x)$ and $z_2(x)$. For the 2x2 case $\mathcal{N}(G_o)$ is an ellipse and $\mathcal{N}(G_N)$ is a straight line. Intuitively it seems obvious that $\mathcal{N}(G_N)$ must be aligned with the major axis of the ellipse $\mathcal{N}(G_o)$ when ε becomes equal to the semi-minor axis of the ellipse (Fig. 2). In fact this turns out to be the best solution for the 2x2 case and is given by

$$G_N = \tilde{A} + \tfrac{1}{2}(\tilde{b}_1+\tilde{b}_m)I, \text{ with } \quad \varepsilon=\tfrac{1}{2}(\tilde{b}_m-\tilde{b}_1) \tag{11}$$

where

$$\tilde{A} = \tfrac{1}{2}[\exp(j\theta)G_o+\exp(-j\theta)G_o^*] \tag{12}$$

and $\tilde{b}_1$, $\tilde{b}_m$ denote the minimum and maximum eigenvalues of

$$\tilde{B} = -j\tfrac{1}{2}[\exp(j\theta)G_o-\exp(-j\theta)G_o^*] \tag{13}$$

note that $\tilde{b}_1$ and $\tilde{b}_m$ together with the corresponding eigenvalues of $\tilde{A}$, $\tilde{a}_1$ and $\tilde{a}_m$ define the corners of the rectangle of Fig. 2 which is known to contain and be tangential to $\mathcal{N}(G_o)$; θ denotes the orientation of the

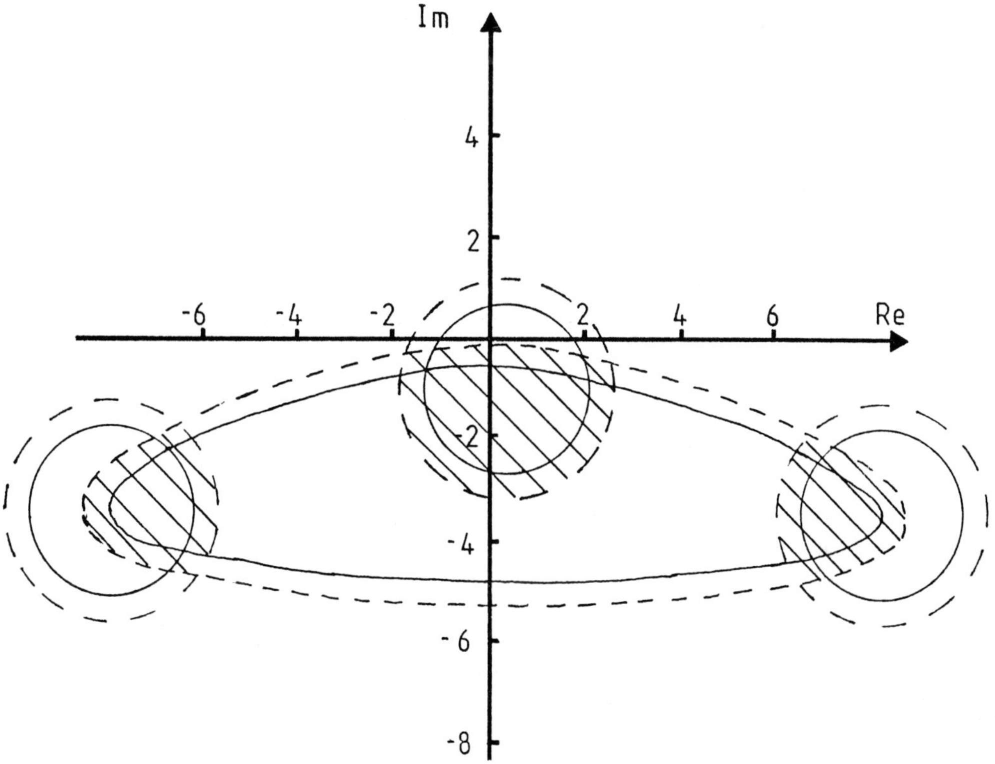

Fig. 1 Eigenvalue Inclusion Regions for $G_o(j\omega)$ of equ. 9

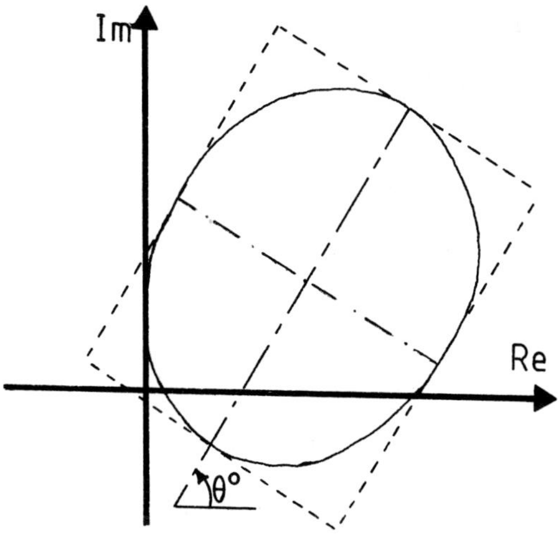

Fig. 2 The Numerical Range of a 2x2 matrix

rectangle with respect to the real axis. Clearly $\varepsilon=\frac{1}{2}(\tilde{b}_m-\tilde{b}_1)$ is one half of the smallest side of the rectangle and thus the best approximation G_N will be obtained for the θ which corresponds to the smallest rectangle that contains and is tangential to $\mathcal{N}(G_o)$.

The same result (eqn. 11) can be used for the mxm case, but the relevant G_N will not be optimum. Indeed, a simple modification of G_N will produce a further reduction in the error ε.(6)

STABILITY AND DESIGN

The necessary and sufficient condition for closed-loop stability is that the Characteristic Loci of $G_p(s)$, $\lambda_i(G_p(jw))$, give the correct number of critical point encirclements. Because of the presence of uncertainty, the $\lambda_i(G_p(s))$ are not known. However by Section 2, it is known that at each frequency $s=jw$, $\lambda_i(G_p(jw))$ will lie in the regions of Result 3 and hence the Characteristic Loci of $G_p(s)$ will lie inside the bands swept by the regions, when these are computed and plotted for all w, $0<w<\infty$. It follows therefore that(6):

Result4. Closed-loop stability in the presence of uncertainty is ensured if the bands formed by the regions of Result 3 when these are plotted for $0<w<\infty$, give a sum of counterclockwise encirclements of the critical point equal to the number of open-loop unstable poles.

A similar result may be stated for the case of a compensated uncertain plant:

$$Q_p' = G_pK = Q_o' + \Delta' = G_oK + \Delta K \tag{14}$$

where K is assumed to be a pre-compensator. However in this form, K is seen to multiply the uncertainty and thus the approach would lead to conservative and complicated results. Instead it is possible to use a loop transformation in order to replace Q_p' by

$$Q_p = Q_o + \Delta = G_o + K^{-1} - I + \Delta \tag{15}$$

where now the uncertainty Δ on the compensated Q_o is the same as for G_o. This way, Result 4 can be repeated to give stability conditions for the compensated uncertain system; the only difference here being that the regions of Result 3 must be plotted for $Q_o=G_o+K^{-1}-I$ and not G_o.

The need for compensation arises out of poor stability margins as demonstrated by the proximity of the eigenvalue inclusion regions to the critical point. This could be due to either large regions when G_o is far from normal or due to inherently poor gain/phase margins. The former calls for the design of a K which makes $Q_o=G_o+K^{-1}-I$ normal and the latter calls for the use of a controller which effects "robust gain /phase sharing". For simplicity, real constant controllers which achieve the above aims at one frequency only are considered here; continuity arguments imply that the results will hold good over a band of

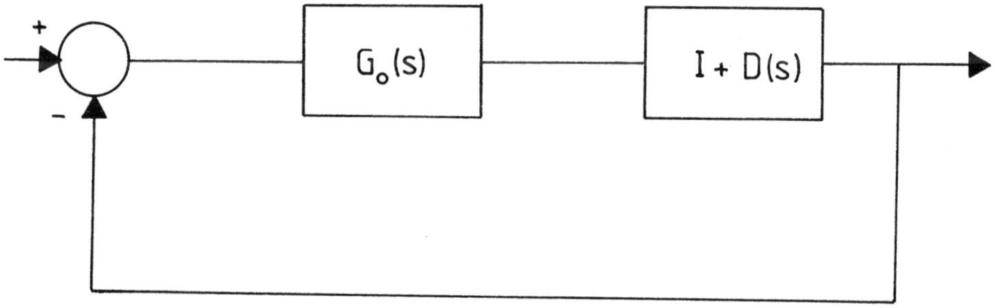

Fig. 3a Feedback Configuration for the Multiplicat ve Case

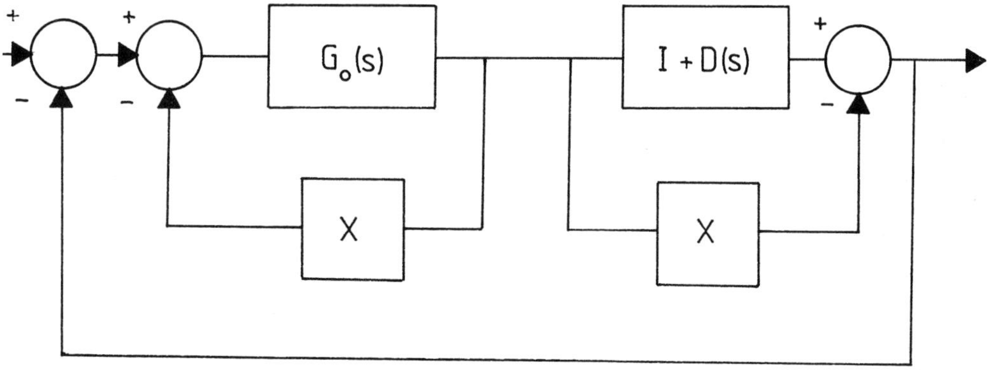

Fig. 3b The Loop Transformation

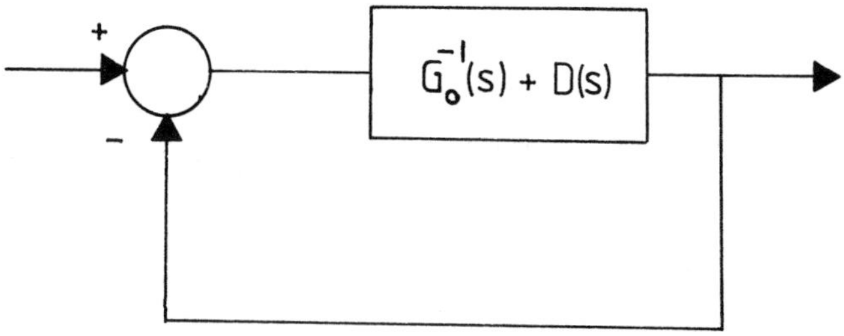

Fig. 3c. The Resulting Feedback Configuration in Additive Form

frequencies. If compensation is required over more than one band of frequencies then several real controllers must be designed and must be integrated together by means of dynamics.

The general philosophy behind the design is to choose K to adjust the matrices $\tilde{A}_Q=\frac{1}{2}(\exp(j\theta)Q_o+\exp(-j\theta)Q_o^*)$ and $\tilde{B}_Q=-j\frac{1}{2}(\exp(j\theta)Q_o-\exp(-j\theta)Q_o^*)$, with Q_o as defined by eqn. 11, in order to manipulate the rectangles of Fig. 2 for an appropriate angle θ. Normality can be attained by "squashing" one of the two dimensions of the rectangle; so long as θ is chosen to be different from 0 to $\pi/2$, there will exist enough degrees of design freedom to effect this operation exactly. Robust gain/phase sharing can be achieved by minimising the spread of $\mathcal{N}(G_o)$ in the two orthogonal directions of the sides of the rectangle of Fig. 2 for a choice of θ which would normally be dictated by the particular example under study; this leads to a least squares problem which admits a solution of closed form based on eigenvalue/eigenvector calculations. Suitable illustrations of the two types of design can be found in Reference 6.

THE MULTIPLICATIVE CASE

Consider now the case where the plant description is given in terms of a transfer function matrix $G_p(s)$ which is related to a nominal model $G_o(s)$ according to the equation

$$G_p(s) = [I+D(s)]G_o(s) \tag{16}$$

where D(s) is bounded with $\bar{\sigma}(D(s))\leqslant d(s)$ but is otherwise assumed to be unstructured. The system is embedded in a unity feedback configuration as shown in Fig. 3a and the problem of stability is raised once again. As before, for d(s)=0 the Characteristic Loci of $G_o(s)$ provide necessary and sufficient stability conditions but the presence of a non-zero d(s) implies that the Characteristic Loci of $G_p(s)$ may differ a good deal from those of $G_o(s)$ and thus can no longer be used to assess stability.

In this section we show that it is possible to convert the multiplicatively perturbed problem to an equivalent additively perturbed problem. This can be achieved by the application to the feedback system of Fig.3a of a stability preserving transformation of the kind depicted in Fig.3b. Such a transformation therefore would enable the introduction of the techniques of the previous section to the multiplicatively perturbed case. Thus consider Fig.3b and choose the operator X such that $G_o(s)$ with the feedback path around it reduces to the identity operator, namely choose X such that

$$G_o(s)\ [I+XG_o(s)]^{-1} = I \tag{17}$$

or

$$X = I - G_o^{-1}(s) \tag{18}$$

Then the multiplicative perturbation term I+D(s) with the feedforward term X around it becomes

$$Q_p = I + D(s) - X = G_o^{-1}(s) + D(s) \tag{19}$$

Thus, the proposed transformation converts the multiplicatively perturbed feedback configuration of Fig. 3a to the additively perturbed configuration of Fig. 3c. Furthermore, the stability properties of the two configurations are equivalent providing the $G_o(s)$ is both stable and minimum phase; note that the theorem below obviates the need for any assumptions concerning the locations of the poles/zeros of $G_o(s)$.

The loop transformation above indicates the type of result that can be derived and the alternative approach, described below, obtains the result in its most general form:

Theorem 1. Let n_z denote the number of right half plane zeros of $G_o(s)$ and let n_o, n_p denote the right half plane poles of $G_o(s)$, $(I+D(s))G_o(s)$ respectively. Then a necessary and sufficient conditions for the stability of the additively perturbed feedback system of Fig. 3a is that the Characteristic Loci of $G_o^{-1}(s)+D(s)$ give a net sum of counterclockwise encirclements of the critical point equal to $n_o-n_p-n_z$.

Proof: Let $p_c(s)$ and $p_o(s)$ denote the closed-loop and open-loop pole polynomials for the system of Fig. 3a and consider the identity

$$\det\,[I+(I+D(s))G_o(s)] = \frac{p_c(s)}{p_o(s)} \tag{20}$$

Then, providing that $G_o(s)$ is not identically singular, we may also write

$$\det\,[I+G_o^{-1}(s)+D(s)] = \frac{p_c(s)}{p_o(s)\det(G_o(s))} \tag{21}$$

Now let s describe the Nyquist contour once clockwise, so that by the Principal of the Argument the determinant above will encircle the origin, or equivalently the Characteristic Loci of $G_o^{-1}(s)+D(s)$ in the net sum will encircle the critical point, in a clockwise direction a number of times equal to

$$n = n_c - n_p - n_z + n_o \tag{22}$$

But for stability, n_c must be zero and this will hold true if and only if the number of clockwise encirclements n, or counterclockwise encirclements -n is

$$-n = n_p + n_z - n_o \tag{23}$$

This completes the proof.

For the case of a stable $D(s)$, the unstable poles of $G_p(s) = (I+D(s))G_o(s)$ will be precisely those of $G_o(s)$ so that $n_p=n_o$ and thus:

Corollary 1. Let $D(s)$ be stable and let n_z denote the number of right half plane zeros of $G_o(s)$. Then a necessary and sufficient condition for the stability of the additively perturbed feedback system of Fig. 3a is that the Characteristic Loci of $G_o^{-1}(s)+D(s)$ give a net sum of counterclockwise encirclements of the critical point equal to n_z.

Clearly $D(s)$ is not known and therefore $G_o^{-1}(s)+D(s)$ and its Characteristic Loci will also be unknown and hence Theorem 1 may not be applied directly. However since $G_o^{-1}(s)+D(s)$ is in an additive form, Result 3 may be applied to $G_o^{-1}(s)$ and $D(s)$, instead of $G_o(s)$ and $\Delta(s)$, in order to yield:

Theorem 2. Let $\hat{G}_N$, $\hat{\varepsilon}$ denote a normal approximation to G_o^{-1} and the maximum singular value of the corresponding error, respectively and consider the bands swept by the intersection of $C[\lambda_i(\hat{G}_N), d+\hat{\varepsilon}]$ and $\mathcal{N}(G_o^{-1},d)$ for $0 \leqslant w < \infty$. Then the closed-loop stability of $G_o(s)$ in the presence of the multiplicative perturbation $I+D(s)$ is ensured if these bands give a net sum of counterclockwise encirclements of the critical point equal to $n_o-n_p-n_z$; n_o, n_p and n_z are as defined in Theorem 1.

It is possible to use Theorem 2 as the basis of a design procedure. However, unlike the additive case where a compensator K had the effect of replacing G_o by $G_o+K^{-1}-I$, a similar controller here would replace G_o^{-1} by $K^{-1}G_o^{-1}$. Thus K^{-1} acts multiplicatively on G_o^{-1} (rather than additively on G_o) and as a result the design procedure is more complicated and will be omitted.

A ROBUST STABILITY STUDY

The results discussed in this paper can be used to appraise the effectiveness of feedback control schemes devised by standard multivariable design techniques such as the Inverse Nyquist Array method(1) and the Characteristic Locus method(2,3).

An example which has been studied by both these methods(8,3) is the model of a chemical reactor described in terms of the nominal transfer function matrix:

$$G_o(s) = \frac{1}{d(s)} \begin{bmatrix} 29.2s+263.3 & -3.146s^3-32.62s^2-89.83s-31.81 \\ 5.679s^3+42.67s^2-68.84s-106.8 & 9.43s+15.15 \end{bmatrix} \tag{24}$$

where

$$d(s) = s^4+11.67s^3+15.75s^2-88.31s+5.154$$

The compensated system is neither open-loop nor closed-loop diagonally dominant and thus the Inverse Nyquist Array method cannot be used to asses its stability margins. The plots of the Characteristic Loci however predict that the system, which is open-loop unstable, becomes stable in the closed-loop if a scalar gain $k>1.5$ is used in each of the feedback loops; the phase margins are assessed to be poor[3]. This describes the situation when $G_p(s)$ is known to be given exactly by the nominal transfer function matrix $G_o(s)$ above.

For the case of additive uncertainty in the model, the plots of Fig. 4a show the bands formed by the eigenvalue inclusion regions and Fig. 4b gives a more detailed description of the situation at high frequencies. The disparity of the Characteristic Gains at low frequencies and the poor eigenvector span (indicated by the large Misalgnment Angles) at high frequencies - both of which are demonstrated by the diagram of Fig. 5 - cause the eigenvalue inclusion regions to become large for most frequencies; as a result the bands of Result 4 include the part of the real axis over which it would have been possible to place the critical point in order to obtain the prerequisite number of critical point encirclements. The need for compensation is therefore obvious.

The controllers proposed by the Inverse Nyquist Array method[8] and the Characteristic Locus method[3] are:

$$K_{INA}(s) = \begin{bmatrix} 0 & 10(s+1)/s \\ -10(s+1)/s & 0 \end{bmatrix} \qquad (25a)$$

$$K_{CL}(s) = \begin{bmatrix} 0 & (10s+9)/s \\ -(10s+21)/s & -8/s \end{bmatrix} \qquad (25b)$$

and produce similar results. This is demonstrated by the bands of Result 4 plotted for $1 \leq w \leq 100$ for $K_{INA}(s)$ (Fig. 6a) and $K_{CL}(s)$ (Fig. 6b). For both controllers, over the frequency range, the regions stay well away from the $-1+j0$ point. However due to large misalignment angles at low frequencies, the region grow larger as $w \to 0$ and at such frequencies $K_{CL}(s)$ gives slightly better stability margins; for this reason the analysis that follows concentrates on $K_{CL}(s)$ alone. Fig. 7 shows the regions of $G(s)K_{CL}(s)$ for low frequencies and the $-1+j0$ point can be seen to be included. A close look of the problem at $w=0.1$ rad/sec produces the regions shown in Fig. 8. Consideration of the controller gains at $w=0.1$ implies that a perturbation on $G(s)K_{CL}(s)$ with $\delta(j0.1)=10$ corresponds approximately to a 0.5%-5% perturbation on $G(s)$, depending on which elements of $G(s)$ are being perturbed. Thus a change of less than 5% on the plant description will cause the inclusion regions to enclose the $-1+j0$ point and will therefore violate the stability requirement of Result 4.

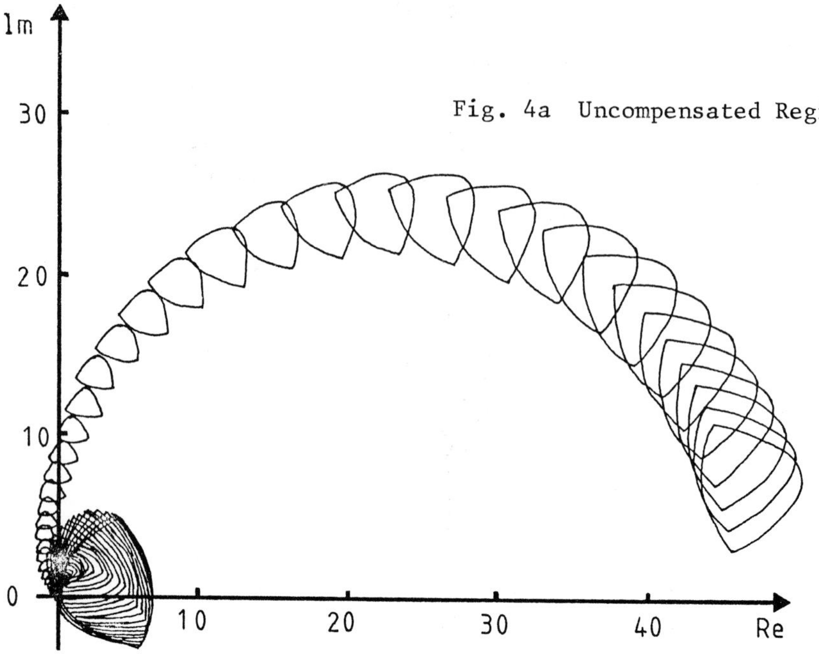

Fig. 4a Uncompensated Regions

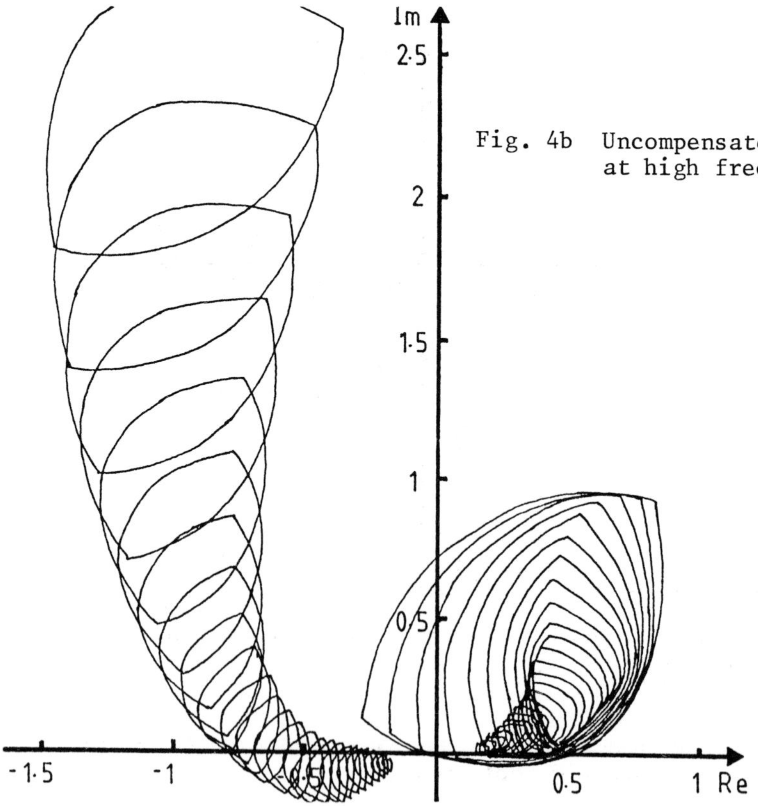

Fig. 4b Uncompensated Regions at high frequencies

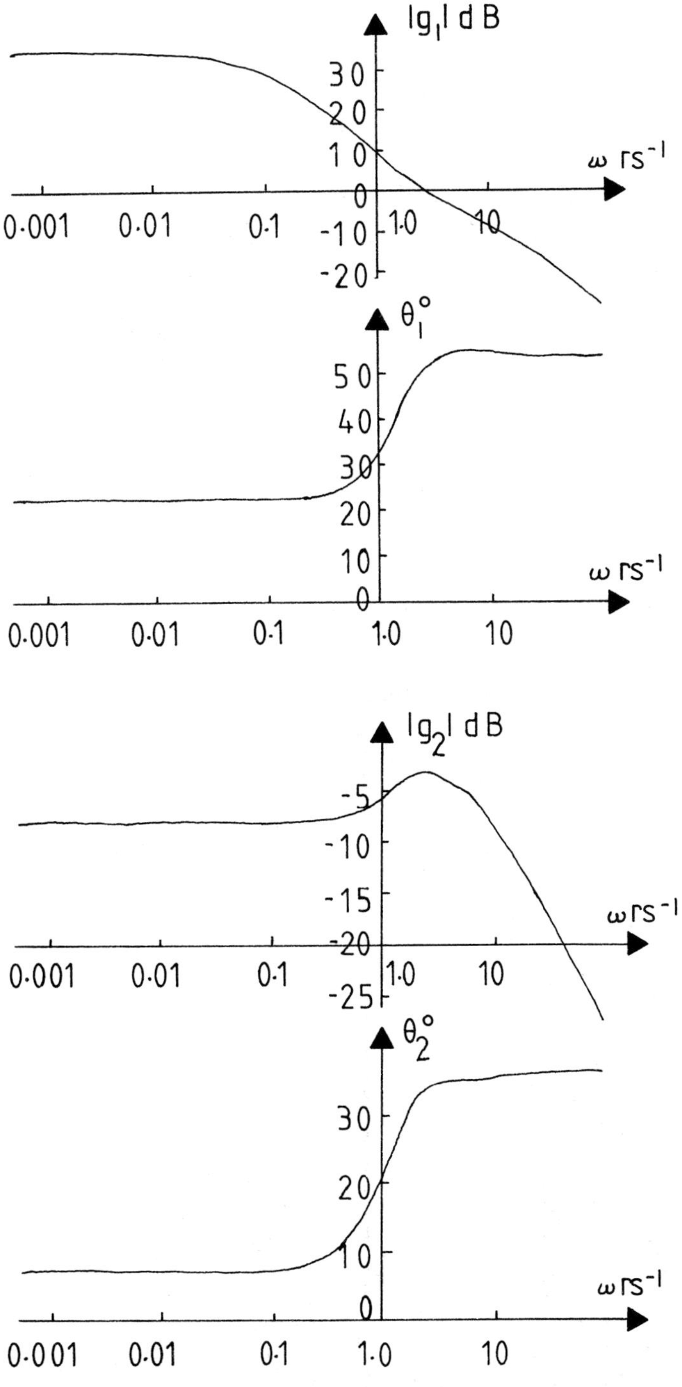

Fig. 5a,b The moduli of the uncompensated Characteristic Loci and Misalignment Angles

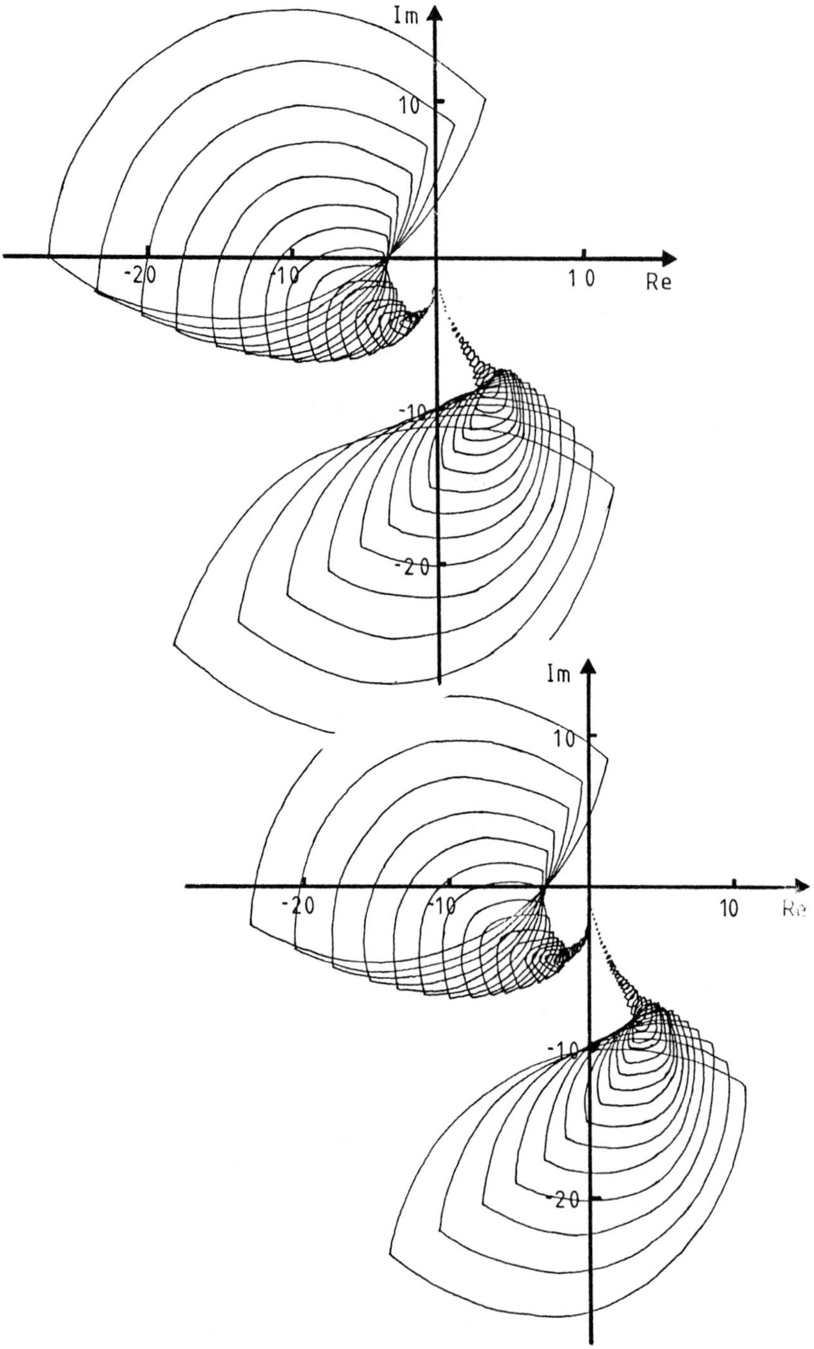

Fig. 6a,b The regions of GK for the CL and the INA controllers

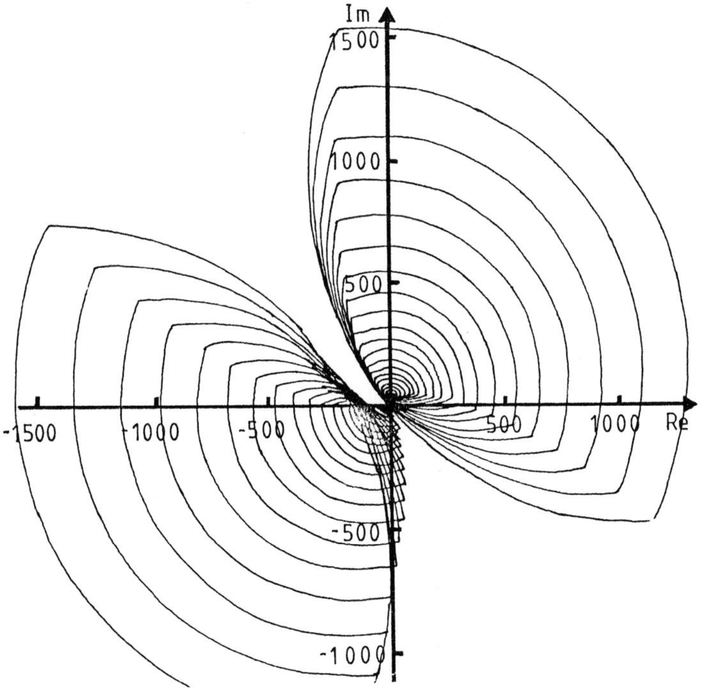

Fig. 7 The regions for GK_{CL} at low frequencies

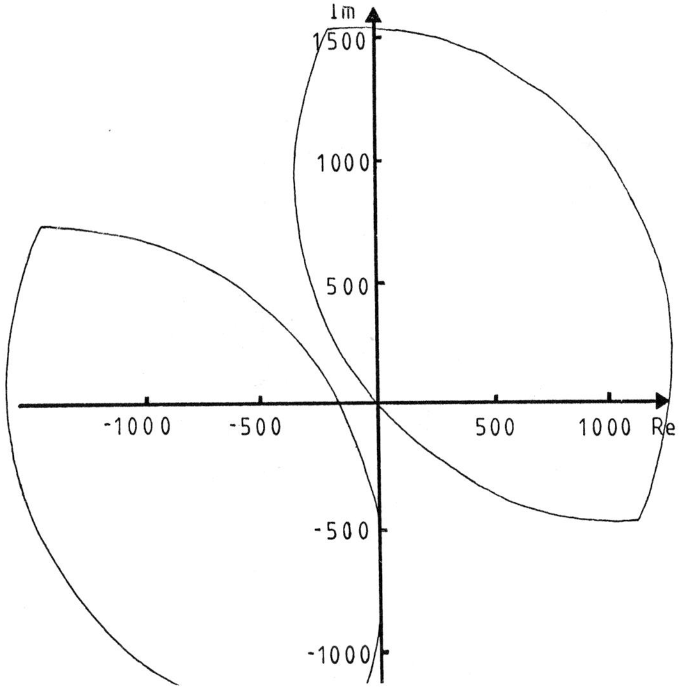

Fig. 8 The regions for GK_{CL} at w=0.1 rad/sec

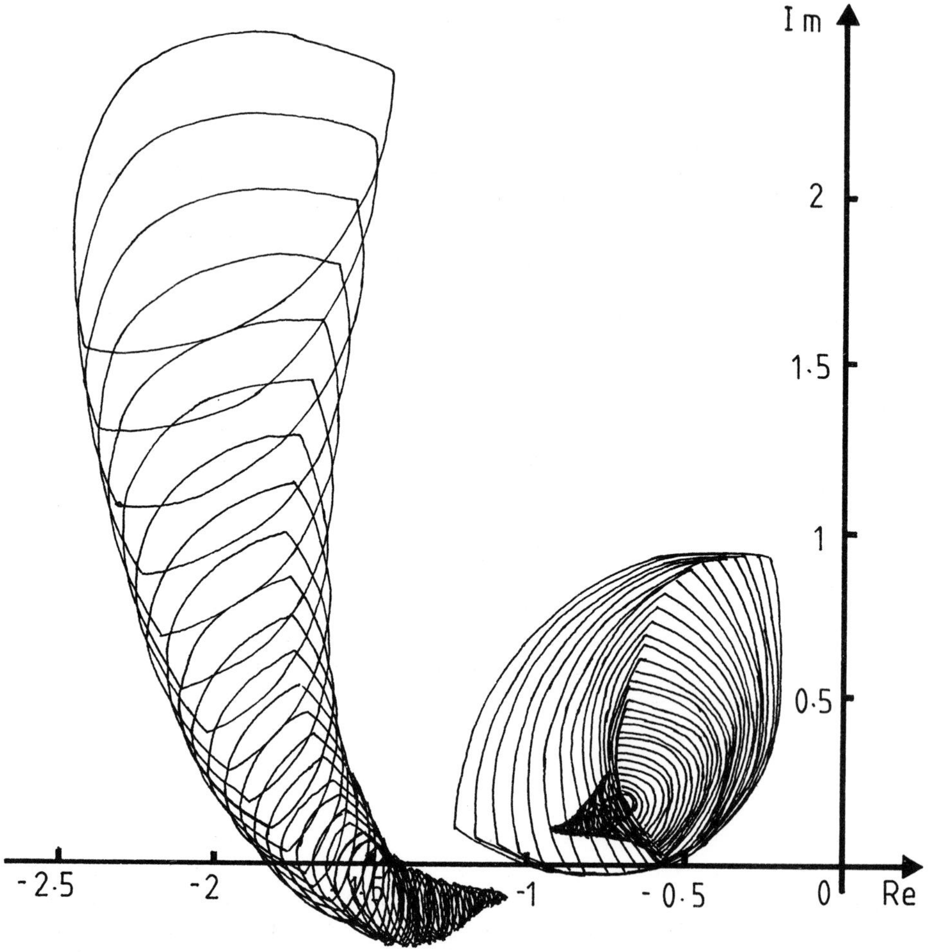

Fig. 9 The regions for $G+K_{CL}^{-1}-I$ for 1<w<100 rad/sec

Thus far we have looked at the inclusion regions for $G_p(s)K(s)$, which under uncertainty is as given earlier. Thus the perturbation on the compensated nominal system $G_o(s)K(s)$, is the product of the controller with $\Delta(s)$; this is clearly inconvenient. The identity

$$\det\,(I+G_p(s)K(s) = \det(G_p(s)+K^{-1}(s))\det(K(s)) = \frac{p_c(s)}{p_o(s)} \qquad (26a)$$

or

$$\det(I+G_p(s)+K^{-1}(s)-I) = \frac{p_c(s)}{\det(K(s))p_o(s)} \qquad (26b)$$

where $p_c(s)$, $p_o(s)$ denote the closed-loop and open-loop pole polynomials, provides an alternative, and possibly more convenient way of assessing the effect of $\Delta(s)$ on the stability properties of the closed-loop system; identity 26 is consistent with the loop transformation used earlier in order to derive eqn. 15. Clearly for an open-loop stable, minimum-phase controller $K(s)$, eqn. 15 implies that the closed-loop system will be stable if and only if the eigenvalues of $G_o(s)+\Delta(s)+K^{-1}(s)-I$ give a net sum of critical point encirclements (in a counterclockwise sense) equal to the number of open-loop unstable poles; 2 in the case of the model of the chemical reactor of eqn. 24. Thus instead of looking at the eigenvalue inclusion regions for $G_o(s)K(s)$ with perturbation $\Delta(s)$ $K(s)$, we can examine the eigenvalue inclusion regions of $G_o(s)+K^{-1}(s)-I$ with perturbation $\Delta(s)$.

Fig. 9 shows the bands of Result 4 for $G_o(s)+K_{CL}^{-1}(s)-I$ at w, $1\leq w\leq 100$; because of the integral action in the controller, at low frequencies $K_{CL}^{-1}(s)$ is small and the regions of $G(s)+K_{CL}^{-1}(s)-I$ tend to those of $G(s)$ shown in Fig. 4a when the latter are translated by -1. Once again the stability margins are found to be very poor; a relatively small perturbation Δ on G_o, will result in regions which enclose the $-1+j0$ point.

CONCLUSIONS

The analysis discussed in this paper can be applied to appraise the robustness of control schemes. The particular case examined was that of an open-loop unstable chemical reactor and it was seen that if the model $G_p(s)$ of the reactor is known exactly, then either the Characteristic Locus or the Inverse Nyquist Array method produce controllers which secure adequate stability margins. However if $G_p(s)$ is subject to some uncertainty, then the analysis of the respective control schemes based on the eigenvalue inclusion regions, shows that the stability margins deteriorate rapidly with increasing levels of uncertainty. If uncertainty were a consideration in this problem therefore, the Characteristic Locus method and the Inverse Nyquist Array method controllers would be unsatisfactory. A modification of the design procedure along the lines of Ref. [6] may well result in an acceptable feedback control configuration. This however forms the topic for future research.

REFERENCES

1. Rosenbrock, H.H: "Computer-Aided Control System Design", 1974, London: Academic Press

2. MacFarlane, A.G.J., and Postlethwaite, I.: "The generalised Nyquist stability criterion and multivariable root loci", 1977, Int. J. Control, 26, 81

3. MacFarlane, A.G.J., and Kouvaritakis, B.: "A design technique for linear multivariable feedback systems", 1977, Int. J. Control, 25, 837

4. Doyle, J.C., and Stein, G.: "Multivariable feedback design: Concepts for a classical/modern synthesis", 1981, AC-26, 4

5. Daniel, R.W., and Kouvaritakis, B.: "Normal Matrix Approximations And Their Use In The Assessment Of Stability In The Presence Of Uncertainty", 1983, Proceedings of the ACC, San Francisco, California June 22-24

6. Daniel, R.W., and Kouvaritakis, B.: "Analysis and design of linear multivariable feedback systems in the presence of additive perturbations", 1984, Int. J. Control, to appear

7. Wilkinson, J.H.: "The algebraic eigenvalue problem", 1965, Oxford University Press

8. Munro, N.: "Design of controllers for open-loop unstable multivariable system using inverse Nyquist array", 1972, Proc. IEE, 119, 1377

ACKNOWLEDGEMENTS

The authors wish to thank the CEGB and SERC for financial support.

CHAPTER 10

A DESIGN TECHNIQUE FOR MULTI-REPRESENTED LINEAR MULTI-VARIABLE DISCRETE-TIME SYSTEMS USING DIAGONAL OR FULL DYNAMIC COMPENSATORS

Nicos M. Christodoulakis
Control Systems Division
Engineering Department
University of Cambridge
England CB2 1RX

ABSTRACT

The aim of the design method is to find a dynamic compensator that gives a "basically non-interactive" closed-loop performance for more than one model, representing different versions of the same plant. In a decoupled closed-loop behaviour outputs respond with small position offset to the corresponding input demand and with small interaction to the others. The idea is to construct a nominal diagonal model, which does not necessarily belong to the family of models, and then to consider the given members as deviations around it. If the sought precompensator is specified to have a diagonal structure, the design can be split into a number of independent scalar loop designs which are easily carried out. An extension to include precompensators with a full structure is also discussed.

1. INTRODUCTION

The assumption of a specific transfer function description of a plant avails the use of very effective design techniques in order to derive precompensators that achieve stability and good performance of the controlled system. In many practical problems however, to assume exact plant description is a rather special case. Deviations from a given nominal description are likely to occur, due either to variation of parameters or any other kind of structural uncertainties that exist in the system. The central aspects of stability and performance must then be examined in the way that controlled systems stand such uncertainties and in the degree of robustness they show. Regarding stability of uncertain systems, the designer usually compares a spectral norm of the maximum deviations occurring with the robustness measure (see [SAF], [DOYL]), or applies the extended concepts of gain and phase margin (see [POST]) to check whether the perturbed system remains closed-loop stable. Maintaining stability is an essential requirement for a designed controller, but still not sufficient to guarantee that some quality of performance is also preserved. To deal with this problem,

S. G. Tzafestas (ed.), Multivariable Control, 179–196.

there have developed ([HOR1]) design techniques in which specifications of performance are considered in such a way as to be kept under any plant description when it is controlled by the derived common feedback structure. Following this approach, a method of finding an optimal - in the minimum square error sense - nominal description for single-input single-output systems has also been developed ([EAST]) in such a way as to minimise the maximum deviation between the nominal and any perturbed version of the open-loop transfer function. What is further considered as a desired closed-loop performance is the familiar specification of a stable system that effectively tracks the reference input and shows low interactions. Under these considerations, the time-domain bounds are nicely translated into frequency-domain constraints that have tp be incorporated in the design procedure.

Reaping the existing results, the present paper describes how similar methods can be derived for multivariable systems as to preserve closed-loop stability and maintain a certain quality of performance despite uncertainties.

Systems discussed here are assumed deterministic, in the discrete-time, and square with m inputs and outputs. Instead of the familiar situation of a single parameter uncertainty, $\alpha \in A$, we adopt here the pattern of a plant being multi-represented by a family of transfer functions, $\{P_n(z),\ n = 1,\dots,r\} \in \mathbb{R}_{m\times m}(z)$. The latter way of characterising uncertainty is obviously equivalent to the former when A is accounted point-wise, and can easily cope with cases of multiple parameter variations or other kinds of structural uncertainties.

The aim of the design is to derive a dynamic controller that satisfactorily serves all the members of the family, in the sense of stability and de-coupled closed-loop performance. The main idea is the construction of a nominal multi-variable model, by minimising the maximum deviations of any member as they are measured by the spectral norms of a robustness measure. The algorithm generates an optimal nominal model with diagonal structure, so that design is split into m independent scalar loop designs. De-coupling specifications of the closed-loop performance for all the models are translated into frequency-domain bounds for each loop of the nominal reference model, and then a diagonal controller is finally derived. Since a diagonal structure could be in many cases considered as an unnecessary restriction of the control action, an extension of the design method is also given to allow for full dynamic compensators. All the members of the family $\{P_n(z)\}$ are assumed to have the same number of open-loop unstable poles, and the same applies for the nominal construction. For convenience, only open-loop stable models are considered in this paper but an extension to unstable cases will become obvious.

2. FREQUENCY-DOMAIN BOUNDS

Specifications that indicate good tracking and low interactions of the closed-loop system can be written in terms of the responses of output i to input j as follows:

$$|\hat{h}_{ij}(t) - \hat{m}_{ij}(t)| \leq |\hat{v}_{ij}(t)| \tag{1}$$

where $\hat{m}_{ij}(t)$ represents a desired mean path, and $\hat{v}_{ij}(t)$ the maximum allowable deviations from it. By making

$$\lim_{t\to\infty} \hat{m}_{ii}(t) = 1 \quad \text{and} \quad \lim_{t\to\infty} \hat{m}_{ij}(t) = 0, \quad \text{for } i \neq j \tag{2}$$

we have the familiar specification for a closed-loop performance without steady-state errors.

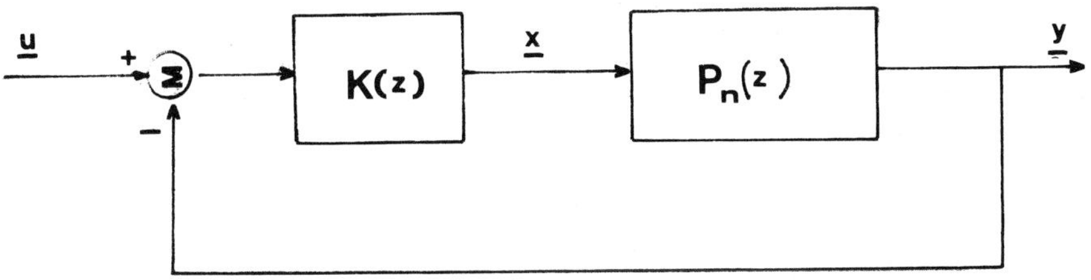

Figure 1. Closed-loop configuration

A weaker form of (1) is one that satisfies the specified bounds in the average over any time period k, i.e.

$$\sum_{t=0}^{k} |\hat{h}_{ij}(t) - \hat{m}_{ij}(t)|^2 \leq \sum_{t=0}^{k} |\hat{v}_{ij}(t)|^2 \tag{3}$$

for any $k = 1,2,\ldots,t,\ldots$

Assuming that all the time-series involved in the above expressions can be viewed as step responses of stable transfer functions, and furthermore that the transfer function corresponding to $\hat{v}_{ij}(t)$ is also minimum phase, then a necessary and sufficient condition for (3) to hold in the time-domain is to have the following relation between transfer functions in the frequency-domain:

$$|H_{ij}(\zeta) - M_{ij}(\zeta)| \leq |V_{ij}(\zeta)| \tag{4}$$

for every $\zeta = e^{j\omega T}$ on the Nyquist contour D_{NYQ} of the complex z-plane; see [KREI], [CHR1]. (T denotes the sampling period and ω the frequency on the imaginary axis of the s-plane.) These properties for $M_{ij}(z)$ and $V_{ij}(z)$ can always be guaranteed by choice, while stability of $H_{ij}(z)$, i.e. closed-loop stability for compensated system $P_n(z)$, has to be ensured by design. Manipulating expression (4) leads to bounds of the return-ratio matrix in the frequency-domain.

Let us now introduce the following notation of transfer functions for systems referred to in Figure 1. All of them are assumed square operators in the space $\mathbb{R}_{m\times m}(z)$.

$P(z)$:= $[P_{ij}(z)]$, generic notation for any member of the set of models $\{P_n, \ell = 1,\ldots,r\}$.
$Q(z)$:= $P^{-1}(z)$:= $[Q_{ij}(z)]$
$K(z)$:= $\mathrm{diag}[K_{ii}(z)]$, diagonal dynamic compensator.
$L(z)$:= $P(z)K(z)$, return-ratio matrix for some compensated member $P = P_n$.
$H(z)$:= $(I+L)^{-1}L$:= $[H_{ii}(z)]$, closed-loop transfer function for some compensated member P.
$M(z)$:= $\mathrm{diag}[M_{ii}(z)]$, diagonal matrix of desired closed-loop transfer functions, whose step responses are the mean paths $\hat{m}_{ii}(k)$ between upper and lower bounds as in Figure 2.
$V(z)$:= $[V_{ij}(z)]$, specified deviations from $M(z)$. Step responses of the off-diagonal elements represent the maximum specified interactions.

The above notation with a subscript N represents the corresponding operators for the nominal reference.

Using this notation we can write a variance of the sensitivity function ([EAST]) that relates the nominal transfer function to those of any member of the given family:

$$I - H_N H^{-1} = (I + L_N)^{-1}(I - P_N P^{-1}) \qquad (5)$$

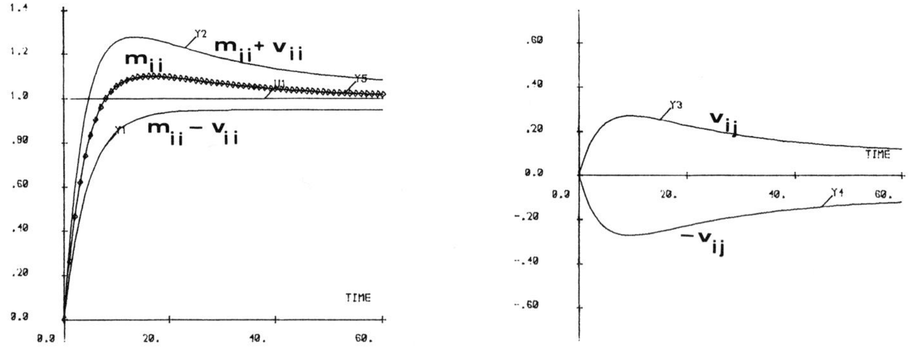

Figure 2. Typical pattern for de-coupled diagonal and off-diagonal responses

When the nominal reference is chosen to be diagonal, the above expression is simplified in a very manageable way, after the introduction of two further assumptions liable to a post-design check:

(i) that the closed-loop step responses of the compensated diagonal nominal model are reasonably close to the mean path $[\hat{m}_{ii}(k)]$ of the specified bounds.

(ii) that the resulting transfer function matrix $H(z)$ for any compensated member of the family $\{P_n(z)\}$ is to be "basically non-interactive" having rather insignificant non-diagonal terms in comparison with the diagonal ones at any frequency $\omega \in D_{NYQ}$. Under this assumption, we can find an approximate expression for the inverse matrix H^{-1}:

$$ii \text{ diagonal element of } H^{-1} \simeq \frac{1}{H_{ii}} \qquad (6a)$$

$$ij \text{ non-diagonal element of } H^{-1} \simeq -\frac{H_{ij}}{H_{ii}H_{jj}} \tag{6b}$$

and then (5) is reduced to a number of scalar expressions

- for diagonal elements

$$1 - \frac{H_{N,ii}}{H_{ii}} \simeq \frac{1}{1 + L_{N,ii}} (1 - P_{N,ii}Q_{ii}) \tag{7a}$$

- and for non-diagonal elements

$$\frac{H_{N,ii}H_{ij}}{H_{ii}H_{jj}} \simeq -\frac{P_{N,ii}Q_{ij}}{1 + L_{N,ii}} \tag{7b}$$

Choosing to satisfy (7a),(7b) when $H_{ii}(\zeta)$ is substituted by the most heavily filtered response $(M_{ii}-V_{ii})$ (see [EAST]) it requires some straightforward manipulations to arrive at the following exclusion bounds for the Nyquist loci of the nominal compensated system $L_N(z)$:

$$|1 + L_{N,ii}(\zeta)| \geq \max_j(r_{ij}) := r_i(\omega) \tag{8}$$

where

$$r_{ii} := \frac{|M_{ii}-V_{ii}|}{|V_{ii}|} \max_n |1 - P_{N,ii}Q_{n,ii}| \tag{9a}$$

$$r_{ij} := \frac{|M_{ii}-V_{ii}||M_{jj}-V_{jj}|}{|V_{ij}||M_{ii}|} \max_n \left|P_{N,ii}Q_{n,ij}\right|, \quad \text{for } i \neq j \tag{9b}$$

for any $i,j = 1,2,\ldots,m$, $n = 1,\ldots,r$ and $\zeta = e^{j\omega T}$.

3. CHOICE OF THE NOMINAL REFERENCE

The nominal diagonal reference is chosen in such a way as to ensure the closed-loop stability of the compensated systems $\{P_n(z)K(z)\}$. By considering model $P_n(z)$ as a multiplicative variation of the diagonal nominal $P_N(z)$ we can write:

$$P_n := (I + \Delta_n)P_N \tag{10}$$

where $\Delta_n := \Delta_n(z) \in \mathbb{R}_{m\times m}(z)$, and then we obtain

$$L_n := (I + \Delta_n)L_N \tag{11}$$

The closed loop stability is maintained for any perturbation $\Delta(z)$ around $P_N(z)$ for which:

$$\sigma_{max}(\Delta) \leq \sigma_{min}(I + L_N^{-1}) \tag{12}$$

where σ_{max} and σ_{min} denote respectively the maximum and minimum singular values of a matrix (see [DOYL]). Since we have that

$$H_N^{-1} = I + L_N^{-1} \tag{13}$$

and

$$\Delta_n = P_n P_N^{-1} - I \tag{14}$$

expression (12) can be written:

$$\sigma_{max}(P_n P_N^{-1} - I) \le \sigma_{min}(H_N^{-1})$$

or

$$\sigma_{max}(P_n Q_N - I) \le \frac{1}{\sigma_{max}(H_N)} \tag{15}$$

Condition (15) can be exploited for choosing Q_N (and then P_N) such as to ensure the minimum level for the expression in the left-hand side, at each frequency $\omega \in \Omega$. Hence the following optimisation problem is formulated:

$$\min_{Q_N} \max_{n} \sigma_{max}(P_n Q_N - I) := \beta(\omega) \tag{16}$$

Solving the above optimisation problem may prove to be a difficult task because of the complexities introduced by the singular value operator. Thus instead of (16) we choose Q_N to optimise the Frobenius norm, i.e.

$$\min_{Q_N} \max_{n} \| PQ_N - I \|_F \tag{17}$$

which by definition is given as:

$$\sum_{j=1}^{m} \{ |P_{n,jj} Q_{N,jj} - 1|^2 + |Q_{N,jj}|^2 \sum_{k \neq j} |P_{n,kj}|^2 \} \tag{18}$$

In the above expression the diagonal elements $Q_{N,jj}$ count only for the j-column of P_n and thus the optimisation problem is finally split into m independent optimal choices:

$$\min_{Q_{N,jj}} \max_{n} \{ |P_{n,jj} Q_{N,jj} - 1|^2 + |Q_{N,jj}|^2 \sum_{k \neq j} |P_{n,kj}|^2 \} := \beta_j(\omega) \tag{19}$$

where $j = 1,\ldots,m$, and $n = 1,\ldots,r$.

This scalar minimax problem can be solved iteratively in the following way (see [CHR1]): Start with some scalar transfer function $Q_{N,jj}(z)$ and on each frequency $\omega \in \Omega$ choose among the j-columns of the models $\{P_n(z)\}$ a column p_j^o that maximises the expression:

$$|P^o_{jj}Q_{N,jj} - 1|^2 + |Q_{N,jj}|^2 \underset{k\neq j}{} |P^o_{jk}|^2$$

By defining

$$\delta(\omega) := \frac{|P^o_{jj}|^2}{\|p^o_j\|^2_F} \tag{20}$$

a new optimal is found through

$$P_{N,jj}(e^{j\omega T}) = \frac{P^o_{jj}(e^{j\omega T})}{\delta(\omega)} \tag{21}$$

and the nominal model is finally constructed by solving a rational synthesis problem for (21). Regarding now the right-hand side of (18), taking into account that matrix $H_N(z)$ is diagonal, and using expression (19) we have that the following condition must be satisfied:

$$|L_{N,ii}|\ \beta(\omega) \leq |1 + L_{N,ii}| \tag{22}$$

for every $i = 1,\ldots,m$.

If we now consider that the bound in expression (8) is satisfied, then a sufficient condition for (22) is:

$$|L_{N,ii}(e^{j\omega T})| \leq \frac{r_i(\omega)}{\beta(\omega)} \tag{23}$$

that provides inclusion bounds for the nominal return ratio loci in order to guarantee closed-loop stability for the family $P_n(z)$, as (8) provides exclusion bounds that ensure de-coupling. Observe that exclusion bounds are necessary and sufficient, while the inclusion ones are only sufficient and thus conservative for variations $P_n(z)$ that are described by a finite set.

Assuming that (8) is satisfied, then it can be easily shown that (23) requires

$$\beta(\omega) < \frac{r_i(\omega)}{-1+r_i(\omega)} \tag{24}$$

for every $\omega \in \Omega$, if $r_i(\omega) > 1$, and is redundant for $r_i(\omega) \leq 1$. This expression is particularly useful because it can be checked for any loop i in advance, after the bounds have been specified.

4. REALISATION OF THE COMPENSATOR

In a lengthy proof [GERA] have shown that an optimal design for $L_{N,ii}(z)$, in the sense of minimum high frequency gain for any fixed excess of poles over zeros, is achieved when (8) is satisfied as an equality. Even with this requirement the choice of $L_{N,ii}(z)$ is not unique and the designer is free to consider other additional aspects to be optimised. The procedure followed here chooses the starting point of the locus $L_{N,ii}(\zeta)$ to lie on the positive real semi-axis with abscissa $r_i(0)$ and the rest of the design steps are determined in order to maximise the phase lag by moving tangentially from one bound to the next, as this enhances the low-frequency gain (see [HOR2], [CHR1]). For higher frequencies we specify a boundary in order to give satisfactory phase margins and ensure the appropriate encirclements of the critical point (here none). This procedure provides a set of points, say $\{f_i(\omega)\}$, through which the locus $L_{N,ii}(\zeta)$ must pass. Using the expression

$$L_{N,ii}(z) = P_{N,ii}(z)K_{ii}(z) \tag{25}$$

we have now to realise a proper and stable controller $K_{ii}(z)$ such that the set $\{f_i(\omega)\}$ is approximated in the minimum square error sense, i.e.

$$\min_{K_{ii}} \sum_{\Omega} |P_{N,ii}(e^{j\omega T})K_{ii}(e^{j\omega T}) - f_i(\omega)|^2 \tag{26}$$

As is common in frequency designs, an adjustment of gains may be required after the fitting in order to improve the shape of the loci over a particular region.

5. AN EXTENSION TO FULL FEEDBACK STRUCTURES

Re-writing the expression for the closed-loop transfer function we get:

$$P_n(z) = [I - H_n^{-1}(z)]^{-1}K^{-1}(z) \tag{27}$$

As was previously stated, the design tries to achieve a $H_n(z)$ as de-coupled as possible in order to have desired dynamic properties for the controlled models P_n, $n = 1,2,\ldots,r$. By fixing a diagonal structure for $K(z)$, the right-hand side in the above relation shows that, for finite values of $H_n(z)$, its achievable de-coupling can only reflect a similar de-coupling in the open-loop transfer functions $P_n(z)$. In other words, if models $P_n(z)$ are far from a state of diagonal dominance, opting for diagonal control action results in highly interactive closed-loop performance. The only case where this assumption is not strong occurs in the vicinity of low-frequencies, given that integral control action has been incorporated in $K(z)$. Then as $\omega \to 0$ we have

$$\lim_{z\to 1} K^{-1}(z) = 0_{m\times m} \quad \text{and} \quad \lim_{z\to 1} H^{-1}(z) = I_{m\times m}$$

and thus expression (27) is reduced to the indefinite expression $0^{-1}0$ that can be matched by a $P_n(z)$ however coupled. In terms of time-domain behaviour, to guarantee zero steady-state interactions rarely amounts to a sufficient quality of performance, and further specifications have to be made in order to reduce severe interactions during the transition period. For this purpose the controller must be allowed for a full structure, $F(z)$, which can always be written as:

$$F(z) = C(z)K(z) \tag{28}$$

with K(z) representing a diagonal transfer function and C(z) a full dynamic pre-compensator. The return-ratio matrix of the compensated system is now

$$L(z) = P_n(z)F(z) = G_n(z)K(z) \tag{29}$$

with

$$G_n(z) = P_n(z)C(z) \tag{30}$$

and an obvious choice for pre-compensator C(z) is such as to achieve a satisfactory de-coupling for the new family $\{G_n(z),\ n=1,\dots,r\}$.

Since in (27) matrix K^{-1} accounts as a column scaling, de-coupling of $G_n(z)$ can be considered only column-wise. A criterion for that is for each column j to have:

$$\sum_{i\neq j} |G_{n,ij}(e^{j\omega T})|^2 < |G_{n,jj}(e^{j\omega T})|^2 \tag{31}$$

and by writing $P_n(z)$ and C(z) as row and column vectors respectively:

$$P_n(z) = \begin{bmatrix} p_1'(z) \\ \vdots \\ p_m'(z) \end{bmatrix} \qquad C(z) = [c_1(z)\ \dots\ c_m(z)]$$

(31) is re-written as:

$$\sum_{i\neq j} |p_i'c_j|^2 < |p_j'c_j|^2 \tag{32}$$

at each frequency $\omega \in \Omega$ and for every model $n=1,\dots,r$. (The exponential arguments and subscript n are omitted for simplicity, and (') is for transpose.)

In the above formulation the problem is essentially the same as that of frame-alignment (solved by the algorithm ALIGN, see [MACF], [EDM]), in which a real precompensator is derived in order to achieve a proportional form of (32) at a specific frequency ω_0, or over a frequency range Ω. A slight modification of this ALIGN algorithm is given below to cover the case of dynamic pre-compensators, and a family of models instead of a specific one.

Since vector $c_j(z)$ is determinable up to constant scaling, we specify

$$||c_j(e^{j\omega T})|| = 1 \tag{33}$$

and then we opt for a proportional form of (32), i.e.

$$\sum_{i=1}^{m} s_{ij} \frac{|p_j' c_j|^2}{\|p_j\|^2} < 1 \tag{34}$$

where $s_{ij} = 1$ for $i \neq j$, and -1 for $i = j$.

Defining for any model n, the hermitian matrix

$$E_j^n(1) = \sum_{i=1}^{m} s_{ij} \frac{\bar{p}_i p_j'}{\|p_j\|^2} \tag{35}$$

the problem finally becomes to find a vector $c_j(z)$ such as

$$\min_{c_j} c_j^* E_j^n(\omega) c_j \tag{36}$$

subject to

$$c_j^* c_j = 1$$

with $(\bar{\ })$ denoting complex conjugate and $(*) = (\bar{\ })'$.

For each model n, vector c_j is determined in the minimum square-error sense as the (generally complex) eigenvector, $\varepsilon_j(\omega)$, of matrix $E_j^n(\omega)$ that corresponds to its minimum eigenvalue. Since E_j^n is hermitian, all eigenvalues are real and the lowest, $\lambda_1(\omega)$, is indeed the optimal value of expression (36).

Confronted with more than one model P_n, one can opt for either of the following modifications:

(i) At each frequency $\omega \in \Omega$, consider a weighted hermitian $\hat{E}(\omega)$ given by:

$$\hat{E}_j(\omega) = \sum_{n=1}^{r} \alpha(n,\omega) E_j^n(\omega) \tag{37}$$

with $\alpha(n,\omega)$ denoting the subjective probability that model P_n is true at frequency ω, and $\Sigma\alpha = 1$. A possible drawback of this approach is that a representation n with low probability $\alpha(n,\omega)$ may well give high values for (36) (see [EAST]) and in this case the above weighting should be checked *ex-post* for the actually achieved de-coupling in each individual model.

(ii) Another approach is to apply an iterative algorithm of the kind

$$\min_{c_j} \max_{n} c_j^* E_j^n c_j \tag{38}$$

that naturally implies a degree of conservatism similar to that involved in the minimax problem (16).

After an optimal c_j has been evaluated at each frequency $\omega \in \Omega$, a dynamic realisation can be performed in the minimum square error sense to yield rational expressions for the pre-compensator C(z). Obviously each scalar entry $c_{ij}(z)$ can be realised individually by solving the problem

$$\min_{c_{ij}} \sum_{\Omega} |c_{ij}(e^{j\omega T}) - \varepsilon_{ij}(\omega)|^2 \tag{39}$$

A complex-curve fitting algorithm can be used here again, and the only specifications to be made on $c_{ij}(z)$ are to exclude unstable poles. The remaining step is to design a dynamic controller K(z) with diagonal structure, for the open-loop systems $G_n(z)$.

6. A DESIGN EXAMPLE

For the illustration of the method, a diagonal controller is sought here in order to serve a family $\{P_n(z),\ n = 1,2,3\}$ consisting of three discrete-time models each of order d = 3 with 2 inputs and 2 outputs. These models are in fact simple, reduced versions of three large linearised econometric models, for which a similar design has been already performed [CHR2], but their complication serves no purpose for the present demonstration. Figure 3 shows the responses of the three models to unit steps; some similarity exists in the pattern of their behavious in the long-run, but their intermediate dynamics reveal serious differences among them. Choosing a logarithmically spaced frequency range Ω of 50 values, minimax problem (19) is solved independently for the two loops and yields two sets of points $\{f_i(\omega),\ i = 1,2\}$ in the complex plane. Then a complex-curve fitting algorithm is used to derive stable rational realisations for the nominal diagonal reference:

$$P_{N,11}(z) = \frac{-.347z^2 + .534z - .336}{z^3 - 2.168z^2 + 1.967z - .70} \tag{40a}$$

$$P_{N,22}(z) = \frac{.505z^2 - .685z + .337}{z^3 - 1.745z^2 + 1.33z - .305} \tag{40b}$$

which are depicted in Figure 4 together with the supplied set of points. In Figure 5 the step responses of the nominal reference demonstrate the intuitively expected resemblance to the responses of the diagonal elements in Figure 3.

The next step is the specification of bounds for the closed-loop responses of the compensated systems, by using functionals

$$\Phi_k(z) = b_k \frac{1 - a_k}{z - a_k}, \quad 0 < a_k < 1,\ 0 < b_k \tag{41}$$

in the following way Members Φ_k are obvious candidates for the lower bounds $(M_{ii}-V_{ii})$, differences $(\Phi_k-\Phi_{k'})$ with $a_k > a_{k'}$ can be used for the terms $V_{ii}(z)$, and the upper bounds $(M_{ii}-V_{ii})$ are expressible by differences $(2\Phi_k-\Phi_{k'})$. As explained in section 5, the design of a satisfactory diagonal controller is facilitated by an existing decoupling of

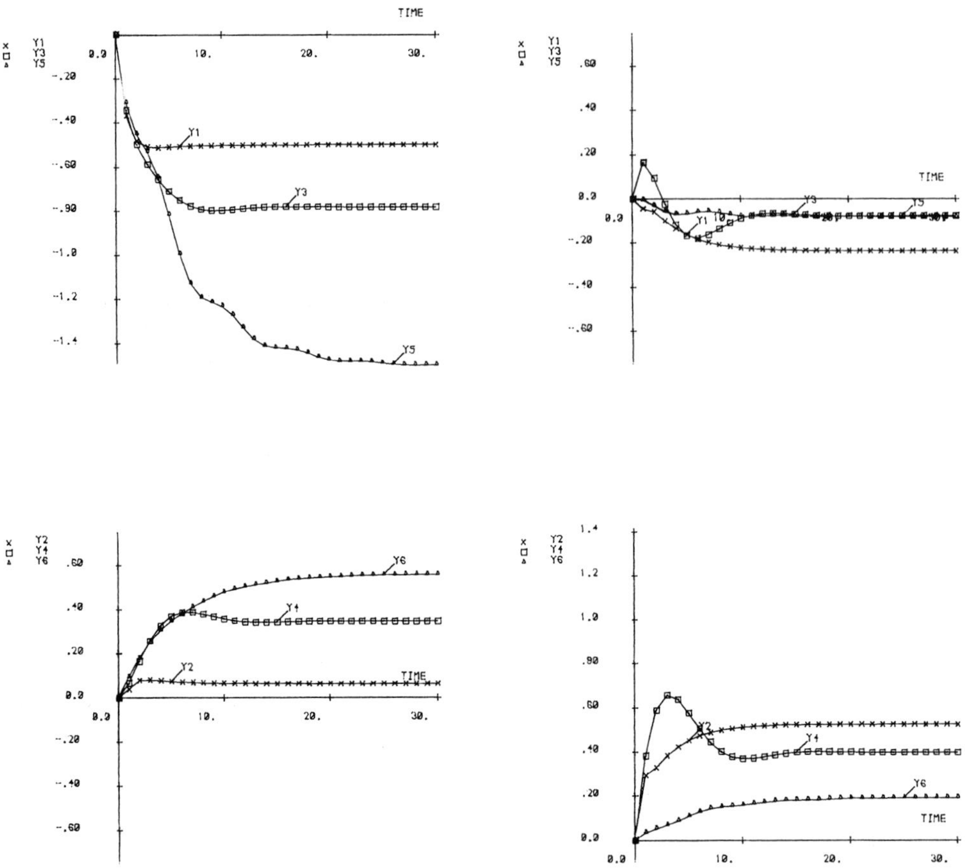

Figure 3. Step responses of the three models, denoted by ×, □, △ respectively

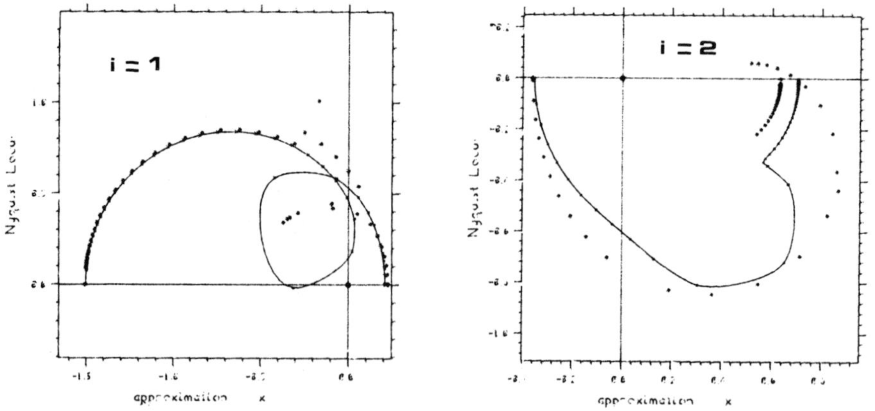

Figure 4. Rational realisation of the diagonal nominal model

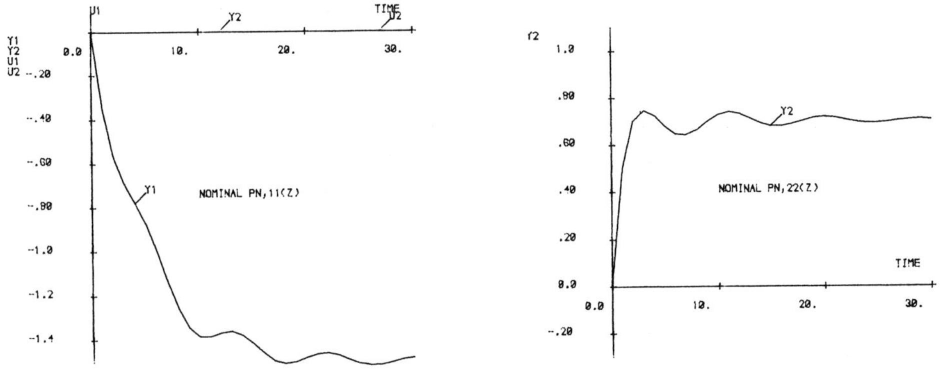

Figure 5. Step responses of the nominal model $P_N(z)$

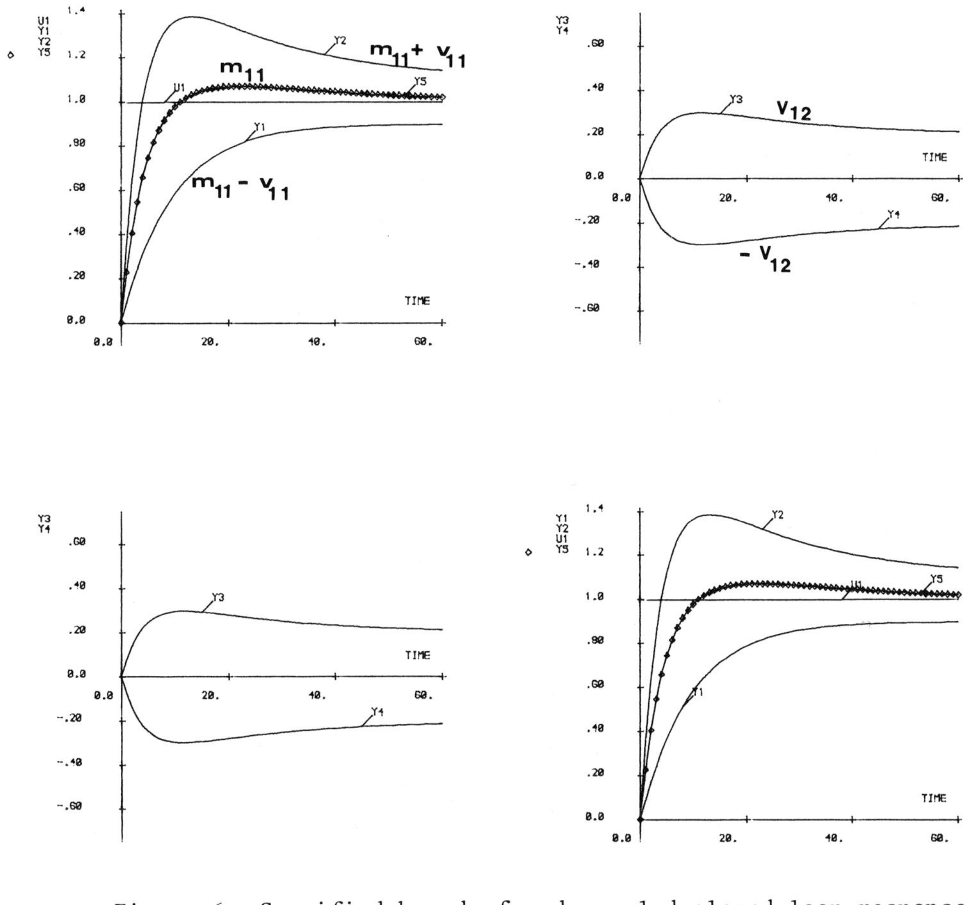

Figure 6. Specified bounds for decoupled closed-loop responses

the open-loop models $P_n(z)$. Looking to Figure 3 one can see that this is the case only for the first column (corresponding to the first loop in the decomposed design), and is not for the second. This situation calls for optimistic bounds of fast and non-overshooting responses for the first loop, and moderate ones for the second; the finally specified bounds for the present problem are shown in Figure 6.

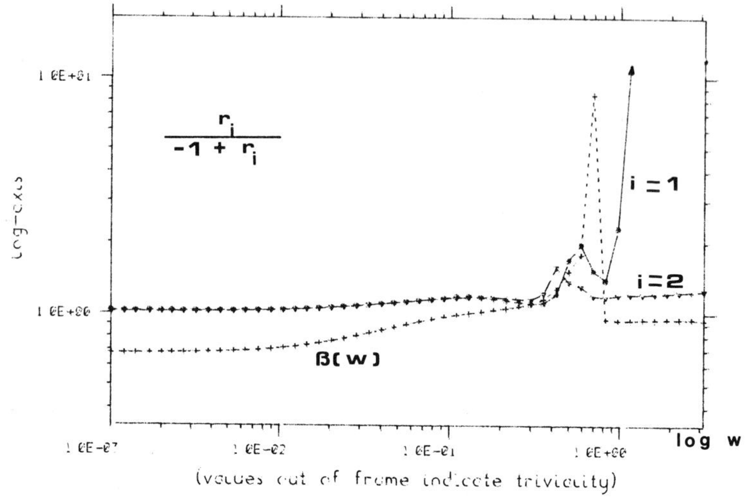

Figure 7. Checking condition (24) for each loop

The sufficient condition (24) for the inclusion bounds (23) is checked in Figure 7, and is found satisfied for the first loop, and slightly violated for the second in the high-frequency region. This however is not inhibiting for the design to proceed, given the conservatism of (24) and the fact that the behaviour of the nominal characteristic loci in high frequencies is dictated by the specified boundary on the phase margin (set here equal to 60°).

Applying the guidelines of section 4, desired nominal loci were first constructed as in Figure 8 and then the following diagonal controller was realised through (26):

$$K_{11}(z) = \frac{-1.896z^2 + .761z + .878}{z^2 - .9805z - .0142}$$

$$K_{22}(z) = \frac{1.90z^2 - .1076z - .678}{z^2 - .916z - .0807}$$

Observe that both expressions include an "almost-integral" action, since the specified bounds of Figure 6 allow for small steady-state offsets. (Introducing exact integrators would simply lead to an ample satisfaction of exclusion bounds (8).) Condition (8) is indeed checked

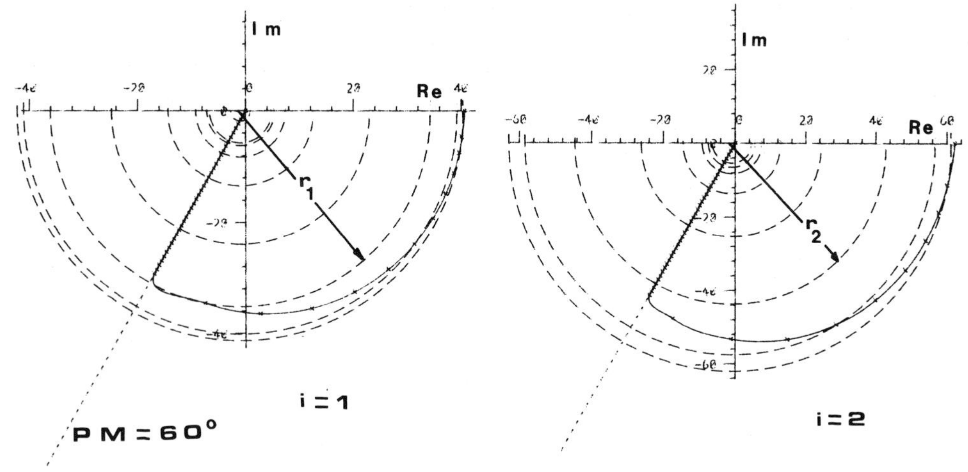

Figure 8. Constructed return-ratios for $L_{N,ii}(z)$

for both loops as in Figure 9, with the finally realised controller.

Closed-loop stability of each model compensated by K(z) is checked through the Generalised Nyquist criterion in Figure 10, that shows good phase and gain margins. Finally, the step responses of the controlled systems are plotted in Figure 11 versus the specified bounds. One can see that bounds are generally kept, in the average sense of expression (3). Observe that whenever bounds are violated, they reflect a corresponding violation in Figure 9.

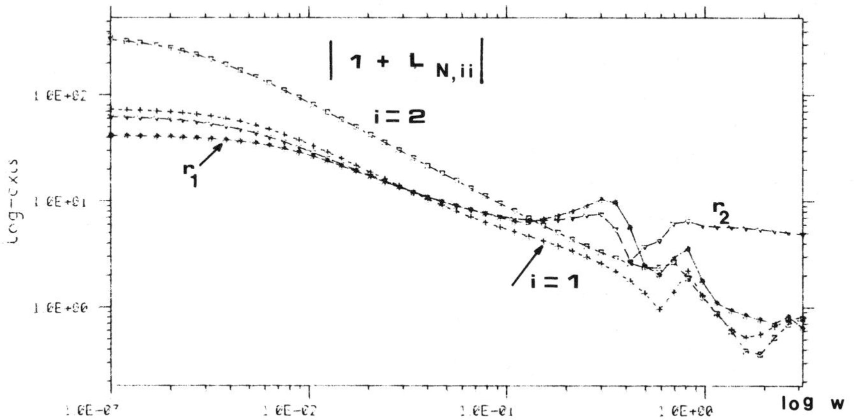

Figure 9. Checking exclusion bounds (8)

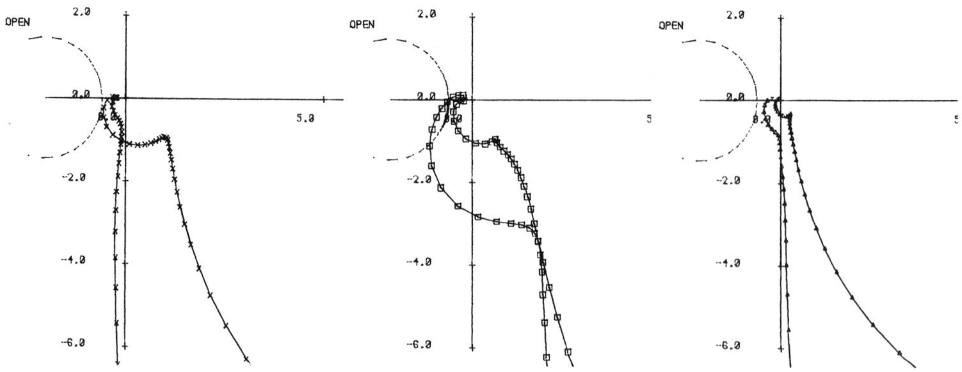

Figure 9. Characteristic loci of the three models $P_n(z)$, compensated by the same $K(z)$

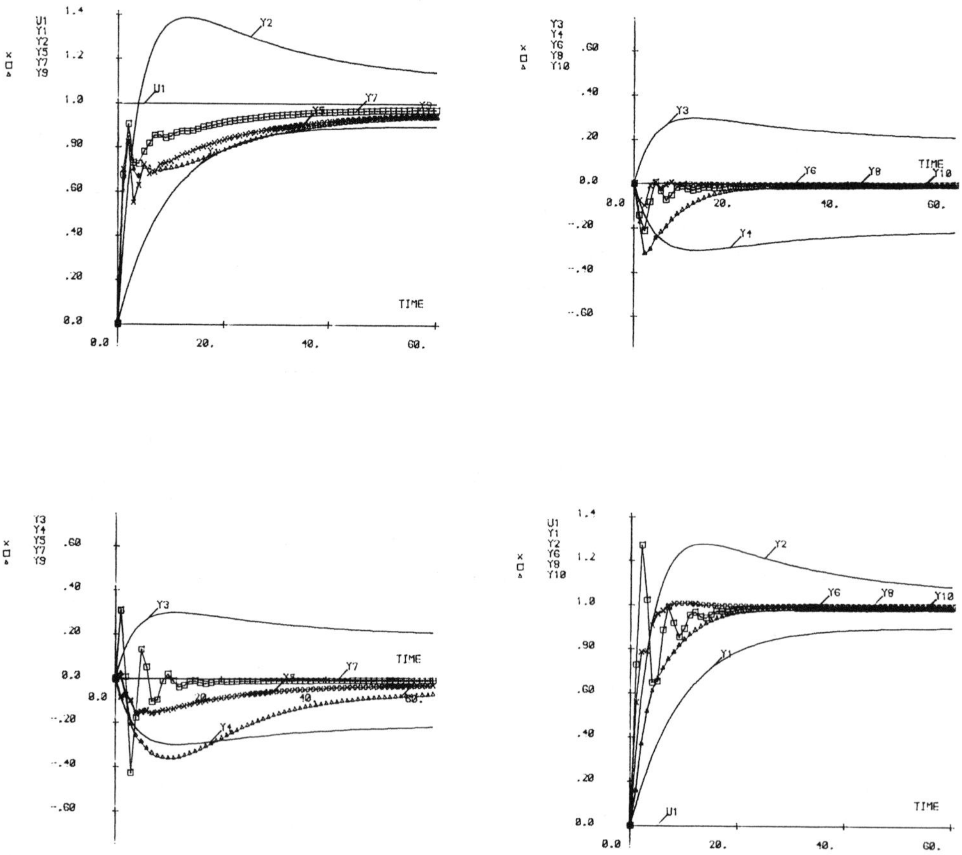

Figure 10. Step responses of the closed-loop systems, $H_n(z)$, versus the specified bounds (Notation as in Fig.3)

ACKNOWLEDGEMENT

The present work is part of the research conducted for a doctoral degree, under a three-year grant from the State Scholarship Foundation of Greece.

The author wishes to thank B. Kouvaritakis for some helpful comments on section 5.

REFERENCES

[CHR1]: Christodoulakis N., 'Designing a De-coupling Controller for Multi-represented Discrete-time MIMO Plants', to appear in *International Journal of Control*.

[CHR2]: Christodoulakis N. and Van de Ploeg F., 'Macro-dynamic Policy Formulation with Conflicting Views of the Economy', *mimeo*, 1984.

[DOYL]: Doyle J. and Stein G., 'Multivariable Feedback Design: Concepts for a Classical/Modern Synthesis', *IEEE*, Trans. AC, 1981, AC-26, 1, pp.4-16.

[EAST]: East D., 'A New Approach to Optimum Loop Synthesis', *I.J.C.*, 1981, 34, 4, pp.731-748.

[EDM]: Edmunds J. and Kouvaritakis B., 'Extensions of the Frame Alignment Technique and their Use in the Characteristic Locus Design Method', *I.J.C.*, 1979, 29, 5, pp.787-796.

[GERA]: Gera A. and Horowitz I., 'Optimisation of the Loop Transfer Function', *I.J.C.*, 1980, 31, 2, pp.389-398.

[HOR1]: Horowitz I., 'Quantitative Synthesis of Uncertain MIMO Plants', *I.J.C.*, 1979, 30, 1, pp.81-106.

[HOR2]: Horowitz I., 'Improved Design Technique for Uncertain MIMO Feedback Systems', *I.J.C.*, 1982, 36, pp.977-988.

[KREI]: Kreidler E., 'On the Definition and Application of the Sensitivity Function', *Journal of the Franklin Institute*, 1968, 285, pp.26-36.

[MACF]: MacFarlane A.G.J. and Kouvaritakis B., 'A Design Technique for Linear Multivariable Feedback Systems', *I.J.C.*, 1977, 25, 6, pp.837-874.

[POST]: Postlethwaite I., Edmunds J.M. and MacFarlane A.G.J., 'Principal Gains and Principal Phases in the Analysis of Linear Multivariable Feedback Systems', *IEEE*, Trans. AC, 26, 1981, pp.32-46.

[SAF]: Safonov M.G., Laub A.J. and Hartman G.L., 'Feedback Properties of Multivariable Systems: The Role and Use of the Return Difference Matrix', *ibid*, pp.37-65.

APPENDIX

The z-transfer function matrices describing the three models used in the example of section 6 are as follows:

Model 1

$$\frac{1}{z^3-.892z^2+.095z+.021}\begin{bmatrix} -.37z^2+.215z+.042 & -.045z^2+.028z-.036 \\ .04z^2+.004z-.03 & .29z^2-.23z+.052 \end{bmatrix}$$

Model 2

$$\frac{1}{z^3-1.75z^2+1.098z-.230}\begin{bmatrix} -.355z^2+.49z-.23 & .162z^2-.35z+.186 \\ .075z^2-.034z-.005 & .384z^2-.46z+.123 \end{bmatrix}$$

Model 3

$$\frac{1}{z^3-1.794z^2+z-.164}\begin{bmatrix} -.315z^2+.430z-.146 & .009z^2-.0688z+.0463 \\ .093z^2-.0778z+.0004 & .083z^2-.10z+.03 \end{bmatrix}$$

CHAPTER 11

MINIMIZING CONSERVATIVENESS OF ROBUSTNESS SINGULAR VALUES

Michael G. Safonov
Department of Engineering-Systems
University of Southern California
Los Angeles, CA 90089-0781
U.S.A.

John C. Doyle
Honeywell, Inc.MN17-2375
2600 Ridgway Pkwy.
Minneapolis, MN 55413
U.S.A.

ABSTRACT

It is proved that the function

$$\bar{\sigma}^2(DMD^{-1})$$

is convex in D for any diagonal scaling matrix $D=\mathrm{diag}(d_1,\ldots,d_n)$, guaranteeing convergence of numerical algorithms for mimimizing $\bar{\sigma}(DMD^{-1})$. Convergence is also proved for more general situations involving certain classes of block diagonal D matrices. The result has application to stability margin analysis for multivariable control systems with structured undertainty. A key intermediate result (Lemma 1) is an expression for the second derivatives of the eigenvalues of a matrix A(t) dependent on a single real parameter t.

INTRODUCTION

In robust control system design as well as in numerical analysis, the issue arises of how to scale a square matrix $M\epsilon C^{nxn}$ via a similarity transformation

$$M \rightarrow XMX^{-1}$$

so as to minimize a given matrix norm [1]-[7]. In most cases the scaling matrix X is restricted to be within a specified class, say $X\epsilon\chi$ which solves

$$\inf_{X\epsilon\chi}\|XMX^{-1}\| \tag{1}$$

for a given matrix norm $\|\cdot\|$.

No analytic solution to this optimization problem is available except for special cases. If $\chi = C^{nxn}$ (the set of nxn complex matrices) and the norm $\|\cdot\|$ is the ℓ_p bound-norm

S. G. Tzafestas (ed.), Multivariable Control, 197–207.

$$\|M\|_{p\to p} := \sup_{\|x\|_p = 1} \|Mx\|_p \ , \quad p \in [1,\infty] \tag{2}$$

where

$$\|x\|_p := \Big(\sum_{i=1}^{n} |x|^p\Big)^{1/p}$$

then it is easily seen that the infimal value of (1) is the spectral radius of M.

Solutions expressed analytically in terms of Perron eigenvectors and eigenvalues are available [1], for the special case where X is restricted to be diagonal, i.e.,

$$\chi = \mathscr{D}_1 := \Big\{ D \,\Big|\, D = \text{diag}\,(d_1,\ldots,d_n) \Big\} ,$$

and either

a) M is a positive matrix, i.e., $M_{ij} \varepsilon R_+ \ \forall\ i,j$, or

b) $p = 1$, or

c) $p = \infty$.

It is also shown in [1] that the minimal values of $\|DMD^{-1}\|_{p\to p}$ are equal for $p = 1$ and $p = \infty$ and that the minimal value for any $p \varepsilon (1,\infty)$ is never greater than for $p = 1$ and for $p = \infty$.

Another norm for which an optimal diagonal $D \varepsilon \mathscr{D}_1$ can be computed reliably is the Frobenius norm

$$\|M\|_F := (\text{Tr}(M^* M))^{1/2} \tag{3}$$

A globally convergent iterative algorithm for computing

$$\inf_{D \varepsilon \mathscr{D}_1} \|DMD^{-1}\|_F$$

has been developed by Osborne[6].

Of more general interest are cases where the set χ may take a block diagonal form such as [2]

$$\chi_\infty := \{\text{diag}\,(\underbrace{\Delta_1,\Delta_1,\ldots,\Delta_1}_{m_1},\underbrace{\Delta_2,\ldots,\Delta_2}_{m_2},\ldots \ \ldots,\underbrace{\Delta_n,\ldots,\Delta_n}_{m_n}) \,|\, \Delta_j \varepsilon C^{k_j \times k_j} \ \forall j\} \tag{4}$$

or

$$\mathscr{D}:= \operatorname{diag}\{d_{11}I_{k_1}, d_{21}I_{k_1},\dots,d_{m_1 1}I_{k_1},d_{12},I_{k_2},\dots,d_{m_2 2}I_{k_2},$$

$$\dots \dots, d_{1n}I_{k_n},\dots,d_{m_n n}I_{k_n} \,|\, d_i \varepsilon R^+ := (0,\infty)\} \;. \tag{5}$$

A problem of special interest in the analysis and design of robustly stable control systems is the computation of Doyle's [2] μ-function

$$\mu(M) := (\inf_{X\,\varepsilon X_\infty}\{\|X\|_{2\to 2} | \det(I+MX) = 0\})^{-1}. \tag{6}$$

This problem is intimately related to the optimal scaling problem via the inequalities [1]-[3]

$$\max_{U\varepsilon} \rho(UM) \leqq \mu(M) \leqq \inf_{D\varepsilon} \| DMD^{-1}\|_{p\to p} \tag{7}$$

where $\rho(\cdot)$ denotes the spectral radius $\rho(M) = \lambda_{max}(M)$ and $\mathscr{U} := \{U \varepsilon X_\infty | U^* U = I\}$. Doyle [2] has proved that the left inequality is in fact an equality and the right inequality becomes an equality when $m_j = 1 \forall j$ and $n \leqq 3$.

When $k_j = 1$ for one or more values of j then it is easily shown that the right hand side of (7) may be replaced by the tighter bound

$$\mu(M) \leqq \inf_{Y\varepsilon\mathscr{Y}} \|YMY^{-1}\|_{p\to p} \tag{8}$$

where

$$\mathscr{Y} = \{\operatorname{diag}(Y_1,\dots,Y_n) | Y_j \varepsilon \mathbb{C}^{m_j x m_j} \text{ if } k_j = 1; \text{ and}$$

$$Y_j = \operatorname{diag}(d_{1j} I_{k_j},\dots,d_{m_j n} I_{k_j}) \text{ if } k_j \geqq 2\}, \tag{9}$$

i.e., the jth block $\operatorname{diag}(d_{1j},\dots,d_{m_j j})$ in (5) may be replaced by any $m_j x m_j$ matrix when $k_j = 1$. More general situations can be handled at the expense of increasingly difficult notation. No analytic expression for the optimal $U\varepsilon\mathscr{U}$, $D\varepsilon\mathscr{D}$ or $Y\varepsilon\mathscr{Y}$ in (7)-(8) is known, but one may resort to nonlinear programming methods based on gradient following to compute the extremal values of U and D. Gradient algorithms reliably converge only when the functional being minimized or maximized does not have disjoint local minima or maxima, respectively The quantity on the left of (7) is known to have multiple local maxima [2] in general and consequently cannot be reliably computed via gradient methods.

In [2] a gradient descent algorithm is presented for finding local minima of

$$\|DMD^{-1}\|_{2\to 2} \tag{10}$$

The principal conclusion of the present paper is a proof of the convexity of $(\|DMD^{-1}\|_{2\to 2})^2$. We make use of the well-known fact that

$$\|M\|_{2\to 2} = \max_i \sigma_i(M) \tag{11}$$

where $\sigma_i(M)$ $(i=1,\ldots,n)$ denote the n singular values of the matrix M. Convexity of $\|DMD^{-1}\|^2_{2\to 2}$ is established by showing that for any $D'\varepsilon\mathscr{D}$ and any $M'\varepsilon C^{nxn}$

$$\frac{d^2}{dt^2}\sigma_i^2\left(e^{tD'}M'e^{-tD'}\right) \geq 0 \tag{12}$$

when $\sigma_i(\cdot)$ is the greatest singular value. This implies gradient descent optimization methods for finding the minimizing D in (10) will reliably converge to a global minimum.

FIRST AND SECOND DERIVATIVES OF SINGULAR VALUES

Consider the nxn complex matrix A(t) analytically dependent on the real parameter t in a neighbourhood of t=0. Thus A(t) has a Taylor series expansion

$$A(t) = A_0 + \dot{A}_0 t + \frac{1}{2}\ddot{A}_0 t^2 + \ldots .$$

We assume that A_0 is diagonable at t=0, i.e., that the Jordan form of A is diagonal. It is well known [8] that the eigenvalues of A(t) may be indexed so that they too are analytic almost everywhere in a neighbourhood of t=0 and each eigenvalue $\lambda(t)$ has Taylor series

$$\lambda(t) = \lambda_0 + \dot{\lambda}_0^i t + \frac{1}{2}\ddot{\lambda}_0^i t^2 + \ldots , \quad i=1,\ldots,m . \tag{13}$$

where m is the multiplicity of the eigenvalue λ_o of A_o.

We assume that the eigenvectors of A(t) corresponding to each eigenvalue can be selected in such a way that they are analytic and non-zero in a neighbourhood of t=0. For example, it is known (e.g.,[8]) that analytic, non-zero eigenvectors can always be selected when A(t) is Hermitian, i.e., when A(t) = A*(t). Let λ_0 be an eigenvalue of A_0 of multiplicity m. Let $P_0\varepsilon C^{nxn}$ be any matrix such that

$$A_0 P_0 = P_0 \begin{bmatrix} \Lambda_0 & 0 \\ 0 & \bar{\Lambda}_0 \end{bmatrix}$$

where $\Lambda_0 \varepsilon C^{m x m}$, $\bar{\Lambda}_0 \varepsilon C^{(n-m) x (n-m)}$

$$\det(Is-\Lambda_0) = (s-\lambda_0)^m \tag{15}$$

$$\det(Is-\bar{\Lambda}_0) = \frac{\det(I_s-A_0)}{(s-\lambda_0)^m} . \tag{16}$$

Partition P_0 as

$$P_0 = [R_0, Q_0] \tag{17}$$

with $R_0 \varepsilon \mathbb{C}^{n x m}$, $Q_0 \varepsilon \mathbb{C}^{n x (n-m)}$. Similarly partition

$$\hat{P}_0^* := P_0^{-1} = \begin{bmatrix} \hat{R}_0^* \\ \hat{Q}_0^* \end{bmatrix} \tag{18}$$

with $\hat{R}_0^* \varepsilon \mathbb{C}^{m x n}$, $\hat{Q}_0^* \varepsilon \mathbb{C}^{(n-m) x n}$.

By the analyticity of $\lambda(t)$ it follows that except at isolated values of t the multiplicity of $\lambda(t)$ remains constant. We shall term such points t_0 where for some $\varepsilon>0$

$$\text{multiplicity}(\lambda(t_0)) = \text{multiplicity}(\lambda(t))\ \forall t \varepsilon (t_0-\varepsilon, t_0+\varepsilon) \tag{19}$$

as <u>regular points</u> of $A(t)$.

<u>Lemma 1</u>. The m first derivatives $\dot{\lambda}_0^i$ $(i=1,\ldots,m)$ of a multiplicity m eigenvalue are the m eigenvalues of the matrix

$$\hat{R}_0^* \dot{A}_0 R_0 . \tag{20}$$

Furthermore, if $t=0$ is a regular point and if λ_0 is nondegenerate (i.e., the Jordan block associated with λ_0 is diagonal) then the second derivatives $\ddot{\lambda}_0^i$ $(i=1,\ldots,m)$ are the eigenvalues of

$$\hat{R}^*(\ddot{A}_0 + 2\dot{A}Q_0(\lambda_0 I - \bar{\Lambda}_0)^{-1}\hat{Q}_0^*\dot{A})R_0. \tag{21}$$

Proof: See appendix.

Note that the restriction in the Lemma to t=0 being a regular point does not cause any insurmountable problems in using the Lemma to compute $\ddot{\lambda}_0^i$. The analyticity of $\lambda(t)$ implies that $\ddot{\lambda}(t)$ is continuous function. Thus, each $\ddot{\lambda}_0^i$ may always be computed by taking the limit of $\ddot{\lambda}^i(t)$ as $t \to 0$.

Lemma 2. (Derivatives of Singular Values): Let $X \in C^{n\times n}$ and let $M(t) = e^{Xt}M_0 e^{-Xt}$. Let

$$\sigma^2(t) = \sigma_0 + t\dot{\sigma}_0^{2^i} + \frac{1}{2}t^2\ddot{\sigma}_0^{2^i} + \ldots \tag{22}$$

be any singular value of $M(t)$. Let V_0 be a matrix whose columns are corresponding right singular vectors of M_0 or, more generally, let V_0 be any matrix satisfying

$$M_0^* M_0 V_0 = \sigma_0^2 V_0 \in \mathbb{C}^{n\times m} \tag{23}$$

where m is the multiplicity of the singular value σ_0. The first derivatives $\dot{\sigma}_0^{2^i}$ are the eigenvalues of

$$V_0^*(M_0^*\dot{M}_0 + \dot{M}_0^* M_0)V_0 \tag{24}$$

where

$$\dot{M}_0 := XM_0 - M_0 X . \tag{25}$$

Furthermore, if $X=X^*$ is hermitian and if t=0 is a regular point of $M^*(t)M(t)$, then the second derivatives $\ddot{\sigma}_0^{2^i}$ are the eigenvalues of

$$2V_0^*\left(X^*(\sigma_0^2 I - A_0)X + 2\dot{M}_0^*\dot{M}_0 + \dot{A}\bar{V}_0(\sigma_0^2 I - \bar{\Sigma}^2)\bar{V}_0^*\dot{A}\right)V_0 \tag{26}$$

where

$$A = M_0^* M_0 \tag{27}$$

$$A = M_0^* M_0 + M_0^* M_0 \tag{28}$$

and $\bar{\Sigma} := \mathrm{diag}(\sigma_{m+1}, \ldots, \sigma_n)$ where $\sigma_{m+1}, \ldots, \sigma_n$ are the singular values of A_0 not equal to σ_0, and $\bar{V}$ is a matrix whose columns are corresponding right singular vectors satisfying $M_0^* M_0 \bar{V} = \bar{V}\bar{\Sigma}$.

Proof: Noting that the squares of the singular values of $M(t)$ are the eigenvalues of $M^*(t)M(t)$, the result follows from Lemma 1 by direct substitution.

MAIN RESULT

Theorem 1: Suppose that either $\chi = \mathscr{Y}$ or $\chi = \mathscr{D}$; see (5) and (9). Then every local minimum of the functional

$$\bar{\sigma}(XMX^{-1}) := \|XMX^{-1}\|_{2\to 2}, \; X \in \chi \tag{29}$$

is also a global minimum. In the case $\chi = \mathscr{D}$, the square of the functional (29) is actually convex in X.

Proof: Let X_1 be any invertible element of χ that is not a global minimum of (29). Choose an invertible $X_0 \in \chi$ such that $\bar{\sigma}(X_0 M X_0^{-1}) < \bar{\sigma}(X_1 M X_1^{-1})$. Without loss of generality we may choose X_0 such that $X_0 X_1^{-1}$ is hermitian and positive definite. Choose H to be the matrix logarithm of $X_0 X_1^{-1}$; so, H is determined by $X_0 X_1^{-1} = e^H$. Note that the matrix H is hermitian. To establish nonexistence of a local minimum, it suffices to prove for $t \in [0,1]$ convexity of the function

$$\bar{\sigma}^2(t) := \max_i \sigma_i^2(t) \tag{30}$$

where $\sigma_i(t), (i=1,\ldots,n)$ denote the n singular values of the matrix

$$e^{Ht} X_1 M X_1^{-1} e^{-Ht} . \tag{31}$$

Noting that the right hand side of (26) is positive definite whenever σ_0 is a maximal singular value, it follows that each function $\sigma_i^2(t)$ is convex over any interval on which $\bar{\sigma}(t) = \sigma_i(t)$. Since the maximum of several convex functions is always convex, we conclude that (30) is convex.† Q.E.D.

†An interesting implication of this is that one can always find a descent direction HX_1 with H hermitian.

DISCUSSION

In multivariable feedback synthesis it is known that the set of realizable closed-loop transfer function matrices can be parameterized in the form

$$M(W) = M_0 + M_1 W M_2 \tag{32}$$

where M_0, M_1 and M_2 are given stable transfer matrices determined by the plant and the matrix W may be any stable transfer function matrix (e.g., [9]-[10]). Thus, an important question is whether or not the gradient descent techniques can be used to minimize

$$\bar{\sigma}\left(D(M_0+M_1WM_2)D^{-1}\right). \tag{33}$$

Clearly this functional is convex in W since $\bar{\sigma}(A+B)\leq\bar{\sigma}(A)+\bar{\sigma}(B)\,\forall\, A,B$. So, for fixed D gradient descent methods will reliably converge to a global minimum. For fixed W, Theorem 1 guarantees convergence likewise. Unfortunately, the functional may still be non-convex as may be easily seen from the example

$$M(W) = \begin{bmatrix} 0 & 1-w \\ 1+w & 0 \end{bmatrix} \tag{34}$$

$$D = \begin{pmatrix} 1 & 0 \\ 0 & d \end{pmatrix} \tag{35}$$

for which

$$\bar{\sigma}(DM(W)D^{-1}) = \max\{|d(1+w)|, |d^{-1}(1-w)|\}. \tag{36}$$

Evidently $\bar{\sigma}$ has <u>two</u> local minima $(w,d) = (1,0)$ and $(w,d)=(-1,\infty)$ plus a saddle point at $(w,d)=(0,1)$.

The situation is no better with the usually well-behaved Frobenius norm (3). For the foregoing example one computes

$$||DM(W)D^{-1}||_F = (d^2(1+w)^2+d^2(1-w)^2)^{\frac{1}{2}} \tag{37}$$

which likewise has two local minima at $(w,d) = (1,0)$ and $(w,d)= (-1,\infty)$.

CONCLUSIONS

Motivated by the need to reliably compute the stability robustness measure $\mu(M)$, expressions for the first and second derivatives of the eigenvalues of a matrix $A(t)$ have been computed. Of special significance is the conclusion that the square of the greatest singular value of $e^{Dt}Me^{-Dt}$ is a convex function of t for any M and any diagonal D. Convexity guarantees the reliable convergence of gradient descent algorithms such as developed by Doyle [2] for computing the optimal diagonal scaling matrix D for bounding $\mu(M)$. More generally, our Theorem 1 establishes the nonexistence of disjoint local minima of $\{\bar{\sigma}(YMY^{-1}) \mid Y \in \mathcal{Y}\}$ ensuring the reliable convergence of gradient descent algorithms for such problems.

REFERENCES

[1] M.G. Safonov, "Stability Margins of Diagonally Perturbed Multivariable Feedback Systems." IEE Proc, Part D, Control Theory and Applications, 129,November 1982.

[2] J.C. Doyle, "Analysis of Feedback Systems with Structured Uncertainties," IEE Proc., Part D, Control Theory and Applications, 129, November 1982.

[3] M.F. Barratt, "Conservatism with Robustness Tests for Linear Feedback Systems," Ph.D. Thesis, Univ. of Minnesota, June 1980.

[4] M.G. Safonov and M. Athans, "A Multiloop Generalization of the Circle Criterion for Stability Margin Analysis,"IEE Trans. on Automatic Control, AC-26,April 1981.

[5] J.H. Wilkenson, The Algebraic Eigenvalue Problem, Oxford: Clarendon Press, 1965.

[6] E.E. Osborne,"On Preconditioning of Matrices," J. Assoc. Comput. Mach., 7, pp. 338-345, 1960.

[7] F.L. Bauer, "Optimally Scaled Matrices," Numer. Math., 5,pp.73-87, 1963.

[8] T. Kato, Perturbation Theory for Linear Operators. NY: Springer-Verlag, 1976.

[9] D.C. Youla, H.A. Jabr, and J.J. Bongiorno, Jr., "Modern Wiener-Hopf Design of Optimal Controllers, Part II: The Multivariable Case." IEEE Trans. on Automatic Control, AC-21,pp. 319-338,1976.

[10] C.A. Desoer, R.W. Liu, J. Murray, and R. Saeks, "Feedback System Design: The Fractional Representation Approach to Analysis and Synthesis," IEEE Trans. on Automatic Control, AC-25,pp. 399-412, 1980.

APPENDIX

Proof of Lemma 1: Transform $A(t)$ into its Jordan form

$$A(t)=P(t)\begin{bmatrix}\Lambda(t) & 0\\ 0 & \bar{\Lambda}(t)\end{bmatrix}\hat{P}^*(t) \tag{38}$$

where

$$\hat{P}^*(t) = P^{-1}(t). \tag{39}$$

and the columns of $P(t)$ and $\hat{P}^*(t)$ are analytic in t near $t = 0$. Without loss of generality, assume

$$P(0) = P_o = [R_o, Q_o]. \tag{40}$$

Partition $P(t)$ and $\hat{P}^*(t)$ as

$$P(t) = [R(t), Q(t)] \tag{41}$$

$$\hat{P}^*(t) = [\hat{R}^*(t), \hat{Q}(t)] \tag{42}$$

and expand in Taylor series about $t = 0$

$$R(t) = R_0 + tR_1 + \frac{1}{2}t^2 R_2 + O(t^3) \tag{43}$$

$$Q(t) = Q_o + tQ_1 + O(t^2) \tag{44}$$

$$\hat{R}^*(t) = \hat{R}^*_o + t\hat{R}^*_1 + \frac{1}{2}t^2\hat{R}^*_2 + O(t^3) \tag{45}$$

$$\hat{Q}^*(t) + \hat{Q}^*_o + t\hat{Q}^*_1 + O(t^2) \tag{46}$$

Note that in general the matrix R_1 can be represented in the form

$$R_1 = Q_o X + R_o Y \tag{47}$$

for some $X \in \mathbb{C}^{(nxm)xm}$ and $Y \in \mathbb{C}^{mxm}$. Thus in general

$$R(t) = R_o(I+tY) + tQ_o X + O(t^2) . \tag{48}$$

Next, make the substitutions

$$R_{new}(t)(I+tY) = R(t) \tag{49}$$

$$(I+tY)^{-1}\hat{R}^*_{new}(t) = \hat{R}^*(t) \tag{50}$$

$$(I+tY)^{-1}\Lambda_{new}(t)\ (I+Y) = \Lambda(t). \tag{51}$$

This leaves (37)-(46) intact, except that $\Lambda_{new}(t)$ is no longer precisely in Jordan form. However, expanding (51) in powers of t and setting like powers of t equal one has:

$$t^0: \quad \Lambda_{new}(0) = \Lambda(0) \tag{52}$$

$$t^1: \quad \dot{\Lambda}_{new}(0) = \dot{\Lambda}(0) \tag{53}$$

$$t^2: \quad \ddot{\Lambda}_{new}(0) = \ddot{\Lambda}(0) + Y\dot{\Lambda}(0) - \dot{\Lambda}Y. \tag{54}$$

Furthermore, when $\dot{\Lambda}(0)$ is diagonal and all its diagonal elements are equal, then $Y\Lambda = \Lambda Y$ and

$$\ddot{\Lambda}_{new}(0) = \ddot{\Lambda}(0). \tag{55}$$

Thus, we may assume $Y = 0$.

Equating like powers of t in the Taylor series expansion of the identity $\hat{P}^*(t)P(t) = I$ yields the useful identities

$$\begin{bmatrix} I & 0 \\ 0 & I \end{bmatrix} = \begin{bmatrix} \hat{R}_0^* R_0 & \hat{R}_0^* Q_0 \\ \hat{Q}_0^* R_0 & \hat{Q}_0^* Q_0 \end{bmatrix} \tag{56}$$

Expanding the identity $0 = R(t)\Lambda(t) - A(t)R(t)$, equating terms involving like powers of t, multiplying on the left by $\hat{R}_0^*$, and simplifying using the identity (56) yields

$$\dot{\Lambda} = \hat{R}_0^* \dot{A} R_0 \tag{57}$$

and

$$\ddot{\Lambda} = \hat{R}_0^* (\ddot{A} R_0 + 2\dot{A} R_1). \tag{58}$$

Doing the same, but this time multiplying on the left by $\hat{Q}_0^*$ instead of $\hat{R}_0^*$ yields

$$R_1 = Q_0 X = Q_0 (\lambda_0 I - \bar{\Lambda}_0)^{-1} \hat{Q}_0^* \dot{A} R_0 \tag{59}$$

from whence it follows that, for equal $\dot{\lambda}^i$ $(i = 1, \ldots, m)$ and (31) diagonal,

$$\ddot{\Lambda}(0) = \hat{R}_0^* (\ddot{A} + 2AQ_0 (\lambda_0 I - \bar{\Lambda}_0)^{-1} \hat{Q}_0^* \dot{A}) R_0. \tag{60}$$

PART III ALGEBRAIC AND OPTIMAL CONTROLLER DESIGN

CHAPTER 12

FREQUENCY ASSIGNMENT PROBLEMS IN LINEAR MULTIVARIABLE SYSTEMS: EXTERIOR ALGEBRA AND ALGEBRAIC GEOMETRY METHODS

N. Karcanias and C. Giannakopoulos
Control Engineering Centre
School of Electrical Engineering and Applied Physics
The City University
Northampton Square, London EC1V 0HB

ABSTRACT

The determinantal assignment problem (DAP) is defined as a generalisation of the pole, zero assignment problems of linear multivariable theory. The multilinear nature of DAP is reduced to a linear problem of zero assignment of polynomial combinants and a standard problem of decomposability of multivectors. The characterisation of the system problems by decomposable polynomial multivectors leads to the definition of the various system Plücker matrices. Necessary conditions for the solvability of frequency assignment problems are given in terms of the new system invariants, the Plücker matrices. The multilinear problem of decomposability is characterised by a minimal set of algebraically independent quadratics, the Reduced Quadratic Plücker Relations (RQPR). The set of RQPRs is used for the study of linearising compensators (feedbacks), and a linearising family of feedbacks superior to that of dyadic feedbacks is defined. A new proof to the pole assignment theorem by state feedback is given. The approach unifies the various problems of frequency assignment, provides a common algebrogeometric framework for their study, and establishes the basis for the development of a common algorithmic procedure for the computation of solution.

1. INTRODUCTION

This paper is concerned with the development of a unifying approach for problems of pole assignment by state, output feedback [1,2] and the problem of zero assignment by "squaring down" [3,4]. These problems are shown to be special cases of the determinantal assignment problem (DAP). The multilinear nature of DAP suggests that the natural framework for its study is that of exterior algebra [5,6] and algebraic geometry [7]. Recent work [8,9,10] has demonstrated the importance of concepts and techniques from exterior algebra [8] and algebraic geometry [9,10] for linear control theory. The present approach relies on the explicit description of a Grassmannian variety (Quadratic Plücker Relations [7]) of a projective space and thus the analysis leads to

S. G. Tzafestas (ed.), Multivariable Control, 211–232.

algorithmic procedures. The general case of decomposability of multivectors is considered, rather than the special case where a multivector is always decomposable [8].

Problems of pole, zero assignment in linear systems are shown to be special cases of a more general problem, the determinantal assignment problem (DAP). The definition of this new problem suggests that a common, unifying approach may be developed for the study of solvability, as well as computation of solution of all frequency assignment problems. The characteristic polynomial of DAP is expressed as inner product of two decomposable multivectors: the system polynomial multivector $\underline{m}(s)\wedge$ and the compensator constant multivector $\underline{h}\wedge$. By setting $\underline{h}\wedge = \underline{k}$, $\underline{k}$ free, the linear part of DAP is defined as a problem of zero assignment of the polynomial combinant [11] $f_M(s,\underline{k}) = \langle \underline{k}, \underline{m}(s)\wedge \rangle$. The nonlinear part of DAP is related to the problem of realisability of solutions of the linear part as compensators, and it is defined by the exterior equation $\underline{h}\wedge = \underline{k}$; the latter equation has a solution when $\underline{k}$ corresponds to a point in the Grassmannian variety of a projective space, or when $\underline{k}$ is decomposable [6]. When $\underline{m}(s)\wedge$ is reduced (the polynomials are coprime) and monic, then it defines a complete invariant of the rational vector space $\mathcal{X}_M$ associated with DAP [17] and the matrix coefficient of $\underline{m}(s)\wedge$, P, is defined as the Plücker matrix of $\mathcal{X}_M$. Some structural properties of P are investigated and the importance of the Plücker matrices in the linear part of DAP is investigated. As a result of this study, necessary conditions for assignability for the various frequency assignment problems are derived. The controllability, observability Plücker matrices are defined and a new criterion for controllability, observability respectively is given in terms of the corresponding Plücker matrices.

The decomposability of multivectors is examined next. The general set of Quadratic Plücker Relations (QPR), characterising the solvability of $\underline{h}\wedge = \underline{k}$, is considered and an independent set of them is defined as the Reduced QPR (RQPR). The set of RQPR is then used to study a family H_l of compensators which completely linearises DAP. Using the H_l family an alternative simple proof to the pole assignment by state feedback theorem [1] is given. It is shown that the family H_d of dyadic compensators also linearises DAP; apart from the unity rank disadvantage, H_d has less degrees of freedom than H_l and thus H_l is superior to the H_d family.

NOTATION

$\mathbb{R}, \mathbb{C}, \mathbb{R}(s)$ denote the fields of real, complex numbers, and rational functions respectively. $\mathbb{R}[s]$ is the ring of polynomials and $\mathbb{R}^n[s]$ is the nx1 vector with elements from $\mathbb{R}[s]$. $\mathbb{R}^n, \mathbb{C}^n, \mathbb{R}^n(s)$ denote vector spaces over $\mathbb{R}, \mathbb{C}, \mathbb{R}(s)$ respectively. If F is a field, or ring, then $F^{m\times n}$ denote the set of matrices with elements from F. Script capital letters denote linear vector spaces and Roman capitals denote linear transformations. The range space of a map H is denoted by $\mathcal{R}(H)$ and its right, left null space by $\mathcal{N}_r(H)$, $\mathcal{N}_\ell(H)$. If $\mathcal{V}$ is a vector space, then we denote by $\underline{v}$ a vector, V a basis and V a basis matrix of $\mathcal{V}$. $Q_{k,n}$ denotes the set of lexicographically ordered, strictly increasing sequences of k integers from 1,2,...,n. If $\{\underline{x}_{i_1},\dots,\underline{x}_{i_k}\}$ is a set of vectors of V, $\omega = (i_1,\dots,i_k) \in Q_{k,n}$, then $\underline{x}_{i_1} \wedge \dots \wedge \underline{x}_{i_k} = \underline{x}_{\omega}\wedge$ denotes their exterior product and by $\wedge^r V$ we denote the r-th exterior power of V. If $H \in F^{m\times n}$ and $r \le \min\{m,n\}$, then by $C_r(H)$ we denote the r-th compound matrix of H [12]. Finally, if a property is said to be true for $i \in \underset{\sim}{n}$, this means that it is true for all $1 \le i \le n$.

2. THE DETERMINANTAL POLE, ZERO ASSIGNMENT PROBLEMS

Consider the linear system described by

$$S(A,B,C,D): \quad \begin{array}{llll} \dot{\underline{x}} = A\underline{x}+B\underline{u} , & A \in \mathbb{R}^{n\times n} , & B \in \mathbb{R}^{n\times \ell} \\ \underline{y} = C\underline{x}+D\underline{u} , & C \in \mathbb{R}^{m\times n} , & D \in \mathbb{R}^{m\times \ell} \end{array} \qquad (2.1)$$

where (A,B) controllable, (A,C) observable, or by the transfer function matrix $G(s) = C(sI-A)^{-1}B+D$, where $\operatorname{rank}_{\mathbb{R}(s)} \{G(s)\} = \min\{m,\ell\}$. In terms of left, right coprime matrix fraction descriptions (LCMFD , RCMFD) G(s) may be represented as

$$G(s) = D_l(s)^{-1}N_l(s) = N_r(s)D_r(s)^{-1} \qquad (2.2)$$

where $N_l(s),N_r(s) \in \mathbb{R}^{m\times \ell}[s], D_l(s) \in \mathbb{R}^{m\times m}[s]$ and $D_r(s) \in \mathbb{R}^{\ell\times \ell}[s]$. The system will be called *square* if $m = \ell$, and *nonsquare* if $m \ne \ell$. A number of frequency assignment problems arising in linear control theory are considered below.

(i) Pole assignment by state feedback: Consider $L \in \mathbb{R}^{n\times \ell}$, where L is a state feedback applied on the system (2.1). The closed loop characteristic polynomial is then given by

$$p_L(s) = \det\{sI-A-BL\} = \det\{B(s)\tilde{L}\} \qquad (2.3)$$

where $B(s) = [sI-A,-B]$ and $\tilde{L} = [I_n,L^t]^t$.

(ii) Design of an n-state observer: Consider the problem of designing an n-state observer for the system of (2.1). The characteristic polynomial of the observer is then defined by

$$p_T(s) = \det\{sI-A-TC\} = \det\{\tilde{T}C(s)\} \qquad (2.4)$$

where $T \in \mathbb{R}^{n\times m}$ is a feedback, $\tilde{T} = [I_n,T]$ and $C(s) = [sI-A^t,-C^t]^t$.

(iii) Pole assignment by constant output feedback: Consider the system described by (2.2) under an output feedback $F \in \mathbb{R}^{m\times \ell}$. The closed-loop characteristic polynomial $p_F(s)$ and the open-loop characteristic polynomial $p_o(s)$ satisfy the relation

$$p_F(s)/p_o(s) = \det\{I_m+G(s)F\} = \det\{I_\ell+FG(s)\} \tag{2.5}$$

Using the LCMFD, RCMFD of G(s), Eqn.(2.5) yields

$$p_F(s) = \det\{D_l(s)+N_l(s)F\} = \det\{D_r(s)+FN_r(s)\} \tag{2.6}$$

By defining the matrices

$$T_l(s)=[D_l(s),N_l(s)]\in\mathbb{R}^{m\times(m+\ell)}[s],\ T_r(s)=\begin{bmatrix}D_r(s)\\N_r(s)\end{bmatrix}\in\mathbb{R}^{(m+\ell)\times\ell}[s]$$

$$\tilde{F}_l=[I_m,F^t]^t\in\mathbb{R}^{(m+\ell)\times m},\ \tilde{F}_r=[I_\ell,F]\in\mathbb{R}^{\ell\times(m+\ell)} \tag{2.7}$$

then,

$$p_F(s) = \det\{T_l(s)\tilde{F}_l\} = \det\{\tilde{F}_r T_r(s)\} \tag{2.8}$$

(iv) Zero assignment by squaring down: For a system with $m>\ell$, we can expect to have independent control over at most ℓ linear combinations of m outputs. If $\underline{c}\in\mathbb{R}^\ell$ is the vector of the variables which are to be controlled, then $\underline{c}=K\underline{y}$, where $K\in\mathbb{R}^{\ell\times m}$ is a *squaring down postcompensator*, and $G'(s)=KG(s)$ is the *squared down transfer function matrix* [3]. A right MFD for G'(s) is defined by $G'(s)=KN_r(s)D_r(s)^{-1}$, where $G(s)=N_r(s)D_r(s)^{-1}$. Finding K such that G'(s) has assigned zeros is defined as the *zero assignment by squaring down* problem. The zero polynomial of $S(A,B,KC,KD)$ is given by

$$z_K(s) = \det\{KN_r(s)\} \tag{2.9}$$

The problems listed above have been defined for regular state-space systems; similar problems, for extended state space, or descriptor type systems [18], may also be formulated in a similar manner. The common formulation of these problems clearly suggest that they are special cases of a more general problem, which is defined below.

The determinantal assignment problem: Let $M(s)\in\mathbb{R}^{p\times r}[s]$, $r\le p$, $\text{rank}_{\mathbb{R}(s)}\{M(s)\}=r$ and let $\mathcal{H}=\{H: H\in\mathbb{R}^{r\times p},\text{rank}\{H\}=r\}$. Finding $H\in\mathcal{H}$ such that the polynomial

$$f_M(s,H) = \det\{HM(s)\} \tag{2.10}$$

has assigned zeros, is defined as the *determinantal assignment problem* (DAP). If $\underline{h}_i^t$, $\underline{m}_i(s)$, $i\in\underset{\sim}{r}$, denote the rows of H, columns of M(s) respectively, then $C_r(H)=\underline{h}_1^t\wedge\ldots\wedge\underline{h}_r^t=\underline{h}^t\wedge\in\mathbb{R}^{1\times\sigma}$ and $C_r(M(s))=$ $=\underline{m}_1(s)\wedge\ldots\wedge\underline{m}_r(s)=\underline{m}(s)\wedge\in\mathbb{R}^\sigma[s]$, $\sigma=\binom{p}{r}$ and by the Binet-Cauchy theorem [12] we have that

$$f_M(s,H)=C_r(H)C_r(M(s))=\langle\underline{h}\wedge,\underline{m}(s)\wedge\rangle=\sum_{\omega\in Q_{r,p}} h_\omega m_\omega(s) \tag{2.11}$$

where $\langle\cdot,\cdot\rangle$ denotes inner product, $\omega=(i_1,\ldots,i_r)\in Q_{r,p}$, and $h_\omega, m_\omega(s)$ are the coordinates of $\underline{h}\wedge,\underline{m}(s)\wedge$ respectively. Note that h_ω is the rxr minor of H which corresponds to the ω set of columns of H and thus h_ω is a multilinear alternating function of the entries h_{ij} of H. The multilinear, skew symmetric nature of DAP suggests that the natural framework for its study is that of exterior algebra [5,6]. The essence of exterior algebra is that it reduces the study of multilinear skew-symmetric functions to the simpler study of linear functions. The study of the zero structure of the multilinear function $f_M(s,H)$ may thus be reduced to a linear subproblem and a standard multilinear algebra problem as it is shown below.

(i) Linear subproblem of DAP: Set $\underline{m}(s)\wedge=\underline{p}(s)\in\mathbb{R}^\sigma[s]$ Determine whether there exists a $\underline{k}\in\mathbb{R}^\sigma$, $\underline{k}\neq\underline{0}$, such that

$$f_M(s,\underline{k}) = \underline{k}^t\underline{p}(s) = \Sigma\, k_i p_i(s) = f(s),\ i\in\underset{\sim}{\sigma},\ f(s)\in\mathbb{R}[s] \tag{2.12}$$

(ii) Multilinear subproblem of DAP: Assume that K is the family of solution vectors $\underline{k}$ of (2.12). Determine whether there exists $H^t=[\underline{h}_1,\ldots,\underline{h}_r]$, where $H^t\in\mathbb{R}^{pxr}$, such that

$$\underline{h}_1\wedge\ldots\wedge\underline{h}_r=\underline{h}\wedge=\underline{k},\ \underline{k}\in K \tag{2.13}$$

Polynomials defined by Eqn.(2.12) are called *polynomial combinants* [11,13] and the zero assignability of them provides necessary conditions for the solution of DAP. The solution of the exterior equation (2.13) is a standard problem of exterior algebra and it is known as *decomposability* of multivectors. Note that notions and tools from exterior algebra play also an important role in the linear subproblem, since $f_M(s,\underline{k})$ is generated by the decomposable multivector $\underline{m}(s)\wedge$. Some of the key notions and results from exterior algebra which are relevant to this approach are summarised next [5,6].

3. THE GRASSMANN REPRESENTATIVE OF A VECTOR SPACE

Let U be a linear vector space over a field F and let $\dim U=p$. The *Grassmannian* $G(r,U)$ is defined as the set of r-dimensional subspaces V of U. Let $V\in G(r,U)$ and let $V=\{\underline{v}_i, i\in\underset{\sim}{r}\}$, $U=\{\underline{u}_j, j\in\underset{\sim}{p}\}$ be bases of V,U respectively. The injection map f: $V\to U$ defined by $f(x)=\underline{x}$, $\underline{x}\in V$, is linear and induces an injection map $\wedge^\tau f$: $\wedge^\tau V\to\wedge^\tau U$. If $\tau=r$, then $\wedge^r V$ is a one-dimensional subspace of $\wedge^r U$ and it is spanned by elements $\underline{v}_1\wedge\ldots\wedge\underline{v}_r$. A matrix representation of f with respect to V,U, induces a matrix representation of $\wedge^r f$ with respect to the induced bases $\wedge^r V$, $\wedge^r U$ of $\wedge^r V$, $\wedge^r U$. The construction of such matrix representation is described by the following commutative diagram:

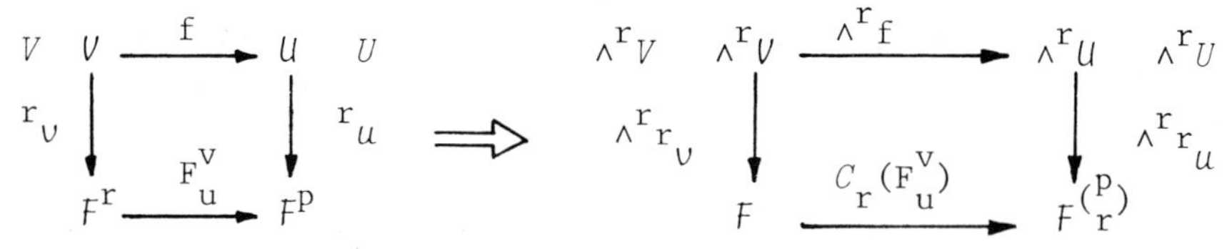

$$
\begin{array}{ccc}
V \quad V \xrightarrow{\ f\ } U \quad U \\
{\scriptstyle r_\upsilon} \downarrow \qquad \downarrow {\scriptstyle r_u} \\
F^r \xrightarrow[\ F^V_u\ ]{} F^p
\end{array}
\implies
\begin{array}{ccc}
\wedge^r V \quad \wedge^r V \xrightarrow{\ \wedge^r f\ } \wedge^r U \quad \wedge^r U \\
{\scriptstyle \wedge^r r_\upsilon} \downarrow \qquad \downarrow {\scriptstyle \wedge^r r_u} \\
F \xrightarrow[\ C_r(F^V_u)\ ]{} F^{\binom{p}{r}}
\end{array}
$$

where r_υ, r_u are representation maps, $\wedge^r r_\upsilon, \wedge^r r_u$ are their r-th exterior powers, $V = F^V_u = [\underline{\upsilon}_1, \ldots, \underline{\upsilon}_r]$, $V \in F^{p \times r}$ is the matrix representation of f and $C_r(F^V_u) = \underline{\upsilon}_1 \wedge \ldots \wedge \underline{\upsilon}_r \in F^\sigma$, $\sigma = \binom{p}{r}$, is the matrix representation of $\wedge^r f$. If $\underline{\upsilon}_1 \wedge \ldots \wedge \underline{\upsilon}_r = [\ldots, c_\omega, \ldots]^t$, $\omega = (i_1, \ldots, i_r) \in Q_{r,p}$ and $\wedge^r U = \{u_\omega = \underline{u}_{i_1} \wedge \ldots \wedge \underline{u}_{i_r}\}$, then

$$\underline{v}^\wedge = \underline{v}_1 \wedge \ldots \wedge \underline{v}_r = \Sigma c_\omega \underline{u}_\omega \wedge, \quad \omega \in Q_{r,p} \tag{3.1}$$

The σ-tuple c_ω are the coordinates of $\wedge^r V$ with respect to the bases V, U and they are known as the *Plücker coordinates* (PC) of V. If V' is another basis of V, $F^{V'}_u$ the matrix representation of f with respect to V', U, then $F^{V'}_u = F^V_u Q^V_{V'}$, where $Q^V_{V'}$ is a coordinate transformation. By the Binet-Cauchy theorem

$$\underline{v}'^\wedge = \underline{v}'_1 \wedge \ldots \wedge \underline{v}'_r = \wedge^r U C_r(F^{V'}_u) = \wedge^r U C_r(F^V_u) C_r(Q^V_{V'}) = a \cdot \underline{v}^\wedge \tag{3.2}$$

where $a = \det\{Q^V_{V'}\} \in F - \{0\}$. Thus, the two sets of PC of V differ only by a scalar. The PCs of V, enumerated in lexicographic order, can be considered as the homogeneous coordinates of a point in the projective space $\mathbb{P}_{\sigma-1}(F)$. However, not every point in $\mathbb{P}_{\sigma-1}(F)$ represents a $V \in G(r,U)$, since $\wedge^r V$ is spanned by $\underline{v}_1 \wedge \ldots \wedge \underline{v}_r$. Vectors of $\wedge^r U$ of the type $\underline{v}_1 \wedge \ldots \wedge \underline{v}_r$ are called *simple* or *decomposable*.

Theorem (3.1): Let $\underline{x}^\wedge = \underline{x}_1 \wedge \ldots \wedge \underline{x}_r$, $\underline{z}^\wedge = \underline{z}_1 \wedge \ldots \wedge \underline{z}_r$ be two nonzero and decomposable vectors of $\wedge^r U$ and let $V_x, V_z \in G(r,U)$ be the associated subspaces. Necessary and sufficient condition for $V_x = V_z$ is that $\underline{x} = a \cdot \underline{z}$, where $a \in F - \{0\}$. □

Corollary (3.1.1): A nonzero, decomposable vector $\underline{v}^\wedge \in \wedge^r U$, $\underline{v}^\wedge = \underline{v}_1 \wedge \ldots \wedge \underline{v}_r$, uniquely defines (modulo $a \in F$) $V = \text{span}\{\underline{v}_i\} \in G(r,U)$. □

Such a multivector $\underline{v}^\wedge$ is called a *Grassmann representative* (GR) of V. The map $\upsilon : V \to \wedge^r V \in \wedge^r U$ expresses a natural injective correspondence between $G(r,U)$ and one-dimensional subspaces of $\wedge^r U$. By associating to every $\wedge^r V$ the Plücker coordinates, the map $p : G(r,U) \to \mathbb{P}_{\sigma-1}(F)$ is defined; p is known as the *Plücker embedding* of $G(r,U)$ in $\mathbb{P}_{\sigma-1}(F)$. The Plücker image of $G(r,U)$ in $\mathbb{P}_{\sigma-1}(F)$ is a $(p-r)r$-dimensional algebraic variety, known as the *Grassmann variety* of $\mathbb{P}_{\sigma-1}(F)$. This variety is defined by a linear system of quadratics, known as *Quadratic Plücker*

Relations (QPR). The GRs, which may be associated with a linear system, determine the solvability of the linear part of DAP, whereas the QPRs determine the realisability of solutions of the latter problem as compensators.

4. PLÜCKER MATRICES AND THE LINEAR SUBPROBLEM OF DAP

Let $T(s) \in \mathbb{R}^{p\times r}(s)$, $T(s) = [\underline{t}_1(s),\dots,\underline{t}_r(s)]$, $p \geq r$, $\text{rank}_{\mathbb{R}(s)}\{T(s)\} = r$ and let $\mathcal{X}_t = \mathcal{R}_{\mathbb{R}(s)}(T(s))$. If $T(s) = M(s)D(s)^{-1}$ is a RCMFD of $T(s)$, then $M(s)$ is a polynomial basis for $\mathcal{X}_t$. If $Q(s)$ is a greatest right divisor of $M(s)$ [14] then $T(s) = \tilde{M}(s)Q(s)D(s)^{-1}$, where $\tilde{M}(s)$ is a least degree polynomial basis of $\mathcal{X}_t$ [15]. A GR for $\mathcal{X}_t$ is defined by

$$\underline{t}(s)\wedge = \underline{t}_1(s)\wedge\dots\wedge\underline{t}_r(s) = \tilde{\underline{m}}_1(s)\wedge\dots\wedge\tilde{\underline{m}}_r(s)\cdot z_t(s)/p_t(s) \tag{4.1}$$

where $z_t(s) = \det\{Q(s)\}$, $p_t(s) = \det\{D(s)\}$ are the zero, pole polynomials of $T(s)$ and $\tilde{\underline{m}}(s)\wedge = \tilde{\underline{m}}_1(s)\wedge\dots\wedge\tilde{\underline{m}}_r(s) \in \mathbb{R}^{\sigma}[s]$, $\sigma = \binom{p}{r}$, is also a GR of $\mathcal{X}_t$. Since $\tilde{M}(s)$ is a least degree polynomial basis of $\mathcal{X}_t$, the polynomials of $\tilde{\underline{m}}(s)\wedge$ are coprime and $\tilde{\underline{m}}(s)\wedge$ will be referred to as a *reduced polynomial GR* (R-$\mathbb{R}$[s]-GR) of $\mathcal{X}_t$. If $\delta = \deg(\tilde{\underline{m}}(s)\wedge)$, then δ is the Forney dynamical order [16] of $\mathcal{X}_t$. $\tilde{\underline{m}}(s)\wedge$ may always be expressed as

$$\tilde{\underline{m}}(s)\wedge = \underline{p}(s) = \underline{p}_0 + s\underline{p}_1 + \dots + s^{\delta}\underline{p}_{\delta} = P_{\delta}\underline{e}_{\delta}(s), \quad P_{\delta} \in \mathbb{R}^{\sigma\times(\delta+1)} \tag{4.2}$$

where P_{δ} is a basis matrix for $\tilde{\underline{m}}(s)\wedge$ and $\underline{e}_{\delta}(s) = [1,s,\dots,s^{\delta}]^t$. It can be readily shown that all R-$\mathbb{R}$[s]-GRs of $\mathcal{X}_t$ differ only by a nonzero scalar factor $a \in \mathbb{R}$. By choosing an $\tilde{\underline{m}}(s)\wedge$ for which $\|\underline{p}_{\delta}\| = 1$, a *monic* R-$\mathbb{R}$[s]-GR is defined; such a GR of $\mathcal{X}_t$ is defined as the *canonical polynomial Grassmann representative* (C-$\mathbb{R}$[s]-GR) of $\mathcal{X}_t$ [17] and shall be denoted by $\underline{g}(\mathcal{X}_t)$. The basis matrix P_{δ} of $\underline{g}(\mathcal{X}_t)$ is defined as the *Plücker matrix* of $\mathcal{X}_t$.

Theorem (4.1) [17]: $\underline{g}(\mathcal{X}_t)$, or the associated Plücker matrix P_{δ}, is a complete (basis free) invariant of $\mathcal{X}_t$. □

Corollary (4.1.1) [17]: Let $T(s) \in \mathbb{R}^{p\times r}(s)$, $p \geq r$, $\text{rank}_{\mathbb{R}(s)}\{T(s)\} = r$, $z_t(s)$, $p_t(s)$ be the monic zero, pole polynomials of $T(s)$ and let $\underline{g}(\mathcal{X}_t) = \underline{p}(s)$ be the C-$\mathbb{R}$[s]-GR of the column space $\mathcal{X}_t$ of $T(s)$. $\underline{t}(s)\wedge$ may be uniquely decomposed as

$$\underline{t}(s)\wedge = c\cdot\underline{p}(s)\cdot z_t(s)/p_t(s), \quad \text{where} \quad c \in \mathbb{R} - \{0\} \tag{4.3}$$

□

If $M(s) \in \mathbb{R}^{p\times r}[s]$, $p \geq r$, $\text{rank}_{\mathbb{R}(s)}\{M(s)\} = r$, then $M(s) = \tilde{M}(s)Q(s)$, where $\tilde{M}(s)$ is a least degree basis and $Q(s)$ is a greatest right divisor of the rows of $M(s)$ and thus

$$\underline{m}(s)\wedge = \tilde{\underline{m}}(s)\wedge\cdot\det\{Q(s)\} = \underline{p}(s)\cdot z_m(s) = P_{\delta}\underline{e}_{\delta}(s)\cdot z_m(s) \tag{4.4}$$

The linear part of DAP is thus reduced to

$$f_M(s,\underline{k}) = \underline{k}^t \underline{p}(s) z_m(s) = \underline{k}^t P_\delta \underline{e}(s) z_m(s) \tag{4.5}$$

Proposition (4.1): The zeros of $M(s)$ are fixed zeros of all combinants of $\underline{m}(s)\wedge$.

□

The zeros of $f_M(s,\underline{k})$ which may be freely assigned are those of the combinant $f_{\tilde{M}}(s,\underline{k}) = \underline{k}^t \tilde{\underline{m}}(s)\wedge$, where $\tilde{\underline{m}}(s)\wedge$ is reduced. Given that the zeros of $f_{\tilde{M}}(s,\underline{k})$ are not affected by scaling with constants, we may always assume that $\tilde{\underline{m}}(s)\wedge = P_\delta \underline{e}_\delta(s)$. In the following, the case of combinants generated by reduced $\tilde{\underline{m}}(s)\wedge$ will be considered. If $a(s) \in \mathbb{R}[s]$ is the polynomial which has to be assigned, then $\max(\deg a(s)) = \delta$, where δ is the Forney dynamical order of $\mathcal{X}_t$. If $a(s) = \underline{a}_\delta^t \underline{e}_\delta(s) = a_o + a_1 s + \ldots + a_\delta s^\delta$, where $\underline{a} \in \mathbb{R}^{\delta+1}$, then the problem of finding $\underline{k}$, $\underline{k} \in \mathbb{R}^\sigma$, such that $f_{\tilde{M}}(s,\underline{k}) = a(s)$ is reduced to the solution of

$$P_\delta^t \underline{k} = \underline{a}\ , \quad P_\delta^t \in \mathbb{R}^{(\delta+1)\times\sigma}\ , \quad \sigma = \binom{p}{r} \tag{4.6}$$

The matrix $M(s) \in \mathbb{R}^{p\times r}[s]$ generating DAP will be called *linearly assignable* (LA), if Eqn.(4.6) has a solution for all $\underline{a}$; otherwise, it will be called *linearly nonassignable* (LNA). $M(s)$ will be called *completely assignable* (CA), if it is LA and Eqn.(2.13) has a solution for at least a solution of the linear problem defined by Eqn.(4.6). Some results characterising the above mentioned properties of $M(s)$ are listed below.

Proposition (4.2): Necessary condition for $M(s)$ to be LA is that $M(s)$ is a least degree matrix (i.e. has coprime rows).

□

Proposition (4.3): Let $M(s) \in \mathbb{R}^{p\times r}[s]$ be a least degree matrix, P_δ be the Plücker matrix of $\mathcal{X}_m$, δ the Forney order of $\mathcal{X}_m$ and let $\pi = \text{rank}\{P_\delta\}$. Necessary and sufficient condition for $M(s)$ to be LA is that $\pi = \delta+1$ (i.e. $\binom{p}{r} \geq \delta+1$ and $\pi = \delta+1$).

□

For LA $M(s)$ the solution of Eqn.(4.6) is defined below.

Proposition (4.4): Let $M(s) \in \mathbb{R}^{p\times r}[s]$ be LA, P_δ be the Plücker matrix of $\mathcal{X}_m$, and let $\tilde{P}^\dagger$ be right inverse of P_δ^t and $\tilde{P}^\perp$ a basis for $N_r(P_\delta^t)$. For every $\underline{a} \in \mathbb{R}^{\delta+1}$, the solution of (4.6) is given by

$$\underline{k} = \tilde{P}^\dagger \underline{a} + \tilde{P}^\perp \underline{c}\ , \text{ where } \underline{c} \in \mathbb{R}^{\sigma-\delta-1} \text{ arbitrary} \tag{4.7}$$

□

For the control problems discussed in section (2), the matrix $M(s)$ has a special structure; thus the matrix coefficient of $\underline{m}(s)\wedge$ has important properties which stem from the properties of the corresponding control problem. A number of Plücker type matrices associated with a linear system are defined below:

(i) For the pair (A,B), $\underline{b}(s)^t\wedge$ denotes the exterior product of the rows of $B(s) = [sI-A,-B]$ and $P(A,B)$ is the $(n+1)\times\binom{n+\ell}{n}$ basis matrix of $\underline{b}(s)^t\wedge$. $P(A,B)$ will be called the *controllability Plücker matrix*.

(ii) For the pair (A,C), $\underline{c}(s)\wedge$ denotes the exterior product of the columns of $C(s) = [sI-A^t, -C^t]^t$ and P(A,C) is the $\binom{n+m}{n}\times(n+1)$ basis matrix of $\underline{c}(s)\wedge$. P(A,C) will be called the *observability Plücker matrix*.

(iii) For the transfer function matrix G(s) represented by the RCMFD, LCMFD of Eqn.(2.2) we define by $\underline{t}_r(s)\wedge$, $\underline{t}_l(s)^t\wedge$ the exterior product of the columns of $T_r(s)$, rows of $T_l(s)$ respectively, where $T_r(s)$, $T_l(s)$ are defined by Eqn.(2.7). By $P(T_r)$ we denote the $\binom{m+\ell}{\ell}\times(n+1)$ basis matrix of $\underline{t}_r(s)\wedge$, and by $P(T_l)$ the $(n+1)\times\binom{m+\ell}{m}$ basis matrix of $\underline{t}_l(s)^t\wedge$. $P(T_r)$, $P(T_l)$ will be referred to as the *right*, *left fractional representation Plücker matrices* respectively.

(iv) For the transfer function G(s), $m\geq\ell$, we denote by $\underline{n}(s)\wedge$ the exterior product of the columns of the numerator $N_r(s)$, of a RCMFD and by P(N) the $\binom{m}{\ell}\times(d+1)$ the basis matrix of $\underline{n}(s)\wedge$. Note that $d=\delta$, the Forney order of $\mathcal{X}_g$, if G(s) has no finite zeros and $d=\delta+\kappa$, where κ is the number of finite zeros of G(s), otherwise. If $N_r(s)$ is least degree (has no finite zeros), then P(N) will be called the *transfer function Plücker matrix*.

The matrices P(A,B), P(A,C), $P(T_r)$, $P(T_l)$ and P(N) play a crucial role in the solution of the corresponding frequency assignment problems of linear systems and some of their properties are discussed next.

It is expected that the matrices $P(T_l)$, $P(T_r)$ associated with LCMFDs, RCMFDs of G(s) to contain the same information about the system. In fact, since

$$[D_l(s), N_l(s)]\begin{bmatrix}0 & I_\ell\\ -I_m & 0\end{bmatrix}\begin{bmatrix}D_r(s)\\ N_r(s)\end{bmatrix} = 0, \text{ or } T_l(s)QT_r(s) = 0 \qquad (4.8)$$

the vector spaces $\bar{\mathcal{X}}_\delta = \text{row-span}_{\mathbb{R}(s)}\{T_l(s)Q\}$, $\mathcal{X}_\delta = \text{col-span}_{\mathbb{R}(s)}\{T_r(s)\}$ are dual and certain relationships hold between their Plücker coordinates [7]. Before we state these relationships we introduce some useful notation. Let $\omega_i = (i_1,\dots,i_\ell)\in Q_{\ell,m+\ell}$, $\rho_j = (j_1,\dots,j_m)\in Q_{m,m+\ell}$, and let $\tilde{\omega}_i = (i_{\ell+1},\dots,i_{\ell+m})\in Q_{m,m+\ell}$, where $\omega_i\cap\tilde{\omega}_i = \emptyset$; let us also denote by $\tau_i = \omega_i\oplus\tilde{\omega}_i = \{i_1,\dots i_\ell, i_{\ell+1},\dots i_{\ell+m}\}\in S_{\ell+m}$ ($S_{\ell+m}$ denotes the permutation group) and by $\sigma(\tau_i)$ the sign of $\tau_i\in S_{\ell+m}$. The relationships between the Plücker coordinates of dual spaces is described by the following result [7].

Lemma (4.1): Let $\bar{T}_l\in\mathcal{F}^{m\times(m+\ell)}$, $T_r\in\mathcal{F}^{(m+\ell)\times\ell}$ be basis matrices of the dual spaces $\bar{\mathcal{X}} = \text{row-span}_{\mathcal{F}}\{\bar{T}_l\}$, $\mathcal{X} = \text{col-span}_{\mathcal{F}}\{T_r\}$ respectively and let us denote by $\bar{t}_{\omega_i}, t_{\rho_j}$ the Plücker coordinates of $\bar{\mathcal{X}},\mathcal{X}$ correspondingly. Then,

$$\bar{t}_{\tilde{\omega}_i} = \alpha\cdot\sigma(\tau_i)t_{\omega_i}, \quad \alpha\in\mathcal{F}-\{0\} \qquad (4.9)$$

for all $\omega_i \in Q_{\ell,m+\ell}$, $\tilde{\omega}_i \in Q_{m,m+\ell}$, such that $\omega_i \cap \tilde{\omega}_i = \emptyset$. □

Lemma (4.1) and equation (4.8) readily yield the following result [17].

Proposition (4.5) Let $P(T_r), P(T_l)$ be the Plücker matrices of the rational vector spaces $\mathcal{X}_\delta = \text{col-span}_{\mathbb{R}(s)}\{T_r(s)\}$, $\mathcal{X}'_\delta = \text{row-span}_{\mathbb{R}(s)}\{T_l(s)\}$ respectively. Then,

$$P(T_l)^t = C_m(Q)^t R P(T_r) \tag{4.10}$$

where $\rho = \binom{m+\ell}{\ell} = \binom{m+\ell}{m}$ and the matrices R,Q are defined by

$$R = \begin{bmatrix} 0 & 0 & \dots & 0 & \sigma(\tau_\rho) \\ 0 & 0 & \dots & \sigma(\tau_\rho - 1) & 0 \\ \vdots & \vdots & & \vdots & \vdots \\ \sigma(\tau_1) & 0 & \dots & 0 & 0 \end{bmatrix} \in \mathbb{R}^{\rho \times \rho}, \quad Q = \begin{bmatrix} 0 & I_\ell \\ -I_m & 0 \end{bmatrix} \tag{4.11}$$

□

The explicit relationship between $P(T_r)$ and $P(T_l)$ established by the above result suggests that either $P(T_r)$ or $P(T_l)$ may be used in the study of pole assignment by constant output feedback. Note that the matrix $C_m(Q^t)R$ is orthogonal and thus $\text{rank}\{P(T_r)\} = \text{rank}\{P(T_l)\}$.

The Plücker matrices P(A,B), P(A,C) characterise the pairs (A,B), (A,C) respectively; thus, it is expected that system controllability, observability should be connected with the properties of P(A,B), P(A,C) correspondingly.

Theorem (4.2): Let P(A,B), P(A,C) be the Plücker matrices associated with the pairs (A,B), (A,C) respectively.

(i) (A,B) is controllable if and only if $\text{rank}\{P(A,B)\} = n+1$.

(ii) (A,C) is observable if and only if $\text{rank}\{P(A,C)\} = n+1$.

Proof The result will be proved for the pair (A,B); the proof for the (A,C) pair is identical.

(i) (Necessity): If (A,B) is controllable, then the matrix pencil $B(s) = [sI-A, -B]$ is characterised by column minimal indices only (there are no decoupling zeros). Let $0 < \varepsilon_1 \le \varepsilon_2 \le .. \le \varepsilon_\ell$ be the set of column minimal indices of B(s) and let (M,N) be a pair of strict equivalence transformations which reduce B(s) to its Kronecker canonical form [20] that is

$$B_k(s) = MB(s)N, \quad M \in \mathbb{R}^{n\times n}, \quad N \in \mathbb{R}^{(n+\ell)\times(n+\ell)}, \quad |M|, \; |N| \neq 0 \tag{4.12}$$

where

$$B_k(s) = \text{block-diag}\{L_{\varepsilon_1}(s);\dots;L_{\varepsilon_\ell}(s)\},\quad L_{\varepsilon_i}(s) = \begin{bmatrix} s & -1 & 0 & \dots & 0 & 0 \\ 0 & s & -1 & \dots & 0 & 0 \\ \vdots & \vdots & \vdots & & \vdots & \vdots \\ 0 & 0 & 0 & \dots & s & -1 \end{bmatrix} \tag{4.13}$$

A basis matrix for $N_r(B_k(s))$ is defined by $E_k(s)$, where

$$E_k(s) = \left[\begin{array}{c|c} \underline{e}_{\varepsilon_1}(s) & 0 \\ \hline 0 & \ddots \quad \underline{e}_{\varepsilon_\ell}(s) \end{array}\right],\quad \underline{e}_{\varepsilon_i} = [1,s,\dots,s^{\varepsilon_i}]^t \tag{4.14}$$

If $\underline{b}_k^t(s)\wedge$, $\underline{e}_k(s)\wedge$ denote the exterior products of the rows of $B_k(s)$, columns of $E_k(s)$ respectively and if $P(B_k)$, $P(E_k)$ denote the matrix coefficients of $\underline{b}_k^t(s)\wedge$, $\underline{e}_k(s)\wedge$ correspondingly, then from $B_k(s)E_k(s) = 0$, Lemma (4.1) and the least degree property of $B_k(s)$, $E_k(s)$ it follows that

$$P(B_k)^t = \gamma\cdot RP(E_k),\quad \gamma \in \mathbb{R} - \{0\} \tag{4.15}$$

where R is a matrix of type defined by eqn.(4.11). By applying the Binet-Cauchy Theorem [12] on eqn.(4.12) we have that $\underline{b}_k(s)\wedge = |M|\cdot\underline{b}(s)\wedge\cdot C_n(N)$ and thus

$$P(B_k) = P(A,B)\cdot C_n(N)\cdot\alpha,\quad \alpha = |M| \tag{4.16}$$

By eqns.(4.15) and (4.16) it follows that

$$P(A,B) = P(E_k)^t R^t C_n(N)^{-1}\beta,\quad \beta = \gamma\alpha^{-1} \in \mathbb{R} - \{0\} \tag{4.17}$$

Given that R and $C_n(N)$ are full rank square matrices it follows that $\text{rank}\{P(A,B)\} = \text{rank}\{P(E_k)^t\} = \text{rank}\{P(E_k)\}$. From the special structure of $E_k(s)$ (echelon type matrix) it is readily seen that $\text{rank}\{P(E_k)\} = n+1$ and thus $\text{rank}\{P(A,B)\} = n+1$.

(ii) (Sufficiency): If $\text{rank}\{P(A,B)\} = n+1$, then $N_\ell\{P(A,B)\} = \{0\}$ and thus there is no vector $\underline{e}_n^t(\lambda) = [1,\lambda,\dots,\lambda^n]$ such that $\underline{e}_n^t(\lambda)P(A,B) = 0$ for all $\lambda \in \mathbb{C}$; thus the polynomials in $\underline{b}^t(s)\wedge$ are coprime and B(s) has no decoupling zeros. □

Note that $\underline{b}^t(s)\wedge = \underline{e}_n^t(s)P(A,B)$ and $\underline{c}(s)\wedge = P(A,C)\underline{e}_n(s)$ where

$$P(A,B) = \left[\begin{array}{c|ccc} \alpha_o & a_o^2 & \dots & a_o^\rho \\ \alpha_1 & a_1^1 & & a_1^\rho \\ \vdots & \vdots & & \vdots \\ \alpha_{n-1} & a_{n-1}^2 & \dots & a_{n-1}^\rho \\ \hline 1 & 0 & \dots & 0 \end{array}\right] = \left[\begin{array}{c|c} \underline{\alpha} & \hat{P}(A,B) \\ \hline 1 & \underline{0}^t \end{array}\right],\quad \rho = \binom{n+\ell}{n} \tag{4.18}$$

$$P(A,C) = \left[\begin{array}{cccc|c} \alpha_o & \alpha_1 & \cdots & \alpha_{n-1} & 1 \\ \hline b_o^2 & b_1^2 & \cdots & b_{n-1}^2 & 0 \\ \vdots & \vdots & & \vdots & \vdots \\ b_o^\tau & b_1^\tau & \cdots & b_{n-1}^\tau & 0 \end{array}\right] = \left[\begin{array}{c|c} \underline{\alpha}^t & 1 \\ \hline \hat{P}(A,C) & \underline{0} \end{array}\right], \quad \tau = \binom{n+m}{n} \tag{4.19}$$

where $[\alpha_o,\alpha_1,\ldots,\alpha_{n-1},1] = [\underline{\alpha}^t,1]$ is the coefficient vector of $|sI-A|$. The matrices $\hat{P}(A,B)$, $\hat{P}(A,C)$ will be referred to as the *reduced controllability, observability Plücker matrices* respectively. Controllability tests may be defined in terms of $\hat{P}(A,B)$, $\hat{P}(A,C)$ as:

Corollary (4.2.1): Let $\hat{P}(A,B)$, $\hat{P}(A,C)$ be the reduced Plücker matrices of the pairs (A,B), (A,C) respectively. Then,

(i) The pair (A,B) is controllable if and only if $\text{rank}\{\hat{P}(A,B)\} = n$.

(ii) The pair (A,C) is observable if and only if $\text{rank}\{\hat{P}(A,C)\} = n$.

Proof: For $\text{rank}\{P(A,B)\} = n+1$, a $(n+1)\times(n+1)$ minor of $P(A,B)$ must exist which is nonzero; because of the structure of $P(A,B)$, this minor must contain the first column of $P(A,B)$ and thus its nonzero value is due to the existence of a nxn nonzero minor in $\hat{P}(A,B)$. □

For the case of pole assignment by state feedback eqn.(4.6) takes the following form

$$\left[\begin{array}{c|c} \underline{\alpha} & \hat{P}(A,B) \\ \hline 1 & \underline{0}^t \end{array}\right] \left[\begin{array}{c} 1 \\ \hline \underline{k} \end{array}\right] = \left[\begin{array}{c} \underline{\gamma} \\ \hline 1 \end{array}\right] \tag{4.20}$$

where $[1,\underline{k}^t]^t$ is the exterior product of the columns of $[I_n,L^t]^t$ and $[\underline{\gamma}^t,1] = [\gamma_o,\gamma_1,\ldots,\gamma_{n-1},1]$ are the coefficients of the polynomial which is to be assigned. A simple manipulation of eqn.(4.20) yields

$$\hat{P}(A,B)\underline{k} = \underline{\gamma} - \underline{\alpha} \tag{4.21}$$

and thus the following result is established.

Corollary (4.2.2): Necessary condition for complete pole assignment by state feedback is that the pair (A,B) is controllable.

The above results apart from providing an alternative characterisation of controllability and observability, have the extra advantage that they are naturally linked to the pole assignment problem by state feedback (design of a full state observer).

The solvability of the linear part of the other assignment problems discussed in section (2), depends on the rank properties of the corresponding Plücker matrices. By specialising the more general results stated for DAP to the case of pole, zero assignment problems we have:

Proposition (4.6): For an observable and controllable proper system, necessary conditions for arbitrary pole assignment by constant output feedback are:

(i) $\text{rank}\{P(T_r)\} = n+1$

(ii) $\binom{m+\ell}{\ell} = \binom{m+\ell}{m} \geq n+1$ □

Proposition (4.7): Let $G(s) \in \mathbb{R}^{m\times\ell}(s)$, $m \geq \ell$, and let $z(s)$ be the zero polynomial for $G(s)$. If $KG(s)$ denotes a transfer function derived from $G(s)$ under squaring down, and $z_k(s)$ the zero polynomial of $KG(s)$, then the following properties hold true:

(i) $z_k(s) = z(s)\tilde{z}(s)$, for all squaring down compensators K.

(ii) The maximal degree of $\tilde{z}(s)$ is equal to the Forney order δ.

(iii) Necessary condition for arbitrary assignment of the zeros of $\tilde{z}(s)$ is that $\binom{m}{\ell} \geq \delta+1$. □

Clearly, the results of this section are only necessary, but not sufficient, since it remains to be shown how from the multivectors $\underline{k}$ we may construct compensators. This is examined in the following section.

5. DECOMPOSABILITY OF MULTIVECTORS: THE REDUCED QUADRATIC PLÜCKER RELATIONS

A sufficient condition for a solution of DAP, is that from the family of solutions (4.7) of Eqn.(4.6) there exists at least a $\underline{k}$ for which Eqn. (2.13) has a solution. A vector $\underline{k} \in \mathbb{R}^\sigma$ defines a point in the projective space $\mathbb{P}_{\sigma-1}(\mathbb{R})$; the points of $\mathbb{P}_{\sigma-1}$ which satisfy for some $H \in \mathbb{R}^{r\times p}$ Eqn.(2.13) are those which belong to the *Grassmann variety* of $\mathbb{P}_{\sigma-1}(\mathbb{R})$ [6,7]. This standard result may be expressed as follows:

Theorem (5.1): Let $\underline{k} \in \mathbb{R}^\sigma$, $\sigma = \binom{p}{r}$ and let k_ω, $\omega = (i_1,\ldots,i_r) \in Q_{r,p}$ be the coordinates of $\underline{k}$ (seen as the Plücker coordinates of a point in $\mathbb{P}_{\sigma-1}(\mathbb{R})$). Necessary and sufficient condition for an $H \in \mathbb{R}^{r\times p}$, $H = [\underline{h}_1,\ldots,\underline{h}_r]^t$, to exist such that

$$\underline{h}\wedge = \underline{h}_1\wedge\ldots\wedge\underline{h}_r = \underline{k} = [\ldots,k_\omega\ldots]^t \tag{5.1}$$

is that the coordinates k_ω satisfy the following quadratic relations

$$\sum_{\kappa=1}^{r+1} (-1)^{\nu-1} k_{i_1,\ldots,i_{r-1},j_\nu} k_{j_1,\ldots,j_{\nu-1},j_{\nu+1},j_{r+1}} = 0 \tag{5.2}$$

where $1 \leq i_1<i_2<\ldots<i_{r-1} \leq n$ and $1 \leq j_1<j_2<\ldots<j_{r+1} \leq n$. □

The set of quadratics defined by Eqn.(5.2) are known [6,7] as *Quadratic Plücker Relations* (QPR) and they define the Grassmann variety of $\mathbb{P}_{\sigma-1}(\mathbb{R})$. Conditions (5.2) clearly reveal the nonlinear nature of DAP. The two main questions which naturally arise are: (1) Given a decomposable $\underline{k}$, which satisfies (5.2), construct the matrix H. (2) Parametrize the set of conditions (5.2). The second question is crucial for the study of compensators which considerably simplify DAP. The following example demonstrates the redundancy in the definition of the QPRs by Eqn. (5.2).

Example (5.1): Let $p = 5$, $r = 3$ and let $(k_0, k_1, k_2, \ldots, k_9)$ be the coordinates of a vector defining a point in the projective space $\mathbb{P}_9$. The set of QPRs describing the Grassmann variety of $\mathbb{P}_9$ (usually denoted by $\Omega(3,5)$) is given by

$$k_0k_5-k_1k_4+k_2k_3=0,\; k_0k_8-k_1k_7+k_2k_6=0,\; k_0k_9-k_3k_7+k_4k_6=0 \tag{5.3}$$

$$k_1k_9-k_3k_8+k_5k_6=0,\; k_2k_9-k_4k_8+k_5k_7=0 \tag{5.4}$$

It may be readily shown that the above set of equations is not minimal; in fact, the set (5.4) may be obtained from the set (5.3) and thus (5.3) is a minimal set of quadratics describing the Grassmann variety $\Omega(3,5)$.

The above example makes clear the need for the definition of a minimal set of quadratics describing $\Omega(r,p)$. The problem of reconstructing H from a decomposable $\underline{k}$ is examined first [7].

Lemma (5.1): Let $\underline{k} = [\ldots, k_\omega, \ldots]^t \in \mathbb{R}^\sigma$, $\sigma = \binom{p}{r}$, be a decomposable vector (satisfying the set of QPRs) and let $k_{\alpha_1,\ldots,\alpha_r}$ be a nonzero coordinate of $\underline{k}$. If we define by

$$h_{ij} = k_{\alpha_1,\ldots,\alpha_{i-1},j,\alpha_{i+1},\ldots,\alpha_r},\quad i \in \underset{\sim}{r},\ j \in \underset{\sim}{p} \tag{5.5}$$

then for the matrix $H = [h_{ij}]$, $C_r(H) = \underline{k}$.

Remark (5.1): Let $\underline{k} = [\ldots, k_\omega, \ldots]^t \in \mathbb{R}^\sigma$, $\sigma = \binom{p}{r}$, be a decomposable vector and let the first coordinate of $\underline{k}$ be nonzero. The H matrix defined by Lemma (5.1) has the form

$$H = [k_\alpha I_r, X^t]^t \in \mathbb{R}^{p\times r},\quad \text{where } k_\alpha = k_{1,2,\ldots,r} \neq 0 \tag{5.6}$$

The reconstruction of H from a decomposable vector is demonstrated by the following example.

Example (5.2): Let $\underline{k} = [k_0, k_1, k_2, k_3, k_4, k_5]^t$ be a point of the Grassmann variety $\Omega(2,4)$ of the projective space $\mathbb{P}^5(\mathbb{R})$. A basis matrix for the vector space $\mathcal{V}$ whose Plücker coordinates are coordinates of the given point is defined by:

$$H_o = \begin{bmatrix} k_o & 0 \\ 0 & k_o \\ -k_3 & k_1 \\ -k_4 & k_2 \end{bmatrix}, \text{ if } k_o \neq 0 \text{ and } H_2 = \begin{bmatrix} k_2 & 0 \\ k_4 & k_o \\ k_5 & k_1 \\ 0 & k_2 \end{bmatrix}, \text{ if } k_2 \neq 0$$

It may be readily shown that $H_2 = H_oQ$, where

$$Q = \begin{bmatrix} k_2/k_o & 0 \\ k_4/k_o & 1 \end{bmatrix}, \quad |Q| \neq 0$$

We may verify that $C_2(H_o) = [k_o^2, k_ok_1, k_ok_2, k_ok_3, k_ok_4, k_1k_4-k_2k_3]^t$ and since $k_ok_5-k_1k_5+k_2k_3 = 0$ we have that $k_1k_4-k_2k_3 = k_ok_5$ and thus $C_2(H) = k_o.\underline{k}$. □

The above procedure for constructing H out of a decomposable $\underline{k}$ also suggests a procedure for writing down an independent set of QPRs which completely describes $\Omega(r,p)$; such a set will be referred to as the *Reduced Quadratic Plücker Relations* (RQPR). It is known [7], that $\dim\Omega(r,p) = r(p-r)$ and thus the number of RQPR is:

$$n_{PQPR} = \binom{p}{r} - 1 - r(p-r) \tag{5.7}$$

The following result [19] provides the means for writing down a set of RQPRs for $\Omega(r,p)$.

Theorem (5.2): Let $\underline{k} = [\ldots,k_\omega,\ldots]^t \in \mathbb{R}^\sigma$, $\sigma = \binom{p}{r}$, $\omega \in Q_{r,p}$, be a decomposable vector and let us assume that $k_{1,2,\ldots,r} \neq 0$. If H_o is the matrix defined with respect to $k_{1,2,\ldots,r}$ by Lemma (5.1), that is

$$H_o = \left[\begin{array}{cccc} k_{1,2,\ldots,r} & 0 & \cdots & 0 \\ 0 & k_{1,2,\ldots,r} & \cdots & 0 \\ \vdots & \vdots & \ddots & \vdots \\ 0 & 0 & & k_{1,2,\ldots,r} \\ \hline (-1)^{r-1}k_{2,\ldots,r,r+1} & (-1)^{r-2}k_{1,3,\ldots,r,r+1}, & \ldots, & k_{1,2,\ldots,r-1,r+1} \\ \vdots & \vdots & & \\ (-1)^{r-1}k_{2,\ldots,r,p} & (-1)^{r-2}k_{1,3,\ldots,r,p} & \ldots, & k_{1,2,\ldots,r-1,p} \end{array}\right] \tag{5.8}$$

then a set of RQPRs is defined by the nontrivial relations of the following vector equation

$$C_r(H_o) = [\ldots,k_\omega,\ldots]^t \, k^{r-1}_{1,2,\ldots,r} \tag{5.9}$$

□

The above result has been stated with respect to the first Plücker coordinate $k_{1,2,\ldots,r}$ which has been assumed to be nonzero. However, a similar result may be stated with respect to any other nonzero Plücker coordinate. The resulting set of RQPRs is then different from the one defined by Eqn.(5.9), but equivalent. An important question in the study of compensators linearising DAP is the classification of the RQPRs in terms of the number of quadratic terms they involve. Some corollaries of Theorem (5.2) which classify the RQPRs are given next [19].

Corollary (5.2.1) Every relation from the set (5.9), which involves a minor of H_o containing at least r-1 rows from the set of the first r rows of H_o is trivial. The number of trivial relations in (5.9) is $r(p-r)+1$.

Corollary (5.2.2) (i) If $p \geq 2r$, then the maximum number of terms in a set of RQPRs is r+1. The set of RQPRs is classified as follows:

$$n_{RQPR} = \binom{p}{r} - r(p-r) - 1 = \underbrace{\binom{r}{r-2}\binom{p-r}{2}}_{\text{3-term relations}} + \ldots + \underbrace{\binom{r}{o}\binom{p-r}{r}}_{\text{(r+1)-term relations}}$$

(ii) If $p < 2r$, then the maximum number of terms in a set of RQPRs is p-r+1. The set of RQPRs is classified as follows:

$$n_{RQPR} = \binom{p}{r} - r(p-r) - 1 = \underbrace{\binom{r}{r-2}\binom{p-r}{2}}_{\text{3-term relations}} + \ldots + \underbrace{\binom{r}{2r-p}\binom{p-r}{p-r}}_{\text{(p-r+1)-term relations}}$$

□

The Grassmann variety $\Omega(r,p)$ characterising the compensators H of DAP will be called the *compensator Grassmann variety*. The parametrization of RQPRs provides the means for the definition of linear subvarieties of $\Omega(r,p)$, which when used in the study of DAP linearise its multilinear nature. Two such linear varieties are discussed next.

6. LINEARISATION OF THE RQPRs AND FEEDBACK

For a LA $M(s) \in \mathbb{R}^{p\times r}[s]$, the solvability of DAP is reduced to a problem of determining whether the family of solutions K of the linear part of DAP defined by

$$K = \{\underline{k}:\underline{k} = \tilde{P}^{\dagger}\underline{a}+\tilde{P}^{\perp}\underline{c}, \text{ where } \underline{c} \in \mathbb{R}^{\sigma-\delta-1} \text{ arbitrary}, \ \sigma = \binom{p}{r}\} \tag{6.1}$$

contains at least one decomposable element $\underline{k}$ for some appropriate choice of $\underline{c}$. This is equivalent to a problem of finding $\underline{c}$ such that $\underline{k}$ satisfies the set of RQPRs. The family of solutions K is a coset of the affine space $\mathbb{R}^{\sigma}$ and thus it defines an irreducible algebraic variety of the projective space $\mathbb{P}_{\sigma-1}(\mathbb{R})$, $\Lambda(K)$, of dimension $\sigma-\delta-1$ and of order one [7]. Within the framework of algebraic geometry, the solvability of DAP is equivalent to a study of real intersection of the irreducible algebraic varieties $\Lambda(K)$ and $\Omega(r,p)$ of $\mathbb{P}_{\sigma-1}(\mathbb{R})$. A simplified version of this problem is to study intersections of the linear variety $\Lambda(K)$ with linear subvarieties of $\Omega(r,p)$. The advantage of this approach is that whenever intersections exist, they are always real; furthermore, the solution may be computed by solving a set of linear equations and not of a mixed set of linear and quadratic equations.

The various techniques which have been developed for the solution of pole assignment problems are based on the choice of special types of feedback (like the dyadic) which linearise the multilinear nature of DAP. Such feedbacks correspond to linear subvarieties $\Lambda(H)$ of $\Omega(r,p)$; the study of $\Lambda(H)$ varieties provides a better understanding of the action of the corresponding feedbacks. Linear subvarieties $\Lambda(H)$ of $\Omega(r,p)$ may be defined by linearising a set of RQPRs. This implies fixing some coordinates of the vector $\underline{k}$ such that the RQPRs become linear. In the following, two procedures for linearising the RQPRs are discussed: the dyadic linearisation and the full rank linearisation. We restrict the investigation to the case of pole assignment problems and thus the first coordinate in $\underline{k}$ must be one. By Theorem (5.2), the set of RQPRs which corresponds to $k_{1,2,\ldots,r} = 1$ is defined by

$$C_r(H) = [\ldots,k_{\omega},\ldots]^t, \quad H = \begin{bmatrix} I_r \\ --- \\ R \end{bmatrix} \in \mathbb{R}^{p\times r} \tag{6.2}$$

where $R \in \mathbb{R}^{(p-r)\times r}$ and its structure is defined in terms of the k_{ω} coordinates as it is shown by Eqn.(5.8). Note that for the pole assignment problems, R represents state, or output feedback.

(α) Dyadic linearisation: The matrix R is written as a dyad.

(i) $p \geq 2r$: $r = \underline{x}\underline{f}^t$, where $\underline{x} \in \mathbb{R}^{p-r}$ is a vector of variables and $\underline{f} \in \mathbb{R}^r$ is a vector of arbitrarily fixed parameters. Then in $\underline{k} = \underline{h}\wedge$ we have one coordinate equal to 1, $r(p-r)$ are expressed linearly in terms of the $(p-r)$ variables and $\binom{p}{r}-r(p-r)-1$ are set zero.

(ii) $p \leq 2r$: $R = \underline{f}\underline{x}^t$, where $\underline{f} \in \mathbb{R}^{p-r}$ is a vector of arbitrarily fixed parameters and $\underline{x} \in \mathbb{R}^r$ is a vector of variables. Then in $\underline{k} = \underline{h}\wedge$ there is one coordinate equal to 1, $r(p-r)$ are expressed linearly in terms of r variables and $\binom{p}{r}-r(p-r)-1$ are set equal to zero.

(β) Full rank linearisation: The matrix R may be chosen to have full rank.

(i) $p \geq 2r$: Fix arbitrarily $(r-1)$ columns of R and leave one column of R as the vector of $(p-r)$ variables. Then in $\underline{k} = \underline{h}\wedge$ we have one coordinate equal to 1, $(r-1)(p-r)$ coordinates arbitrarily chosen, $(p-r)$ coordinates as variables, and the rest $\binom{p}{r}-r(p-r)-1$ are expressed as linear functions of the p-r variables.

(ii) $p \leq 2r$: Fix arbitrarily $(p-r-1)$ rows of R and leave one row of R as the vector of r variables. Then in $\underline{k} = \underline{h}\wedge$ we have one coordinate equal to 1, $(p-r-1)r$ coordinates arbitrarily chosen, r coordinates as variables and the rest $\binom{p}{r}-r(p-r)-1$ are expressed as linear functions of the r variables.

If we denote by μ_f, μ_d the degrees of freedom of the families of full rank and dyadic linearising feedbacks respectively, then

$$\mu_f = (r-1)(p-r), \; \mu_d = r, \; \text{if } p \geq 2r \tag{6.3}$$

$$\mu_f = (p-r-1)r \;, \; \mu_d = p-r, \;\; \text{if } p \leq 2r \tag{6.4}$$

from which it is clear that $\mu_f > \mu_d$. Both families have the same number of free variables; the superiority of the full rank linearising family is due to the greater number of degrees of freedom and to the full rank property. The definition and properties of the two families of feedbacks are illustrated by the following example.

Example (6.1): Let $p = 5$, $r = 3$, $\underline{k} = [1,k_1,\ldots,k_9]^t \in \mathbb{R}^{10}$. Then,

$$H = \begin{bmatrix} 1 & 0 & 0 \\ 0 & 1 & 0 \\ 0 & 0 & 1 \\ k_6 & -k_3 & k_1 \\ k_7 & -k_4 & k_2 \end{bmatrix} = \begin{bmatrix} I_3 \\ R \end{bmatrix}, \; R = \begin{bmatrix} k_6 & -k_3 & k_1 \\ k_7 & -k_4 & k_2 \end{bmatrix}$$

and the set of RQPRs is defined from the nontrivial relations obtain from equation $C_3(H) = \underline{k}$, that is

$$1=1, \; k_1=k_1, \; k_2=k_2, \; k_3=k_3, \; k_4=k_4, \; k_6=k_6, \; k_7=k_7 \;\; \text{(trivial)}$$

$$k_5=-k_2k_3+k_1k_4, \; k_8=-k_2k_6+k_1k_7, \; k_9=k_6k_4+k_3k_7 \;\; \text{(nontrivial)}$$

(i) Dyadic linearisation: $p < 2r$ and thus $R = \underline{f}\underline{x}^t$, $\underline{f} = [\alpha,\beta]^t$, $\underline{x}^t = [x_1,x_2,x_3]$. Then the corresponding multivector becomes

$$[1,\underline{k}_d^t]^t = [1,\alpha x_3,\beta x_3,-\alpha x_2,-\beta x_2,0,\alpha x_1,\beta x_1,0,0]^t$$

(ii) Full rank linearisation: $p < 2r$ and by fixing the elements of the first row of R, by setting $k_6 = \alpha$, $k_3 = -\beta$, $k_1 = \gamma$, we have

$$[1,\underline{k}_f^t]^t = [1,\gamma,k_2,-\beta,k_4,\beta k_2+\gamma k_4,\alpha,k_7,-\alpha k_2+\gamma k_7,-\alpha k_4-\beta k_7]^t$$

□

The linearisation procedures defined above result in decomposable vectors $\underline{k}$ of the following type:

(i) Dyadic linearisation:

$$\underline{k} = \begin{bmatrix} 1 \\ --- \\ \underline{k}_d \end{bmatrix} = \underline{e}_1 + \begin{bmatrix} \underline{0}^t \\ --- \\ \hat{M} \end{bmatrix} \underline{x} = \underline{e}_1 + M\underline{x} \tag{6.5}$$

where $\underline{x} \in \mathbb{R}^{\nu}$ is the vector of variables of the dyadic feedback, $\hat{M} \in \mathbb{R}^{(\sigma-1)x\nu}$ is a canonical structure matrix with zeros and the elements of $\underline{f}$ only and $\underline{e}_1$ the first standard basis vector of $\mathbb{R}^{\sigma}$.

(ii) Full rank linearisation:

$$\underline{k} = \begin{bmatrix} 1 \\ --- \\ \underline{k}_f \end{bmatrix} = \underline{e}_1 + \begin{bmatrix} 0 \\ -- \\ \hat{\underline{\tau}} \end{bmatrix} + \begin{bmatrix} \underline{0}^t \\ --- \\ \hat{N} \end{bmatrix} \underline{x} = \underline{e}_1 + \underline{\tau} + N\underline{x} \tag{6.6}$$

where $\underline{x} \in \mathbb{R}^{\nu}$ is the vector of variables in R, $\hat{\underline{\tau}} \in \mathbb{R}^{(\sigma-1)}$ is a vector with zeros and the arbitrarily fixed parameters in R and $\hat{N} \in \mathbb{R}^{(\sigma-1)x\nu}$ is a canonical structure matrix with zeros, ones, and the arbitrarily fixed parameters in R.

Using the types of feedback defined above the general solution of DAP is reduced to the solution of the following system of linear equations:

Dyadic: $$P_{\delta}^t M\underline{x} = \underline{a} - P_{\delta}^t \underline{e}_1 \tag{6.7}$$

Full rank: $$P_{\delta}^t N\underline{x} = \underline{a} - P_{\delta}^t \{\underline{e}_1 + \underline{\tau}\} \tag{6.8}$$

Proposition (6.1): DAP has a dyadic linearising solution if and only if there is a choice of the vector of variables $\underline{f}$ such that

$$\{\underline{a} - P_{\delta}^t \underline{e}_1\} \in \mathcal{R}\{P_{\delta}^t \hat{M}\} \tag{6.9}$$

□

Proposition (6.2) DAP has a full rank linearising solution if and only if there is a selection of the vector of variables $\underline{\tau}$ such that

$$\{\underline{a} - P_{\delta}^t \{\underline{e}_1 + \underline{\tau}\}\} \in \mathcal{R}\{P_{\delta}^t N\} \tag{6.10}$$

□

For the case of pole assignment by state feedback Eqn.(6.8) is reduced to

$$\hat{P}(A,B)\hat{N}\underline{x} = \underline{\gamma} - \underline{\alpha} - \hat{P}(A,B)\hat{\underline{\tau}} = \underline{\xi} \tag{6.11}$$

where $\underline{\gamma}, \underline{\alpha}$ are the vectors of coefficients of the closed-loop and open-loop polynomials respectively, $\hat{P}(A,B) \in \mathbb{R}^{nx(\rho-1)}$, $\rho = \binom{n+\ell}{n}$, is the reduced Plücker matrix of (A,B) and $\hat{N} \in \mathbb{R}^{(\rho-1)xn}$ is the canonical matrix of the full rank linearisation. For Eqn.(6.11) to have a solution for all $\underline{\xi} \in \mathbb{R}^n$, the nxn matrix $\hat{P}(A,B)\hat{N}$ must have full rank for at least one matrix $\hat{N}$ characterising the full rank linearising feedback. By the Binet-Cauchy theorem $\det\{\hat{P}(A,B)\cdot\hat{N}\} = C_n(\hat{P}(A,B))C_n(\hat{N})$ and since the system is assumed controllable the multivector $C_n(\hat{P}(A,B))$ has at least one non-zero component. By inspection of the canonical structure of $\hat{N}$ it may be readily seen that there is always a full rank linearising feedback for which $C_n(\hat{N})$ has a nonzero component at the same position with the nonzero coordinate of $C_n(\hat{P}(A,B))$; thus, the matrix P(A,B)N may become full rank and this provides an alternative formulation and proof of the pole assignment theorem by state feedback [1].

Theorem (6.1): The system described by the pair (A,B) is CA, if and only if rank$\{\hat{P}(A,B)\} = n$.

Corollary (6.1.1): If rank$\{\hat{P}(A,B)\} = n$, there exists a full rank linearising state feedback which assigns the characteristic polynomial of A + BL.

The advantage of the full rank linearising feedback in comparison to the dyadic, may be demonstrated by the next example.

Example (6.2): Consider the system $S = \{\dot{x}_1 = u_1,\ \dot{x}_2 = x_3,\ \dot{x}_3 = u_2\}$. By following the procedure suggested in this section, it can be readily shown that although $\hat{P}(A,B)$ has full rank, there is no dyadic feedback which assigns the closed-loop poles; by choosing a full rank linearising state feedback, such as

$$L = \begin{bmatrix} \alpha & \beta & \gamma \\ \ell_1 & \ell_2 & \ell_3 \end{bmatrix}, \text{ with } \beta \neq \gamma\alpha \tag{6.13}$$

it can be readily verified that arbitrary assignment is always possible.

7. CONCLUSIONS

The aim of the paper was to provide a unifying framework for the study of solvability, as well as the computation of solutions for the frequency assignment problems defined on linear multivariable systems. The formulation of DAP as the unifying problem and its natural splitting to a linear problem of zero assignment of polynomial combinants and a standard problem of decomposability of multivectors, provides the means for the reduction of DAP to a standard problem of algebraic geometry: the study of intersections of the Grassmann variety $\Omega(r,p)$ of $\mathbb{P}_{\sigma-1}$ with the linear variety defined by the solution of the linear subproblem. For the realisability of compensators, the intersections have to be real points. The study of real intersections depends on the properties of the Plücker matrix P_δ; the rank properties of the various Plücker matrices arising in control problems, as well as the conditions under which real interesections are guaranteed are under investigation. The study of linearising feedbacks yields an explicit description of the linear subvarieties of $\Omega(r,p)$, which when deployed may provide sufficient conditions for the solvability, as well as computation of solutions of DAP. The advantage of the present approach when compared to those in [9],[10] is that it relies on the explicit description of $\Omega(r,p)$ in terms of the PQPRs and on the properties of new system invariants, defined by the Plücker matrices; thus it may provide algorithms for the design of feedbacks and not just existence results. The crucial role played by the Plücker matrices suggests that the solvability of the various versions of DAP may depend on the structural properties of P_δ and not just on the numbers of inputs, states and outputs.

REFERENCES

[1] Wonham, W.M., 1979. "*Linear multivariable control: A geometric approach*". Springer Verlag, New York.
[2] Kimura , H., 1975. "Pole assignment by gain output feedback". *IEEE Trans.Aut. Control*, AC-20, 509-516.
[3] Kouvaritakis, B. and MacFarlane, A.G.J., 1976. "Geometric approach to analysis and synthesis of system zeros. Part II: Non-square systems". *Int.J. Control*, 23, 167-181.
[4] Karcanias, N. and Kouvaritakis, B., 1979. "The output zeroing problem and its relationship to the invariant zero structure: a matrix pencil approach". *Int.J. Control*, 30, 395-415.
[5] Greub, W.H., 1967. "*Multilinear Algebra*", Springer Verlag, New York.
[6] Marcus, M., 1973. "*Finite dimensional multilinear algebra*" (in two parts). Marcel Deker, New York.
[7] Hodge, W.V.D. and Pedoe, P.D., 1952. "*Methods of algebraic geometry*". Vol.2, Cambridge Univ. Press.
[8] Sain, M.K., 1976. "The growing algebraic presence in systems engineering". *IEEE Proc.*, Vol.64, No.1, 96-111.
[9] Martin, C. and Hermann, R., 1978. "Applications of algebraic geometry to systems theory: The MacMillan degree, and Kronecker indices...". *SIAM J. Control and Opt.*, 16, 743-.
[10] Brockett, R.W. and Byrnes, C.I., 1981. "Multivariable Nyquist Criterion, Root Loci and Pole placement: A geometric viewpoint". *IEEE Trans.Aut. Control*, AC-26, 271-283.
[11] Karcanias, N., Giannakopoulos, C. and Hubbard, M., 1983. "Almost zeros of a set of polynomials of $\mathbb{R}[s]$". To appear in *Int.J. Control*.
[12] Marcus, M. and Minc, H., 1964. "*A survey of matrix theory and matrix inequalities*". Allyn and Bacon, Boston.
[13] Giannakopoulos, C., Karcanias, N. and Kalogeropoulos, G., 1983. "Polynomial combinants, almost zeros and zero assignment". *Proc. MECO 83*, Athens.
[14] Wolovich, W.A., 1974. "*Linear multivariable systems*". Appl.Math. Sc., 11, Springer Verlag, New York.
[15] Rosenbrock, H.H., 1979. "Order, degree and complexity". *Int.J. Control*, 19, 323-331.
[16] Forney, G.D., 1975. "Minimal bases of rational vector spaces". *SIAM J. Control*, 13, 493-520.
[17] Karcanias, N. and Giannakopoulos, C., 1983. "Grassmann invariants, almost zeros and the determinantal zero, pole assignment problems of linear multivariable systems". The City Univ., Dept. of Systems Science Res. Report, DSS/NK-CG/236.
[18] Verghese, G., 1978. "Infinite frequency behaviour in generalised dynamical systems", Ph.D. Thesis, Stanford University, U.S.A.

[19] Giannakopoulos, C., Kalogeropoulos, G. and Karcanias, N., 1984. "The Grassmann variety of nondynamic compensators and the determinantal assignment problem of linear systems". Control Engin. Centre, The City University, Res. Rep., CEC/CG-GK-NK/4, U.K.
[20] Gantmacher, G.,1959. *Theory of Matrices*, Vol. 2, Chelsea, New York.

CHAPTER 13

ON THE STABLE EXACT MODEL MATCHING AND STABLE MINIMAL DESIGN PROBLEMS

Antonis I.G. Vardulakis
Department of Mathematics
Faculty of Sciences
Aristotle University of Thessaloniki
Thessaloniki
Greece

Nicos Karcanias
Control Engineering Centre
School of Electrical Engineering and Applied Physics
The City University
Northampton Square
London EC1V 0HB, England

ABSTRACT

A number of results on the module structure of the set M* of all *proper* rational vectors which have no poles inside a "forbidden" region Ω of the finite complex plane and which are also contained in a given rational vector space $T(s)$ are surveyed. The structure of the various bases of M* is examined and the notion of a "*simple*" basis of M* is introduced. The existence and construction of *simple proper and Ω-stable* bases of $T(s)$ having minimal MacMillan degree among all other proper bases of $T(s)$ is established. In the light of these concepts and results a number of linear multivariable control algebraic synthesis problems are examined. Thus, necessary and sufficient conditions for the solvability of the "stable exact model matching problem" (SEMMP) are derived and the family of all "proper and stable" solutions to the SEMMP is characterized. Also the minimal design and the stable minimal design problems (MDP)(SMDP) are examined.

1. INTRODUCTION

Recent algebraic synthesis methods for linear multivariable control problems are based on what is known as the "fractional representation" of rational matrices (Vidyasagar et al. [8][12][14], Desoer et al. [9], Francis and Vidyasagar [10], Saeks and Murray [11][13]. These fractional representations which are defined over the ring of *proper and stable* rational functions reveal the importance of the set M* of "*proper and stable*" rational vectors which are contained in a given rational vector space $T(s)$.

Two other important sets of rational vectors contained in a rational vector space $T(s)$ are the set N* of *polynomial vectors* and the set P* of *proper rational vectors*.

The algebraic structure of N* has been thoroughly investigated using polynomial matrix theory by many authors (Rosenbrock [7], Rosenbrock and Hayton [23], Wolovich [22][24], Wang and Davison [25], Forney [18] and references therein). Through these studies, which heavily

S. G. Tzafestas (ed.), Multivariable Control, 233–263.

relied on the theory of (finite) Smith forms of polynomial matrices, it has been established that N^* has the structure of a free $\mathbb{R}[s]$-module [26]. Forney's work [18] is primarily concerned with the classification of polynomial bases of $T(s)$ with special interest given to those called *minimal polynomial bases*. The algebraic structure of *proper* input-output maps from the (state) feedback equivalence point of view has been investigated by Hautus and Heymann [27], and Hammer and Heymann [28] have examined connections between what they term "causal factorizations" of proper rational maps and linear feedback theory. In particular the notion of proper bases may be found there. The counterpart of the work of Rosenbrock, Wolovich and Forney on polynomial bases of rational vector spaces has been recently developed for the case of the set P^* in [16] where *proper* bases of rational vector spaces were examined and classified.

The study of the algebraic structure of the set of proper rational vectors in $T(s)$ with poles in a prescribed region of the complex plane $\mathbb{C}$ was initiated by Morse [3] and later used by Hung and Anderson [4] for design purposes. Verghese and Kailath [17] further examined the structure of a rational matrix and its associated vector space by making use of valuation theory and of Smith-MacMillan forms at any point s_0 of the extended complex plane $\mathbb{C}^* = \mathbb{C} \cup \{\infty\}$.

In this paper a number of results on the structure of *proper and "stable"* bases of rational vector spaces are surveyed first. The algebraic structure of the set of all *proper* rational vectors which have *no poles inside a "forbidden" region* Ω of the finite complex plane and which are also contained in a given rational vector space (s), is shown to be that of a Noetherian $\mathbb{R}_P(s)$-module M^* ($\mathbb{R}_P(s)$: the Euclidean ring of rational functions which have *no poles* in $P := \Omega \cup \{\infty\}$, i.e. of proper and "$\Omega$-stable" rational functions). The proper submodules M_i of M^* form an ascending chain of submodules partially ordered by an invariant of M_i, defined as the "stathm in P" of M_i. The structure of the various "proper and Ω-stable" bases of the (maximal) Noetherian $\mathbb{R}_P(s)$-module M^* is examined and the notion of a "*simple*" basis $T_1(s)$ of M^* is introduced. These *simple* bases are shown to be (proper and Ω-stable) bases of M^* having the property that their MacMillan degree is given by the sum of the MacMillan degrees of their columns taken separately. Based on these results the existence and construction of *"simple", proper and Ω-stable* bases of $T(s)$ having minimal MacMillan degree among all other proper bases of $T(s)$ is established [35]. These proper bases have been defined [35] as simple, minimal MacMillan degree, proper and Ω-stable bases of $T(s)$ (SMMD-$\mathbb{R}_P(s)$ bases) and it is shown that this notion is the counterpart to Forney's concept of a minimal polynomial basis of $T(s)$, for the case of the $\mathbb{R}_P(s)$-module M^*.

In the light of these concepts and results a number of multivariable control algebraic synthesis problems are examined. Firstly, necessary and sufficient conditions for the solvability of the *stable exact model matching problem* (SEMMP) are derived and the family of all "proper and stable" solutions to the SEMMP is characterised. These results are used in order to express a number of equivalent necessary and sufficient conditions for the existence of proper and stable left or right inverses of proper transfer function matrices. The existence and construction of a proper (but not necessarily stable) solution to the exact model matching problem (EMMP) having also minimal MacMillan degree (a problem known as the minimal design problem (MDP)) is investigated and solved using the notion of a simple, proper, minimal MacMillan degree $\mathbb{R}_{pr}(s)$-basis. Finally the problem of the existence and construction of minimal MacMillan degree solutions to the SEMMP (known as the stable minimal design problem (SMDP)) is examined.

2. BACKGROUND

Let $\mathbb{R}$ be the field of reals, $\mathbb{R}[s]$ the ring of polynomials with coefficients in $\mathbb{R}$ and $\mathbb{R}(s)$ the field of rational functions: $t(s) = n(s)/d(s)$, $n(s), d(s) \in \mathbb{R}[s]$, $d(s) \not\equiv 0$. Define the map $\delta_\infty: \mathbb{R}(s) \to \mathbb{Z} \cup \{\infty\}$ [1] ($\mathbb{Z}$ the ring of integers), via $\delta_\infty(t(s)) = \deg.d(s) - \deg.n(s)$, $\delta_\infty(0) = \infty$. The map $\delta_\infty(\cdot)$ is a "discrete valuation" on $\mathbb{R}(s)$ [2] and every $t(s) = n(s)/d(s) \in \mathbb{R}(s)$ can be written as:

$$t(s) = \left(\frac{1}{s}\right)^{q_\infty} n'(s)/d'(s)$$

where $q_\infty := \delta_\infty(t(s))$ and $\deg.n'(s) = \deg.d'(s)$. If $q_\infty > 0$, we say that $t(s)$ has a zero at $s = \infty$ of order q_∞ while if $q_\infty < 0$, then we say that $t(s)$ has a pole at $s = \infty$ of order $|q_\infty|$. If $t(s) \in \mathbb{R}(s)$ has $\delta_\infty(t(s)) \geq 0$ then $t(s)$ is called proper rational function and if the inequality is strict, then $t(s)$ is called strictly proper. It can be easily verified (e.g. see [1]) that the set of proper rational functions, which we denote by $\mathbb{R}_{pr}(s)$, is a Euclidean ring with "degree" given by the map $\delta_\infty(\cdot)$. Thus $\mathbb{R}_{pr}(s)$ is a principal ideal ring and both $\mathbb{R}[s]$ and $\mathbb{R}_{pr}(s)$ are subrings of $\mathbb{R}(s)$. The elements of $\mathbb{R}[s]$ can be regarded as rational functions with no poles in $\mathbb{C}$ (the finite complex plane) while the elements of $\mathbb{R}_{pr}(s)$ can be regarded as rational functions with no poles at $s = \infty$. The units in $\mathbb{R}_{pr}(s)$ are proper rational functions $u(s)$ for which $\delta_\infty(u(s)) = 0$ (i.e. having also no zeros at $s = \infty$) and they are called *biproper* rational functions [1]. We denote by $\mathbb{R}^{p\times m}(s)$ the set of $p\times m$ matrices with elements in $\mathbb{R}(s)$ and by $\mathbb{R}^{p\times m}[s]$, $\mathbb{R}^{p\times m}_{pr}(s)$ the subsets of $\mathbb{R}^{p\times m}(s)$ consisting of $p\times m$ matrices with elements respectively in $\mathbb{R}[s]$ and $\mathbb{R}_{pr}(s)$. A $T(s) \in \mathbb{R}^{p\times p}_{pr}(s)$ is called $\mathbb{R}_{pr}(s)$-unimodular or *biproper* if there exists a $\bar{T}(s) \in \mathbb{R}^{p\times p}_{pr}(s)$ such that $T(s)\bar{T}(s) = I_p$.

In order to generalise the above, let Ω be a region in the finite complex plane $\mathbb{C}$, symmetrically located with respect to the real axis $\mathbb{R}$ and which excludes at least one point α on the real axis, and let Ω^c be the complement of Ω with respect to $\mathbb{C}$ (i.e. $\mathbb{C} = \Omega \cup \Omega^c$). Let $t(s) \in \mathbb{R}(s)$ and factorise it as

$$t(s) = t_\Omega(s)\hat{t}(s) = \frac{n_\Omega(s)}{d_\Omega(s)} \frac{\hat{n}(s)}{\hat{d}(s)} \tag{2.1}$$

where $n_\Omega(s)$, $d_\Omega(s)$ are coprime polynomials *with all their zeros in* Ω and $\hat{n}(s)$, $\hat{d}(s)$ are coprime polynomials *with all their zeros outside* Ω. We define now the map $\delta_\Omega: \mathbb{R}(s) \to \mathbb{Z} \cup \{\infty\}$ via

$$\delta_\Omega(t(s)) = \begin{cases} \deg.\hat{d}(s) - \deg.\hat{n}(s) \in \mathbb{Z}, & t(s) \not\equiv 0 \\ \infty \quad , & t(s) \equiv 0 \quad (*) \end{cases} \tag{2.2}$$

Then from the definition of the "valuation at $s = \infty$" we have that

$$\delta_\Omega(t(s)) = q_\infty + \deg.n_\Omega(s) - \deg.d_\Omega(s) \tag{2.3}$$

Consider now the subset of $\mathbb{R}(s)$ consisting of all rational functions that satisfy the following two requirements: (i) are proper and (ii) have no poles in Ω (such rational functions we will call Ω-stable) and denote this set by $\mathbb{R}_P(s)$, i.e. let

$$\mathbb{R}_P(s) = \{t(s) \in \mathbb{R}(s):\ t(s) \text{ has no poles in } P := \Omega \cup \{\infty\}\}$$

The set $\mathbb{R}_P(s)$ endowed with the operations of addition and multiplication forms a commutative ring with unity element (the real number 1) and no zero divisors and thus it is an integral domain. If $t(s) \in \mathbb{R}_P(s)$, then $t(s) = n_\Omega(s)\hat{n}(s)/\hat{d}(s)$ and since $\deg.[n_\Omega(s)\hat{n}(s)] \le \deg.\hat{d}(s)$ it follows that $\delta_\Omega(t(s)) := \deg.\hat{d}(s) - \deg.\hat{n}(s) \ge 0$; thus, $\delta_\Omega(\cdot)$ for the nonzero elements of $\mathbb{R}_P(s)$ may serve as a "degree" function. The algebraic structure of $\mathbb{R}_P(s)$ has been examined initially by Morse [3] and subsequently by Hung and Anderson [4], and it has been shown that with $\delta_\Omega(\cdot)$ as degree, $\mathbb{R}_P(s)$ is a Euclidean ring and therefore a principal ideal domain (PID). In the sequel, the function δ_Ω when restricted to the subdomain $\mathbb{R}_P(s) \subset \mathbb{R}(s)$ will be denoted by $\delta_P: \mathbb{R}_P(s) \to \mathbb{Z} \cup \{\infty\}$.

The units of $\mathbb{R}_P(s)$ are biproper rational functions which have no poles and no zeros in $P = \Omega \cup \{\infty\}$; equivalently, $t(s) \in \mathbb{R}_P(s)$ is a unit if and only if $\delta_P(t(s)) = 0$.

Remark 2.1: If Ω coincides with the closed right half plane $\mathbb{C}_+ := \{s \in \mathbb{C},\ \mathrm{Re}(s) \ge 0\}$ then $P \equiv \mathbb{C}_+ \cup \{\infty\} =: \bar{\mathbb{C}}_+$, and $\mathbb{R}_{\bar{\mathbb{C}}_+}(s)$ is the Euclidean ring of "proper and stable" rational functions. The units in $\mathbb{R}_{\bar{\mathbb{C}}_+}(s)$ are biproper, stable and "minimum phase" rational functions.

From (2.3) it simply follows that if $t(s) \in \mathbb{R}_P(s)$ then

$$q := \delta\ (t(s)) = q_\infty + \deg.n_\Omega(s) \tag{2.4}$$

where now $q_\infty \ge 0$ gives the order of the zero at $s = \infty$ of $t(s) \in \mathbb{R}_P(s)$ and $\deg.n_\Omega(s)$ gives the number of finite zeros of $t(s)$ inside Ω. We examine now some factorizations of rational functions. Let $t(s) \in \mathbb{R}(s)$, then $t(s)$ can be written as

$$t(s) = \frac{n_\Omega(s)}{d_\Omega(s)} \frac{1}{(s+\alpha)^q} \left[\frac{\hat{n}(s)\ (s+\alpha)^q}{\hat{d}(s)}\right] \tag{2.5}$$

where $-\alpha \in \mathbb{R}$ is outside Ω and otherwise arbitrary, $q := \delta_\Omega(t(s)) = \deg.\hat{d}(s) - \deg.\hat{n}(s)$ and $\hat{n}(s)(s+\alpha)^q/\hat{d}(s)$ is a unit in $\mathbb{R}_P(s)$. (From the

---

*Notice that if $t(s) = n(s)/d(s) \equiv 0$, $d(s) \not\equiv 0$, then $n(s) = n_\Omega(s)\hat{n}(s) \equiv 0$ and if $d(s) = d_\Omega(s)\hat{d}(s)$ then $\delta_\Omega(0) = \deg.\hat{d}(s) - \deg.(0) = \deg.\hat{d}(s) - (-\infty) = \infty$.

above factorization it becomes clear why the region Ω must "exclude at least one point $-\alpha \in \mathbb{R}$"). The term $n_\Omega(s)/d_\Omega(s).1/(s+\alpha)^q$ gives the pole-zero structure of t(s) in $P = \Omega \cup \{\infty\}$. Thus the zeros of n (s) give the (finite) zeros of t(s) in Ω and the zeros of $d_\Omega(s)$ give the (finite) poles of t(s) in Ω. Furthermore if

$$q_\infty := q + \deg.d_\Omega(s) - \deg.n_\Omega(s) > 0$$

then t(s) has a zero at $s = \infty$ of order q_∞ while if $q_\infty < 0$ then t(s) has a pole at $s = \infty$ of order $|q_\infty|$ [1]. From the above we see that every $t(s) \in \mathbb{R}_P(s)$ can be written as

$$t(s) = \frac{n_\Omega(s)}{(s+\alpha)^q} u(s) \tag{2.6}$$

where $n_\Omega(s)$ has no zeros outside Ω, $-\alpha \in \mathbb{R}$ is outside Ω, $q := \delta_P(t(s))$ and u(s) is a unit in $\mathbb{R}_P(s)$.

Denote now by $\mathbb{R}_P^{p\times m}(s)$ the set of all p×m matrices with elements in $\mathbb{R}_P(s)$. These matrices we call proper and "Ω-stable" rational matrices (in the sense that they have no poles at infinity (proper) and also no poles inside Ω). If $\Omega \equiv \mathbb{C}_+$ then $\mathbb{R}_P^{p\times m}(s)$ represents the set of proper and "stable" rational matrices. A matrix $T(s) \in \mathbb{R}_P^{p\times p}(s)$ is called $\mathbb{R}_P(s)$-unimodular if there exists a $\hat{T}(s) \in \mathbb{R}_P^{p\times p}(s)$ such that $T(s)\hat{T}(s) = I_p$. A direct implication of the above is that $T(s) \in \mathbb{R}_P^{p\times p}(s)$ is $\mathbb{R}_P(s)$-unimodular if it has also no zeros at $s = \infty$ and no finite zeros in Ω (i.e. T(s) has no poles or zeros in $P := \Omega \cup \{\infty\}$). A system theoretic interpretation of an $\mathbb{R}_P(s)$-unimodular matrix $T(s) \in \mathbb{R}_P^{p\times p}(s)$ is the following. Let $T(s) \in \mathbb{R}_P^{p\times p}(s)$ be $\mathbb{R}_P(s)$-unimodular and let $n := \delta_M(T(s)) \geq 0$ be the MacMillan degree of T(s). If now $A \in \mathbb{R}^{n\times n}$, $B \in \mathbb{R}^{n\times p}$, $C \in \mathbb{R}^{p\times n}$, $E \in \mathbb{R}^{p\times p}$ is a canonical (minimal) realization of T(s) then we have that:

1) spectrum (A) is outside Ω, i.e. if $n > 0$ then spectrum (A) $\subset \Omega^c$ (T(s) has no (finite) poles in Ω)
2) $\text{rank}_{\mathbb{R}} E = p$ (T(s) has no zeros at $s = \infty$)
3) spectrum $(A-BE^{-1}C) \subset \Omega^c$ (T(s) has no finite zeros in Ω)

In the particular case when $\Omega \equiv \mathbb{C}_+$, then an $\mathbb{R}_P(s)$-unimodular matrix $T(s) \in \mathbb{R}_P^{p\times p}(s)$ represents a square, biproper, stable and "minimum phase" transfer function matrix. Elementary row and column operations on a $T(s) \in \mathbb{R}_P^{p\times m}(s)$ are now defined as follows: (i) interchange any two rows (columns) of T(s), (ii) multiply row (column) i of T(s) by a *unit* $u(s) \in \mathbb{R}_P(s)$ and (iii) add to row (column) i of T(s) a multiple by $t(s) \in \mathbb{R}_P(s)$ of row (column) j. These elementary operations can be accomplished by multiplying the given T(s) on the left (right) by "elementary" $\mathbb{R}_P(s)$-unimodular matrices obtained by performing the above operations on the identity matrix $I_{p(m)}$. It can also be shown that every $\mathbb{R}_P(s)$-unimodular matrix may be represented as a product of a finite number of elementary $\mathbb{R}_P(s)$-unimodular matrices [5].

<u>Definition 2.1</u>: Let $T_1(s) \in \mathbb{R}^{p\times m}(s)$, $T_2(s) \in \mathbb{R}^{p\times m}(s)$. Then $T_1(s)$ and $T_2(s)$ are called "equivalent in P" if there exist $\mathbb{R}_P(s)$-unimodular matrices $T_L(s) \in \mathbb{R}^{p\times p}(s)$, $T_R(s) \in \mathbb{R}^{m\times m}(s)$ such that:

$$T_L(s)T_1(s)T_R(s) = T_2 \tag{2.7}$$

If $T_L(s) \equiv I_p \in \mathbb{R}^{p\times p}(s)$ $(T_R(s) \equiv I_m \in \mathbb{R}_P^{m\times m}(s))$ then $T_1(s),T_2(s)$ are called "column (row) equivalent in P".

Eq.(2.7) defines an equivalence relation on $\mathbb{R}^{p\times m}(s)$, which we denote by E^P, and if $T_1(s),T_2(s)$ are equivalent in P we denote this fact by writing: $(T_1(s),T_2(s)) \in E^P$. The E^P equivalence class or the "orbit" of a fixed $T(s) \in \mathbb{R}^{p\times m}(s)$ we denote by $[T(s)]_{E^P}$. Let $T(s) \in \mathbb{R}^{p\times m}(s)$ with $\operatorname{rank}_{\mathbb{R}(s)} T(s) = r$ and consider the quotient of $\mathbb{R}^{p\times m}(s)$ by E^P, i.e. the set (denoted by) $\mathbb{R}^{p\times m}(s)/E^P$ of E^P-equivalence classes $[T(s)]_{E^P}$ when $T(s)$ runs through the elements of $\mathbb{R}^{p\times m}(s)$. We can characterise these equivalence classes by determining complete sets of invariants and canonical forms. We have:

Theorem 2.1: (Smith-MacMillan form of a rational matrix in $P := \Omega \cup \{\infty\}$) [6]. Let $T(s) \in \mathbb{R}^{p\times m}(s)$ with $\operatorname{rank}_{\mathbb{R}(s)} T(s) = r$. Then $T(s)$ is equivalent in P to a diagonal matrix $S^P_{T(s)}$ having the form:

$$S^P_{T(s)} = [\operatorname{diag}\{\varepsilon_1(s)\psi_1(s)^{-1},\varepsilon_2(s)\psi_2(s)^{-1},\ldots,\varepsilon_r(s)\psi_r(s)^{-1}\},0_{p-r,m-r}] \tag{2.8}$$

where

$$\varepsilon_i(s) = \frac{\varepsilon_{i\Omega}(s)}{(s+\alpha)^{p_i}} \in \mathbb{R}_P(s) \;, \quad \psi_i(s) = \frac{\psi_{i\Omega}(s)}{(s+\alpha)^{\ell_i}} \in \mathbb{R}_P(s)$$

are coprime in P [6], $i \in \underline{r}$; $\varepsilon_{i\Omega}(s)$, $\psi_{i\Omega}(s) \in \mathbb{R}[s]$ have their zeros not outside Ω, $-\alpha \in \mathbb{R}$ is outside Ω and otherwise arbitrary, and

$$0 \le \delta_\Omega(\varepsilon_i(s)) := p_i \le p_{i+1} =: \delta_\Omega(\varepsilon_{i+1}(s)),\ \varepsilon_i(s)\,|\,\varepsilon_{i+1}(s),\ i \in \underline{r-1}$$

$$0 \le \delta_\Omega(\psi_{i+1}(s)) := \ell_{i+1} \le \ell_i =: \delta_\Omega(\psi_i(s)),\ \psi_{i+1}(s)\,|\,\psi_i(s),\ i \in \underline{r-1}$$

The matrix $S^P_{T(s)}$ can also be written as:

$$S^P_{T(s)} = E^P(s)\Psi^P_R(s)^{-1} = \Psi^P_L(s)^{-1}E^P(s) \tag{2.9}$$

where

$$E^P(s) = [\operatorname{diag}\{\varepsilon_1(s),\varepsilon_2(s),\ldots,\varepsilon_r(s)\},0_{p-r,m-r}] \tag{2.10}$$

$$\Psi^P_R(s) = \operatorname{diag}.\{\psi_1(s),\psi_2(s),\ldots,\psi_r(s),I_{m-r}\} \tag{2.11}$$

$$\Psi^P_L(s) = \operatorname{diag}.\{\psi_1(s),\psi_2(s),\ldots,\psi_r(s),I_{p-r}\} \tag{2.12}$$

and as

$$S^{P}_{T(s)} = [\text{diag.}\{\frac{\varepsilon_{1\Omega}(s)}{\psi_{1\Omega}(s)}(s+\alpha)^{q_1},\ldots,\frac{\varepsilon_{1\Omega}(s)}{\psi_{r\Omega}(s)}(s+\alpha)^{q_r}, 0_{p-r,m-r}] \quad (2.13)$$

where $q_i := p_i - \ell_i \in \mathbb{Z}$, $i \in \underline{r}$. □

Corollary 2.1 [6]: The rational functions $\varepsilon_i(s)\psi_i(s)^{-1} \in \mathbb{R}(s)$ form a complete (mod.α) set of invariants for E^P on $\mathbb{R}^{p\times m}(s)$ and they will be referred to as "the invariant rational functions of T(s) in P". Also $S^{P}_{T(s)}$ is a (unique mod.α) canonical form for E^P on $\mathbb{R}^{p\times m}(s)$. □

Remark 2.2: The finite zeros of $\varepsilon_i(s) = \varepsilon_{i\Omega}(s)/(s+\alpha)^{p_i} \in \mathbb{R}_P(s)$, i.e. the (finite) zeros of $\varepsilon_{i\Omega}(s) \in \mathbb{R}[s]$, $i \in \underline{r}$, give the (finite) zeros of T(s) inside Ω, while the $q^i_{z\infty} := p_i - \text{deg.}\varepsilon_{i\Omega}(s) \geq 0$ give the orders of the zeros at $s = \infty$ of T(s). Also the finite zeros of $\psi_i(s) = \psi_{i\Omega}(s)/(s+\alpha)^{\ell_i}$ $\in \mathbb{R}_P(s)$, i.e. the (finite) zeros of $\psi_{i\Omega}(s) \in \mathbb{R}$ s , $i \in \underline{r}$, give the (finite) poles of T(s) inside Ω, while the $q^i_{p\infty} := \ell_i - \text{deg.}\psi_{i\Omega}(s) \geq 0$ give the orders of the poles at $s = \infty$ of T(s).

If $T(s) \in \mathbb{R}_P^{p\times m}(s)$ then

$$S^{P}_{T(s)} = [\text{diag.}\{\frac{\varepsilon_{1\Omega}(s)}{(s+\alpha)^{q_1}},\ldots,\frac{\varepsilon_{r\Omega}(s)}{(s+\alpha)^{q_r}}\}, 0_{p-r,m-r}] \equiv E^P(s) \in \mathbb{R}_P^{p\times m}(s)$$

and then $S^{P}_{T(s)}$ is called the *Smith form* of T(s) in P. Otherwise, i.e. if $T(s) \notin \mathbb{R}_P^{p\times m}(s)$, then some of the $\psi_i(s)$'s will be different than 1 and then $S^{P}_{T(s)}$ is called the *MacMillan form* of T(s) in P.

Remark 2.3: If Ω coincides with the whole $\mathbb{C}$ plane except a unique point $-\alpha \in \mathbb{R}$, i.e. $\Omega = \mathbb{C}-\{-\alpha\}$, then $S^{P}_{T(s)}$ will give the Smith-MacMillan form of T(s) in all, except one (i.e. except at $s = -\alpha$), points of the "extended" complex plane $\mathbb{C}^\infty := \mathbb{C} \cup \{\infty\}$. If Ω is the empty set, i.e. $\Omega = \emptyset$, then $\varepsilon_{i\Omega}(s) = \psi_{i\Omega}(s) \equiv 1$, and (2.13) gives the Smith-MacMillan form of T(s) at $s = \infty$, $S^{\infty}_{T(s)}$ [1].

3. COPRIMENESS IN P OF PROPER AND Ω-STABLE RATIONAL MATRICES

We describe now the notions of right or left coprimeness of rational matrices in the set $P := \Omega \cup \{\infty\}$. From the definition of the zeros of a $T(s) \in \mathbb{R}^{p\times m}(s)$ inside P via its Smith-MacMillan form $S^{P}_{T(s)}$ in P we have [6]:

Proposition 3.1: Let $T(s) \in \mathbb{R}^{p\times m}(s)$, $\text{rank}_{\mathbb{R}(s)} T(s) = r$. Then the following statements are equivalent:

(i) T(s) has no zeros in $P = \Omega \cup \{\infty\}$.

(ii) $\varepsilon_i(s) = 1$, $i \in \underline{r}$ <=> $\varepsilon_{i\Omega}(s) = 1$ *and* $p_i = 0$, $i \in \underline{r}$.

(iii) $S^{P}_{T(s)} = [\text{diag.}\{\frac{1}{\psi_1(s)},\ldots,\frac{1}{\psi_r(s)}\},0_{p-r,m-r}]$

(i.e. T(s) has possibly only *poles* in P) □

Definition 3.1: Given two rational matrices $A(s) \in \mathbb{R}^{\ell\times m}(s)$, $B(s) \in \mathbb{R}^{t\times m}(s)$ with $p := \ell+t \geq m$ and $\text{rank}_{\mathbb{R}(s)}\begin{bmatrix}A(s)\\B(s)\end{bmatrix} = m$, then we say that (the row of) A(s) and B(s) are right coprime in $P = \Omega \cup \{\infty\}$ if $T(s) := \begin{bmatrix}A(s)\\B(s)\end{bmatrix} \in \mathbb{R}^{(\ell+t)\times m}(s)$ has no zeros in P.

If we restrict ourselves to matrices that are proper and Ω-stable then we have:

Proposition 3.2 [6]: Let $A(s) \in \mathbb{R}_P^{\ell\times m}(s)$, $B(s) \in \mathbb{R}_P^{t\times m}(s)$ (i.e. proper and Ω-stable) with $p := \ell+t \geq m$. Then the following statements are equivalent:

(i) A(s) and B(s) are right coprime in $P := \Omega \cup \{\infty\}$.

(ii) The proper and Ω-stable matrix $T(s) := \begin{bmatrix}A(s)\\B(s)\end{bmatrix} \in \mathbb{R}_P^{p\times m}(s)$ has no zeros in P.

(iii) There exists a $\mathbb{R}_P(s)$-unimodular matrix $T_L(s) \in \mathbb{R}_P^{p\times p}(s)$ such that

$$T_L(s)T(s) = \begin{bmatrix}I_m\\0_{p-m,m}\end{bmatrix} = S^{P}_{T(s)}$$

(iv) There exist proper, Ω-stable rational matrices $X(s) \in \mathbb{R}_P^{m\times \ell}(s)$, $Y(s) \in \mathbb{R}_P^{m\times t}(s)$ such that $[X(s),Y(s)]\begin{bmatrix}A(s)\\B(s)\end{bmatrix} = I_m$

(v) There exist proper, Ω-stable rational matrices $C(s) \in \mathbb{R}_P^{\ell\times(p-m)}(s)$, $D(s) \in \mathbb{R}_P^{t\times(p-m)}(s)$ such that the rational matrix $\begin{bmatrix}A(s) & C(s)\\B(s) & D(s)\end{bmatrix} \in \mathbb{R}_P^{p\times p}(s)$ is $\mathbb{R}_P(s)$-unimodular.

(vi) $\text{rank}_{\mathbb{C}}\begin{bmatrix}A(s_o)\\B(s_o)\end{bmatrix} = m \;\forall\; s_o \in \Omega$ *and* $\lim_{s\to\infty}\begin{bmatrix}A(s)\\B(s)\end{bmatrix} =: E \in \mathbb{R}^{p\times m}$ with $\text{rank}_{\mathbb{R}} E = m$ □

Definition 3.2: A proper and Ω-stable rational matrix $T(s) \in \mathbb{R}_P^{p\times m}(s)$ $(p \geq m)$ satisfying the equivalent conditions of Proposition 3.2 is defined as a $\mathbb{R}_P(s)$-left unimodular rational matrix. ($\mathbb{R}_P(s)$-right unimodular rational matrices can be defined in an analogous manner.)

Notice that from Proposition 3.2 it simply follows that an $\mathbb{R}_P$-left unimodular rational matrix $T(s) \in \mathbb{R}_P^{p\times m}(s)$ may have zeros only in Ω^c. We examine now closely the notions of right (common) divisors in P and of greatest (common) right divisors in P of (the rows of two or more) rational matrices having the same number of columns. Left (common) and greatest left (common) divisors in P can be defined analogously. From the Smith-MacMillan form of T(s), the following factorization of a rational matrix (not necessarily proper) is readily derived.

Proposition 3.3: Any rational matrix $T(s) \in \mathbb{R}^{p\times m}(s)$ with $p \geq m$ and $\text{rank}_{\mathbb{R}(s)} T(s) = m$ can be factorized (in a non-unique way) as

$$T(s) = T_1(s)T_{GR}(s) \tag{3.1}$$

where $T_1(s) \in \mathbb{R}_P^{p\times m}(s)$ is $\mathbb{R}_P(s)$-left unimodular and $T_{GR}(s) \in \mathbb{R}^{m\times m}(s)$ has pole-zero structure in $P = \Omega \cup \{\infty\}$ the same with that of $T(s)$. □

Corollary 3.1: If $T(s) \in \mathbb{R}_{pr}^{p\times m}(s)$ ($\in \mathbb{R}_P^{p\times m}(s)$) then it can be factorized as in (3.1) where $T_{GR}(s) \in \mathbb{R}_{pr}^{p\times m}(s)$ ($\in \mathbb{R}_P^{p\times m}(s)$). In such a case $T(s)$ and $T_{GR}(s)$ have no poles at $s = \infty$, the same pole structure in Ω, and the same zero structure in $P := \Omega \cup \{\infty\}$ (no poles in P and the same zero structure in $P := \Omega \cup \{\infty\}$). □

Definition 3.3: Let the proper and Ω-stable rational matrices $T(s) \in \mathbb{R}_P^{p\times m}(s)$, $T_1(s) \in \mathbb{R}_P^{p\times m}(s)$, $T_R(s) \in \mathbb{R}_P^{m\ m}(s)$ be related via

$$T(s) = T_1(s)T_R(s) \tag{3.2}$$

The $T_R(s)$ is called a *right divisor in* P of $T(s)$.

Definition 3.4: Let $T(s) \in \mathbb{R}_P^{p\times m}(s)$ with $p \geq m$ and $\text{rank}_{\mathbb{R}(s)} T(s) = m$. Then any rational matrix $T_{GR}(s) \in \mathbb{R}_P^{m\times m}(s)$ that satisfies (3.1) for some $\mathbb{R}_P(s)$-left unimodular rational matrix $T_1(s) \in \mathbb{R}_P^{p\times m}(s)$ is called a *greatest (common) right divisor in* P of (the rows of) $T(s)$.

Remark 3.1: If $T_{GR}(s) \in \mathbb{R}_P^{m\times m}(s)$ is a g.(c.)r.d. in of (the rows of) $T(s) \in \mathbb{R}_P^{p\times m}(s)$ then from Proposition 3.3 it follows that $T_{GR}(s)$ "contains" *all* the zeros of $T(s)$ in $P = \Omega \cup \{\infty\}$ (i.e. the finite ones in Ω and the infinite ones, if any). If $T_{GR}(s) \in \mathbb{R}_P^{m\times m}(s)$ happens to be $\mathbb{R}_P(s)$-unimodular then $T(s)$ has also no zeros in P and its rows are said to be *right coprime in* P. Notice that in such a case $T(s)$ might have only finite zeros outside Ω, i.e. in Ω^c.

Finally the next proposition describes the fact that every rational matrix $T(s)$ can be represented as a ratio of (coprime in P) proper and Ω-stable rational matrices. This representation of rational matrices was firstly introduced by Vidyasagar [8] and later used by Desoer et al. [9], Francis and Vidyasagar [10], Saeks and Murray [11], and forms the basis of what is known as the "fractional representation approach" to analysis and synthesis of linear multivariable control algebraic problems (see also [12][13][14]).

Proposition 3.4: Let $T(s) \in \mathbb{R}^{p\times m}(s)$. Then $T(s)$ can always be represented (in a non-unique way) as

$$T(s) = B_2(s)A_2(s)^{-1} = A_1(s)^{-1}B_1(s) \tag{3.3}$$

where $B_2(s) \in \mathbb{R}_P^{p\times m}(s)$, $A_2(s) \in \mathbb{R}_P^{m\times m}(s)$ are right coprime in P and

$A_1(s) \in \mathbb{R}_P^{p\times p}(s)$, $B_1(s) \in \mathbb{R}_P^{p\times m}(s)$ are left coprime in P.

□

A systematic procedure for deriving the family of coprime fractional representations is based on the Smith-MacMillan form of T(s) in P and it is given in [6].

A pair $(B_2(s),A_2(s))$ $((B_1(s),A_1(s))$ satisfying Proposition 3.4 is defined as a right (left) coprime in P $\mathbb{R}_P(s)$-matrix fraction description of T(s) ($\mathbb{R}_P(s)$-MFD).

3.1 The Function $\delta_P(\cdot)$ for Rational Matrices

We generalise now the definition of the function δ_P introduced in section 2 for the case of proper and Ω-stable rational matrices. Let $T(s) \in \mathbb{R}^{p\times m}(s)$, $\mathrm{rank}_{\mathbb{R}(s)}T(s) = r$. We define the map $\delta_P : \mathbb{R}_P^{p\times m}(s) \to \mathbb{Z} \cup \{\infty\}$ via:

$$\delta_P(T) = \begin{cases} \min\left\{\begin{array}{l}\delta_P(\cdot) \text{ among the } \delta_P(\cdot)\text{'s of all r-th} \\ \text{order (non-zero) minors of } T(s)\end{array}\right. & \text{if } r > 0 \\ +\infty & \text{if } r = 0 \end{cases}$$

We give now a number of properties regarding $\delta_P(\cdot)$. These properties are stated for the case of a rational matrix $T(s) \in \mathbb{R}_P^{p\times m}(s)$ with $p \geq m$ and similar ones can be stated for the case $p \leq m$.

<u>Proposition 3.1.1</u> [6]: Let $T(s) \in \mathbb{R}_P^{p\times m}(s)$. Then $\delta_P(T) \geq 0$. Moreover if $p = m = \mathrm{rank}_{\mathbb{R}(s)}T(s)$ then $\delta_P(T) = 0$ iff T(s) is $\mathbb{R}_P(s)$-unimodular.

The following proposition describes the $\delta_P(\cdot)$ of a product of two proper and Ω-stable rational matrices having special sizes.

<u>Proposition 3.1.2</u> [6]: Let $T_1(s) \in \mathbb{R}_P^{p\times m}(s)$, $p \geq m$, $\mathrm{rank}_{\mathbb{R}(s)}T(s) = m$ and $T_2(s) \in \mathbb{R}_P^{m\times m}(s)$, $\mathrm{rank}_{\mathbb{R}(s)}T_2(s) = m$ and let $T(s) = T_1(s)T_2(s)$, then

$$\delta_P(T) = \delta_P(T_1) + \delta_P(T_2) \tag{3.1.2}$$

□

<u>Corollary 3.1.1</u>: Let $T(s) \in \mathbb{R}_P^{p\times m}(s)$, $p \geq m$, $\mathrm{rank}_{\mathbb{R}(s)}T(s) = m$ and let $T_R(s) \in \mathbb{R}_P^{m\times m}(s)$ be a (common) right divisor in P (greatest (common) right divisor in P) of (the rows of) T(s), i.e. let $T(s) = T_1(s)T_R(s)$ for some ($\mathbb{R}_P(s)$-left unimodular) rational matrix $T_1(s) \in \mathbb{R}_P^{p\times m}(s)$. Then $\delta_P(T) \geq \delta_P(T_1)$ with equality holding iff $T_R(s)$ is $\mathbb{R}_P(s)$-unimodular; i.e. $\delta_P(\cdot)$ is an invariant of the column $\mathbb{R}_P(s)$-equivalence class of the rational matrix T(s).

4. PROPER AND Ω-STABLE, MINIMAL MACMILLAN DEGREE BASES OF RATIONAL VECTOR SPACES

We will examine now the algebraic structure of the set of all proper and Ω-stable rational vectors $t(s) \in \mathbb{R}_P^{p\times 1}(s)$ which are contained in the rational vector space $\mathcal{T}(s)$ spanned by the columns $t_j(s) \in \mathbb{R}_P^{p\times 1}(s)$, $j \in \underline{m}$, of a general rational matrix $T(s) \in \mathbb{R}^{p\times m}(s)$.

Firstly the existence of proper and Ω-stable bases for $\mathcal{T}(s)$ follows directly from Proposition 3.4, i.e. if $T(s) \in \mathbb{R}^{p\times m}(s)$ is a basis for $\mathcal{T}(s)$ ($p \geq m$, $\mathrm{rank}_{\mathbb{R}(s)} T(s) = m$) and is expressed as a (right coprime in $\mathcal{P}$) $\mathbb{R}_{\mathcal{P}}(s)$-MFD: $T(s) = B(s)A(s)^{-1}$ where $B(s) \in \mathbb{R}_{\mathcal{P}}^{p\times m}(s)$, $A(s) \in \mathbb{R}_{\mathcal{P}}^{m\times m}(s)$, then clearly $B(s)$ is a proper and Ω-stable basis of $\mathcal{T}(s)$.

Let now $T_1(s) \in \mathbb{R}_{\mathcal{P}}^{p\times m}(s)$ be a basis for $\mathcal{T}(s)$ and consider the set of all proper and Ω-stable rational vectors that can be obtained as linear combinations of the columns $t_j(s) \in \mathbb{R}_{\mathcal{P}}^{p\times 1}(s)$, $j \in \underline{m}$, of $T_1(s)$, with coefficients in the ring $\mathbb{R}_{\mathcal{P}}(s)$. This set is a free $\mathbb{R}_{\mathcal{P}}(s)$-module M_1 and any other basis for M_1 can be obtained from $T_1(s)$ by post-multiplying $T_1(s)$ by a $\mathbb{R}_{\mathcal{P}}(s)$-unimodular matrix, i.e. if $T_R(s) \in \mathbb{R}_{\mathcal{P}}^{m\times m}(s)$ is $\mathbb{R}_{\mathcal{P}}(s)$-unimodular and we define

$$\bar{T}_1(s) := T_1(s)T_R(s) \tag{4.1}$$

then $\bar{T}_1(s) \in \mathbb{R}_{\mathcal{P}}^{p\times m}(s)$ is also a basis for M_1.

In such a case, from Corollary 3.1.1 we have that $\delta_{\mathcal{P}}(\bar{T}_1) = \delta_{\mathcal{P}}(T_1)$, i.e. all the bases of M_1 have the same $\delta_{\mathcal{P}}(\cdot)$ and thus the $\delta_{\mathcal{P}}(\cdot)$ of any basis $T_1(s)$ is an *invariant* of the $\mathbb{R}_{\mathcal{P}}(s)$-module M_1 which is generated by its columns, and which we call the "*stathm in $\mathcal{P}$*" of M_1 and denote by $\delta_{\mathcal{P}}(M_1)$.

Assume now that $T_1(s) \in \mathbb{R}_{\mathcal{P}}^{p\times m}(s)$ is *not* $\mathbb{R}_{\mathcal{P}}(s)$-left unimodular and let $T_{1G}(s) \in \mathbb{R}_{\mathcal{P}}^{m\times m}(s)$ be a non-$\mathbb{R}_{\mathcal{P}}(s)$-unimodular right divisor in $\mathcal{P}$ of $T_1(s)$ (not necessarily a g.c.r.d. in $\mathcal{P}$ of $T_1(s)$), i.e. assume that

$$T_1(s) = T_2(s)T_{1G}(s) \tag{4.2}$$

for some (not necessarily $\mathbb{R}_{\mathcal{P}}(s)$-left unimodular) $T_2(s) \in \mathbb{R}_{\mathcal{P}}^{p\times m}(s)$ with $\delta_{\mathcal{P}}(T_1) \geq \delta_{\mathcal{P}}(T_2)$ (Corollary 3.1.1) so that $T_{1G}(s)$ contains some of the zeros of $T_1(s)$ in $\mathcal{P}$. If we now consider the $\mathbb{R}_{\mathcal{P}}(s)$-module M_2 which is generated by (the columns of) $T_2(s)$, then M_1 is a proper submodule of M_2 i.e.

$$M_1 \subset M_2 \tag{4.3}$$

In general if $T_{1G}(s), T_{2G}(s), \ldots, T_{iG}(s) \in \mathbb{R}_{\mathcal{P}}^{m\times m}(s)$ are (non $\mathbb{R}_{\mathcal{P}}(s)$-unimodular) right divisors in $\mathcal{P}$ of $T_1(s)$ such that

$$0 < \delta_{\mathcal{P}}(T_{1G}) < \delta_{\mathcal{P}}(T_{2G}) < \ldots < \delta_{\mathcal{P}}(T_{iG}) \tag{4.4}$$

and

$$T_1(s) = T_{i+1}(s)T_{iG}(s) \qquad i = 1,2,\ldots \tag{4.5}$$

for some (not necessarily $\mathbb{R}_{\mathcal{P}}(s)$-left unimodular) rational matrix $T_{i+1}(s) \in \mathbb{R}_{\mathcal{P}}^{p\times m}(s)$, then the $\mathbb{R}_{\mathcal{P}}(s)$-modules M_{i+1}, $i = 1,2,\ldots$ generated by (the columns of) $T_{i+1}(s)$, $i = 1,2,\ldots$ form an ascending sequence of submodules [15]:

$$M_1 \subset M_2 \subset \ldots \subset M_{i+1} \qquad i = 1,2,\ldots \tag{4.6}$$

and

$$\delta_P(M_1) > \delta_P(M_2) > \ldots > \delta_P(M_{i+1}) \qquad i = 1,2,\ldots \tag{4.7}$$

If now for some $i = 1,2,\ldots$ $T_{iG}(s) =: T_{GR}(s)$ is a *greatest right divisor in* P of $T_1(s)$, so that

$$T_1(s) = \bar{T}(s)T_{GR}(s) \tag{4.8}$$

for some $\mathbb{R}_P(s)$-left unimodular matrix $\bar{T}(s) \in \mathbb{R}_P^{p\,m}(s)$, then the $\mathbb{R}_P(s)$-module generated by (the columns of) $\bar{T}(s)$, and which we denote by M^*, satisfies an ascending chain condition on submodules [15], i.e. we have:

$$M_1 \subset M_2 \subset \ldots \subset M_{i+1} \equiv M^* \tag{4.9}$$

for some $i = 1,2,\ldots$ and coincides with the set of all proper and Ω-stable rational vectors which are contained in the rational vector space $\mathcal{T}(s)$.

In the sequel we examine the structure of the various ($\mathbb{R}_P(s)$-left unimodular) bases of the "maximal" $\mathbb{R}_P(s)$-module M^*. As we show below, these bases can be further classified according to properties of their MacMillan degree. In order to proceed with this investigation we will need a number of definitions and results from the theory of proper bases of rational vector spaces [16].

We start with

Definition 4.1 [16]: Let $T(s) \in \mathbb{R}_{pr}^{p\times m}(s)$, $p \geq m$, $\text{rank}_{\mathbb{R}(s)}T(s) = m$, with column vectors $t_j(s)$ and denote by $\delta_M(t_j)$ the MacMillan degree of $t_j(s)$. Then the *column MacMillan degree complexity* of $T(s)$, denoted by $C_M^c(T)$, is the sum of the MacMillan degrees of its columns, i.e.

$$C_M^c(T) := \sum_{j=1}^{m} \delta_M(t_j)$$

The following proposition [16] states under what conditions the column MacMillan degree complexity $C_M^c(T)$ of a *column reduced at* $s = \infty$ [16][17] proper rational matrix $T(s)$ coincides with its MacMillan degree $\delta_M(T)$.

Proposition 4.1 [16]: Let $T(s) \in \mathbb{R}_{pr}^{p\times m}(s)$, $p \geq m$, $\text{rank}_{\mathbb{R}(s)}T(s) = m$, be a *column reduced at* $s = \infty$ rational matrix. Let $d_j(s) \in \mathbb{R}[s]$ be the (monic) least common multiple of the (p) denominators which appear in the j-th column t_j of $T(s)$ and write: $t_j(s) = n_j(s).1/d_j(s)$, $n_j(s) \in \mathbb{R}^{p\times 1}[s]$, $j \in \underline{m}$. Let $N(s) := [n_1(s),\ldots,n_m(s)] \in \mathbb{R}^{p\times m}[s]$ and $D(s) := \text{diag}(d_1(s),\ldots,d_m(s))$. Then we have:

$$\sum_{j=1}^{m} \delta_M(t_j) \geq \delta_M(T)$$

with equality holding iff N(s),D(s) are right coprime. □

Definition 4.2 [16]: A column reduced at $s = \infty$ proper rational matrix $T(s) \in \mathbb{R}_{pr}^{p\times m}(s)$ with $p \geq m$ and $\text{rank}_{\mathbb{R}(s)} T(s) = m$, which satisfies:

$$\sum_{j=1}^{m} \delta_M(t_j) = \delta_M(T)$$

is defined as a "*simple*" basis of the $\mathbb{R}_{pr}(s)$-module M generated by its columns $t_j(s) \in \mathbb{R}_{pr}^{p\times 1}(s)$.

Remark 4.1: It can be easily proved that given a *column reduced at* $s = \infty$ basis $T(s) \in \mathbb{R}_{pr}^{p\times m}(s)$ of M then we can always determine an $\mathbb{R}_{pr}(s)$-unimodular (biproper) rational matrix $U_R(s) \in \mathbb{R}_{pr}^{m\times m}(s)$ such that $\bar{T}(s) := T(s)U_R(s)$ is a column reduced at $s = \infty$ and simple basis of M with $\delta_M(\bar{T}) = \delta_M(T)$.

The other two results needed for our investigation are two lemmas. The first lemma describes a particular case of a more general Theorem which concerns the relationship between (i) the total number of poles and zeros (finite and infinite ones) of a general rational matrix $T(s) \in \mathbb{R}^{p\times m}(s)$ and (ii) the sum of the invariant dynamical indices [18] of the left and right null spaces of T(s). The general result appeared originally in [19] and as Theorem 2.1 in [20] or as Theorem 3 in [21] (see also Corollary 8 in [16]).

Lemma 4.1: Let $T(s) \in \mathbb{R}_{pr}^{p\times m}(s)$, $p \geq m$, $\text{rank}_{\mathbb{R}(s)} T(s) = m$ and $\lim_{s\to\infty} T(s) = E \in \mathbb{R}^{p\ m}$ with $\text{rank}_{\mathbb{R}} E = m$. Let $v_j \geq 0$, $j \in \underline{m}$ the (Forney) invariant dynamical indices of the rational vector space $\mathcal{T}(s)$ spanned by the columns of T(s), $\text{ord}_F(\mathcal{T}(s)) := \sum_{j=1}^{m} v_j$ the (Forney) invariant dynamical order of T(s) and $z_f(T)$ the number of finite zeros of T(s). Then:

$$\delta_M(T) = \text{ord}_F(\mathcal{T}(s)) + z_f(T) \tag{4.10}$$

Proof: $T(s) \in \mathbb{R}_{pr}^{p\times m}(s)$ and $\text{rank}_{\mathbb{R}} E = m$ imply respectively that T(s) has no poles and zeros at $s = \infty$ [1][16]. Hence $\delta_M(T)$ is the number of finite poles of T(s). The result then follows as a particular case of the more general Theorem in [19][20][21]. □

Lemma 4.2: Let $T_i(s) \in \mathbb{R}_{pr}^{p\times m}(s)$, $i = 1,2$, with $\lim_{s\to\infty} T(s) =: E_i \in \mathbb{R}^{p\times m}$ and $\text{rank}_{\mathbb{R}} E_i = m$. If there exists a $Q(s) \in \mathbb{R}^{m\times m}(s)$, $\text{rank}_{\mathbb{R}(s)} Q(s) = m$, such that $T_1(s) = T_2(s)Q(s)$, then $Q(s) \in \mathbb{R}_{pr}^{m\times m}(s)$ and $\lim_{s\to\infty} Q(s) =: Q_o \in \mathbb{R}^{m\times m}$ with $\text{rank}_{\mathbb{R}} Q_o = m$.

Proof: Consider the Laurent expansion of $T_1(s)$, $T_2(s)$, $Q(s)$ at $s = \infty$:

$$T_i(s) = E_i + \sum_{j=1}^{\infty} \frac{1}{s^j} C_{ij}, \quad i = 1,2, \quad Q(s) = Q_o s^r + Q_1 s^{r-1} + \ldots, \quad Q_0 \neq 0$$

Then

$$[E_1 + \frac{1}{s} C_{11} + \ldots] = [E_2 + \frac{1}{s} C_{21} + \ldots][Q_o s^r + Q_1 s^{r-1} + \ldots]$$

(i) If $r > 0$ then $E_2 Q_o = 0 \Rightarrow Q_o = 0$.

(ii) If $r < 0$ then $E_1 = 0$.

Thus $r = 0$ and $E_1 = E_2 Q_o$. Given that E_1, E_2 have full rank it follows that $\det Q_o \neq 0$. □

Having introduced the above concepts and results we return now to the examination of the various $\mathbb{R}_P(s)$-left unimodular bases of the maximal $\mathbb{R}_P(s)$-module M* [35]. Firstly (and unlike the proper submodules M_i of M*) all bases of M* are column reduced at $s = \infty$, since (by the definition of M*) they are all $\mathbb{R}_P(s)$-left unimodular (see Proposition 9 in [16]). Secondly, and due to the above fact, it also follows from Remark 4.1 that if $T(s) \in \mathbb{R}_P^{p\times m}(s)$ is a basis of M* then we can always determine an $\mathbb{R}_P(s)$-unimodular matrix $U_R(s) \in \mathbb{R}_P^{m\times m}(s)$ such that $\bar{T}(s) := T(s)U_R(s) \in \mathbb{R}_P^{p\times m}(s)$ is a *simple* basis of M* which satisfies: $\delta_M(\bar{T}) = \delta_M(T)$. The following Theorem which is the main result of this section proves the fact that given a basis T(s) of M* we can always determine an $\mathbb{R}_P(s)$-unimodular matrix $U_R(s) \in \mathbb{R}_P^{m\times m}(s)$ such that $\bar{T}(s) := T(s)U_R(s)^{-1}$ is a *simple* basis of M* which has *desired* poles (in Ω^c) and whose MacMillan degree $\delta_M(\bar{T})$ is *minimum* among the MacMillan degrees of all other proper or proper and Ω-stable bases of the rational vector space $\mathcal{T}(s)$ spanned by the columns of T(s), i.e. that $\delta_M(\bar{T}) \leq \delta_M(T)\ \forall\ T(s) \in \mathbb{R}_{pr}^{p\times m}(s)$ basis of $\mathcal{T}(s)$. We thus prove that if $\mathcal{T}(s)$ is a rational vector space then among the proper and Ω-stable bases of $\mathcal{T}(s)$ which have no zeros in $\mathcal{P}$ (i.e. they are $\mathbb{R}_P(s)$-left unimodular) there is a subfamily of *simple* $\mathbb{R}_P(s)$-left unimodular bases which have any desired set of poles and whose MacMillan degrees are minimum among the MacMillan degrees of all other proper bases of $\mathcal{T}(s)$. This result is formally stated in the following:

Theorem 4.1 [35]: Let $T(s) \in \mathbb{R}_P^{p\times m}(s)$, $p \geq m$, $\operatorname{rank}_{\mathbb{R}(s)} T(s) = m$ be $\mathbb{R}_P(s)$-left unimodular and let M* be the $\mathbb{R}_P(s)$-module generated by its columns $t_j(s)$. Then T(s) can always be factorized (in a non-unique way) as:

$$T(s) = \bar{T}(s)U_R(s) \tag{4.11}$$

where $\bar{T}(s) = [\bar{t}_1(s),\ldots,\bar{t}_m(s)] \in \mathbb{R}_P^{p\times m}(s)$ is $\mathbb{R}_P(s)$-left unimodular and *simple* basis of M* which has no *finite zeros* and $U_R(s) \in \mathbb{R}_P^{m\times m}(s)$ is $\mathbb{R}_P(s)$-unimodular and the set of its finite zeros* contains as a subset the set of finite zeros (which, if any, are in Ω^c) of T(s). Furthermore if $v_j \geq 0$, $j \in \underline{m}$, are the (Forney) *invariant dynamical indices* of the rational vector space $\mathcal{T}(s)$ spanned by the columns of T(s) (and also of

*Notice that a $\mathbb{R}_P(s)$-unimodular matrix has (possibly) only finite poles and zeros not outside Ω^c.

$\bar{T}(s)$), $\text{ord}_F(\mathcal{T}(s)) := \sum_{j=1}^{m} v_j$ is the (Forney) *invariant dynamical order* of $\mathcal{T}(s)$ then:

(i) $\delta_M(\bar{t}_j) = v_j$, $j \in \underline{m}$

(ii) $\delta_M(\bar{T}) = \sum_{j=1}^{m} \delta_M(\bar{t}_j) = \sum_{j=1}^{m} v_j = \text{ord}_F(\mathcal{T}(s))$

and $\delta_M(\bar{T})$ is minimum among the MacMillan degrees of all other proper bases of $\mathcal{T}(s)$.

Proof: Let $T(s) = N(s)D(s)^{-1}$, $N(s) \in \mathbb{R}^{p\times m}[s]$, $D(s) \in \mathbb{R}^{m\times m}[s]$ be a right coprime MFD of $T(s)$ (where due to the assumption that $T(s)$ is $\mathbb{R}_p(s)$-left unimodular, the finite zeros of $N(s)$ and $D(s)$ are confined in Ω^c), and factorise $N(s)$ as $N(s) = N_L(s)N_R(s)$ where $N_R(s) \in \mathbb{R}^{m\times m}[s]$ is a g.c.r.d. of (the rows of) $N(s)$ and $N_L(s)$ is a minimal polynomial basis of the rational vector space $\mathcal{T}(s)$ spanned by the columns of $T(s)$ [18] (i.e. $N_L(s)$ is (i) column proper and (ii) has relatively right prime rows [18][22]). If $n_j(s) = [n_{\ell j}(s),\dots,n_{pj}(s)]^T \in \mathbb{R}^{p\times 1}[s]$, $j \in \underline{m}$, are the columns of $N_L(s)$ then by definition [18]: $\deg.n_j(s) := \max_{i\in\underline{p}}\{\deg.n_{ij}(s)\} = v_j$, $j \in \underline{m}$. Let now $\hat{D}(s) := \text{diag}(\hat{d}_1(s),\dots,\hat{d}_m(s))$ with $\deg.\hat{d}_j(s) = v_j$ and $\hat{d}_j(s)$ arbitrary monic polynomials with zeros in Ω^c. Then $T(s)$ can be written as

$$T(s) = [N_L(s)\,\hat{D}(s)^{-1}][\hat{D}(s)\,N_R(s)\,D(s)^{-1}] := \bar{T}(s)U_R(s) \qquad (4.12)$$

Now since $p \geq m$ and $N_L(s)$ is a minimal basis: $\text{rank}_{\mathbb{R}}[N_L(s)]_c^h = m$ (*) and $\text{rank}_{\mathbb{C}}\begin{bmatrix} N_L(s) \\ \hat{D}(s) \end{bmatrix} = \text{rank}_{\mathbb{C}} N_L(s) = m$ for every $s \in \mathbb{C}$, i.e. $N_L(s),\hat{D}(s)$ are right coprime. By construction $\bar{T}(s) := N_L(s)\hat{D}(s)^{-1}$ is proper and *simple* and $\lim_{s\to\infty} \bar{T}(s) = [N_L(s)]_c^h$. Therefore $\bar{T}(s)$ is proper and has no zeros in $\mathbb{C} \cup \{\infty\}$. Also by construction $\bar{T}(s)$ has no poles in Ω, therefore $\bar{T}(s) \in \mathbb{R}_p^{p\times m}(s)$ is $\mathbb{R}_p(s)$-left unimodular with no finite zeros at all. From Lemma 4.2 it follows that $U_R(s) := \hat{D}(s)N_R(s)D(s)^{-1}$ has $\lim_{s\to\infty} U_R(s) =: Q_o \in \mathbb{R}^{m\times m}$ with $\text{rank}_{\mathbb{R}} Q_o = m$. However $U_R(s)$ has all its finite zeros and poles in Ω^c and so $U_R(s) \in \mathbb{R}_p^{m\times m}(s)$ is $\mathbb{R}_p(s)$-unimodular and some of its zeros, i.e. the zeros of $N_R(s)$ are zeros of $T(s)$. Now $\delta_M(\bar{t}_j) := \deg.\hat{d}_j(s) = v_j$ and $\sum_{j=1}^{m} \delta_M(\bar{t}_j(s)) = \sum_{j=1}^{m} v_j =: \text{ord}_F(\mathcal{T}(s)) = \deg.\det \hat{D}(s) =: \delta_M(\bar{T})$. Finally to see that $\delta_M(\bar{T})$ is a minimum notice that from Lemma 4.1 we have that for every proper basis $T(s)$ of $\mathcal{T}(s)$ with $\lim_{s\to\infty} T(s) = E$ and $\text{rank}_{\mathbb{R}} E = m$

(*) By $[N(s)]_c^h$ we denote the "highest column degree coefficient matrix of $N(s) \in \mathbb{R}^{p\times m}[s]$, [22].

$$\delta_M(T) = \mathrm{ord}_F(\mathcal{T}(s)) + z_f(T)$$

where now $z_f(T) \geq 0$ denotes the number of finite zeros (if any) of T(s) in Ω^c). Hence from (ii) it follows that $\delta_M(\bar{T}) \leq \delta_M(T)$ for every proper basis T(s) of $\mathcal{T}(s)$.

□

Remark 4.2: If p = m then the factorization of T(s) in (4.11) is trivial with $\bar{T}(s) = I_m$ and $U_R(s) = T(s)$. If p > m the factorization is not unique. The above theorem gives rise to the following.

Definition 4.3 [35]: *A* $\mathbb{R}_P(s)$-*left unimodular and simple* basis $\bar{T}(s) \in \mathbb{R}_P^{p\times m}(s)$ of M* which has no finite zeros (and thus it satisfies (i) and (ii) of Theorem 4.1) is defined as a *simple, minimal Macmillan degree, proper* and Ω-*stable basis* of the rational vector space $\mathcal{T}(s)$ spanned by its columns (SMMD-$\mathbb{R}_P(s)$ basis).

Remark 4.3: From the above analysis it becomes clear that any proper and Ω-stable rational matrix $T(s) \in \mathbb{R}_P^{p\times m}(s)$ with $p \geq m$ ($p \leq m$) and $\mathrm{rank}_{\mathbb{R}(s)} T(s) = m(p)$ with no zeros in $\mathbb{C} \cup \{\infty\}$, but not necessarily "*simple*", is a *minimal MacMillan degree* $\mathbb{R}_P(s)$-*basis* of the rational vector space $\mathcal{T}(s)$ spanned by its columns (rows) (MMD-$\mathbb{R}_P(s)$ basis).

Notice that given any SMMD $\mathbb{R}_P(s)$-basis we can rearrange its columns (rows) by post (pre) multiplying it by a constant permutation (and so $\mathbb{R}_P(s)$-unimodular) matrix $Q \in \mathbb{R}_P^{m\times m}$ ($\in \mathbb{R}^{p\times p}$) so that the resulting matrix $\hat{T}(s) := T(s)Q$ ($QT(s)$) has columns $\hat{t}_j(s)$ (rows $\check{t}_i(s)$) that are arranged in order of ascending MacMillan degrees, i.e. $\delta_M(\hat{t}_1) \leq \delta_M(\hat{t}_2) \leq \ldots \leq \delta_M(t_M)$ etc.

Example: Let $\Omega \equiv \mathbb{C}_+$ and consider the rational matrix

$$T(s) = \begin{bmatrix} \dfrac{s-2}{(s+1)\ (s+2)} & \dfrac{s-2}{(s+1)^2} \\ 0 & \dfrac{s-1}{(s+1)^2} \\ \dfrac{s-1}{(s+2)^2} & \dfrac{(s-1)^2}{(s+1)\ (s+2)} \end{bmatrix} \in \mathbb{R}_P^{3\times 2}(s)$$

which has a finite sero at s = 1 and two zeros at s = ∞ each one of order $q^i_{z\infty} = 1$, $i = 1,2$. Extracting a greatest right divisor in P from T(s) we can write

$$T(s) = \begin{bmatrix} \dfrac{s-2}{s+1} & 0 \\ 0 & \dfrac{1}{s+1} \\ \dfrac{s-1}{s+2} & \dfrac{s-2}{s+2} \end{bmatrix} \begin{bmatrix} \dfrac{1}{s+2} & \dfrac{1}{s+1} \\ 0 & \dfrac{s-1}{s+1} \end{bmatrix} = T_1(s)T_{GR}(s)$$

Now $T_1(s) \in \mathbb{R}_p^{3\times2}(s)$ has no zeros in $\mathbb{C} \cup \{\infty\}$ and therefore it is a proper and stable minimal MacMillan degree basis of the rational vector space spanned by its columns with $\delta_M(T_1) = 3$. Notice that its *column MacMillan degree complexity* is: $C_M^c(T_1) = \delta_M(t_{11}) + \delta_M(t_{12}) = 2 + 2 = 4 > \delta_M(T_1)$, *i.e.* $T_1(s)$ *is not simple*. Multiplying $T_1(s)$ on the right by the $\mathbb{R}_p(s)$-unimodular matrix

$$U_R(s) = \begin{bmatrix} 1 & 0 \\ -3/4 & 1 \end{bmatrix} \in \mathbb{R}_p^{2\times2}(s)$$

we obtain

$$T_1(s)U_R(s) = \begin{bmatrix} \frac{s-2}{s+1} & 0 \\ \frac{-3/4}{s+1} & \frac{1}{s+1} \\ 1/4 & \frac{s-2}{s+2} \end{bmatrix} =: \bar{T}(s)$$

which is also a proper and stable minimal MacMillan degree basis which is simple since $\delta_M(\bar{T}) = 3 = \delta_M(\bar{t}_1) + \delta_M(\bar{t}_2) = 1 + 2$.

Summarizing the various factorizations obtained in Propositions 3.3, 3.4 and Theorem 4.1 we have:

Theorem 4.2 : Let $T(s) \in \mathbb{R}^{p\times m}(s)$, $p \geq m$, $\text{rank}_{\mathbb{R}(s)} T(s) = m$. Then $T(s)$ can be factorized (in a non unique way) as

$$T(s) = T_1(s)T_2(s)T_3(s)T_4(s) \tag{4.12}$$

where:

(i) $T_1(s) \in \mathbb{R}_p^{p\times m}(s)$ is a proper and Ω-stable minimal MacMillan degree basis of $\mathcal{T}(s)$.

(ii) $T_2(s) \in \mathbb{R}_p^{m\times m}(s)$ is $\mathbb{R}_p(s)$-unimodular and the set of its finite zeros (in Ω^c) contains as a subset the set of finite zeros (in Ω^c) of $T(s)$ (if any).

(iii) $T_3(s) \in \mathbb{R}_p^{m\times m}(s)$ and has the same zero structure in $P = \Omega \cup \{\infty\}$ as that of $T(s)$.

(iv) $T_4(s)^{-1} \in \mathbb{R}_p^{m\times m}(s)$ and has the same pole structure in P as that of $T(s)$.

Proof: Express $T(s)$ as a right coprime in P, $\mathbb{R}_p(s)$-MFD: $T(s) = B(s)A(s)^{-1}$ and set $T_4(s) := A(s)^{-1}$. Then factorize $B(s)$ according to Proposition 3.3 as: $B(s) = \hat{T}(s)T_3(s)$ where $\hat{T}(s) \in \mathbb{R}_p^{p\times m}(s)$ is $\mathbb{R}_p(s)$-left unimodular and $T_3(s) \in \mathbb{R}_p^{m\times m}(s)$ is a g.c.r.d. in P of $B(s)$. Finally factorize $\hat{T}(s)$ according to Theorem 4.1 as $\hat{T}(s) = T_1(s)T_2(s)$ where $T_1(s)$ is a MMD-$\mathbb{R}_p(s)$ basis and $T_2(s)$ is $\mathbb{R}_p(s)$-unimodular.

□

5. STABLE EXACT MODEL MATCHING AND STABLE MINIMAL DESIGN PROBLEMS

In the light of the concepts and results discussed in the previous sections we are now able to address a number of questions arising in linear multivariable control algebraic synthesis problems. These are the stable exact model matching (SEMMP) and the stable minimal design problems (SMDP).

5.1 Existence and Characterization of Proper and Stable Solutions to the Exact Model Matching Problem

In this section we determine two equivalent necessary and sufficient conditions for the existence of proper and stable solutions to the exact model matching problem. The ordinary* exact model matching problem (EMMP) is the following. Given $T(s) \in \mathbb{R}_{pr}^{p\times m}(s)$, $M(s) \in \mathbb{R}_{pr}^{q\times q}(s)$ determine under what conditions the equation

$$T(s)K(s) = M(s) \tag{5.1}$$

has a proper solution $K(s) \in \mathbb{R}_{pr}^{m\times q}(s)$. This problem has been the subject of numerous investigations (e.g. see [22][29][34]). It is known that if $p \geq m$ then EMMP has either no solution or a unique solution which can easily be determined [22]. Normally then, it is assumed that $m > p = \text{rank}_{\mathbb{R}(s)} T(s)$. In applications $M(s)$ will belong to $\mathbb{R}_{P}^{q\times q}(s)$ where e.g. $P = \mathbb{C}_+ \cup \{\infty\}$, i.e. the "model" will be *proper and stable*. With the further restriction that $K(s)$ is required to be also proper *and stable* the above is known as the *stable exact model matching problem* (SEMMP) and it has been examined by Anderson and Scott [30].

In the sequen we examine SEMMP for the case when $\text{rank}_{\mathbb{R}(s)} T(s) = p<m$, $M(s) \in \mathbb{R}_{P}^{p\times q}(s)$, $\Omega \equiv \mathbb{C}_+$ and generalizations for the case of $P = \Omega \cup \{\infty\}$ for arbitrary Ω are obvious.

Let $T_R(s) \in \mathbb{R}^{m\times m}(s)$, $\mathbb{R}_P(s)$-unimodular and such that

$$T(s)T_R(s) = [T_{GL}(s), 0_{p,m-p}] \tag{5.2}$$

where $T_{GL}(s) \in \mathbb{R}_{pr}^{p\times p}(s)$ (see (dual) proof in Proposition 3.3 and Corollary 3.1).

We can now state

<u>Theorem 5.1</u>: Let $T(s) \in \mathbb{R}_{pr}^{p\times m}(s)$, $M(s) \in \mathbb{R}_{P}^{p\times q}(s)$ $\text{rank}_{\mathbb{R}(s)} T(s) = p < m$ and $P = \mathbb{C}_+ \cup \{\infty\}$. Then the SEMMP in (5.1) has a proper and stable solution $K(s) \in \mathbb{R}_{P}^{m\times q}(s)$ iff

$$M_o(s) := T_{GL}(s)^{-1} M(s) \in \mathbb{R}_{P}^{p\times q}(s) \tag{5.3}$$

---

*i.e. without the constraint of the stability of the (proper) solution K(s).

Proof: (if) (i.e. if (5.3) is satisfied then (5.1) has a solution $K(s) \in \mathbb{R}_p^{m\times q}(s)$). Partitioning the $\mathbb{R}_p(s)$-unimodular matrix $T_R(s) \in \mathbb{R}_p^{m\times m}(s)$ in (5.2) as

$$T_R(s) = [T_{R1}(s), T_{R2}(s)] \tag{5.4}$$

where $T_{R1}(s) \in \mathbb{R}_p^{m\times p}(s)$, $T_{R2}(s) \in \mathbb{R}_p^{m\times (m-p)}(s)$ (and from Proposition 3.2 (v) both $\mathbb{R}_p(s)$-left unimodular), from (5.2) we have:

$$T(s)T_{R1}(s) = T_{GL}(s) \tag{5.5}$$

$$T(s)T_{R2}(s) = 0_{p,m-p} \tag{5.6}$$

Assuming now (5.3), i.e. that $M(s) = T_{GL}(s)M_o(s)$ for some $M_o(s) \in \mathbb{R}_p^{p\times q}(s)$, and multiplying (5.5) on the right by $M_o(s)$ we have:

$$T(s)T_{R1}(s)M_o(s) = T_{GL}(s)M_o(s) = M(s)$$

Hence $K(s) := T_{R1}(s)M_o(s) \in \mathbb{R}_p^{m\times q}(s)$ is a solution.

(only if) (i.e. if (5.1) has a solution $K(s) \in \mathbb{R}_p^{m\times q}(s)$ then (5.3) is satisfied) Let $K(s) \in \mathbb{R}_p^{m\times q}(s)$ be a solution of (5.1) and partition it as $K(s) = \begin{bmatrix} K_1(s) \\ K_2(s) \end{bmatrix}$ where $K_1(s) \in \mathbb{R}_p^{p\times q}(s)$, $K_2(s) \in \mathbb{R}_p^{(m-p)\times q}(s)$. Now consider the $\mathbb{R}_p(s)$-unimodular matrix $T_R(s)$ in (5.2) let $T_R(s)^{-1} =: \hat{T}(s) \in \mathbb{R}_p^{m\times m}(s)$ (and $\mathbb{R}_p(s)$-unimodular) and partition it as

$$\hat{T}(s) = \begin{bmatrix} \hat{T}_1(s) & \hat{T}_2(s) \\ \hat{T}_3(s) & \hat{T}_4(s) \end{bmatrix} \begin{matrix} p \\ m-p \end{matrix}$$

(column blocks: p, $m-p$)

Then (5.2) gives

$$T(s) = [T_{GL}(s), 0_{p,m-p}]\hat{T}(s) = T_{GL}(s)[\hat{T}_1(s), \hat{T}_2(s)]$$

and by assumption:

$$\begin{aligned} T(s)K(s) &= T_{GL}(s)[\hat{T}_1(s), \hat{T}_2(s)]\begin{bmatrix} K_1(s) \\ K_2(s) \end{bmatrix} \\ &= T_{GL}(s)[\hat{T}_1(s)K_1(s) + \hat{T}_2(s)K_2(s)] = M(s) \end{aligned}$$

i.e. (5.3) is satisfied for some $M_o(s) := \hat{T}_1(s)K_1(s) + \hat{T}_2(s)K_2(s) \in \mathbb{R}_p^{p\times q}(s)$. □

Corollary 5.1: If $T(s) \in \mathbb{R}_p^{p\times m}(s)$ is $\mathbb{R}_p(s)$-right unimodular (see Definition 3.2 and Proposition 3.2) then the SEMMP (5.1) has always a solution for every $M(s) \in \mathbb{R}_p^{p\times q}(s)$.

Corollary 5.2: If the SEMMP (5.1) has a proper and stable solution K(s) (i.e. if (5.3) is satisfied) then the family of all proper and stable solutions is given by:

$$\bar{K}(s) = K(s) + T_{R2}(s)Z(s) \tag{5.7}$$

where $T_{R2}(s) \in \mathbb{R}_P^{m\times(m-p)}(s)$ is the $\mathbb{R}_P(s)$-left unimodular block of $T_R(s)$, indicated in (5.4) and $Z(s) \in \mathbb{R}_P^{(m-p)\times q}(s)$ is arbitrary.

Proof: Let $K(s) \in \mathbb{R}_P^{m\times q}(s)$ be a solution of (5.1). Combining eqs.(5.1) and (5.6) in matrix form we have:

$$T(s)[K(s), T_{R2}(s)] = [M(s), 0_{p,m-p}] \tag{5.8}$$

Multiplying the above equation on the right by the proper and stable rational matrix $\begin{bmatrix} I_q & 0_{q,m-p} \\ Z(s) & I_{m-p} \end{bmatrix}$ we obtain the general proper and stable solution of (5.1) given by (5.7).

□

(A similar result to the above has been obtained by Anderson and Scot [30])

Another form of the necessary and sufficient condition for the solvability of the SEMMP can be expressed in terms of a property of any MMD-$\mathbb{R}_P(s)$ basis of $\ker[T(s),-M(s)]$ (when $P = \Omega \cup \{\infty\}$ and $\Omega \equiv \mathbb{C}_+$) and is contained in the following

Theorem 5.2: Let $T(s) \in \mathbb{R}_{pr}^{p\times m}(s)$, $\text{rank}_{\mathbb{R}(s)} T(s) = p < m$, $M(s) \in \mathbb{R}_P^{p\times q}(s)$ and $P = \mathbb{C}_+ \cup \{\infty\}$. Let $Q(s) \in \mathbb{R}_P^{(m+q)\times(m+q-p)}(s)$ be a minimal MacMillan degree $\mathbb{R}_P(s)$-basis for $\ker[T(s),-M(s)]$ and let $Q(s) = \begin{bmatrix} B(s) \\ A(s) \end{bmatrix}$ where $B(s) \in \mathbb{R}_P^{m\times(m+q-p)}(s)$, $A(s) \in \mathbb{R}_P^{q\times(m+q-p)}(s)$. Then the SEMMP has a solution $K(s) \in \mathbb{R}_P^{m\times q}(s)$ if and only if $A(s)$ is $\mathbb{R}_P(s)$-right unimodular (see Proposition 3.2).

Proof: (if) (i.e. if there exists a $K(s) \in \mathbb{R}_P^{m\times q}(s)$ solving (5.1) then $A(s)$ is $\mathbb{R}_P(s)$-right unimodular) By assumption we have that a $K(s) \in \mathbb{R}_P^{m\times q}(s)$ satisfies

$$[T(s),-M(s)]\begin{bmatrix} K(s) \\ I_q \end{bmatrix} = 0 \tag{5.9}$$

Now since

$$\begin{bmatrix} K(s) \\ I_q \end{bmatrix} \in \mathbb{R}_P^{(m+q)\times q}(s) \quad \text{and} \quad \text{rank}_{\mathbb{C}} \begin{bmatrix} K(s_o) \\ I_q \end{bmatrix} = q \ \forall\ s_o \in \mathbb{C} \cup \{\infty\}, \quad \begin{bmatrix} K(s) \\ I_q \end{bmatrix}$$

is a MMD-$\mathbb{R}_P(s)$ basis of the rational vector space that it spans and which vector space, due to (5.9), is a subspace of $\ker[T(s),-M(s)]$ for which by assumption a MMD-$\mathbb{R}_P(s)$ basis is given by $Q(s)$. Therefore there must exist a $C(s) \in \mathbb{R}^{(m+q-p)\times q}(s)$ with $\text{rank}_{\mathbb{R}(s)} C(s) = q$ and such that

$$\begin{bmatrix} K(s) \\ I_q \end{bmatrix} = \begin{bmatrix} B(s) \\ A(s) \end{bmatrix} C(s) \tag{5.10}$$

Since both $\begin{bmatrix} K(s) \\ I_q \end{bmatrix}$ and $\begin{bmatrix} B(s) \\ A(s) \end{bmatrix}$ are MMD-$\mathbb{R}_{\mathcal{P}}(s)$ bases, if $\lim_{s\to\infty} \begin{bmatrix} K(s) \\ I_q \end{bmatrix} =: E_1 = \begin{bmatrix} E_K \\ I_q \end{bmatrix}$ $\in \mathbb{R}^{(m+q)\times q}$, then $\mathrm{rank}_{\mathbb{R}} E_1 = q$. Also if $\lim_{s\to\infty} \begin{bmatrix} B(s) \\ A(s) \end{bmatrix} =: E_Q \in \mathbb{R}^{(m+q)\times(m+q-p)}$ then $\mathrm{rank}_{\mathbb{R}} E_Q = m+q-p$. Therefore taking limits in (5.10) and letting $\lim_{s\to\infty} C(s) =: E_C \in \mathbb{R}^{(m+q-p)\times q}$, we have that $\mathrm{rank}_{\mathbb{R}} E_C = q$, i.e. that $C(s) \in \mathbb{R}_{pr}^{(m+q-p)\times q}(s)$ and must have no zeros at $s = \infty$. Similarly for evergy $s_o \in \Omega \equiv \mathbb{C}_+$ we have:

$$\mathrm{rank}_{\mathbb{C}} \begin{bmatrix} K(s_o) \\ I_q \end{bmatrix} = q \quad \text{and} \quad \mathrm{rank}_{\mathbb{C}} \begin{bmatrix} B(s_o) \\ A(s_o) \end{bmatrix} = m+q-p$$

therefore from the above and (5.10) we conclude that we must have that $C(s_o)$ is finite and that $\mathrm{rank}_{\mathbb{C}}\, C(s_o) = q$ for every $s_o \in \Omega \equiv \mathbb{C}_+$, i.e. $C(s)$ must have no poles or zeros in $\mathbb{C}_+$. The above imply that $C(s) \in \mathbb{R}_{\mathcal{P}}^{(m+q-p)\times q}(s)$ where $\mathcal{P} = \mathbb{C}_+ \cup \{\infty\}$ (*). Finally from (5.9) $I_q = A(s)C(s)$ and since $A(s) \in \mathbb{R}_{\mathcal{P}}^{q\times(m+q-p)}(s)$ from Proposition 3.2(iv) we conclude that $A(s)$ must be $\mathbb{R}_{\mathcal{P}}(s)$-right unimodular. (only if) (i.e. if $A(s)$ is $\mathbb{R}_{\mathcal{P}}(s)$-right unimodular then there exists a $K(s) \in \mathbb{R}_{\mathcal{P}}^{m\times q}(s)$ solving (5.1)). From Proposition 3.2, $A(s)$ being $\mathbb{R}_{\mathcal{P}}(s)$-right unimodular implies that there exists a $T_R(s) \in \mathbb{R}_{\mathcal{P}}^{(m+q-p)\times(m+q-p)}(s)$ and $\mathbb{R}_{\mathcal{P}}(s)$-unimodular such that

$$A(s)T_R(s) = [I_q, 0_{q,m-p}] \tag{5.11}$$

hence

$$\begin{bmatrix} B(s) \\ A(s) \end{bmatrix} T_R(s) =: \begin{array}{c} \overset{\leftarrow q \rightarrow \;\; \leftarrow m-p \rightarrow}{\left[\begin{array}{c|c} B_1(s) & B_2(s) \\ \hline I_q & 0 \end{array}\right]} \end{array} \begin{array}{l} \updownarrow m \\ \updownarrow q \end{array} \tag{5.12}$$

which implies that $K(s) := B_1(s) \in \mathbb{R}_{\mathcal{P}}^{m\times q}(s)$ is a solution of (5.1). □

As a necessary condition for a $T(s) \in \mathbb{R}_{\mathcal{P}}^{p\times m}(s)$ $(p \le m)$ to be $\mathbb{R}_{\mathcal{P}}(s)$-right unimodular is that $T(s)$ is $\mathbb{R}_{pr}(s)$-right unimodular (or right biproper [16]), i.e. that $\lim_{s\to\infty} T(s) = E \in \mathbb{R}^{p\times m}$ and $\mathrm{rank}_{\mathbb{R}} E = p$ (see Definition 3.2 and 2nd condition in (vi) of Proposition 3.2), from Theorem 5.2 we obtain a necessary and sufficient condition for the solvability of the ordinary EMMP which is stated in

<u>Corollary 5.3</u>: Let $Q(s) = \begin{bmatrix} B(s) \\ A(s) \end{bmatrix} \in \mathbb{R}_{pr}^{(m+q)\times(m+q-p)}(s)$ be a *proper minimal MacMillan degree* basis of $\ker[T(s),-M(s)]$ (not necessarily stable) [16]. Then a necessary and sufficient condition for the existence of a proper (not necessarily stable) solution $K(s) \in \mathbb{R}_{pr}^{m\times q}(s)$ for the EMMP is that $A(s) \in \mathbb{R}_{pr}^{q\times(m+q-p)}(s)$ is *right biproper* (or equivalently: if

---

(*) Notice that $C(s)$ might have zeros outside Ω, i.e. in Ω^c.

$\lim\limits_{s\to\infty} A(s) =: E_A \in \mathbb{R}^{q\times(m+q-p)}$ then $\operatorname{rank}_{\mathbb{R}} E_A = q$, see [16]).

5.2 Existence and Characterization of Proper, Stable and "Minimum Phase" Left or Right Inverses of Rational Matrices

In the particular case when $\Omega \equiv \mathbb{C}_+$, a $\mathbb{R}_P(s)$-left (right) unimodular rational matrix $T(s) \in \mathbb{R}_P^{p\times m}(s)$ represents a proper, stable and "minimum phase" transfer function matrix that satisfies also: $\lim\limits_{s\to\infty} T(s) = E$ and $\operatorname{rank}_{\mathbb{R}} E = m\ (= p)$ (i.e. $T(s)$ has also no zeros at $s = \infty$). In such a case Proposition 3.2 and Theorems 5.1 and 5.2 for $M(s) = I_p$ (and their duals) can be viewed as expressing a number of equivalent necessary and sufficient conditions for the existence of proper, stable and "minimum phase" left (right) inverses of a $T(s) \in \mathbb{R}_{pr}^{p\times m}(s)$ with $p \geq m$ $(p \leq m)$ and $\operatorname{rank}_{\mathbb{R}(s)} T(s) = m\ (= p)$.

If a $T(s) \in \mathbb{R}_{pr}^{p\times p}(s)$ satisfies these conditions then the family of all proper, stable and "minimum phase" left inverses of $T(s)$ can be obtained as follows. Let $T_L(s) \in \mathbb{R}_P^{p\times p}(s)$ be a $\mathbb{R}_P(s)$-unimodular rational matrix which reduces $T(s)$ to its *Smith* form in P, i.e. let

$$T_L(s)T(s) = \begin{bmatrix} I_m \\ 0 \end{bmatrix} = S_{T(s)}^{P} \tag{5.13}$$

and partition $T_L(s)$ as $T_L(s) = \begin{array}{c} \begin{array}{cc} \leftarrow \ell \rightarrow & \leftarrow t \rightarrow \end{array} \\ \left[\begin{array}{c|c} X(s) & Y(s) \\ \hline B(s) & A(s) \end{array}\right] \end{array} \begin{array}{l} \updownarrow m \\ \updownarrow p-m \end{array}$, $\ell + t = p$. Then $T_{LI}(s) := [X(s), Y(s)] \in \mathbb{R}_P^{m\times p}(s)$ is a proper, stable and "minimum phase" left inverse of $T(s)$. Multiplying (5.13) on the left by the $\mathbb{R}_P(s)$-unimodular rational matrix $\begin{bmatrix} I_m & Q(s) \\ 0 & I_{p-m} \end{bmatrix} \in \mathbb{R}_P^{p\times p}(s)$, where $Q(s) \in \mathbb{R}_P^{m\times(p-m)}(s)$ and otherwise arbitrary, we easily see that the family of all proper, stable, and "minimum phase" left inverses of $T(s)$ is given by

$$[X(s) + Q(s)B(s)\ ,\ Y(s) + Q(s)A(s)] \in \mathbb{R}_P^{m\times p}(s) \tag{5.14}$$

5.3 Stable Minimal Design Problem (SMDP)

If we consider the ordinary EMMP in (5.1) i.e. without the constraint of the stability of the proper solution $K(s)$, but with the added constraint that the proper solution $K(s)$ has MacMillan degree which is minimum among the MacMillan degrees of all other proper solutions, then this problem is known as the *minimal design problem* (MDP) [25][18]. Obviously a necessary condition for the solution of the MDP is that the corresponding EMMP is solvable. The MDP together with the stability constraint of the proper and minimal MacMillan degree solution is known as the *stable minimal design problem* (SMDP) and it has been studied by Wolovich et al. [31] and Scott and Anderson [32].

In the following we investigate solvability of the MDP employing the theory of proper (not necessarily stable) minimal MacMillan degree bases of rational vector spaces [16] and this analysis is then used in order to examine solvability of the SMDP.

We start by stating some facts regarding the structure of different simple, minimal MacMillan degree, proper bases (SMMD-$\mathbb{R}_{pr}(s)$ bases) of a rational vector space. Let $T_i(s) \in \mathbb{R}_{pr}^{p\times m}(s)$, $i=1,2$ $(p \geq m)$ be any two *column ordered* SMMD-$\mathbb{R}_{pr}(s)$ bases of a rational vector space $\mathcal{T}(s)$, i.e. with their columns $t_{ij}(s) \in \mathbb{R}_{pr}^{p\times 1}(s)$, $i=1,2$; $j \in \underline{m}$, arranged in order of ascending MacMillan degrees $\delta_M(t_{ij}) = v_j$. Let also that

$$v_1 = v_2 = \ldots = v_{r_1} < v_{r_1+1} = \ldots = v_{r_1+r_2} < v_{r_1+r_2+1} = \ldots \leq v_m$$

i.e. that there are in $T_i(s)$ η groups $(1 \leq \eta \leq m)$ of columns with equal MacMillan degrees, each group having r_i columns, $i \in \underline{\eta}$ $(\sum_{i=1}^{\eta} r_i = m)$. Then we have

Lemma 5.1:

$$T_1(s) = T_2(s)U(s) \tag{5.15}$$

where $U(s) \in \mathbb{R}_{pr}^{m\times m}(s)$ is $\mathbb{R}_{pr}(s)$-unimodular and has the structure

$$U(s) = \hat{D}_2(s)V(s)\hat{D}_1(s)^{-1} \tag{5.16}$$

where $T_i(s) = N_{iL}(s)\hat{D}_i(s)^{-1}$, $N_{iL}(s) \in \mathbb{R}^{p\times m}[s]$, $\hat{D}_i(s) = \text{diag}[d_{i1}(s),\ldots,d_{im}(s)] \in \mathbb{R}^{m\times m}[s]$ is a coprime MFD of $T_i(s)$, $\deg.d_{ij}(s) = v_j$, $i=1,2$; $j \in \underline{m}$ (see proof of Theorem 4.1) and $V(s) \in \mathbb{R}^{m\times m}[s]$ is a "v-form" $\mathbb{R}[s]$-unimodular matrix corresponding to the indices v_j of $\mathcal{T}(s)$ and having the following structure

$$V(s) = \begin{bmatrix} V_1 & W_{12}(s) & W_{13}(s) & \ldots & W_{1\eta}(s) \\ 0 & V_2 & W_{23}(s) & \ldots & W_{2\eta}(s) \\ \vdots & & & & \\ 0 & 0 & 0 & & V_\eta \end{bmatrix} \tag{5.17}$$

$V_i \in \mathbb{R}^{r_i\times r_i}$, $\det V_i \neq 0$ and the entries of the $W_{ij}(s) \in \mathbb{R}^{r_i\times r_j}[s]$ having degrees less than or equal to $v_{\sigma_j} - v_{\sigma_i}$, $\sigma_i = \sum_{j=1}^{i-1} r_j + 1$, $i=1,2,\ldots,\eta$ (see [32][33]).

Proof: The lemma is a direct consequence of the results in [32][33] and the fact that $N_{1L}(s), N_{2L}(s)$ are "column ordered" minimal polynomial bases of $\mathcal{T}(s)$. □

Corollary 5.4: If $T_i(s) \in \mathbb{R}_p^{p\times m}(s)$ are column ordered SMMD-$\mathbb{R}_p(s)$ bases of $\mathcal{T}(s)$ then $U(s) \in \mathbb{R}_p^{m\times m}(s)$ is $\mathbb{R}_p(s)$-unimodular.

Now let $Q^i(s) = \begin{bmatrix} B^i(s) \\ A^i(s) \end{bmatrix} \in \mathbb{R}_{pr}^{(m+q)\times(m+q-p)}(s)$, $i = 1,2$, be any two column ordered SMMD-$\mathbb{R}_{pr}(s)$ bases for $\ker[T(s),-M(s)]$ and assume that the condition of Corollary 5.3 for the solvability of the EMMP is satisfied. Partition $A^i(s) \in \mathbb{R}_{pr}^{q\times(m+q-p)}(s)$, $i = 1,2$, as

$A^i(s) = [A_1^i(s), A_2^i(s), \ldots, A_\eta^i(s)]$ where $A_k^i(s) \in \mathbb{R}_{pr}^{q\times r_k}(s)$, $k \in \underline{\eta}$, are blocks of r_k columns corresponding to columns of $Q^i(s)$ having equal MacMillan degrees. Let $A_{\cdot j_1}^1(s), A_{\cdot j_2}^1(s), \ldots, A_{\cdot j_q}^1(s)$ be the first q columns of $A^1(s)$ (enumerated from left to right) that are such that the matrix $\bar{A}^1(s) := [A_{\cdot j_1}^1(s), \ldots, A_{\cdot j_q}^1(s)] \in \mathbb{R}_{pr}^{q\times q}(s)$ is $\mathbb{R}_{pr}(s)$-unimodular (biproper) * and assume that the q columns of $\bar{A}^1(s)$ involve ρ_1 columns from block $A_1^1(s)$, ρ_2 columns from block $A_2^1(s), \ldots, \rho_\eta$ columns from block $A_\eta^1(s)$, $(0 \le \rho_i \le \min(r_i, q)$, $i \in \underline{\eta}$, $\rho_1+\rho_2+\ldots+\rho_\eta = q)$. Then we have

Lemma 5.2: If $A_{\cdot j_1}^2(s), A_{\cdot j_2}^2, \ldots, A_{\cdot j_q}^2(s)$ are the first q columns of $A^2(s) \in \mathbb{R}_{pr}^{q\times(m+q-p)}(s)$ (enumerated also from left to right) that compose an $\mathbb{R}_{pr}(s)$-unimodular matrix $\bar{A}^2(s) := [A_{\cdot j_1}^2(s), \ldots, A_{\cdot j_q}^2(s)] \in \mathbb{R}_{pr}^{q\times q}(s)$, then $\bar{A}^2(s)$ will also involve exactly ρ_1 columns from block $A_1^2(s)$, ρ_2 columns from block $A_2^2(s), \ldots, \rho_\eta$ columns from block $A_\eta^2(s)$.

Proof: From Lemma 5.1

$$[A_1^1(s), A_2^1(s), \ldots, A_\eta^1(s)] = [A_1^2(s), A_2^2(s), \ldots, A_\eta^2(s)] \begin{bmatrix} V_1 & W_{12}(s) & \ldots & W_{1\eta}(s) \\ 0 & V_2 & \ldots & W_{2\eta}(s) \\ \vdots & \vdots & & \vdots \\ 0 & 0 & & V_\eta \end{bmatrix} \tag{5.18}$$

Considering the first block eq. of (5.18)

$$A_1^1(s) = A_1^2(s)V_1 \tag{5.19}$$

then $\rho_1 = \mathrm{rank}_{\mathbb{R}} A_1^1(\infty)$ and because of (5.19) it follows that

$$\mathrm{rank}_{\mathbb{R}} A_1^2(\infty) = \rho_1 \tag{5.20}$$

Considering the second block equation of (5.18) then from $\rho_1 + \rho_2 = \mathrm{rank}_{\mathbb{R}}[A_1^1(\infty), A_2^1(\infty)]$ it follows that $\mathrm{rank}_{\mathbb{R}}[A_1^2(\infty), A_2^2(\infty)] = \rho_1 + \rho_2$ and because of (5.20) we have that $\mathrm{rank}_{\mathbb{R}} A_2^2(\infty) = \rho_2$. Continuing in this way with the remaining block equations of (5.18) and using similar arguments we can prove Lemma 5.2. □

If we now denote by $\bar{B}^i(s) := [B_{\cdot j_1}^i(s), B_{\cdot j_2}^i(s), \ldots, B_{\cdot j_q}^i(s)] \in \mathbb{R}_{pr}^{m\times q}(s)$ the matrix composed from the q corresponding columns of $B^i(s)$, then $\bar{Q}^i(s) := \begin{bmatrix} \bar{B}^i(s) \\ \bar{A}^i(s) \end{bmatrix} \in \mathbb{R}_{pr}^{(m+q)\times q}(s)$ is also a simple, proper, minimal MacMillan degree basis whose columns span a subspace of $\ker[T(s), -M(s)]$

*Under the assumption of condition in Corollary 5.3 such a matrix always exists.

and by virtue of Lemma 5.2 the MacMillan degrees of the columns $Q^i_{j_k}(s) \in \mathbb{R}^{(m+q)\times 1}_{pr}(s)$, $k \in \underline{q}$, of $\bar{Q}^i(s)$ are independent of the particular choice of $Q^i(s)$. It readily follows now that a proper solution of the corresponding EMMP will be given by

$$\bar{K}(s) := \bar{B}^i(s)\bar{A}^i(s)^{-1} \in \mathbb{R}^{m\times q}_{pr}(s) \tag{5.21}$$

Moreover $\bar{K}(s)$ will have MacMillan degree which will be minimal among the MacMillan degrees of all other proper solutions of (5.1), i.e. $\bar{K}(s)$ will be a solution of the MDP with

$$\delta_M(\bar{K}) = \delta_M(\bar{B}^i(s)\bar{A}^i(s)^{-1}) = \delta_M(\bar{Q}^i(s)) = \sum_{k=1}^{q} \delta_M(Q_{j_k}(s)) \tag{5.22}$$

and these facts follow directly from the construction of $\bar{Q}^i(s)$ and the following.

Proposition 5.1: Let $\bar{Q}(s) = \begin{bmatrix}\bar{B}(s)\\ \bar{A}(s)\end{bmatrix} \in \mathbb{R}^{(m+q)\times q}_{pr}(s)$ be a proper minimal MacMillan degree basis (not necessarily stable) and let $\bar{A}(s) \in \mathbb{R}^{q\times q}_{pr}(s)$ be $\mathbb{R}_{pr}(s)$-unimodular (biproper). Then

$$\delta_M(\bar{B}(s)\bar{A}(s)^{-1}) = \delta_M(\bar{Q}) \tag{5.23}$$

Proof: Let $\bar{Q}(s) = \hat{N}(s)\hat{D}(s)^{-1}$, $\hat{N}(s) \in \mathbb{R}^{(m+q)\times q}[s]$, $\hat{D}(s) \in \mathbb{R}^{q\times q}[s]$ be a right coprime polynomial MFD and partition $\hat{N}(s)$ as $\hat{N}(s) = \begin{bmatrix}N(s)\\ D(s)\end{bmatrix}$ where $N(s) \in \mathbb{R}^{m\times q}[s]$, $D(s) \in \mathbb{R}^{q\times q}[s]$ are right coprime since $\bar{Q}(s)$ is a proper MMD basis. Then

$$\delta_M(\bar{Q}) = \deg|\hat{D}(s)| =: \deg.\hat{D}(s) \tag{5.24}$$

Now since $\bar{A}(s) = D(s)\hat{D}(s)^{-1} \in \mathbb{R}^{q\times q}_{pr}(s)$ is $\mathbb{R}_{pr}(s)$-unimodular

$$\delta_\infty(\bar{A}(s)) := \deg.\hat{D}(s) - \deg.D(s) = 0$$

(see [16]), i.e. $\deg.\hat{D}(s) = \deg.D(s) = \delta_M(\bar{Q})$. Now

$$\bar{K}(s) := \bar{B}(s)\bar{A}(s)^{-1} = N(s)\hat{D}(s)^{-1}[D(s)\hat{D}(s)^{-1}]^{-1} = N(s)D(s)^{-1} \in \mathbb{R}^{m\times q}_{pr}(s) \tag{5.25}$$

and since $N(s),D(s)$ are right coprime it follows that

$$\delta_M(\bar{K}) = \deg.D(s) = \delta_M(\bar{Q}) \tag{5.26}$$ □

Of course the above analysis of the MDP is a restatement in the "language" of proper, minimal MacMillan degree bases of rational vector spaces of the results obtained by Forney [18]. Nevertheless this analysis gives some hints for the investigation of the solvability of the SMDP.

Thus let $Q(s) = \begin{bmatrix}B(s)\\ A(s)\end{bmatrix} \in \mathbb{R}^{(m+q)\times(m+q-p)}(s)$ be a SMMD-$\mathbb{R}_p(s)$ basis for ker[T(s),-M(s)] whose columns $Q_j(s) \in \mathbb{R}_P^{(m+q)\times 1}(s)$, $j \in \underline{m+q-p}$, are arranged in order of ascending MacMillan degrees and assume that the corresponding SEMMP is solvable, i.e. that the condition in Theorem 5.2 is satisfied. Assume also that among the columns $A_j(s) \in \mathbb{R}_P^{q\times 1}$ of $A(s) \in \mathbb{R}_P^{q\times(m+q-p)}(s)$ there exist q columns $A_{j_1}(s),\dots,A_{j_q}(s)$ (selected from left to right) such that the matrix $\bar{A}(s) := [A_{j_1}(s),\dots,A_{j_q}(s)] \in \mathbb{R}_P^{q\times q}(s)$ is $\mathbb{R}_P(s)$-unimodular. If now $\bar{B}(s) := [B_{j_1}(s),\dots,B_{j_q}(s)] \in \mathbb{R}_P^{m\times q}(s)$ is the matrix composed from the q corresponding columns of B(s), then

$$\bar{K}(s) := \bar{B}(s)\bar{A}(s)^{-1} \in \mathbb{R}_P^{m\times q}(s) \tag{5.27}$$

will be a solution of the corresponding SEMMP and by arguments similar to the ones in Lemma 5.2 the MacMillan degree of $\bar{K}(s)$ will be minimum among the MacMillan degrees of all other proper and stable solutions of eq.(5.1). Also, and again by virtue of Lemma 5.2 and Proposition 5.1, the (minimal) MacMillan degree of $\bar{K}(s)$ will be given by

$$\delta_M(\bar{K}) = \delta_M(\bar{B}(s)\bar{A}(s)^{-1}) = \delta_M(\bar{Q}) = \sum_{k=1}^{q} \delta_M(Q_{j_k}(s)) \tag{5.28}$$

where $\bar{Q}(s) := [Q_{j_1}(s),Q_{j_2}(s),\dots,Q_{j_q}(s)] = \begin{bmatrix}\bar{B}(s)\\ \bar{A}(s)\end{bmatrix} \in \mathbb{R}_P^{(m+q)\times q}(s)$ is a column ordered SMMD-$\mathbb{R}_P(s)$ basis.

Of course given an $\mathbb{R}_P(s)$-right unimodular matrix $A(s) \in \mathbb{R}_P^{q\times(m+q-p)}(s)$ then by no means it follows that among its m+q-p columns there always exists a subset of q columns which forms an $\mathbb{R}_P(s)$-unimodular matrix, and this fact constitutes the main difficulty in the construction of minimal MacMillan degree solutions to the SEMMP. In such a case the construction of a minimal MacMillan degree solution of the SEMMP (i.e. a solution of the SMDP) can be investigated by firstly determining (using Lemma 5.1 and Corollary 5.4) a parametric expression for all column ordered SMMD-$\mathbb{R}_P(s)$ bases for ker[T(s),-M(s)] and then examining whether there exist parameters such that q leftmost columns of a parametric $\mathbb{R}_P(s)$-right unimodular matrix $A(s) \in \mathbb{R}_P^{q\times(m+q-p)}(s)$ constitute an $\mathbb{R}_P(s)$-unimodular matrix.

Example 1: Consider the SEMMP with $\Omega \equiv \mathbb{C}_+$ and $T(s) = [\frac{s-1}{(s-2)(s+2)}\ \frac{s-2}{s+2}] \in \mathbb{R}_{pr}^{1\times 2}(s)$, $M(s) = \frac{1}{s+2} \in \mathbb{R}_P(s)$. To investigate solvability of SEMMP we apply Theorem 5.1. We firstly write the elements $t_1(s),t_2(s)$ of T(s) as quotients of coprime (in P) elements of $\mathbb{R}_P(s)$

$$t_1(s) = \frac{s-1}{(s-2)(s+2)} = \frac{s-1}{(s+2)(s+\alpha)}\left[\frac{s-2}{s+\alpha}\right]^{-1} = a_1(s)b_1(s)^{-1}, \quad \alpha > 0$$

$$t_2(s) = \frac{s-2}{s+2} = \frac{s-2}{s+2}\,1^{-1} = a_2(s)b_2(s)^{-1}$$

We then write: $T(s) = d(s)^{-1}N(s)$, where $d(s) \in \mathbb{R}_P(s)$ is the least common multiple of the "denominators" $b_1(s), b_2(s) \in \mathbb{R}_P(s)$ and $N(s) \in \mathbb{R}_P^{1\times 2}(s)$. Thus

$$T(s) = \left[\frac{s-2}{s+\alpha}\right]^{-1} \left[\frac{s-1}{(s+2)(s+\alpha)}, \frac{(s-2)^2}{(s+2)(s+\alpha)}\right] = d(s)^{-1}N(s)$$

It can be easily seen that $N(s)$ is $\mathbb{R}_P(s)$-right unimodular hence its "Smith" form in P is: $S^P_{N(s)} = [1\ 0]$ and therefore the MacMillan form of $T(s)$ in P is $S^P_{T(s)} = [\varepsilon_1(s)\psi_1(s)^{-1}, 0] = [\frac{s+\alpha}{s-2}, 0]$ with $\varepsilon_1(s) = 1$, $\psi_1(s) = \frac{s-2}{s+\alpha} \in \mathbb{R}_P(s)$. Therefore from the dual of Proposition 3.3 we have that $T_{GL}(s) = \frac{s+\alpha}{s-2} \in \mathbb{R}_{pr}(s)$ and since $T_{GL}(s)^{-1}M(s) = \frac{s-2}{s+\alpha}\frac{1}{s+2} \in \mathbb{R}_P(s)$ from Theorem 5.1 it follows that the SEMMP has solution. In order to find a solution we do not need to determine the $\mathbb{R}_P(s)$-unimodular matrix $T_R(s)$ in eq.(5.2). Instead we can use Theorem 5.2. We firstly determine a column ordered SMMD-$\mathbb{R}_P(s)$ basis for $\ker[T(s), -M(s)]$. Such a basis is given by

$$Q^1(s) = \begin{bmatrix} \frac{s-2}{s+2} & 0 \\ 0 & \frac{1}{s+2} \\ \hline \frac{s-1}{s+2} & \frac{s-2}{s+2} \end{bmatrix} = \begin{bmatrix} B^1(s) \\ A^1(s) \end{bmatrix} = N_{1L}(s)\hat{D}_1(s)^{-1}$$

(Notice that $A^1(s) = [\frac{s-1}{s+2}\ \ \frac{s-2}{s+2}] \in \mathbb{R}^{1\times 2}_P(s)$ is $\mathbb{R}_P(s)$-right unimodular as expected from Theorem 5.2). Now as no element of $A^1(s)$ is a unit in $\mathbb{R}_P(s)$, from $Q^1(s)$ we generate the parametric family of all column ordered SMMD-$\mathbb{R}_P(s)$ bases using Lemma 5.1 as:

$$Q^2(s) = N_{1L}(s)V_1\hat{D}_2(s)^{-1} = \begin{bmatrix} s-2 & 0 \\ 0 & 1 \\ s-1 & s-2 \end{bmatrix} \begin{bmatrix} v_{11} & v_{12} \\ v_{21} & v_{22} \end{bmatrix} \begin{bmatrix} s+\alpha & \\ 0 & s+\beta \end{bmatrix}^{-1}$$

$$= \begin{bmatrix} \frac{v_{11}(s-2)}{s+\alpha} & \frac{v_{12}(s-2)}{s+\beta} \\ \frac{v_{21}}{s+\alpha} & \frac{v_{22}}{s+\beta} \\ \frac{v_{11}(s-1)+v_{21}(s-2)}{s+\alpha} & \frac{v_{12}(s-1)+v_{22}(s-2)}{s+\beta} \end{bmatrix}$$

where $v_{ij} \in \mathbb{R}$, $\det V_1 \neq 0$, $\alpha > 0$, $\beta > 0$ and arbitrary. For $v_{11} = 1$, $v_{21} = -3/4$, $v_{12} = 0$, $v_{22} = 1$

$$Q^2(s) = \begin{bmatrix} \frac{s-2}{s+\alpha} & 0 \\ \frac{-3/4}{s+\alpha} & \frac{1}{s+\beta} \\ \hline \frac{1/4(s+2)}{s+\alpha} & \frac{s-2}{s+\beta} \end{bmatrix} = \begin{bmatrix} B^2(s) \\ A^2(s) \end{bmatrix}$$

and since $\bar{A}^2(s) := \frac{1/4(s+2)}{s+\alpha}$ is a unit in $\mathbb{R}_p(s)$, a minimal MacMillan degree solution is

$$\bar{K}(s) = \bar{B}^2(s)\bar{A}^2(s)^{-1} = \begin{bmatrix} \frac{s-2}{s+\alpha} \\ \frac{-3/4}{s+\alpha} \end{bmatrix} \left[\frac{1/4\cdot(s+2)}{s+\alpha}\right]^{-1} = \begin{bmatrix} \frac{4(s-2)}{s+2} \\ \frac{-3}{s+2} \end{bmatrix} \in \mathbb{R}_p^{2\times 1}(s)$$

Example 2: Let $\Omega \equiv \mathbb{C}_+$ and consider the SEMMP with [31]

$$T(s) = \begin{bmatrix} \frac{s}{(s+1)(s+2)} & 0 & \frac{s^2+2s+2}{(s+1)(s+2)} \\ \frac{2s+1}{s+2} & \frac{s-1}{s+2} & 0 \end{bmatrix} \in \mathbb{R}_p^{2\times 3}(s), \quad M(s) = I_2$$

Since (i) $\lim_{s\to\infty} T(s) = \begin{bmatrix} 0 & 0 & 1 \\ 2 & 1 & 0 \end{bmatrix} =: E_T$ with $\operatorname{rank}_{\mathbb{R}} E_T = 2$ and (ii) $T(s)$ has no finite zeros, $T(s)$ is $\mathbb{R}_p(s)$-right unimodular and thus from Corollary 5.1 and the discussion in section 5.2 there exists a family of stable, proper and "minimum phase" right inverses of $T(s)$. According to [31] a minimal polynomial basis for $\ker[T(s),-M(s)]$ is given by

$$N_L(s) = \begin{bmatrix} 1 & 0 & 0 \\ -1 & s+2 & 0 \\ 1 & 0 & (s+1)(s+2) \\ \hline 1 & 0 & s^2+2s+2 \\ 1 & s-1 & 0 \end{bmatrix}$$

with $v_1 = 0$, $v_2 = 1$, $v_3 = 2$. Taking $\hat{D}(s) = \operatorname{diag}[\hat{d}_1(s),\hat{d}_2(s),\hat{d}_3(s)]$ with $\deg.\hat{d}_i(s) = v_i$, $i = 1,2,3$, and $\hat{d}_i(s)$ arbitrary Hurwitz polynomials e.g. $\hat{d}_1(s) = 1$, $\hat{d}_2(s) = s+1$, $\hat{d}_3(s) = (s+1)(s+2)$, from Theorem 4.1 a column ordered SMMD-$\mathbb{R}_p(s)$ basis for $\ker[T(s),-M(s)]$ is given by

$$Q(s) = N_L(s)\hat{D}(s)^{-1} = \begin{bmatrix} 1 & 0 & 0 \\ -1 & \frac{s+2}{s+1} & 0 \\ 1 & 0 & 1 \\ \hline 1 & 0 & \frac{s^2+2s+2}{(s+1)(s+2)} \\ 1 & \frac{s-1}{s+1} & 0 \end{bmatrix} = \begin{bmatrix} B(s) \\ A(s) \end{bmatrix}$$

Now the matrix consisting from the 1st and 3rd columns of A(s)

$$\bar{A}(s) = \begin{bmatrix} 1 & \dfrac{s^2+2s+2}{(s+1)(s+2)} \\ 1 & 0 \end{bmatrix} \in \mathbb{R}_P^{2\times 2}(s)$$

is $\mathbb{R}$ (s)-unimodular, hence a proper, stable right inverse of T(s) of least degree is given by

$$T_{RI}(s) = \bar{B}(s)\bar{A}(s)^{-1} = \begin{bmatrix} 1 & 0 \\ -1 & 0 \\ 1 & 1 \end{bmatrix} \begin{bmatrix} 1 & \dfrac{s^2+2s+2}{(s+1)(s+2)} \\ 1 & 0 \end{bmatrix}^{-1} = \begin{bmatrix} 0 & 1 \\ 0 & -1 \\ \dfrac{(s+1)(s+2)}{s^2+2s+2} & \dfrac{-s}{s^2+2s+2} \end{bmatrix}$$

and has MacMillan degree $\delta_M(T_{RI}) = v_1 + v_3 = 2$.

CONCLUSIONS

In this paper we have surveyed a number of results on the algebraic structure of the set M* of all rational vectors which are contained in a given rational vector space $\mathcal{T}(s)$ and which also have no poles in a set $P := \Omega \cup \{\infty\}$, i.e. they are "proper and Ω-stable". Relying on the notions of "equivalence in P" : E^P and "Smith-MacMillan form in P" of rational matrices, the existence of proper and Ω-stable bases for $\mathcal{T}(s)$ has been established. Starting from any proper and Ω-stable basis $T_1(s)$ of $\mathcal{T}(s)$ we then focused attention on the $\mathbb{R}_P(s)$-module M_1 generated by its columns. It was shown that all bases of M_1 belong to the same E^P-equivalence class and that the notion of the "stathm in P" of every basis of M_1 is an invariant. We then considered the ascending chain of submodules $M_1 \subset M_2 \subset M_3$ ... whose respective bases $T_2(s), T_3(s), \ldots$ are obtained from $T_1(s)$ by "extracting right divisors in P" (which loosely speaking are square, proper and Ω-stable rational matrices containing zeros in P of $T_1(s)$). It was demonstrated that by extracting from $T_1(s)$ a "greatest right divisor in P" we arrive at a proper and Ω-stable basis $\bar{T}(s)$ of $\mathcal{T}(s)$ which has no zeros in P and that the $\mathbb{R}_P(s)$-module generated by any such basis coincides with the set M* of all proper and Ω-stable rational vectors contained in $\mathcal{T}(s)$. By investigating the structure of the various bases of the $\mathbb{R}_P(s)$-module M* we have established the notion of a "simple, proper, Ω-stable, and minimal MacMillan degree basis of $\mathcal{T}(s)$ as the counterpart to Forney's concept of a minimal polynomial basis of $\mathcal{T}(s)$, for the case of the $\mathbb{R}_P(s)$-module M*. Finally the applicability of these concepts and results to the resolution of algebraic control problems which involve questions of properness, stability and/or minimality of solutions of rational matrix equations has been demonstrated. In particular necessary and sufficient conditions for the solvability of the stable exact model matching problem have been derived and the problem of construction of solutions to the stable minimal design problem has been elucidated.

REFERENCES

[1] A.I.G. Vardulakis, D.J.N. Limebeer and N. Karcanias, "Structure and Smith-MacMillan form of a rational matrix at infinity", Int.J. Contr., Vol.35, pp.701-725, 1982.

[2] O. Zariski and P. Sammuel, Commutative Algebra, Vol.II, Springer, 1960.

[3] A.S. Morse, "System invariants under feedback and cascade control", in Lecture Notes in Economics and Mathematical Systems, Vol.131, Springer (Proc.Int.Symp., Udine, Italy), pp.61-74, 1975.

[4] N.T. Hung and B.D.O. Anderson, "Triangularization technique for the design of multivariable control systems", IEEE Trans.Automat. Contr., Vol.AC-24, pp.455-460, 1979.

[5] R.F. Gantmacher, The Theory of Matrices, Vol.1,2, Chelsea Publishing Co., New York, 1960.

[6] A.I.G. Vardulakis and N. Karcanias, "Structure, Smith-MacMillan form and coprime MFD's of a rational matrix inside a region $P = \Omega \cup \{\infty\}$", Int.J.Contr., Vol.38, No.5, pp.927-957, 1983.

[7] H.H. Rosenbrock, State Space and Multivariable Theory, London, Nelson, 1970.

[8] M. Vidyasagar, "On the use of right-coprime factorizations in distributed feedback systems containing unstable subsystems", IEEE Trans. Circuits Syst., Vol.CAS-25, pp.916-921, 1978

[9] C.A. Desoer, R.W. Liu, J. Murray and R. Saeks, "Feedback system design: the fractional representation approach to analysis and synthesis", IEEE Trans.Automat.Contr., Vol.AC-25, No.3, pp.399-412, 1980.

[10] B.A. Francis and M. Vidyasagar, "Algebraic and topological aspects of the Servo problem for lumped linear systems", S & IS Report No.8003, Nov. 1980.

[11] R. Saeks and J. Murray, "Feedback system design: the tracking and disturbance rejection problems", IEEE Trans.Automat.Contr., Vol. AC-26, pp.203-217, 1981.

[12] M. Vidyasagar, H. Schneider and B.A. Francis, "Algebraic and topological aspects of feedback stabilization", IEEE Trans.Automat. Contr., Vol.AC-27, pp.880-894, 1982.

[13] R. Saeks and J. Murray, "Fractional representation, algebraic geometry and the simultaneous stabilization problem", IEEE Trans. Automat.Contr., Vol.Ac-27, pp.895-903, 1982.

[14] M. Vidyasagar and N. Viswanadham, "Algebraic design for reliable stabilization", IEEE Trans.Automat.Contr., Vol.AC-27, pp.1085-1095, 1982.

[15] S. MacLane and G. Birkhof, Algebra, MacMillan Pub.Co.Inc., New York, 1979.

[16] A.I.G. Vardulakis and N. Karcanias, "Classification of proper bases of rational vector spaces: minimal MacMillan degree bases", Int.J.Contr., Vol.38, No.4, pp.779-809.

[17] G. Verghese and T. Kailath, "Rational matrix structure", IEEE Trans.Automat.Contr., Vol.Ac-26, pp.434-439, 1981.

[18] G.D. Forney, "Minimal bases of rational vector spaces with applications to multivariable linear systems", SIAM J. Control, Vol.13, pp.943-520, 1975.

[19] G. Verghese, Ph.D. Dissertation, Electrical Engineering Department, Stanford University, California, U.S.A., 1978.

[20] S. Kung and T. Kailath, Some notes on valuation theory in linear systems, IEEE Conf. on Decision and Control, San Diego, U.S.A., 1979.

[21] G. Verghese, P. Van Dooren and T. Kailath, Properties of the system matrix of a generalised state-space system, In.T.Contr., Vol.30, pp.235-243, 1979.

[22] W.A. Wolovich, Linear Multivariable Systems, Springer, New York, 1974.

[23] H.H. Rosenbrock and G.E. Hayton, "Dynamical indices of a transfer function matrix", Int.J.Contr., Vol.9, pp.97-106, 1974.

[24] W.A. Wolovich, "The determination of state-space representations for linear multivariable systems", Automatica, Vol.9, pp.97-106, 1973.

[25] S.H. Wang and E.J. Davison, A minimization algorithm for the design of linear multivariable systems, IEEE Trans.Automat.Contr., Vol.AC-18, pp.220-225, 1973.

[26] T. Kousiouris, Ph.D. Thesis, UMIST Manchester, U.K., 1977.

[27] M.L.J. Hautus and M. Heymann, "Linear feedback - an algebraic approach", SIAM J. Control and Optimization, Vol.16, pp. , 1978.

[28] J. Hammer and M. Heymann, "Causal factorization and linear feedback", SIAM J. Control and Optimization, Vol.19, pp. , 1981.

[29] B.C. Moore and L.M. Silvermann, "Model matching by state feedback and dynamic compensation", IEEE Trans.Automat.Contr., Vol.AC-17, pp.491-497, 1972.

[30] B.D.O. Anderson and R.W. Scott, "Parametric solution of the stable exact model matching problem", IEEE Trans.Automat.Contr., Vol. AC-22, pp.137-138, 1977.

[31] W.A. Wolovich, P. Antsaklis and H. Elliot, "On the stability of solutions to the minimal and nonminimal design problems", IEEE Trans.Automat.Contr., Vol.AC-22, pp.88-94, 1977.

[32] R.W. Scott and B.D.O. Anderson, "Least order, stable solution of the exact model matching problem", Automatica, Vol.14, pp.481-492, 1978.

[33] W.A. Wolovich, "Equivalence and invariants in linear multivariable systems", Proc. 1974 JACC, pp.177-185.

[34] L. Pernebo, An algebraic theory for the design of controllers for linear multivariable systems, Parts I and II, IEEE Trans.Automat. Contr., Vol.AC-26, pp.171-190.

[35] A.I.G. Vardulakis and N. Karcanias, "Proper and stable, minimal MacMillan degree bases of rational vector spaces", To appear in IEEE Trans.Automat.Contr., Nov. 1984.

CHAPTER 14

POLE PLACEMENT IN DISCRETE MULTIVARIABLE SYSTEMS BY TWO AND THREE-TERM CONTROLLERS

H. Seraji
Department of Electrical Engineering
Tehran University of Technology
P.O. Box 3406
Tehran, Iran

ABSTRACT. The paper presents simple methods for the design of digital two-term (PI, PD) and three-term (PID) controllers for pole placement in discrete-time multivariable systems, where the controllers act directly on the available system outputs. It is shown that for an m-input ℓ-output system of order n, the PID controller can in general place up to $3m+\ell-1$ poles of the $(n+2\ell)$th order closed-loop system arbitrarily. The PI and PD controllers can each place up to $2m+\ell-1$ poles of the $(n+\ell)$th order closed-loop system. The PID and PI controllers also ensure that in the steady-state, the system outputs follow step references in the presence of inaccessible constant disturbances and in the face of system parameter variations. The PID, PI and PD controllers can be "tuned" to improve closed-loop transient responses.

1. INTRODUCTION

It is a well-known empirical fact that complex industrial processes are often controlled satisfactorily by means of surprisingly simple three-term (PID) controllers. According to an extensive survery [1], 34 out of 37 listed industrial analogue controllers are of PID type. Since the introduction of digital computers in control loops, many analogue three-term controllers have been replaced by digital PID algorithms in direct digital control (DDC). The advent of microcomputers in recent years and the consequent ease of digital control implementation have made digital PID controllers far more attractive from a practical viewpoint [2].

The purpose of this paper is to develop simple methods for the design of digital multivariable three-term (PID) and two-term (PI, PD) controllers for linear discrete-time multivariable systems. These controllers are designed to achieve specified closed-loop pole positions so as to ensure stability and desirable transient performance. In the steady-state, the inherent features of PID and PI controllers ensure that the system outputs follow step references in the presence of inaccessible constant disturbances and in the face of system parameter variations.

S. G. Tzafestas (ed.), Multivariable Control, 265–279.

2. DESIGN OF PID CONTROLLERS

The multivariable system to be controlled is described by the linear discrete-time state model

$$\begin{aligned} x(k+1) &= A\,x(k)+Bu(k)+E\,d(k) \\ y(k) &= C\,x(k)+F\,d(k) \end{aligned} \qquad \ldots(1)$$

where x is the nx1 state vector, u is the mx1 control input vector, d is the νx1 inaccessible disturbance input vector and y is the ℓx1 vector of available outputs to be controlled. The scalar parameter k specifies the sampling instant and assumes the integer values k=0,1,2,.... where t=kT, and the sampling period T is defined as the unit of time and is omitted for simplicity. The system (1) is assumed to be controllable and observable. The system matrix A is either cyclic or is made cyclic by the initial application of an arbitrary constant output feedback matrix [3]; and the matrices B and C have full ranks.

In many practical multivariable control problems, the basic design requirements can be stated as follows:

(i) In the steady-state, the output vector y follows the ℓx1 step reference vector $y_r(k)=Y_r$ in the presence of the inaccessible constant disturbance vector d(k)=D and in the face of system parameter variations.

(ii) The closed-loop system has acceptable transient responses and stability margin as expressed in terms of specified pole locations inside the unit circle in the z-plane.

(iii) The controller must have a simple structure, be easy to tune and should act directly on the available system outputs.

In order to meet the design requirements, in theory it is sufficient to employ a purely integral controller acting on the tracking-error $e(k) = y_r(k) - y(k)$, where $y_r(k)$ is the ℓx1 reference (setpoint) vector. This controller meets the steady-state specification (i) at the cost of reduction in speed of response and aggravation of system stability. The proportional and derivative controllers enhance both speed of response and system stability at the same time. Thus the PID controller combines the speed of response and the stabilizing effect due to P and D actions with the steady-state output control due to I action. The proposed digital PID control law

$$u(k) = -\,Py(k) + Q\,w(k) - Rv(k) \qquad \ldots(2)$$

is applied to the system (1), where P, Q and R are constant mxℓ proportional, integral and derivative gain matrices respectively and w(k) and v(k) are ℓx1 state vectors related to the integral and the derivative terms in the controller respectively. The ℓx1 vector w(k) is the integral of the tracking-error e(k) in discrete form and is defined by the difference equation

$$w(k) = w(k-1) + e(k) \qquad \ldots(3)$$

On taking z-transform from equation (3), w and e are related by the transfer-function model $W(z) = \frac{z}{z-1} E(z)$. The ℓx1 vector v(k) is the derivative of the output in discrete form and is defined by the difference equation

$$v(k) = y(k) - y(k-1) \qquad \ldots(4)$$

Taking z-transform from equation (4) gives the transfer-function relationship $V(z) = \frac{z-1}{z} Y(z)$. From equations (2)-(4), it is seen that the implementation of the digital PID control law adds 2ℓ new state-variables to the system through the additional state vectors w(k) and v(k) and hence increases the order of the system by 2ℓ. The integral state vector w(k) statisfies the discrete-time state model

$$\begin{aligned} w(k+1) &= w(k) + e(k+1) = w(k) + y_r(k+1) - y(k+1) \\ &= w(k) + y_r(k+1)-CAx(k)-CBu(k)-CEd(k)-Fd(k+1) \qquad \ldots(5) \end{aligned}$$

The derivative state vector v(k) satisfies the discrete-time state model

$$\begin{aligned} v(k+1) &= y(k+1)-y(k) = Cx(k+1)+Fd(k+1)-Cx(k)-Fd(k) \\ &= (CA-C)x(k)+CBu(k)+(CE-F)d(k)+Fd(k+1) \qquad \ldots(6) \end{aligned}$$

Expanding the system model to incorporate the additional state vectors w(k) and v(k), from equations (1), (5) and (6) we obtain the state model of the $(n+2\ell)$th order augmented open-loop system as

$$\begin{bmatrix} x(k+1) \\ w(k+1) \\ v(k+1) \end{bmatrix} = \begin{bmatrix} A & 0 & 0 \\ -CA & I & 0 \\ CA-C & 0 & 0 \end{bmatrix} \begin{bmatrix} x(k) \\ w(k) \\ v(k) \end{bmatrix} + \begin{bmatrix} B \\ -CB \\ CB \end{bmatrix} u(k) + \begin{bmatrix} 0 \\ I \\ 0 \end{bmatrix} y_r(k+1) + \begin{bmatrix} E \\ -CE \\ CE-F \end{bmatrix} d(k)$$

$$+ \begin{bmatrix} 0 \\ -F \\ F \end{bmatrix} d(k+1)$$

$$\begin{bmatrix} y(k) \\ w(k) \\ v(k) \end{bmatrix} = \begin{bmatrix} C & 0 & 0 \\ 0 & I & 0 \\ 0 & 0 & I \end{bmatrix} \begin{bmatrix} x(k) \\ w(k) \\ v(k) \end{bmatrix} + \begin{bmatrix} F \\ 0 \\ 0 \end{bmatrix} d(k) \qquad \ldots(7)$$

The PID control law (eq.2) is now in the form of a constant output feedback law for the augmented system. Thus, the requirements for the existence of (P,Q,R) for pole placement are that the augmented system be controllable and observable. Furthermore, since unity-rank gain matrices will be used for simplicity, the augmented system must also be cyclic. These requirements are now discussed:

(i) Controllability: The augmented system (7) is controllable if and only if [4]:

1- The original system (A,B) is controllable; i.e. rank $[B \vdots AB \vdots \cdots \vdots A^{n-1}B] = n$.

2- $\text{rank} \begin{bmatrix} A-I & B \\ C & 0 \end{bmatrix} = n+\ell$

3- $\text{rank} \begin{bmatrix} A & B \\ C & 0 \end{bmatrix} = n+\ell$

It is seen that the controllability of the original system is a necessary but not a sufficient condition for the controllability of the augmented system. Furthermore, conditions 2 and 3 imply that $m \geq \ell$, i.e. the original system must have at least as many control inputs as there are outputs.

(ii) Observability: It can readily be shown that a necessary and sufficient condition for observability of the augmented system (7) is that the original system (C,A) be observable; i.e. rank $[C^T \vdots A^T C^T \vdots \cdots \vdots (A^T)^{n-1} C^T] = n$.

(iii) Cyclicity: A necessary and sufficient condition for cyclicity of the augmented system (7) is that the polynomial matrix $\phi(z) = (zI-A_0)^{-1}$ be irreducible where A_0 is the augmented system matrix [3]. It can be shown [4] that for multi-output systems ($\ell>1$), the factor $z^{\ell-1}(z-1)^{\ell-1}$ is cancelled out in $\phi(z)$ and hence for multi-output system, the augmented system matrix is always non-cyclic.

2.1. Design Method

In this section, a method is described for the design of the PID controller gain matrices P,Q and R (eq. 2) to place up to $3m+\ell-1$ poles of the $(n+2\ell)$th order augmented system (eq. 7) at the desired locations $\lambda_1,\ldots,\lambda_{3m+\ell-1}$ in the z-plane subject to complex pairing. The design is carried out in three steps. In Step One, the augmented system is made cyclic so that unity-rank gain matrices can be used in subsequent steps. In Step Two, $\ell-1$ poles of the cyclic augmented system are moved to desired locations; while in Step Three the assigned poles are preserved and up to 3m additional poles are moved to specified positions. Steps One and Two are omitted for single-output systems.

(i) Step One: The augmented system (7) is made cyclic by applying the integral-derivative feedback law

$$u(k) = \hat{u}(k) + \hat{Q}w(k) - \hat{R}v(k) \qquad \ldots(8)$$

where $\hat{Q}$ and $\hat{R}$ are arbitrary full-rank $m\times\ell$ integral and derivative gain matrices respectively and $\hat{u}(k)$ is the $m\times 1$ control vector. This results in the new system matrix $A_1^* = \begin{bmatrix} A & B\hat{Q} & -B\hat{R} \\ -CA & I-CB\hat{Q} & CB\hat{R} \\ CA-C & CB\hat{Q} & -CB\hat{R} \end{bmatrix}$ which has distinct eigenvalues and is therefore cyclic [3].

(ii) Step Two: The unity-rank integral control law

$$\hat{u}(k) = \bar{u}(k) + q_1 k_1 w(k) \qquad \ldots(9)$$

is applied to the cyclic augmented system (C^*, A_1^*, B^*) so as to move $\ell-1$ poles to the distinct* specified locations $\lambda_1,\ldots,\lambda_{\ell-1}$ subject to complex pairing, where q_1 and k_1 are $m\times 1$ and $1\times\ell$ vectors, $\bar{u}(k)$ is the $m\times 1$ control vector, $B^* = \begin{bmatrix} B \\ -CB \\ CB \end{bmatrix}$ and $C^* = (C\ 0\ 0)$. On applying (9) we obtain the closed-loop characteristic polynomial [4]:

$$H_2(z) = H_1(z) + \frac{z}{z-1} k_1 G_1(z) q_1 \qquad \ldots(10)$$

where $H_1(z) = |zI-A_1^*|$ and $G_1(z) = C^* \text{ adj } (zI-A_1^*)B^*$. The vector q_1 is specified arbitrarily such that the single-input system $(A_1^*, B^* q_1)$ is controllable. The vector k_1 is then found to move $\ell-1$ poles to $\lambda_1,\ldots,\lambda_{\ell-1}$ by solving the $\ell-1$ linear equations

$$H_1(\lambda_i) + \frac{\lambda_i}{\lambda_i-1} k_1 G_1(\lambda_i) q_1 = 0 \ ; \quad i=1,\ldots,\ell-1 \qquad \ldots(11)$$

---

* The $\ell-1$ poles assigned in Step Two must be distinct so that they can be preserved in Step Three. Therefore, if necessary, the specified locations $\lambda_1,\ldots,\lambda_{3m+\ell-1}$ are perturbed slightly so as to obtain $\ell-1$ distinct locations.

The resulting system matrix $A_2^* = \begin{bmatrix} A & B(\hat{Q}+q_1k_1) & -B\hat{R} \\ -CA & I-CB(\hat{Q}+q_1k_1) & CB\hat{R} \\ CA-C & CB(\hat{Q}+q_1k_1) & -CB\hat{R} \end{bmatrix}$ has $\ell-1$

eigenvalues at $\lambda_1, \ldots, \lambda_{\ell-1}$.

(iii) Step Three: The unity-rank PID control law

$$\bar{u}(k) = -pk_2y(k)+qk_2w(k)-rk_2v(k) \qquad (12)$$

where p,q,r and k_2 are mx1, mx1, mx1 and 1xℓ vectors respectively, is applied to the system (C^*,A_2^*,B^*) so as to move 3m additional poles to the specifed locations $\lambda_\ell,\ldots,\lambda_{3m+\ell-1}$ while preserving the ℓ-1 poles assigned in Step Two. On applying (12), we obtain the closed-loop characteristic polynomial [4]:

$$H_3(z)=H_2(z)+k_2G_2(z)p+\frac{z}{z-1}k_2G_2(z)q+\frac{z-1}{z}k_2G_2(z)r \quad \ldots(13)$$

where $H_2(z) = |zI-A_2^*|$ and $G_2(z) = C^* \operatorname{adj}(zI-A_2^*)B^*$. The vector k_2 is used to perserve the ℓ-1 assigned poles at $\lambda_1,\ldots,\lambda_{\ell-1}$, whereas the vectors p,q and r are employed to move 3m additional poles to $\lambda_\ell \ldots$, $\lambda_{3m+\ell-1}$ subject to complex pairing.

In order to preserve the ℓ-1 assigned poles at $\lambda_1,\ldots,\lambda_{\ell-1}$ irrespective of p,q and r, from equation (13) the vector k_2 must satisfy the equation

$$k_2G_2(\lambda_i) = 0 \qquad i=1,\ldots,\ell-1$$

Each ℓxm matrix $G_2(\lambda_i)$ is of unity-rank [5] and has only one linearly independent column denoted by g_i. Thus the vector k_2 is found from the ℓ-1 linear equations

$$k_2g_i = 0 \qquad i=1,\ldots,\ell-1 \qquad (14)$$

Once k_2 is determined, the vectors p,q and r are found to place 3m additional poles at $\lambda_\ell,\ldots,\lambda_{3m+\ell-1}$ by solving the 3m linear equations

$$H_2(\lambda_i)+k_2G_2(\lambda_i)p+\frac{\lambda_i}{\lambda_i-1}k_2G_2(\lambda_i)q+\frac{\lambda_i-1}{\lambda_i}k_2G_2(\lambda_i)r=0; \; i=\ell,\ldots,3m+\ell-1 \quad (15)$$

Note that for systems with $(3m+\ell-1) > (n+2\ell)$, $(3m-\ell-n-1)$ elements of the vectors p, q or r are not used for pole placement and can be specified arbitrarily. In this case, equations (15) are solved for

$i=\ell,\ldots, n+2\ell$.

Finally, the digital PID control law for the original system is given by $u(k)=-Py(k)+Qw(k)-Rv(k)$ where $P=pk_2$ is the unity-rank proportional gain matrix and $Q=\hat{Q}+q_1k_1+qk_2$ and $R=\hat{R}+rk_2$ are the full-rank integral and derivative gain matrices respectively. The state model of the final $(n+2\ell)$th order closed-loop system is given by

$$\begin{bmatrix} x(k+1) \\ w(k+1) \\ v(k+1) \end{bmatrix} = \begin{bmatrix} A-BPC & BQ & -BR \\ -C(A-BPC) & I-CBQ & CBR \\ C(A-I-BPC) & CBQ & -CBR \end{bmatrix} \begin{bmatrix} x(k) \\ w(k) \\ v(k) \end{bmatrix} + \begin{bmatrix} 0 \\ I \\ 0 \end{bmatrix} y_r(k+1) + \begin{bmatrix} E-BPF \\ -C(E-BPF) \\ C(E-BPF)-F \end{bmatrix} d(k) + \begin{bmatrix} 0 \\ -F \\ F \end{bmatrix} d(k+1)$$

$$y(k)=(C \quad 0 \quad 0)\begin{bmatrix} x(k) \\ w(k) \\ v(k) \end{bmatrix} +F\, d(k) \qquad \ldots(16)$$

The closed-loop system has $3m+\ell-1$ poles at the specified locations $\lambda_1,\ldots,\lambda_{3m+\ell-1}$. For systems with $n < 3m-\ell$, all $n+2\ell$ closed-loop poles can in general be placed arbitrarily by a PID controller using the above method* and hence closed-loop stability is ensured. When $n \geq 3m-\ell$, $(n-3m+\ell+1)$ closed-loop poles move to unspecified locations. If these pole locations are undesirable, the matrices $\hat{Q}$, $\hat{R}$ and the vector q_1 are altered and the design procedure is repeated. However, for some systems with $n \geq 3m-\ell$, the remaining $(n-3m+\ell+1)$ unassignable poles may dominate the closed-loop responses or even cause instability.

2.2. Controller Robustness

Consider now the steady-state characteristics of the closed-loop system (16). Assuming stability, when the closed-loop system is subjected simultaneously to the step reference vector $y_r(k) = Y_r$ and the step disturbance vector $d(k) = D$, the augmented state vector $x^*(k)$ reaches a constant value in the steady-state. Hence $w(k) = w(k+1)$ in the

---

* There are singular cases where eqs. (15) are linearly dependent and hence $3m+\ell-1$ poles cannot be placed by a PID controller. This, however, is not a limitation of the present method and is an inherent feature of certain singular systems, e.g. for systems with $C^*B^*=CB=0$ all closed-loop poles cannot be placed by a PID controller (see, e.g.,[6]).

steady-state, that is $e(k) = 0$ and $y(k) = y_r(k) = Y_r$. Thus in the steady-state, the output vector $y(k)$ follows the step reference vector $y_r(k) = Y_r$ in the presence of the step disturbance vector $d(k) = D$.

An important feature of the PID controller is "robustness", namely the preservation of the above steady-state characteristics in the face of variations or uncertainties in the system parameters. Let the system parameters (A,B,C,E,F,n) change from their nominal design values to arbitrary new values (A',B',C',E',F',n') provided only that the closed-loop system remains stable. By repeating the above argument, it can be shown that the system outputs still follow step references despite constant disturbances. This feature is of vital importance in practical applications where system parameters are not known exactly or are subject to large variations.

It must be noted that the achievement of the steady-state characteristics of the system is solely due to the integral term in the PID controller. The inherent property of integral term in robust reference tracking and disturbance rejection makes this term an essential part of most industrial controllers.

2.3. Controller Tuning

Although the placement of closed-loop poles is a means for achieving acceptable transient responses, pole locations by themselves do not completely determine the transient behaviour of the system. Therefore, once the PID controller (P,R,Q) is determined for pole placement, the designer must simulate the open-loop system and the controller on a computer and observe the performance of the resulting closed-loop system. At this stage, the designer may improve the transient responses by "tuning" each term in the controller independently as

$$u(k) = -\alpha P\, y(k) + \beta Q w(k) - \gamma R\, v(k)$$

where α, β and γ are scalar tuning parameters. By varying α, β and γ independently, the designer can study the effect of the proportional, integral and derivative terms on the transient responses separately and "tune" each term to improve the transient performance. A significant advantage of this tuning method is that only three parameters need to be tuned on simulation, even though the system may have many inputs and outputs. In general, it is undesirable to have a design method which requires tuning of a large number of parameters depending on the number of inputs and outputs or the order of the system.

3. DESIGN OF PI CONTROLLERS

In some applications, the three basic design requirements stated in Section 2 can be meet satisfactorily using a proportional-integral controller without the need for the derivative term. Furthermore, in certain cases where the system outputs are contaminated by noise or the output fluctuations are excessive, the use of derivative is undesirable.

Consider the system described by equation (1). Suppose that the digital proportional-integral (PI) control law

$$u(k) = -Py(k) + Qw(k) \qquad \ldots(17)$$

is applied to the system (1) to achieve pole placement and steady-state output control, where the symbols are as defined before and the integral state vector w(k) satisfies equation (5). On combining equations (1) and (5), we obtain the state model of the $(n+\ell)$th order augmented open-loop system as

$$\begin{bmatrix} x(k+1) \\ w(k+1) \end{bmatrix} = \begin{bmatrix} A & 0 \\ -CA & I \end{bmatrix} \begin{bmatrix} x(k) \\ w(k) \end{bmatrix} + \begin{bmatrix} B \\ -CB \end{bmatrix} u(k) + \begin{bmatrix} 0 \\ I \end{bmatrix} y_r(k+1) + \begin{bmatrix} E \\ -CE \end{bmatrix} d(k) + \begin{bmatrix} 0 \\ -F \end{bmatrix} d(k+1)$$

$$\begin{bmatrix} y(k) \\ w(k) \end{bmatrix} = \begin{bmatrix} C & 0 \\ 0 & I \end{bmatrix} \begin{bmatrix} x(k) \\ w(k) \end{bmatrix} + \begin{bmatrix} F \\ 0 \end{bmatrix} d(k) \qquad \ldots(18)$$

Equation (17) is now in the form of a constant output feedback law for the augument system (18). Hence for the design of unity-rank P and Q, the augumented system must be controllable, observable and cyclic.

(i) Controllability: The augmented system (18) is controllable if and only if [4]:

1- The original system (A,B) is controllable.

2- $\text{rank} \begin{bmatrix} A-I & B \\ C & 0 \end{bmatrix} = n+\ell$

(ii) Observability: It can readily be shown that the augmented system (18) is observable provided that the original system (C,A) is observable.

(iii) Cyclicity: As in Section 2, the augmented system matrix

$A^*_0 = \begin{bmatrix} A & 0 \\ -CA & I \end{bmatrix}$ is always non-cyclic for multi-output systems ($\ell > 1$).

3.1. Design Method

In this section, a method is described for the design of the PI controller (P,Q) so as to shift up to $2m+\ell-1$ poles of the $(n+\ell)$th order augmented system (18) to the specified locations $\lambda_1,\ldots,\lambda_{2m+\ell-1}$. The method proceeds in three steps, but for single-output systems, Steps One and Two are omitted. Since the design steps are similar to Section 2.1, the description here will be brief.

(i) Step One: The augmented system matrix A_0^* is made cyclic by applying the integral feedback law $u(k) = \hat{u}(k) + \hat{Q}w(k)$, where $\hat{Q}$ is an arbitrary full-rank $m\times\ell$ matrix. This results in the cyclic augmented system matrix $A_1^* = \begin{bmatrix} A & B\hat{Q} \\ -CA & I-CB\hat{Q} \end{bmatrix}$.

(ii) Step Two: As in Section 2.1 (ii), the unity-rank integral control law $\hat{u}(k) = \bar{u}(k)+q_1k_1w(k)$ is applied to the system (C^*,A_1^*,B^*) where $B^* = \begin{bmatrix} B \\ -CB \end{bmatrix}$ and $C^* = (C \quad 0)$. The $m\times 1$ vector q_1 is specified arbitrarily while the $1\times\ell$ vector k_1 is found to shift $\ell-1$ eigenvalues of A_1^* to the distinct locations $\lambda_1,\ldots,\lambda_{\ell-1}$ by solving the $\ell-1$ linear equations (eq. 11)

$$H_1(\lambda_i) + \frac{\lambda_i}{\lambda_i-1}\, k_1G_1(\lambda_i)\, q_1 = 0; \qquad i=1,\ldots,\ell-1$$

where $H_1(z) = |zI-A_1^*|$ and $G_1(z) = C^*\,\mathrm{adj}(zI-A_1^*)B^*$. The resulting system matrix $A_2^* = \begin{bmatrix} A\,, & B(\hat{Q}+q_1k_1) \\ -CA, & I-CB(\hat{Q}+q_1k_1) \end{bmatrix}$ has $\ell-1$ eigenvalues at $\lambda_1,\ldots,\lambda_{\ell-1}$.

(iii) Step Three: As in Section 2.1 (iii), the unity-rank PI control law $\bar{u}(k) = -pk_2y(k)+qk_2w(k)$ is applied to the system (C^*,A_2^*,B^*). The $1\times\ell$ vector k_2 is used to preserve the $\ell-1$ assigned eigenvalues of A_2^* at $\lambda_1,\ldots,\lambda_{\ell-1}$ irrespective of p and q by solving the $\ell-1$ linear equations (eq. 14):

$$k_2g_i = 0\,; \qquad i=1,\ldots,\ell-1$$

where g_i is any non-zero column of $G_2(z) = C^*\,\mathrm{adj}(zI-A_2^*)B^*$ for $z = \lambda_i$. The $m\times 1$ vectors p and q are found to shift 2m additional eigenvalues of

A_2^* to $\lambda_\ell,\ldots,\lambda_{2m+\ell-1}$ by solving the 2m linear equations (eq. 15 with r=0)

$$H_2(\lambda_i)+k_2\,G_2(\lambda_i)p+\frac{\lambda_i}{\lambda_i-1}\,k_2G_2(\lambda_i)q = 0; \qquad i=\ell,\ldots,2m+\ell-1$$

where $H_2(z) = |zI-A_2^*|$. When $(2m+\ell-1) > (n+\ell)$, $(2m-n-1)$ elements of p or q are specified and the above equations are solved for $i=\ell,\ldots,n+\ell$.

The digital PI control law for the original system is $u(k)=-Py(k)+Qw(k)$, where $P=pk_2$ and $Q=\hat{Q}+q_1k_1+qk_2$ are the unity-rank proportional and the full-rank integral gain matrices respectively. The resulting $(n+\ell)$th order closed-loop system is given by

$$\begin{bmatrix} x(k+1) \\ w(k+1) \end{bmatrix} = \begin{bmatrix} A-BPC & -BQ \\ -C(A-BPC) & I-CBQ \end{bmatrix}\begin{bmatrix} x(k) \\ w(k) \end{bmatrix} + \begin{bmatrix} 0 \\ I \end{bmatrix} y_r(k+1) + \begin{bmatrix} E-BPF \\ -C(E-BPF) \end{bmatrix} d(k) + \begin{bmatrix} 0 \\ -F \end{bmatrix} d(k+1)$$

$$y(k) = (C \quad 0)\begin{bmatrix} x(k) \\ w(k) \end{bmatrix} + Fd(k) \qquad \ldots(19)$$

and has $2m+\ell-1$ poles at $\lambda_1,\ldots,\lambda_{2m+\ell-1}$. Note that when $n < 2m$, all closed-loop poles can in general be placed arbitrarily by a PI controller (i.e. except in singular cases, e.g. when $C^*B^* = CB = 0$, [6]).

Following Section 2.2 and assuming stability, the output vector y follows the step reference vector $y_r(k) = Y_r$ in the steady-state in the presence of the inaccessible constant disturbance vector $d(k) = D$. Furthermore, the PI controller is "robust" since the steady-state characteristics are insensitive to variations in system parameters provided stability is maintained.

Finally, as in Section 2.3, to improve the transient responses, the acquired controller can be "tuned" as $(\alpha P, \beta Q)$, where the scalar tuning parameters α and β are varied separately to study the effect of the proportional and integral terms on the closed-loop transient responses independently.

4. DESIGN OF PD CONTROLLERS

When steady-state output control is not required, the integral term is removed from the controller. In this case, a proportional-derivative controller is used to obtain acceptable transient responses and stability margin through pole placement.

Consider the system described by equation (1). Suppose that the digital proportional-derivative (PD) control law

$$u(k) = u_c(k) - Py(k) - Rv(k) \qquad \ldots(20)$$

is applied to the system (1) to obtain specified pole positions, where $u_c(k)$ is the mx1 command vector and the derivative state vector $v(k)$ satisfies equation (6). The state model of the $(n+\ell)$th order augmented open-loop system is obtained by combining equations (1) and (6) as

$$\begin{bmatrix} x(k+1) \\ \\ v(k+1) \end{bmatrix} = \begin{bmatrix} A & 0 \\ \\ CA-C & 0 \end{bmatrix} \begin{bmatrix} x(k) \\ \\ v(k) \end{bmatrix} + \begin{bmatrix} B \\ \\ CB \end{bmatrix} u(k) + \begin{bmatrix} E \\ \\ CE-F \end{bmatrix} d(k) + \begin{bmatrix} 0 \\ \\ F \end{bmatrix} d(k+1)$$

$$\ldots(21)$$

$$\begin{bmatrix} y(k) \\ \\ v(k) \end{bmatrix} = \begin{bmatrix} C & 0 \\ \\ 0 & I \end{bmatrix} \begin{bmatrix} x(k) \\ \\ v(k) \end{bmatrix} + \begin{bmatrix} F \\ \\ 0 \end{bmatrix} d(k)$$

Equation (20) is in effect a constant output feedback law for the augmented system (21). Thus the controllability, observability and cyclicity of the augmented system are required for the design of unity-rank P and R.

(i) Controllability: The augmented system (21) is controllable if and only if [4]:

1- The original system (A,B) is controllable.

2- $\text{rank}\begin{bmatrix} A & B \\ C & 0 \end{bmatrix} = n+\ell$

(ii) Observability: It can readily be shown that the augmented system (21) is observable provided that the original system (C,A) is observable.

(iii) Cyclicity: As in Section 2, for multi-output systems ($\ell > 1$), the augmented system matrix $A_0^* = \begin{bmatrix} A & 0 \\ CA-C & 0 \end{bmatrix}$ is always non-cyclic .

4.1. Design Method

In this section, a method is described for the design of the PD controller (P,R) to shift up to $2m+\ell-1$ poles of the $(n+\ell)$th order augmented system (21) to the specified locations $\lambda_1,\ldots,\lambda_{2m+\ell-1}$. The design is carried out in three steps, but for single-output systems Steps One and Two are omitted. Since the design procedure is similar to Section 2.1, the description here will be brief.

(i) Step One: The augmented system matrix A_0^* is made cyclic by applying the derivative feedback law $u(k) = u_c(k) + \hat{u}(k) - \hat{R}v(k)$, where $\hat{R}$ is an arbitrary full-rank $m\times\ell$ matrix. The new system matrix $A_1^* = \begin{bmatrix} A & -B\hat{R} \\ CA-C & -CB\hat{R} \end{bmatrix}$ is now cyclic.

(ii) Step Two: The unity-rank derivative control law $\hat{u}(k) = \bar{u}(k)-r_1k_1v(k)$ is applied to the system (C^*,A_1^*,B^*), where $B^* = \begin{bmatrix} B \\ CB \end{bmatrix}$ and $C^* = (C \quad 0)$. The $m\times1$ vector r_1 is chosen arbitrarily, whereas the $1\times\ell$ vector k_1 is determined to shift $\ell-1$ eigenvalues of A_1^* to the distinct locations $\lambda_1,\ldots,\lambda_{\ell-1}$ by solving the $\ell-1$ linear equations [eq. 15 with p=q=0]

$$H_1(\lambda_i) + \frac{\lambda_i-1}{\lambda_i} k_1G_1(\lambda_i)\, r_1 = 0 \;; \qquad i=1,\ldots,\ell-1$$

where $H_1(z) = |zI-A_1^*|$ and $G_1(z) = C^*\,\mathrm{adj}(zI-A_1^*)B^*$. The resulting system matrix $A_2^* = \begin{bmatrix} A & -B(\hat{R}+r_1k_1) \\ CA-C & -CB(\hat{R}+r_1k_1) \end{bmatrix}$ has $\ell-1$ eigenvalues at $\lambda_1,\ldots,\lambda_{\ell-1}$.

(iii) Step Three: The unity-rank PD control law $\bar{u}(k) = -pk_2y(k)-rk_2v(k)$ is applied to the system (C^*,A_2^*,B^*). As in Section 2.2 (iii), the $1\times\ell$ vector k_2 is used to preserve the $\ell-1$ assigned eigenvalues of A_2^* at $\lambda_1,\ldots,\lambda_{\ell-1}$ irrespective of p and r by solving the $\ell-1$ linear equations (eq.[14])

$$k_2 g_i = 0 \quad ; \qquad i=1,\ldots,\ell-1$$

where g_i is any non-zero column of $G_2(z) = C^* \operatorname{adj}(zI-A_2^*)B^*$ for $z = \lambda_i$.

The mx1 vectors p and r are determined to shift 2m additional eigenvalues of A_2^* to $\lambda_\ell,\ldots,\lambda_{2m+\ell-1}$ by solving the 2m linear equations [eq. 15 with q = 0]

$$H_2(\lambda_i)+k_2G_2(\lambda_i)p+\frac{\lambda_i-1}{\lambda_i}k_2G_2(\lambda_i)r = 0 \quad ; \qquad i=\ell,\ldots,2m+\ell-1$$

where $H_2(z) = |zI-A_2^*|$. When $(2m+\ell-1) > (n+\ell)$, $(2m-n-1)$ elements of p or r are specified and above equations are solved for $i = \ell,\ldots, n+\ell$.

The digital PD control law for the original system is $u(k) = u_c(k)-Py(k)-Rv(k)$, where $P = pk_2$ and $R = \hat{R}+r_1k_1+rk_2$ are the unity-rank proportional and the full-rank derivative gain matrices respectively. The resulting $(n+\ell)$th order closed-loop system is given by

$$\begin{bmatrix} x(k+1) \\ v(k+1) \end{bmatrix} = \begin{bmatrix} A-BPC & -BR \\ C(A-I-BPC) & -CBR \end{bmatrix} \begin{bmatrix} x(k) \\ v(k) \end{bmatrix} + \begin{bmatrix} B \\ CB \end{bmatrix} u_c(k) + \begin{bmatrix} E \\ CE-F \end{bmatrix} d(k) + \begin{bmatrix} 0 \\ F \end{bmatrix} d(k+1) \qquad \ldots(22)$$

$$y(k) = (C \quad 0) \begin{bmatrix} x(k) \\ v(k) \end{bmatrix} + Fd(k)$$

and has $2m+\ell-1$ poles at $\lambda_1,\ldots,\lambda_{2m+\ell-1}$. Note that when $n < 2m$, all closed-loop poles can in general be placed arbitrarily by a PD controller (i.e. except in singular cases, e.g. when $C^*B^* = CB = 0$, [6]).

Finally, as in Section, 2.3, to improve the closed-loop responses the acquired controller can be "tuned" as $(\alpha P,\gamma R)$, where the scalar tuning parameters α and γ are varied separately to study the effect of the proportional and derivative terms on responses independently.

5. CONCLUSIONS

The design of digital two and three-term controllers to achieve pole placement in discrete-time multivariable systems is discussed in this paper. These controllers are simple and practical in that they act directly on the available system outputs and do not require the complications of state reconstruction and implementation of state feedback. Furthermore, the 2 and 3-term controllers lead to extremely simple and efficient algorithms which can be implemented on micro-computers for digital control applications.

Finally, it is important to note that the computational times involved in the design methods described in this paper are very short and thus the methods form powerful tools for CAD studies. Using the CAD facility, the designer can iterate on the pole locations and other design parameters while rapidly displaying the resulting responses until the desired performance is obtained. It is believed that this combining of the qualitative feel of pole positions together with a rapid display of accurate quantitative response information leads to quite a powerful and general design technique for discrete-time multivariable systems.

6. REFERENCES

1. Merritt, R.: *"Electronic Controller Survey"*, Instrum. Technol., 1977, 24(5), pp. 43-62.

2. Auslander, D.M., Takahashi, Y. and Tomizuka, M.: *"Direct Digital Process Control: Practice and Algorithms for Microprocessor Application"*, Proceedings IEEE, 1978, 66(2), pp. 199-208.

3 Seraji, H.: *"Cyclicity of Linear Multivariable Systems"*, Int. J. Control, 1975, 21(3), pp. 497-504.

4. Seraji, H.: *"Design of Digital Two and Three-Term Controllers for Discrete-Time Multivariable Systems"*, Int. J. Control, 1983, 38(4), pp. 843-865.

5. Seraji, H.: *"A New Method for Pole Placement Using Output Feedback"*, Int. J. Control, 1978, 28(1), pp. 147-155.

6. Tarokh, M.: *"Necessary Conditions for Stabilization and Pole Placement with Output Feedback"*, Elect. Lett., 1977, 13(5) pp. 148-149.

CHAPTER 15

LINEAR QUADRATIC REGULATORS WITH PRESCRIBED EIGENVALUES FOR A FAMILY OF LINEAR SYSTEMS

F. Heger, P. M. Frank
Mess- und Regelungstechnik
Universität -GH- Duisburg
Postfach 10 16 29
4100 Duisburg, FRG

ABSTRACT

In this paper a new design procedure is proposed that combines linear quadratic optimal control (LQ) with pole assignment. In LQ-problems the quadratic weights are usually found by trial and error to get good time behavior. In contrast the proposed method modifies a given state weighting matrix such that all closed-loop poles are located in a specified region of the complex s-plane. It is shown that this can be done with the smallest possible increase of the quadratic performance index under consideration. The procedure is extended to the design of robust control systems where the plant is described by a family of linear systems according to different operational conditions. The control of a F4E aircraft demonstrates the achievements of the new procedure.

1. INTRODUCTION

The problem of optimally controlling linear plants has been extensively studied /1/2/3/. The closed loop system constructed by solving an optimal regulator problem with a quadratic cost functional has some remarkable properties such as infinite gain margin and at least 60° phase margin for each control channel /4/. Furthermore there exists a constant upper bound for the overshoot of the outputs /5/. These properties do not depend on the special choice of the weighting matrices.

Another popular method to modify the dynamic response of a linear time invariant system is the assignment of the closed loop poles at arbitrary locations in the complex plane. There are many authors who have tried to combine the merits of both approaches by a special choice of the weighting matrices such that all closed loop poles are located at desired positions /7/. However, it is often sufficient to place all poles in a suitable region of the complex plane instead of placing them to their respective positions. A method of deciding weighting matrices in an LQ-problem to locate all poles in a specified region was given in /8/. The main disadvantage of this procedure is that it totally neglects the increase of the originally given performance in-

S. G. Tzafestas (ed.), Multivariable Control, 281–292.

dex thus leading to closed loop poles lying far left in the complex plane that are combined with large values of the control variables /9/.

In this paper a new procedure is introduced that modifies a given quadratic cost functional such that all closed loop poles come to lie in a suitable region of the complex plane while the increase of the original cost functional is kept as small as possible. It is shown that by repeated application of the procedure robust controllers for a family of linear systems can be obtained.

2. PROBLEM FORMULATION

Consider a controllable linear multivariable system

$$\dot{x} = A\,x + B\,u \tag{1}$$

and a quadratic cost functional

$$J = \int_0^\infty (x^T Q\,x + u^T R\,u)dt \tag{2}$$

with the n-dimensional state vector x, the m-dimensional input vector u and the matrices

$A \in \mathbb{R}^{n.n}$; $B \in \mathbb{R}^{n.m}$; $Q \in \mathbb{R}^{n.n}$, $Q \geq 0$; $R \in \mathbb{R}^{m.m}$, $R > 0$.

Additionally it is required that all poles of the closed loop system should be located in a specified region of the complex plane as shown in Fig. 1.

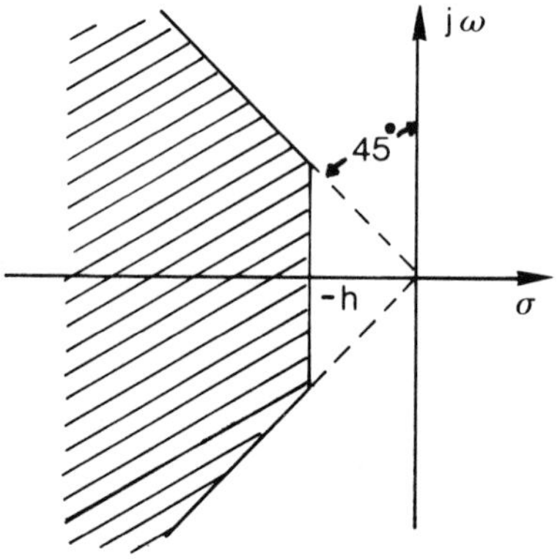

Fig. 1: A desired region where the poles of the closed loop system are to be located for good responses

It is well known that in this case the system responses converge at appropriate speed and any vibrating modes are well damped. Therefore a modification of the weighting matrix Q is sought that provides a suitable pole shifting while the originally given cost functional is increased by a minimal amount. Since a degree of stability of h can be guaranteed if the weighting matrices are multiplied by e^{2ht} /2/ in the following only the case h = 0 is treated.

3. SEQUENTIAL POLE PLACEMENT

If the feedback

$$u^1 = K^1 x \tag{3}$$

found by minimization of the original performance index

$$J^{10} = \int_0^\infty (x^T Q^1 x + u^T R u)dt$$

is implemented the closed loop system takes the form

$$\dot{x} = A^1 x + B u \tag{4}$$

with

$$A^1 = A - B K^1 . \tag{5}$$

Now the state weighting matrix Q is modified according to

$$Q^2 = Q^1 + \Delta Q^1 \tag{6}$$

and the feedback matrix K^2 is obtained. The control

$$u^2 = -K^2 x = -(K^1 + \Delta K^1)x \tag{7}$$

found in this way is optimal with respect to the new performance index

$$J^{2n} = \int_0^\infty (x^T Q^2 x + u^T R u)dt \tag{8}$$

while it is suboptimal with respect to the original performance index evaluated using the new control

$$J^{20} = \int_0^\infty (x^T Q^1 x + u^{2^T} R u^2)dt = \int_0^\infty x^T (Q^1 +$$

$$+ K^{2^T} R K^2) x \, dt \quad . \tag{9}$$

The main idea now is to choose the positive semidefinite matrix ΔQ^1 such that

- all the eigenvalues inside the allowed region remain unchanged
- one pair of complex conjugate eigenvalues is shifted into the admissible region such that the performance index is increased no more than is necessary, i.e.

$$\Delta J = J^{20} - J^{10} \stackrel{!}{=} \min \quad . \tag{10}$$

Therefore consider the eigenvalues of the closed loop system matrix A_1. It can be shown /9/ that keeping n-2 of those eigenvalues unchanged weighting matrices for a quadratic performance index can be found that shift one pair of complex conjugate eigenvalues into the hatched region of Fig. 2. This region is edged by a circle and a hyperbola with the respective pair of complex conjugate eigenvalues at the intersection.

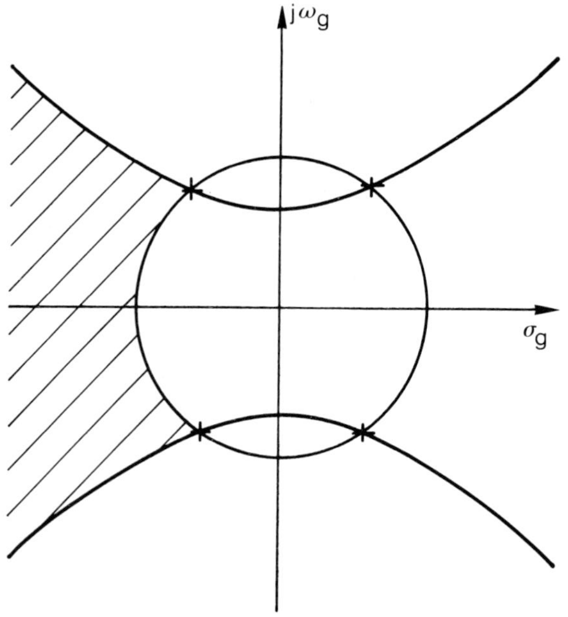

Fig. 2: Region where one pair of complex conjugate eigenvalues can be shifted by LQ-control

To determine the matrix ΔQ_1 a procedure given by Solheim /10/ is used. If the closed loop feedback matrix A^1 is transformed to real modal form

$$T^{-1}A^1T = \Lambda = \begin{bmatrix} \lambda_1 & & & & & & \\ & \ddots & & & & & \\ & & \lambda_{i-1} & & & & \\ & & & -\sigma_{io} & \omega_{io} & & \\ & & & -\omega_{io} & -\sigma_{io} & & \\ & & & & & \lambda_{i+2} & \\ & & & & & & \ddots \\ & & & & & & & \lambda_n \end{bmatrix} \tag{11}$$

the pair of complex conjugate eigenvalues $-\sigma_{oi} \pm j\omega_{oi}$ can be shifted to $-\sigma_{ci} \pm j\omega_{ci}$ by the choice

$$\tilde{\Delta Q}^1 = \begin{bmatrix} 0 & & & & & \\ & 0 & & & & \\ & & \tilde{\Delta q}_{ii} & & & \\ & & & \tilde{\Delta q}_{i+1\,i+1} & & \\ & & & & 0 & \\ & & & & & 0 \end{bmatrix} \tag{12}$$

with

$$\tilde{\Delta q}_{ii} = \tilde{\Delta q}_{i+1\,i+1} = \frac{2(\sigma_{ci}^2 - \omega_{ci}^2) - 2(\sigma_{oi}^2 - \omega_{oi}^2)}{h_{ii} + h_{i+1\,i+1}} \tag{13}$$

where $h_{ii} > 0$, $h_{i+1\,i+1} > 0$ are the corresponding elements of the matrix

$$H = T^{-1}\, B\, R^{-1}\, B^T (T^{-1})^T \quad . \tag{14}$$

Transforming back to the original state variables the necessary modification

$$\Delta Q^1 = T^T\, \tilde{\Delta Q}^1\, T \tag{15}$$

is found.

The method of Solheim thus determines a modification of the state weighting matrix to shift one pair of complex conjugate eigenvalues to desired locations. However, the question how to determine these locations inside the allowed region to minimize ΔJ is still open.

It can be shown that the value of a quadratic performance index is a monotonically increasing function of the diagonal elements q_{ii} of the state weighting matrix Q and the diagonal elements r_{ii} of the input weighting matrix R /11/.

Since the elements r_{ii} are not changed by Solheims procedure ΔJ is minimized by the choice

$$\sigma_{ci} = \omega_{ci} \tag{16}$$

leading to

$$\tilde{\Delta q}_{iimin} = \min_{\sigma_{ci},\ \omega_{ci}} \tilde{\Delta q}_{ii} = \frac{-2(\sigma_{oi}^2 - \omega_{oi}^2)}{h_{ii}+h_{i+1\ i+1}} . \tag{17}$$

Thus the minimal increase ΔJ is obtained if the eigenvalues are shifted to the border of the admissible region.

The next pair of complex conjugate eigenvalues can now be shifted by the same procedure using the closed loop system matrix

$$A^2 = A-BK^2 = A-B(K^1 + \Delta K^1) \tag{18}$$

obtaining ΔQ^2 and ΔK^2. This procedure has to be repeated till finally all closed loop eigenvalues are shifted into the admissible region with

$$Q = Q^1 + \sum_j \Delta Q^j \tag{19}$$

and

$$K = K^1 + \sum_j \Delta K^j . \tag{20}$$

Note that as a characteristic of Solheims procedure the exact location on the border cannot be predicted. The modifications of the feedback matrix ΔK^j can be calculated by the solution of second order Riccati equations.

4. EXAMPLE

To compare the new design method with the procedure given by Kawasaki and Shimemura /8/ a fourth order system with

	Kawasaki/Shimemura	Heger/Frank
Q^1	0_4	0_4
k^2	53.76 51.84 17.82 17.66	12.08 18.92 9.82 1.85
s_{c1} s_{c2} $s_{c3/4}$	-9.660 -5.660 $-2 \pm j$	$-1.926 \pm j\ 1.926$ $-2 \pm j$
J^{20}	2543	243.49
Q^1	I_4	I_4
k^2	133.9 162.8 70.9 11.0	11.56 18.35 9.62 1.85
s_{c1} s_{c2} $s_{c3/4}$	-9.480 -3.707 $-2.054 \pm j\ 0.995$	$-1.873 \pm j\ 1.873$ $-2.054 \pm j\ 0.995$
J^{20}	5671	293.15
Q^1	$1000\ I_4$	$1000\ I_4$
k^2	27.78 87.17 81.26 28.63	19.33 67.02 70.44 28.31
s_{c1} s_{c2} $s_{c3/4}$	-31.606 - 1.119 $-1.578 \pm j\ 0.473$	-31.606 1 1.119 $-0.792 \pm j\ 0.792$
J^{20}	26230	25313

Table 1: Comparison of the two design procedures

$$\underline{A} = \begin{bmatrix} 0 & 1 & 0 & 0 \\ 0 & 0 & 1 & 0 \\ 0 & 0 & 0 & 1 \\ -25 & -30 & -18 & -6 \end{bmatrix} \qquad \underline{b} = \begin{bmatrix} 0 \\ 0 \\ 0 \\ 1 \end{bmatrix}$$

and the eigenvalues

$$s_{o_{1/2}} = -1 \pm j2$$

$$s_{o_{3/4}} = -2 \pm j1$$

is considered. For both methods the obtained feedback vectors k^T the values of the performance indices J^{20} and the closed loop poles s_c are given in Table 1. Note that in all cases the new procedure leads to feedback vectors with a smaller norm that shift the respective pair of complex conjugate eigenvalues to the border of the allowed region (i.e. real and imaginary part are equal).

5. APPLICATION TO SYSTEMS WITH LARGELY VARYING PARAMETERS

Consider now a system with largely varying parameters that can be described by a family of linear state space models

$$\dot{x}_i = A_i x_i + B_i u_i \qquad i = 1,\dots,r \tag{21}$$

due to r different operational conditions. The above described procedure can now be applied successively to all members of the family of linear state space models. After the last member of the family has been treated it is possible that the resulting controller no longer satisfies the pole region requirements for the other previously considered members. Therefore starting with the first operational condition new modifications are necessary until finally in one run through all operational conditions no modification is made. Thus one fixed robust controller is found that satisfies the pole region requirements for all operational conditions.

6. DESIGN EXAMPLE

The proposed design procedure was applied to the control of an airplane F4E with canards /12/. The linearized state equations written in sensor coordinates are given by

$$\dot{x} = A\,x + b\,u$$

$$y = c\,x$$

with the state vector and input defined as

$$x = \begin{bmatrix} N_z & q & \delta_e \end{bmatrix}^T$$

N_z normal acceleration

q pitch rate

δ_e deviation of elevator deflection

u elevator deflection command

and the matrices

$$A = \begin{bmatrix} a_{11} & a_{12} & a_{13} \\ a_{21} & a_{22} & a_{23} \\ 0 & 0 & -14 \end{bmatrix} ; \; b = \begin{bmatrix} b_1 \\ 0 \\ 14 \end{bmatrix} ;$$

$$c = \begin{bmatrix} 1 & 12.46 & 0 \end{bmatrix} .$$

The corresponding values of the parameters for four different flight conditions and for the fictitious nominal case are given in Table 2. In this example the regions of the pole locations are not only specified by the above mentioned requirements but also by military specifications for flying qualities of piloted airplanes. The desired closed loop eigenvalue locations are given in Fig. 3. In addition the step response normalized upon its stationary value should not leave a tolerance band as shown in Fig. 5 (dotted lines).

Mach Alt.	Case 1 0.5 5000'	Case 2 0.85 5000'	Case 3 0.9 35000'	Case 4 1.5 35000'	nominal values
a_{11}	-0.9896	-1.702	-0.6607	-0.5162	-0.967
a_{12}	17.41	50.72	18.11	26.96	28.3
a_{13}	96.15	96.15	84.34	178.9	155
a_{21}	0.2648	0.2648	0.08201	0.6896	-0.003
a_{22}	-0.8512	-1.418	-0.6587	-1.225	-1.038
a_{23}	-11.39	-31.99	-10.81	-30.38	-21.14
b_1	-97.78	-272.2	-85.09	-175.6	-157

Table 2: Aerodynamic data of the system

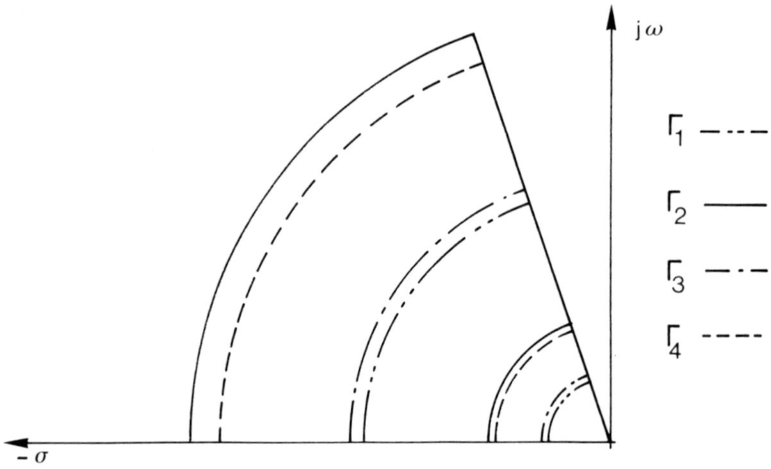

Fig. 3: Preassigned pole regions for the four operating points

It is assumed that the weighting matrices

$$Q^1 = \begin{bmatrix} 0.4 & 0 & 0 \\ 0 & 28 & 0 \\ 0 & 0 & 0.4 \end{bmatrix}, \quad R = 100$$

are given leading to a feedback vector

$$(k^1)^T = \begin{bmatrix} -0.043 & -0.553 & 0.276 \end{bmatrix} \quad .$$

Only one modification was necessary where

$$Q^2 = \begin{bmatrix} 0.680 & 1.155 & 2.809 \\ 1.155 & 37.130 & 6.153 \\ 2.809 & 6.153 & 32.200 \end{bmatrix}$$

and

$$(k^2)^T = \begin{bmatrix} -0.053 & -0.662 & 0.263 \end{bmatrix}$$

was obtained while the value of the performance index has been increased from $J^{10} = 15.310$ to $J^{20} = 31.798$. The resulting pole locations and step responses for the four flight conditions are shown in Fig. 4 and Fig. 5.

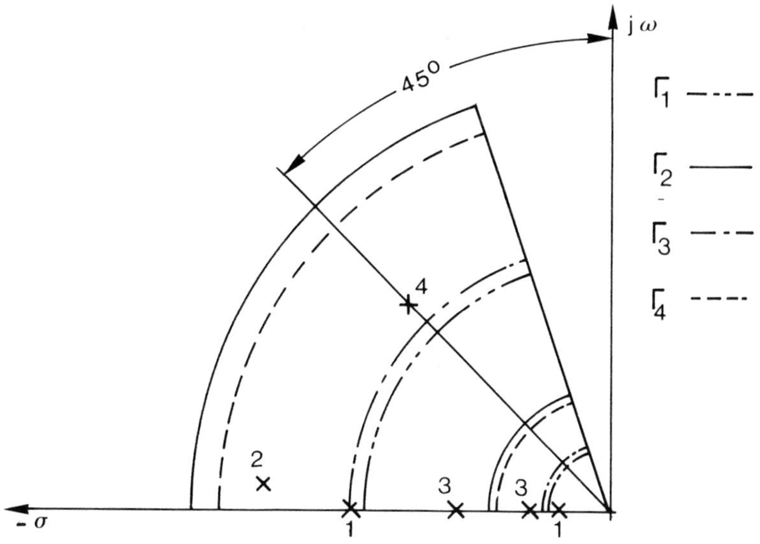

Fig. 4: Closed loop eigenvalues

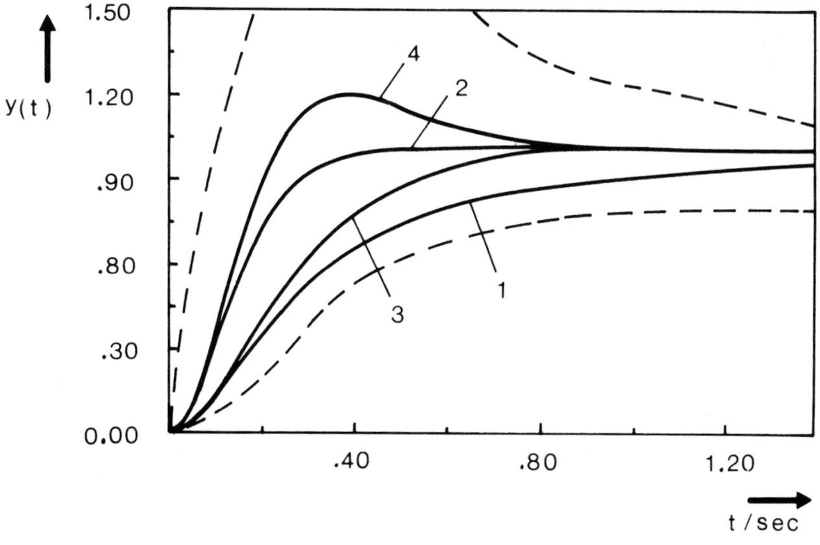

Fig. 5: Step responses for four different flight conditions

7. Conclusions

The paper presents a new design procedure that combines the merits of the LQ problem with the pole assignment approach. By modification of the state weighting matrix Q a feedback matrix K is found such that all closed loop poles come to lie in a given sector of the complex s-plane while the increase of the originally given performance index is minimized. By repeated application of the procedure robust controllers for a family of linear systems can be obtained.

8. REFERENCES

/1/ H. Kwakernaak; R. Sivan, Linear Optimal Control Systems, New York, Wiley Interscience, 1972.

/2/ B.D.O. Anderson; J.B. Moore, Linear Optimal Control, Prentice Hall Inc., 1971.

/3/ R.E. Kalman, 'When is a Linear Control System Optimal?' J. Basic Eng., Vol. 86, 1964.

/4/ M.G. Safonov; M. Athans, 'Gain and Phase Margin for Multiloop LQR Regulators', IEEE Trans. Autom. Contr. Vol. AC-22, 1977.

/5/ H. Kobayashi; E. Shimemura, 'Some Propertis of Optimal Regulators and their Applications', Int. J. Contr., 1981.

/6/ J. Ackermann, 'Entwurf durch Polvorgabe', Regelungstechnik 25, 1977.

/7/ D. Graupe, 'Derivation of Weighting Matrices Towards Satisfying Eigenvalue Requirements', Int. J. Contr., Vol. 16, 1972.

/8/ N. Kawasaki; E. Shimemura, 'A Method of Deciding Weighting Matrices in an LQ-problem to Locate all Poles in the Specified Region', Proc. of the 8th IFAC World Congress, Kyoto, 1981.

/9/ F. Heger, Entwurf robuster Regelungen für Strecken mit großen Parametervariationen, Ph.D. Thesis, University Duisburg, 1983.

/10/ O.A. Solheim, 'Design of Optimal Control Systems with Prescribed Eigenvalues', Int. J. Contr., Vol. 15, 1972.

/11/ K. Heym, 'The Influence of Weghting Matrices on the Optimal Regulator', Proc. of the 2nd IFAC-Symp. on "Multivariable Technical Control Systems". Düsseldorf, 1971.

/12/ K.P. Sondergeld, 'A Collection of Plant Models and Design Specifications for Robust Control', DFVLR Institute for Flight System Dynamics, Oberpfaffenhofen, 1982.

CHAPTER 16

SENSITIVITY REDUCTION OF THE LINEAR QUADRATIC OPTIMAL REGULATOR

C. Verde, P. M. Frank
Mess- und Regelungstechnik
Universität -GH- Duisburg
Postfach 10 16 29
4100 Duisburg, FRG

ABSTRACT

In this paper the power of using sensitivity theory together with the linear optimal regulator theory for the design of robust fixed state feedback for plants with non-infinitesimal parameter variations is discussed. The plant model is assumed to be given by a family of r state equations due to r sets of values of the parameters under consideration. In order to meet the requirements of robustness with respect to the given finite parameter variations, a fixed insensitive linear quadratic optimal regulator is designed by modification of the Q matrix in the performance index J, such that a certain compromise between an L_2 norm of the trajectory sensitivity vector and J is obtained. As nominal model of the plant a fictive state equation containing averaged values of the parameters is used. It is shown that a constant output feedback for the attitude control of an aircraft F4E can be found so that the pole locations in preassigned regions are maintained for the extreme flight conditions under consideration.

1. INTRODUCTION

A realistic design of the feedback control system has to take into consideration the unavoidable deviations between the parameters of the actual physical process and its nominal mathematical model utilized for the design. These deviations may be caused by identification inaccuracies, manufacturing tolerances, environmental or operational influences, or by mathematical simplification /1/. This problem has long been discussed in modern control system design and many contributions have been published in the last decades /2-3/. In principle two distinct methodologies have been proposed to design a control system which compensates for the effects of parameter deviations and uncertainties. One is to design an adaptive controller, i.e., to sense the parameter variations and adjust the controller accordingly. The other classical though not less efficient approach is to design a fixed controller, so that the overall system is low sensitive to parameter deviations. In recent papers /4-5/, a new conception to design a fixed controller for finite

S. G. Tzafestas (ed.), Multivariable Control, 293–306.

known parameter variations has been developed without using sensitivity theory.

On the other hand, it is important to note that two different terms are used today to characterize a control system which compensates for parameter variations. The term robustness refers to tolerances of large disturbances (lying within specified bounds) and the term sensitivity reflects the tendency of the nominal system to parameter variations. Reference /6/ discusses the relationship between sensitivity and robustness.

One of the most powerful approaches to multivariable feedback system design regarding robustness and being straighforward is the linear optimal regulator (LOR). An excellent description of the LOR is given in /7-8/.

Consider a time-invariant linear system whose state vector x(t) (an n vector), control vector u(t) (an m vector) and output vector y(t) (an l vector) are related by

$$\dot{x}(t) = A(\alpha)\ x(t) + B(\alpha)\ u(t) \tag{1.1}$$

$$x(0) = x_o$$

$$y(t) = C\ x(t),$$

where $A(\alpha)$, $B(\alpha)$ and C are matrices of appropriate dimensions and α is a parameter vector (an j vector) with nominal value α_o.

Minimizing the quadratic performance index

$$J^* = \int_o^\infty (x'Q + u'R\ u)dt \tag{1.2}$$

(where Q is a given (nxn) constant symmetric positive semidefinite matrix and R is a given (mxm) constant symmetric positive definite matrix) leads to the optimal linear control law

$$u^* = u(t) = -K\ x(t) = -R^{-1}B_o'\ P\ x(t), \tag{1.3}$$

where P is the positive semidefinite matrix solution of the matrix Riccati equation

$$A_o'\ P + P\ A_o - P\ B_o\ R^{-1}B_o'\ P + Q = 0 \tag{1.4}$$

with

$$A_o = A(\alpha_o),\quad B_o = B(\alpha_o).$$

It is well known that the LOR guarantees, independent of the weighting matrices Q and R in eq. (1.2), the following robustness properties /9/.

(1) The absolute value of the return difference is larger than 1 for all frequencies.
(2) The gain margin is infinite and the gain reduction margin is 0.5.
(3) The phase margin is at least 60°.

Despite these inherent robustness properties various authors /10-15/ have proposed methods to further reduce the sensitivity of the

LQR. Kreindler /10/, Rao e.a. /11/, Fleming e.a. /12/, Elmetwally e.a. /13/, Byrne e.a. /14/ and Rillings e.a. /15/ augment the quadratic performance index by a term of the trajectory sensitivity vector to reach a certain compromise between state and trajectory sensitivity vectors.

$$J_t = \int_0^\infty (x' Q x + u' R u + \sigma' Q_s \sigma) dt \tag{1.5}$$

Several difficulties arise from this new augmented problem:
(1) The standard LQR design is no longer applicable.
(2) The order of the optimization problem is increased.
(3) In /10,11,12,13/ the resulting control system requires an additional feedback of the trajectory sensitivity vector which implies a substantial increase of the order of the control system.

In contrast Subbayyan e.a. /16/ and Verde e.a. /17/ suggest to simply modify the matrix Q in eq. (1.2), so that a trade off of sensitivity reduction versus increase of the cost is achieved. In this paper the approach proposed in /17/ for the reduction of parameter sensitivity of LQR is used to design a robust fixed controller for the attitude control of an aircraft F4E with horizontal canards. In this example the robustness requirement is given by the regions of the pole locations for four flight conditions.

2. SENSITIVITY REDUCTION BY Q MODIFICATION

Consider as a sensitivity measure the L_2 norm of the trajectory sensitivity functions defined by

$$J_s = \int_0^\infty \sum_{i=1}^{j} (\sigma_i' Q_{si} \sigma_i) dt \tag{2.1}$$

where Q_{si} is a positive definite matrix and σ_i represents the trajectory sensitivity function of the closed loop system with the feedback gain matrix K governed by

$$\dot{\sigma}_i(t) = (A_{\alpha i} - B_{\alpha i} K) x(t) + (A_o - B_o K) \sigma_i(t) \tag{2.2}$$

$$\sigma_i(0) = 0$$

with $A_{\alpha i} = \partial A / \partial \alpha_i \big|_{\alpha_o}$ and $B_{\alpha i} = \partial B / \partial \alpha_i \big|_{\alpha_o}$ for $i = 1, \ldots, j$.

The goal is now to find a gain feedback matrix $\tilde{K}$ generated by LQR design, so that the sensitivity measure eq. (2.1) is reduced, i.e., to determine the condition to satisfy the inequality

$$\int_0^\infty \sum_{i=1}^{j} (\tilde{\sigma}_i' Q_{si} \tilde{\sigma} i) dt < \int_0^\infty \sum_{i=1}^{j} (\sigma_i Q_{si} \sigma_i) dt \tag{2.3}$$

where $\tilde{\sigma}_i$ is the trajectory sensitivity function of the closed-loop system with the feedback matrix $\tilde{K}$. In /3/ it has been demonstrated that for a single input plant with A and B in phase variable cononical form and $B_{\alpha i} = 0$, the condition to satisfy (2.3) is reduced to

$$|1 + \tilde{k}\, \Phi(j\omega)\, b_o|^2 > |1 + k\, \Phi(j\omega)\, b_o|^2 \tag{2.4}$$

where $\Phi(j\omega) = (j\omega I - A_o)^{-1}$.

Using the fact that for an LOR design /7/

$$|1 + k\, \Phi(j\omega)\, b_o|^2 = 1 + b_o\, \Phi(-j\omega)\, Q\, \Phi(j\omega)\, b_o \tag{2.5}$$

the inequality (2.4) implies

$$\tilde{Q} - Q > 0. \tag{2.6}$$

Thus for a completely controllable single input system one can always find two optimal solutions using $\tilde{Q} > Q$ with the property (2.3). However, this property does not generally hold for arbitrary state variable representations.

If the control $\tilde{u} = -\tilde{K}\, x$ is implemented instead of $u^* = -K\, x$, the control system is optimal with respect to $\tilde{Q}$ but suboptimal with respect to Q. The corresponding suboptimal performance index with respect to Q

$$\tilde{J} = \int_0^\infty (x'Q\, x + \tilde{u}'R\, \tilde{u})dt \tag{2.7}$$

becomes

$$\tilde{J} = x_o'\, F\, x_o \tag{2.8}$$

where F is the solution of the matrix Liapunov equation

$$(A_o - B_o\, \tilde{K})'F + F(A_o - B_o\, \tilde{K}) = -Q - \tilde{K}'R\, \tilde{K} \tag{2.9}$$

If $\tilde{Q} - Q$ is a positive definite matrix

$$\tilde{J} > J^* \quad .$$

In other words, the matrix $\tilde{Q}$ causes a sensitivity reduction but at the same time an increase in the cost J.

The question is how to choose a suitable matrix $\tilde{Q}$ that yields a satisfactory sensitivity reduction in terms of $\tilde{J}_S$, eq. (2.1) without a substantial increase of the performance index $\tilde{J}$, eq. (2.7).

Subbayyan e.a. /16/ have suggested to choose the matrix $\tilde{Q} = \beta Q$ and to determine a value of $\beta > 1$, such that a desired compromise is obtained. However, this method restricts themself to an equal increase of all state variables which limits the achievements of the sensitivity reduction. Another possiblility is to take into consideration the sensitivity of each state variable for the change of the Q matrix, i.e., to find a positive semidefinite matrix S that represents a measure of the sensitivity of the state variables and then use S as an increment of Q, thus penalizing the state variables due to their sensitivity. This means that the matrix Q is changed due to

$$\tilde{Q} = Q + \beta S \tag{2.10}$$

where β determines the strength of the change in $\tilde{Q}$. To find the matrix S the following procedure is proposed:
First determine the sensitivity measure L_2 with the optimal control u

$$J_s = x_o' H_{11} x_o$$

with

$$H_{11} = \sum_{i=1}^{j} H_{11i} \tag{2.11}$$

where the matrix H_{11} is a submatrix of the matrix

$$H_i = \begin{bmatrix} H_{11i} & H_{12i} \\ H_{21i} & H_{22i} \end{bmatrix} \tag{2.12}$$

which can be found as a solution of the matrix Liapunov eq.

$$A'_{mi} H_i + H_i A_{mi} + W_i = 0 \tag{2.13}$$

with

$$A_{mi} = \begin{bmatrix} A_o - B_o K & 0 \\ A_{\alpha i} - B_{\alpha i} K & A_o - B_o K \end{bmatrix}, \tag{2.14}$$

$$W_i = \begin{bmatrix} 0 & 0 \\ 0 & Q_{si} \end{bmatrix}. \tag{2.15}$$

From eq. (2.11) it is seen that the elements of the positive semidefinite matrix H_{11} define the influence of the corresponding state variable upon the sensitivity index J_s. From this it follows that the matrix H_{11} can be interpreted as a measure of the sensitivity in form of a matrix and can be used to determine the increment of the Q matrix, i.e.

$$S \triangleq H_{11} \tag{2.16}$$

In order to prove this statement in a mathematical way, consider the stable controllable system which for simplicity of interpretation and calculation, may be given in Jordan form

$$\dot{x}^*(t) = \lambda(\alpha)\, x^*(t) \tag{2.17}$$

$$x^*(0) = x_o^*$$

Therefore, for a single parameter α, the trajectory sensitivity equation becomes

$$\dot{\sigma}^*(t) = \lambda_\alpha\, x^*(t) + \lambda(\alpha_o)\sigma^*(t) \tag{2.18}$$

$$\sigma^*(0) = 0$$

with

$$\lambda_\alpha = \mathrm{diag}\left[\frac{\partial\lambda_i}{\partial\alpha}\bigg|_{\alpha_o}, \ldots, \frac{\partial\lambda_n}{\partial\alpha}\bigg|_{\alpha_o}\right] . \tag{2.19}$$

Solving the differential eq. (2.18), we find that

$$\sigma^*_i(t) = \lambda_{\alpha i}\{\exp(\lambda_i t) \int_o^\infty \exp(-\lambda_i\zeta)x^*_i(\zeta)d\zeta\} \tag{2.20}$$

for i = 1,...,n.
This equation shows that for a fixed eigenvalue λ_i the sensitivity of the variable x^*_i depends only on the partial derivative $\lambda_{\alpha i}$.

Now assuming that in the L_2 norm, eq. (2.1),

$$Q_s = \mathrm{diag}\left[q_1, q_2, \ldots, q_n\right] , \tag{2.21}$$

it follows that the sensitivity measure, eq. (2.1), is given by

$$J_s = x^{*\prime}_o H^*_{11} x^*_o \tag{2.22}$$

where

$$H^*_{11} = -\frac{1}{4}\,\mathrm{diag}\left[q_1\frac{\lambda^2_{\alpha 1}}{\lambda^3_1}, \ldots, q_n\frac{\lambda^2_{\alpha n}}{\lambda^3_n}\right] . \tag{2.23}$$

Substituting H^*_{11} in J_s yields

$$J_s = -\frac{1}{4}\sum_{i=1}^{n} q_i(x^*_i(0)\lambda_{\alpha i})^2/\lambda^3_i . \tag{2.24}$$

From eq. (2.24) it is evident that the coefficient q_i determines the weight of the trajectory sensitivity component i to be considered in the L_2 norm measure; it also follows that it is not necessary to solve the differential equation (2.18) to know the sensitivity influence of the variable x_i. The term $h^*_{11}(ii) = \lambda^2_{\alpha i}/\lambda^3_i$ for a fixed eigenvalue preserves the sensitivity information. This means the elements of the matrix H_{11} may be regarded as a measure of the sensitivity of the state variables.

2.1 EXAMPLE

In order to illustrate the efficiency of the described procedure to reduce the sensitivity by the Q modification a comparison of Subbayyan's method with the new method is made. Considered is the third order system treated by Kreindler /18/. For the design the value of the weighting matrix Q proposed by Kreindler is taken as an initial value in the compared methods.

The feedback matrix K and the matrix Q obtained after nine iterations i for the two procedures are given in Table 1. This table also shows the maximal value of the control u for $x_o' = (1.\ 0.\ 0.)$.

Method	Subbayyan	Verde
Algorithm	$Q_{i+1} = Q_i + 0.3\ Q_o$	$Q_{i+1} = Q_i + 0.3\ H_{11}$
Matrix K	1.9 4.6 -2.1	0.4 3.1 -1.6
u_{max} for $x_o' = (1\ 0\ 0)$	1.9	0.4
Matrix Q	$\begin{bmatrix} 400 & 0 & 0 \\ 0 & 20 & 0 \\ 0 & 0 & 0 \end{bmatrix}$	$\begin{bmatrix} 156 & 3.7 & .48 \\ 3.7 & 8.7 & -.2 \\ .48 & -.2 & .3 \end{bmatrix}$

Table I: Results obtained after nine iterations in Q according to Subbayyan and Verde procedure

The values of the cost integral rate

$$\bar{J}^*(i) = \int_o^\infty (x_i' Q_o x_i + u_i' R u_i) dt / J^*(0) \tag{2.25}$$

and the values of the sensitivity measure rate

$$\bar{J}_s(i) = \int_o^\infty (\sigma_i' Q_s \sigma_i) dt / J_s(0) \tag{2.26}$$

for the nine iterations are plotted in Fig. 1. A plot of the root-locus for a variation of Q for both methods is given in Fig. 2.

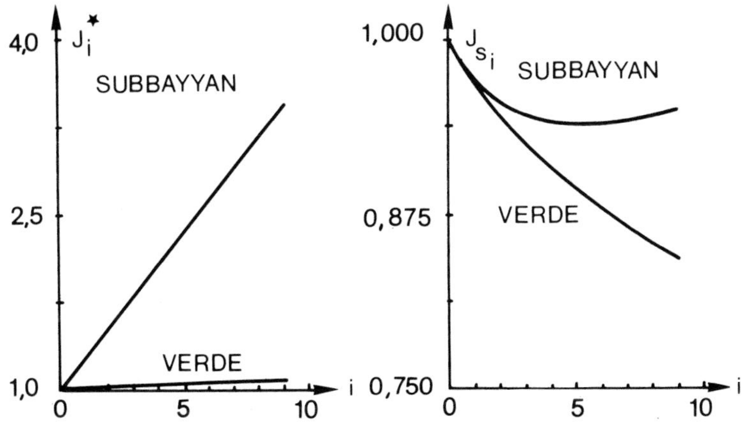

Fig. 1: Variations of the rates for the compared methods

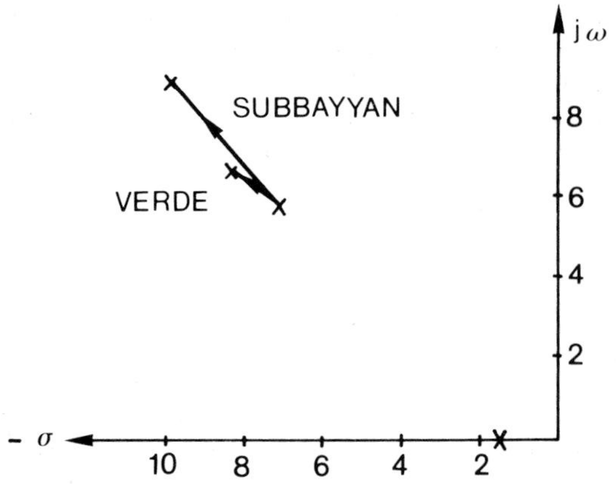

Fig. 2: Root-locus plot as Q is varied for the compared methods

It is evident that the new procedure provides a larger reduction in the sensitivity rate with lower increase in the cost rate and smaller variations of the closed-loop poles than in Subbayyan's approach.

3. EXTENSION FOR A FAMILY OF PLANT EQUATIONS

The basic concept used in the approach introduced above is the sensitivity function which is valid for a first order approximation at small parameter variations. Nevertheless, we shall show that this sensitivity approach can also be a quite useful design procedure to achieve robustness properties for finite parameter variations.

Assume that in the controllable system, eq. (1.1) the parameter vector $\alpha \varepsilon R^J$ may have some typical values

$$\alpha = \alpha(i) \qquad i = 1,\ldots,r \;, \tag{3.1}$$

according to different operating points of the system. Then the matrices $A(\alpha)$ and $B(\alpha)$ have to be replaced by a set of r matrices

$$A(i) = A(\alpha(i)),\; B(i) = B(\alpha(i)) \text{ for } i = 1,\ldots,r \tag{3.2}$$

and the system, eq. (1.1) can be replaced by a family of plants

$$\dot{x}(t) = A(i)\, x(t) + B(i)\, u(t) \quad \text{for } i = 1,\ldots,r \tag{3.3}$$

$$x(0) = x_o$$

$$y(t) = C\, x(t) \quad .$$

The design objective is to find a fixed feedback $u = -K\,x$ that satisfies the following requirements:

(1) it assures the stability of the system for all operating points

(2) it maintains desired system specification in preassigned limits (step-response boundary, region in the eigenvalue plane, etc.).

To use the sensitivity approach it is necessary to build a nominal plant and the corresponding sensitivity model. Then as a nominal system we propose to define the average value over all sets of plant parameters (denoted "fictitious plant"). This means

$$A_o = \frac{1}{r}\sum_{i=1}^{r} A(i) \ , \quad B_o = \frac{1}{r}\sum_{i=1}^{r} B(i) \ . \tag{3.4}$$

For the sensitivity model it is assumed that the parameter vector α consists of the variable coefficients of the matrices A(i) and B(i).

The feedback matrix K can now be calculated with the aid of the sensitivity approach. It is important to note that the approach has only to assert the stability of the nominal or fictive system. Consequently it is necessary, after the calculation of the feedback matrix K, to prove the stability of all the family members. If the robustness requirements are not fulfilled it is necessary to change again the weighting matrix Q according to eq. (2.10). This can be easily done with the aid of a digital computer by an iterative procedure.

3.1 EXAMPLE

Consider the short period longitudinal mode of the flight control system of an airplane F4E with canards treated by Ackermann /4/. The linearized state equations are given by

$$\dot{x}(t) = A\, x(t) + b\, u(t) \tag{3.5}$$

$$y(t) = c\, x(t) \tag{3.6}$$

with the state vector and input defined as

$$x(t) = \begin{bmatrix} N_z & q & \delta_e \end{bmatrix}' \tag{3.7}$$

N_z normal acceleration

q pitch rate

δ_e deviation of elevator deflection

u elevator deflection command

and the matrices

$$A = \begin{bmatrix} a_{11} & a_{12} & a_{13} \\ a_{21} & a_{22} & a_{23} \\ 0 & 0 & -14 \end{bmatrix} \qquad b = \begin{bmatrix} b_1 \\ 0 \\ 14 \end{bmatrix}$$

$$c = \begin{bmatrix} 1 & 12.46 & 0 \end{bmatrix} .$$

The corresponding values of the parameters for four different flight conditions and for the fictitious nominal case are given in Table II. The allowed regions of the closed loop eigenvalues of the short period mode are specified by military requirements for flying qualities of piloted airplanes and are shown in Fig. 3.

Mach Alt.	Case 1 0.5 5000'	Case 2 0.85 5000'	Case 3 0.9 35000'	Case 4 1.5 35000'	nominal values
a_{11}	-0.9896	-1.702	-0.6607	-0.5162	-0.967
a_{12}	17.41	50.72	18.11	26.96	28.3
a_{13}	96.15	263.5	84.34	178.9	155.
a_{21}	0.2648	0.220	0.0820	-0.689	-0.03
a_{22}	-0.8512	-1.418	-0.6587	-1.225	-1.038
a_{23}	-11.89	-31.99	-10.81	-30.38	-21.14
b_1	-97.78	-272.2	-85.09	-175.6	-157.0
λ_1	1.23	1.78	0.56	-0.87 $\pm$	-1.0 $\pm$
λ_2	-3.07	-4.9	-1.87	j [illegible]3	j 2.9
λ_3	-14.0	-14.0	-14.0	-14.0	-14.0

Table II: Aerodynamic data and eigenvalues of the system

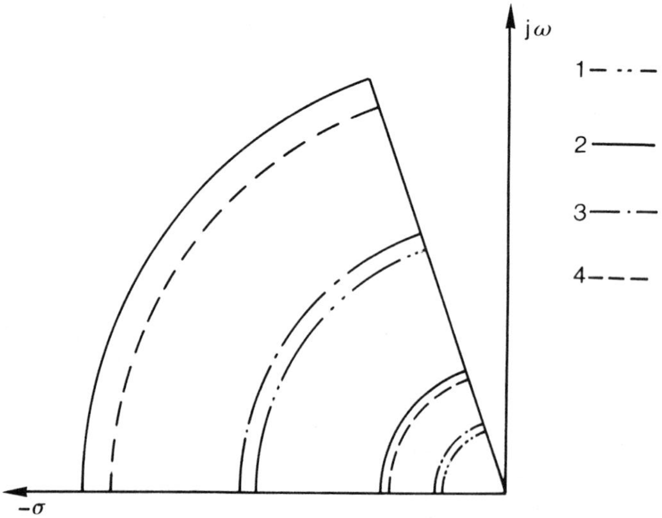

Fig. 3: Preassigned pole regions for the four operating points

From Table II it is seen that there are seven variable coefficients in the model. Considering these coefficients as the parameters of interest, the parameter vector is defined by

$$\alpha = \begin{bmatrix} a_{11} & a_{12} & a_{13} & a_{21} & a_{22} & a_{23} & b_1 \end{bmatrix} . \qquad (3.8)$$

If a total sensitivity vector

$$\sigma = \sum_{i=1}^{j} \sigma_i \qquad (3.9)$$

is introduced, the sensitivity eq. (2.2) becomes

$$\dot{\sigma}(t) = (A_\alpha - B_\alpha K)\, x(t) + (A_o - B_o K)\, \sigma(t) \qquad (3.10)$$

$$\sigma(0) = 0$$

where summations of the partial derivatives of the matrices A and b with respect to the parameters of the vector α are written as

$$A_\alpha = \sum_{i=1}^{7} \frac{\partial A}{\partial \alpha_i} = \begin{bmatrix} 1 & 1 & 1 \\ 1 & 1 & 1 \\ 0 & 0 & 0 \end{bmatrix} ; \quad b_\alpha = \sum_{i=1}^{7} \frac{\partial b}{\partial \alpha_i} = \begin{bmatrix} 1 \\ 0 \\ 0 \end{bmatrix} . \qquad (3.11)$$

The starting weighting matrix Q for the state x, the weighting matrix R for the input u in the quadratic performance index, eq. (1.2), and the weighting matrix Q_s in the L_2 norm of sensitivity functions, eq. (2.1), are chosen as

$$Q = \begin{bmatrix} 0.4 & 0 & 0 \\ 0 & 28 & 0 \\ 0 & 0 & 0.4 \end{bmatrix} ; \quad Q_s = \begin{bmatrix} 2 & 0 & 0 \\ 0 & 2 & 0 \\ 0 & 0 & 2 \end{bmatrix} \quad ; \quad r = 100.$$

Note that it is no longer possible to choose different weighting matrices Q_{si} if the total sensitivity vector σ is used. For the design it is assumed that the deviation of the elevator deflection δ_e is not easy to measure and is therefore not used for feedback. On account of this constraint in the form of the control u(t), the solution of the matrix Riccati equation can not be used to determine the feedback gain K. Therefore the Levine & Athans algorithm /19/ was implemented to minimize J for output feedback.

After 14 iterations (i.e. 14 changes of Q according to eq. (2.10)) with $\beta = 1$ (and Levine e.a. algorithm) the following feedback matrix was obtained

$$\tilde{K} = \begin{bmatrix} -0.1174 & -0.8812 & 0.0 \end{bmatrix} .$$

This matrix was found minimizing J with the weighting matrix

$$\tilde{Q} = \begin{bmatrix} 1.68 & 2.31 & 10.9 \\ 2.31 & 35.9 & 18.7 \\ 10.9 & 18.7 & 95.0 \end{bmatrix} .$$

The resulting pole locations for the four flight conditions are shown in Fig. 4. Note here that the robustness requirement is achieved for the four cases. In order to illustrate the quality of the system responses numerical simulations for the four cases have been made. The step responses y(t) and the step responses normalized upon their stationary values ($y^*(t) = y(t)/y(\infty)$) are respectively plotted in Fig. 5 and 6. It is seen, that the settling time is

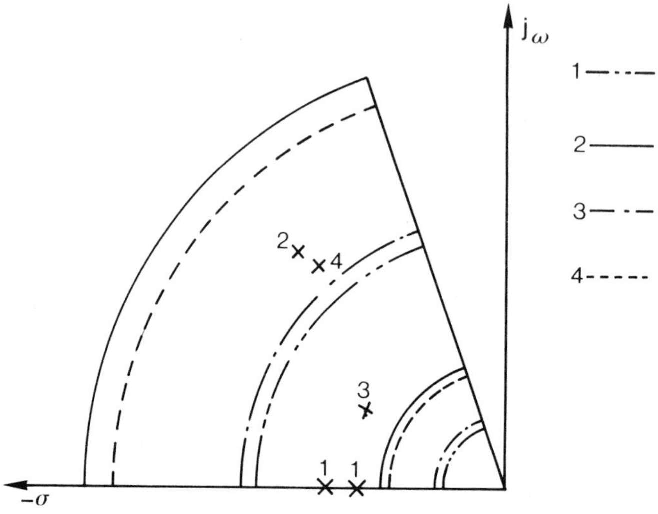

Fig. 4: Closed-loop eigenvalues of the short period mode for the four flight conditions

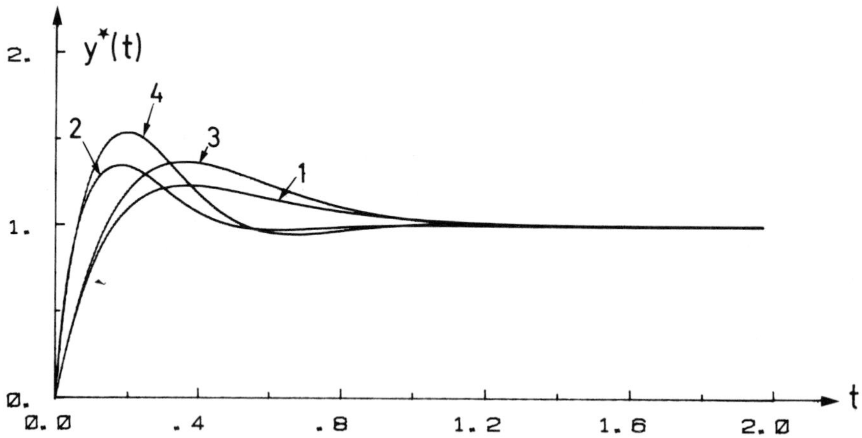

Fig. 5: Step responses $y^*(t) = y(t)/y(\infty)$ for the four flight conditions

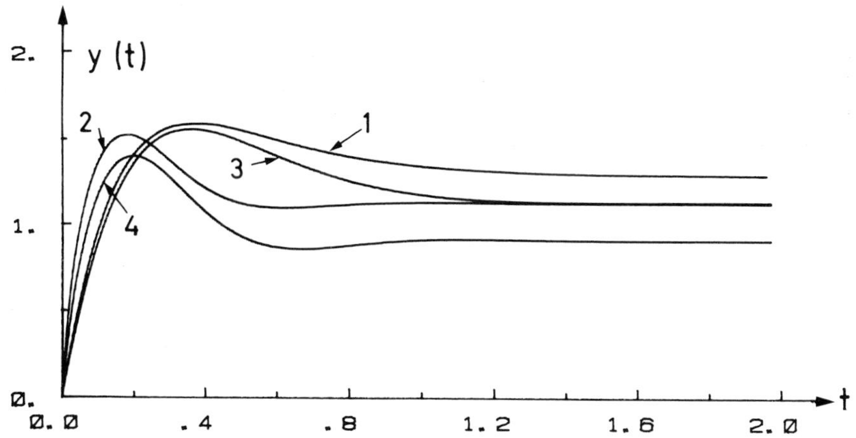

Fig. 6: Step responses y(t) for the four flight conditions

similar for all the cases and the overshoot is only considerable for the cases 2 and 4. This is evident because the complex conjugate poles in these cases have smaller damping rate. It is important to remark from Fig 6 that the system can not follow the step function input with zero steady-state error because one considered only the sensitivity of the state regulator problem. However, this is not a limitation. Encouraging results were also obtained for this example using Davison structure /20/ together with the described procedure (see /3/).

4. CONCLUSIONS

This paper discusses the possibility to design robust control systems at finite parameter variations using sensitivity theory together with the theory of linear optimal regulator. It is shown that the state weighting matrix Q in an LQR design can be used to adjust the sensitivity of the closed-loop system. A generalized procedure for the determination of the matrix Q is proposed. This procedure allows to achieve a desired compromise between sensitivity reduction on the one side and the increase of cost on the other side while retaining the adventages of the LQR design. It should be noticed that the method is not restricted to complete state variable feedback. The results obtained from simulation of an aircraft attitude control show the usefulness of the described design method.

5. REFERENCES

/1/ P.M. Frank, Introduction to System Sensitivity Theory, Academic Press, New York, San Francisco, London, 1978.

/2/ Y.N. Andreev, 'Algebraic Methods of State Space in Linear Object Control Theory';(Survey of Foreign Literature), Translated from Avtomatika i Telemekhanika, No. 3 pp 3-50, March 1977. UDC-62-501.12

/3/ M.C. Verde Rodarte, 'Empfindlichkeitsreduktion bei linearen optimalen Regelungen'. Doctoral Thesis, Univ.-GH-Duisburg, FRG, 1983.
/4/ J. Ackermann, 'Robust Control System Design', Lecture Series No. 109 Fault Tolerance Design and Redundancy Management Techniques, North Atlantic Treaty Organization.
/5/ F. Heger; P.M. Frank, 'Computer-Aided Pole Placement for the Design of Robust Control Systems', IFAC Symp. on CAD of Multivariable Technological Systems, West Lafayette, Indiana USA, 1982.
/6/ P.M. Frank, 'Robustness and Sensitivity: A Comparison of two methods', ACI, IASTED Symposium Copenhagen, Denmark, 1983.
/7/ B.D.O. Anderson; J.B. Moore, Linear Optimal Control; Prentice Hall, Inc. 1971.
/8/ H. Kwakernaak; R.Sivan, Linear Optimal Control Systems; Wiley Interscience, New York, 1972.
/9/ M.G. Safonov; M. Athans, 'Gain and Phase Margin for Multiloop LQR Regulators', IEEE Trans. Autom. Contr., Vol. AC-22, No. 2, April 1977.
/10/ E. Kreindler, 'On minimization of Trajectory Sensitivity', Int. J. Contr., Vol. 8 No. 1, pp 89-96, 1968.
/11/ S.G. Rao; A.C. Soudack, 'Synthesis of Optimal Control Systems with Near Sensitivity Feedback', IEEE Trans. Autom. Contr., Vol. AC-16, No. 2, pp 194-196, 1971.
/12/ P.J. Fleming; M.M. Newman, 'Design Algorithms for a Sensitivity Contrained Suboptimal Controller', Int. J. Contr., 25 No. 6, pp 965-978, 1977.
/13/ M.M. Elmetwally; N.D. Rao, 'Design of Low Sensitivity Optimal Regulators for Synchronous Machines', Int. J. Contr., 19 No. 3, pp 593-607, 1974.
/14/ P.C. Bryne; M. Burke, 'Optimaization with Trajectory Sensitivity Considerations', IEEE Trans. Autom. Contr., Vol. AC-21, pp 282-283, 1976.
/15/ J.H. Rillings; R.J. Roy, 'Analog Sensitivity Design of Saturn V Launch Vehicle; IEEE Trans. Autom. Contr., Vol. AC-15, No. 4, pp 437-442, 1970.
/16/ R. Subbayyan; V.V. Sarma; M.C. Vaithilingam, 'An Approach for Sensitivity Design of Linear Regulators', Int. J. Syst. SCL, Vol. 9, No. 4, pp 65-74, 1978.
/17/ C. Verde; P.M. Frank, 'A Design Procedure for Robust Linear Suboptimal Regulators with Preassigned Trajectory Sensitivity', 21st IEEE Conf. on Decision and Contr., Orlando, Florida, 1982.
/18/ E. Kreindler, 'Closed-Loop Sensitivity Reduction of Linear Optimal Control', Vol. AC-13 No. 3, June 1968.
/19/ W.S. Levine; M. Athans, 'On the Determination of the Optimal Constant Output Feedback Gains for Linear Multivariable Systems', IEEE Trans. Autom. Contr., Vol. AC-15, No. 1, February 1970.
/20/ E.J. Davison, 'The Output Control of Linear Time-Invariant Multivariable Systems with Unmeasureable Arbitrary Disturbances', IEEE Trans. Autom. Contr., Vol. AC-17, No. 5, October 1972.

CHAPTER 17

DESIGN OF LOW-ORDER DELAYED MEASUREMENT OBSERVERS FOR DISCRETE TIME LINEAR SYSTEMS

P. Stavroulakis
Mediterranean College
Athens, Greece

S. G. Tzafestas
Patras University
Patras, Greece

ABSTRACT: This paper considers the development of a discrete-time low order delayed-measurement observer for a discrete time-invariant linear system. The low-order delayed measurement observer developed has several unique features. It utilizes discrete time delayed measurements as part of its inputs and it is an rth order observer for an nth order linear system with q linearly independent outputs where r<n-q. The purpose of this observer is to implement the control directly without estimating the state first. It is shown that under certain conditions the dimension of the observer is much lower than the standard observer of dimension n-q. The procedure employed here presents an approach which can be very important in large scale control implementation.

1. INTRODUCTION

The design and implementation of a linear optimal control system using state techniques often requires the availability of all state variables associated with the system [1-4]. However, in practice, not all state variables in a system are accessible for direct measurement nor is it economical to measure all state variables directly. Furthermore, the measurement data may be contaminated by measurement errors. The design problem caused by the unknown and/or inaccessible state riables may be overcome by replacing the unknown and/or inaccessible state variables by their estimated values. For linear systems, there are two well known filters which may be used to generate estimates of the unknown and/or inaccessible state variables, namely, the Kalman-Bucy filter [5] and the Luenberger observer [6-7]. The Kalman-Bucy filter uses noise contaminated measurements to reconstruct estimates of the unkown and/or inaccessible state variables. The order of the filter is the same as the order of the associated system. On the other hand, when the measurements are perfect, i.e. contain no measurement

S. G. Tzafestas (ed.), Multivariable Control, 307–326.

errors, and there are no random disturbances acting on the system, a Luenberger observer may be used to generate the desired estimates of the unknown and/or inaccessible state variables. The order of a Luenberger observer is generally less than that of the associated system [6-9]; specifically, the n-m unknown and/or inaccessible state variables of an nth order linear system with m linearly independent outputs may be constructed by a minimal-order observer of order n-m (see in particular, [9]).

Since the pioneering work of Luenberger [6-7], observer theory has been studied extensively in the literature [10-21], where in [14-21], observer theory has been extended to stochastic systems. Observer theory has also played an important role in the design of disturbance accommodating control systems [22-26], where various minimal-order Luenberger observers have been developed to provide estimates of various unknown system disturbances which either have a specific waveform structure or can be approximated by a specific waveform structure.

In this chapter, a new reduced-order observer for discrete-time linear systems will be developed. The observer developed utilizes time delayed measurements to generate estimates of an unknown input function which depends linearly on the inaccessible state variables and will be called a low-order-*delayed-measurement observer*. The delayed-measurement observer has an important feature: it is an rth order observer, $r<(n-m)$, for an nth order system with m linearly independent outputs. Hence, the dimension of a delayed-measurement observer is lower than that of the corresponding (n-m)-minimal-order Luenberger observer. Furthemore, r may be varied and may be reduced by using more delayed-measurements. When enough delayed measurements are used, the delayed-measurement observer becomes a pseudo-observer or an observer with no dynamics which reconstructs the present values of the unknown input function (control) instantaneously. Up to now, control via state estimation using delayed measurements have been considered in [27-30]. In [27-29], the results obtained correspond to a pseudo-observer, and in [30], microprocessor implementations of a delayed-measurement observer and a pseudo-observer in an actual optimal control system are carried out. We shall show that a similar formulation could be implemented to derive an observer of reduced order for the estimation of the control variables directly.

2. STATEMENT OF THE PROBLEM

Consider a discrete-time linear system described by

$$x(k+1)=Ax(k)+Bu(k),\ x(0)=x_0 \tag{1}$$

with measurements given by

$$y(k)=Hx(k)=\left[H_{11} \,\vdots\, 0_{m\times(n-m)}\right]x(k), \tag{2}$$

where $x(k)\varepsilon R^n$, $u(k)\varepsilon R^p$, $y(k)\varepsilon R^m$ are, respectively, the state, input and output vectors; A, B, H and H_{11} are, respectively, nxn, nxr, nxn and mxm matrices, and H_{11} is assumed to be nonsingular*.

From (1) and (2), we obtain by using d time delayed measurements with k_d,

$$\begin{bmatrix} y(k) \\ y(k-1) \\ \vdots \\ y(k-d) \end{bmatrix} = \begin{bmatrix} H \\ HA^{-1} \\ \vdots \\ HA^{-d} \end{bmatrix} x(k) - \begin{bmatrix} 0_{mxr} & 0_{mxr} & \ldots & 0_{mxr} \\ 0_{mxr} & HA^{-1}B & \ldots & 0_{mxr} \\ \vdots & & & \vdots \\ 0_{mxr} & HA^{-d}B & \ldots & HA^{-1}B \end{bmatrix} \begin{bmatrix} u(k) \\ u(k-1) \\ \vdots \\ u(k-d) \end{bmatrix} \tag{3}$$

where A is assumed to be invertible**.

Defining

$$y_d^{*T}(k) \stackrel{\Delta}{=} \left[y^T(k-1)\, y^T(k-2) \ldots y^T(k-d)\right] \tag{4}$$

$$u_d^T(k) \stackrel{\Delta}{=} \left[u^T(k-1)\, u^T(k-2) \ldots u^T(k-d)\right] \tag{5}$$

$$W_d^T \stackrel{\Delta}{=} \left[H^T \,\vdots\, (A^{-1})^T H^T \,\vdots\, \ldots \,\vdots\, (A^{-d})^T H^T\right] \tag{6}$$

$$B_d \stackrel{\Delta}{=} \begin{bmatrix} 0_{mxr} & 0_{mxr} & \ldots & 0_{mxr} \\ 0_{mxr} & HA^{-1}B & \ldots & 0_{mxr} \\ \vdots & \vdots & & \vdots \\ 0_{mxr} & HA^{-d}B & \ldots & HA^{-1}B \end{bmatrix} \tag{7}$$

we obtain from (3)-(7)

---

* Without loss of generality, H may be assumed to be of this form.

** The matrix A is invertible if (1) is the discretized version of a continuous-time system, since in that case, A is a nonsingular transition matrix.

$$\begin{bmatrix} y(k) \\ \cdots \\ y_d^*(k) \end{bmatrix} = W_d x(k) - B_d \begin{bmatrix} u(k) \\ \cdots \\ u_d(k) \end{bmatrix} \tag{8}$$

where $y_d^*(k)$ is an md-dimensional delayed measurement vector, $u_d(k)$ is an pd-dimensional delayed input vector; W_d and B_d are, respectively, $m(d+1)xn$ and $m(d+1)xr(d+1)$ matrices. Equation (8) yields

$$\begin{bmatrix} y(k) \\ \cdots \\ y_d(k) \end{bmatrix} \begin{bmatrix} y(k) \\ \cdots \\ y_d^*(k) \end{bmatrix} + B_d \begin{bmatrix} u(k) \\ \cdots \\ u_d(k) \end{bmatrix} . \tag{9a}$$

$$= W_d x(k), \tag{9b}$$

where $y_d(k)$ as defined is known for all $k \geq d$, since the right-side of (9a) is known for all $k \geq d$. We note that rank $[W_d] = q$, where $m \leq q \leq n$ and $q \leq m(d+1)$. Furthermore, if rank $[W_d] = n$, then W_d becomes a constructibility matrix [31] and is equivalent to an observability matrix when A is nonsingular [34]-[35]. If rank $[W_d] = n$, then (9) may be written as

$$\begin{bmatrix} y(k) \\ \cdots \\ y_d(k) \end{bmatrix} = \begin{bmatrix} H_{11} & \vdots & 0_{mx(n-m)} \\ \cdots & \vdots & \cdots \\ M_{21} & \vdots & M_{22} \end{bmatrix} \begin{bmatrix} x_m(k) \\ \cdots \\ x_{n-m}(k) \end{bmatrix}, \tag{10}$$

In [31], it is shown that if rank $(W_d) = n$, the entire state $\underline{x}(k)$ can be estimated using the known measurement vector

$$\hat{\underline{y}}(k) = \begin{bmatrix} y(k) \\ \cdots \\ y_d(k) \end{bmatrix}.$$

If rank $[W_d] = q < n$, it is further shown that an observer $\underline{z}(k)$ of dimension $(n-q)$ can be used to find an estimate $\hat{\underline{x}}(k)$ of $\underline{x}(k)$ where

$$\underline{x}(k) = P\underline{z}(k) + V\hat{\underline{y}}(k) \tag{11a}$$

and

$$\underline{z}(k+1) = F\underline{z}(k) + G\hat{\underline{y}}(k) + D\underline{u}(k) \tag{11b}$$

$$\underline{z}(0)=T\underline{\alpha} \tag{11c}$$

such that F,G,D,P,V,T and $\underline{\alpha}$ are appropriately chosen.

In many (control/estimation) problems, however what is really desired to generate is a linear combination of the state components, i.e.

$$\underline{u}(k)=K\underline{x}(k) \tag{11d}$$

where K is a constant (pxn) matrix and p is usually less than n. The objective of the development that follows is the estimation of u(k) *directly* using the measurement $\hat{\underline{y}}(k)$ and an observer $\underline{z}(k)$ which is of dimension r such that $r<n-q$ and

$$\underline{u}(k)=M\underline{z}(k)+\hat{H}\hat{\underline{y}}(k) \tag{12}$$

Consider (1) and (9). Since we have assumed rank $[W_d]=q$ and the dimensions of W_d are m(d+1)xn, there are m(d+1)-q redundant or linearly dependent output variables in the md-vector $y_d(k)$. Upon collecting (q-m) linearly independent output variables in $\underline{y}_d(k)$, say

$$\underline{y}_d(k)=\begin{bmatrix}\underline{y}_{1d}(k)\\ \\ \underline{y}_{2d}(k)\end{bmatrix}\begin{matrix}q-m\\ \\ md-q-m\end{matrix}$$

equation (9) becomes

$$\begin{bmatrix}\underline{y}(k)\\ \underline{y}_{1d}(k)\\ \underline{y}_{2d}(k)\end{bmatrix}=\left[\begin{array}{c|c|c}H_{11} & 0 & 0\\ \hline W_{11A} & W_{11B} & W_{11c}\\ \hline W_{21A} & W_{21B} & W_{21c}\end{array}\right]\begin{bmatrix}\underline{x}_m(k)\\ \underline{x}_{q-m}(k)\\ \underline{x}_{n-q}(k)\end{bmatrix} \tag{13}$$

If we use the nonsingular transformation

$$T=\left[\begin{array}{c|c|c}I_n & 0 & \\ \hline 0 & I_{q-m} & -W_{11B}^{-1}W_{11c}\\ \hline 0 & 0 & I_{n-q}\end{array}\right] \tag{14}$$

and set $\underline{x}(k)=T^{-1}\underline{x}(k)$, then equations (1) and (13) become

$$\tilde{\underline{x}}(k)=T^{-1}AT\tilde{\underline{x}}(k)+T^{-1}B\underline{u}(k) \tag{15}$$

$$\begin{bmatrix} \underline{y}(k) \\ \underline{y}_{1d}(k) \\ \underline{y}_{2d}(k) \end{bmatrix} = \begin{bmatrix} W_{11} & 0 \\ \hline W_{21} & 0 \end{bmatrix} \begin{bmatrix} \underline{x}_1(k) \\ \hline \underline{x}_2(k) \end{bmatrix} \tag{16}$$

Since T is nonsingular, we could have assumed that we started with the system of eq. (15) therefore we can drop the "$\sim$" from $\underline{x}(k)$.
Hence we can write in general

$$\begin{bmatrix} \underline{x}_1(k+1) \\ \hline \underline{x}_2(k+1) \end{bmatrix} = \begin{bmatrix} A_{11} & \vdots & A_{12} \\ \hline A_{21} & \vdots & A_{22} \end{bmatrix} \begin{bmatrix} x_1(k) \\ \hline x_2(k) \end{bmatrix} + \begin{bmatrix} B_{11} \\ \hline B_{21} \end{bmatrix} \underline{u}(k) \tag{17}$$

and

$$\begin{bmatrix} \underline{y}_1(k) \\ \hline \underline{y}_2(k) \end{bmatrix} = \begin{bmatrix} W_{11} & \vdots & 0 \\ \hline W_{21} & \vdots & 0 \end{bmatrix} \begin{bmatrix} \underline{x}_1(k) \\ \hline \underline{x}_2(k) \end{bmatrix} \tag{18}$$

where

$$x_1(k) \quad \left.\begin{bmatrix} \underline{x}_m(k) \\ \\ \underline{x}_{q-m}(k) \end{bmatrix}\right\} q \qquad x_2(k) = \left[\underline{x}_{n-q}(k)\right]$$

$$W_{11} \quad \begin{bmatrix} W_{11} & \vdots & 0 \\ \hline W_{11A} & \vdots & Q_{11B} \end{bmatrix}$$

$$W_{21} \quad \left[W_{21A} \vdots W_{21B}\right] \quad \text{and} \quad \underline{y}_1(k) = \begin{bmatrix} \underline{y}(k) \\ \underline{y}_{1d}(k) \end{bmatrix}$$

$$\underline{y}_2(k) = \left[\underline{y}_{2d}(k)\right]$$

From (18) we observe that

$$\begin{aligned} \underline{y}_1(k) &= W_{11}\underline{x}_1(k) = \left[W_{11} \vdots 0\right] \underline{x}(k) \\ \underline{y}_2(k) &= W_{21}x_1(k) = \left[W_{21} \vdots 0\right] \underline{x}(k) \end{aligned} \tag{19}$$

Since W_{11} is nonsingular (we have assumed both H_{11} and W_{11B}

nonsingular) we can write

$$\underline{x}_1(k)=W_{11}^{-1}\underline{y}_1(k)$$
$$\underline{y}_2(k)=W_{21}W_{11}^{-1}\underline{y}_1(k) \tag{20}$$

Thus $\underline{y}_2(k)$ and $\underline{y}_1(k)$ are *not* independent but linearly related. Equation (12) then becomes

$$\underline{u}(k)=M\underline{z}(k)+H_1\underline{y}_1(k) \tag{21}$$

where

$$H=\left[H_{11}+H_{12}W_{21}W_{11}^{-1}\right] \quad \text{and} \quad \hat{H}=[\overbrace{H_{11}}^{q} \,\vdots\, H_{12}]$$

3. LOW-ORDER-DELAYED MEASUREMENT OBSERVER

Assume that the control is

$$\underline{u}(k)=K\underline{x}(k)+\underline{e}(k) \tag{22}$$

where $\underline{e}(k)$ is the error due to the constraint that only $\underline{y}_1(k)$ is available.
Consider the possibility of designing the controller so that

$$\underline{e}(k)=M\underline{v}(k) \tag{23}$$

where $\underline{v}(k)$ has dimension r and satisfies the equation

$$\underline{v}(k+1)=F\underline{v}(k) \tag{24}$$

If we set

$$\underline{z}(k)=\underline{v}(k)+W\underline{x}(k) \tag{25}$$

then (25) becomes

$$\underline{z}(k+1)=\underline{v}(k+1)+W\underline{x}(k+1) \tag{26}$$

Equations (24) and (1), substituted into (26), yield

$$\underline{z}(k+1)=F\underline{v}(k)+WA\underline{x}(k)+WB\underline{u}(k)$$
$$=F(\underline{z}(k)-W\underline{x}(k))+WA\underline{x}(k)+WB\underline{u}(k)$$

Hence

$$\underline{z}(k+1)=F\underline{z}(k)+(WA-FW)\underline{x}(k)+WB\underline{u}(k) \tag{27}$$

Equation (22) becomes

$$\begin{aligned}\underline{u}(k) &= K\underline{x}(k)+M\underline{v}(k)\\ &= K\underline{x}(k)+M(\underline{z}(k)-W\underline{x}(k))\end{aligned} \quad (28)$$

Hence

$$\underline{u}(k)=(K-MW)\underline{x}(k)+M\underline{z}(k) \quad (29)$$

From (1),(19), (27) and (29) we can easily see that if

$$\left.\begin{aligned}WA-FW&=GC\\ K-MW&=H_1C\end{aligned}\right\} \quad (30)$$

where $C=[W_{11}\,\vdots\,0]$

then (17) and (18) reduce to

$$\underline{z}(k+1)=F\underline{z}(k)+G\underline{y}_1(k)+WB\underline{u}(k) \quad (31)$$

and

$$\underline{u}(k)=M\underline{z}(k)+H_1\underline{y}_1(k) \quad (32)$$

Thus our problem reduces to finding a simultaneous solution for the set of equations (30).

Case I. Single-input plant (u(k)=scalar)

It is shown in [32] and [33] that for the system of equations (30), the dimension of the matrix F can be r such that

$$\frac{n-q}{q}\leq r\leq n-q \quad (33)$$

provided that the pair $[A,(W_{11}\,\vdots\,0)]$ is an observable pair.

For example if n=10 and q=5, then an observer as low as first order could be found to approximate $\underline{u}(k)$ where as the regular observer solution (estimating $\underline{x}(k)$) would require an observer of fifth order.

Case II. Single-output plant (y(k)=scalar)

It is shown in [33] that if the output of the system is a scalar quantity, i.e. if $\underline{y}(k)$ is a scalar, an observer of order r where $r\leq n-1$ can be found.

Case III. Multi-input/multi-output plant

Let's assume that (see (22))

$$\underline{u}(k)=K\underline{x}(k)+\underline{e}(k) \tag{34}$$

where K is of full rank and dimension pxn. Equation (17) can be written as follows:

$$\underline{x}(k+1)=A\underline{x}(k)+B\underline{u}(k) \tag{35}$$

and (18) as

$$\underline{y}_1(k)=\left[W_{11} \,\vdots\, \underline{0}\right]\underline{x}(k) \tag{36}$$

where

$$A\begin{bmatrix} A_{11} & \vdots & A_{12} \\ \hline A_{21} & \vdots & A_{22} \end{bmatrix}, \quad B\begin{bmatrix} B_{11} \\ \hline B_{21} \end{bmatrix} \tag{37}$$

Equation (35) can also be written as

$$\underline{x}(k+1)=A\underline{x}(t)+\sum_{i=1}^{P}\underline{b}_i u_i(k) \tag{38}$$

where $\underline{b}_i$ are the column vectors of B, $u_i(k)$ are the components of the vector $\underline{u}(k)$ and equations (27) and (34) yield

$$\underline{z}(k+1)=F\underline{z}(k)+(WA-FW)\underline{x}(k)+W\sum_{i=1}^{P}\underline{b}_i u_i(k) \tag{39}$$

and

$$u_i(k)=\underline{k}_i^T\underline{x}(k)+e_i(k) \tag{40}$$

where $\underline{k}_i^T$ is the ith row of the matrix K and e_i is the ith component of the vector $\underline{e}$. Similarly equation (30) gives

$$WA-FW=GC \tag{41}$$

$$\underline{k}_i^T-m_i^T W=\underline{h}_i^T C,\quad i=1,2,\ldots,p \tag{42}$$

where $\underline{m}_i^T$ and $\underline{h}_i^T$ are the ith row of the matrices M and H respectively. Using a procedure similar to that described in [32] it is shown in the appendix that the lowest allowable range for the observer is that given by r where $\frac{n-q}{q}<r<n-q$ iff $\{M,F\}$ and $\{A,C\}$ form observable pairs. In the illustrative example that follows n=4, q=2, thus an observer of dimension as low as $\frac{4-2}{2}=1$ can be used to implement the control whereas if we first estimated $\underline{x}(k)$ we would require an observer of dimension 4-2=2. In large scale problems, this reduction could be quite important.

4. EXAMPLE

The design of a low-order-delayed measurement observer will be illustrated through the following simple example.

Consider the following time-invariant plant whose state equations are given by

$$\underline{x}(k+1)=A\underline{x}(k)+B\underline{u}(k) \tag{43}$$

where

$$A \begin{bmatrix} 0 & 1 & 0 & 0 \\ 0 & 0 & 1 & 0 \\ 0 & 0 & 0 & 1 \\ 1 & 0 & 1 & 0 \end{bmatrix}, \quad B=\begin{bmatrix} 0 & 0 \\ 0 & 1 \\ 0 & 0 \\ 1 & 0 \end{bmatrix}, \quad \underline{u}(k) \begin{bmatrix} u_1(k) \\ u_2(k) \end{bmatrix}$$

$$y(k)=C\underline{x}(k) \tag{44}$$

$$\underline{u}(k)=K\underline{x}(k) \tag{45}$$

where

$$C=\begin{bmatrix} 1 & 0 & 0 & 0 \end{bmatrix}, \quad K=\begin{bmatrix} 0 & 0 & 1 & 1 \\ 1 & 1 & 0 & 1 \end{bmatrix}$$

Suppose that one delayed-measurement is used, then equation (44) gives

$$y(k-1)=\begin{bmatrix} 1 & 0 & 0 & 0 \end{bmatrix}\underline{x}(k-1) \tag{46}$$

and equation (43) gives

$$\underline{x}(k)=A\underline{x}(k-1)+B\underline{u}(k-1) \tag{47}$$

From (47) we obtain

$$\underline{x}(k-1)=A^{-1}\underline{x}(k)-A^{-1}B\underline{u}(k-1) \tag{48}$$

where

$$A^{-1} = \begin{bmatrix} 0 & -1 & 0 & 1 \\ 1 & 0 & 0 & 0 \\ 0 & 1 & 0 & 0 \\ 0 & 0 & 1 & 0 \end{bmatrix} \quad (49)$$

and from (45)

$$\underline{u}(k-1) = \begin{bmatrix} 0 & 0 & 1 & 1 \\ 1 & 1 & 0 & 1 \end{bmatrix} \underline{x}(k-1) \quad (50)$$

Substituting (49) and (50) into (48) we obtain

$$\underline{x}(k-1) + A^{-1}Bk\underline{x}(k-1) = A^{-1}\underline{x}(k)$$

or

$$x(k-1) = \begin{bmatrix} -1 & -2 & 1 & -1 \\ 0 & 1 & 0 & 0 \\ 1 & 1 & 0 & 0 \\ 0 & 0 & 0 & 1 \end{bmatrix} \underline{x}(k) \quad (51)$$

and

$$y(k-1) = \begin{bmatrix} 1 & 0 & 0 & 0 \end{bmatrix} \underline{x}(k-1) = \begin{bmatrix} 1 & 0 & 0 & 0 \end{bmatrix} \begin{bmatrix} -1 & -2 & 1 & -1 \\ 0 & 1 & 0 & 0 \\ 1 & 1 & 0 & 0 \\ 0 & 0 & 0 & 1 \end{bmatrix} \underline{x}(k) \quad (52)$$

i.e.

$$y(k-1) = \begin{bmatrix} -1 & -2 & 1 & -1 \end{bmatrix} \underline{x}(k)$$

Thus

$$\begin{bmatrix} y(k) \\ y(k-1) \end{bmatrix} = \begin{bmatrix} 1 & 0 & 0 & 0 \\ -1 & -2 & 1 & -1 \end{bmatrix} \quad (53)$$

or

$$\underline{y}_1(k) = \begin{bmatrix} 1 & 0 & 0 & 0 \\ -1 & -2 & 1 & -1 \end{bmatrix} \underline{x}(k) \quad (54)$$

Hence for the augmented vector $\underline{y}_1(k)$, the matrix which relates $y_1(k)$ and $\underline{x}(k)$ is given by

$$C = \begin{bmatrix} 1 & 0 & 0 & 0 \\ -1 & -2 & 1 & -1 \end{bmatrix}$$

If we form the matrix

$$\left[C^T \,\vdots\, A^T C^T \,\vdots\, \dots\dots \,\vdots\, A^{T(\alpha-1)} C^{T(\alpha-1)}\right]$$

we can easily show that the observability index of A,C is two or $\alpha=2$. Considering that the plant dimension is n=4 and the measurement dimension q=2 we can design an observer of dimension $\frac{n-q}{q}=\frac{4-2}{2}=1$.

Equation (31) then becomes

$$z(k+1)=fz(k)+\left[g_1 \;\; g_2\right]\underline{y}_1(k)+\left[\omega_1 \;\; \omega_2 \;\; \omega_3 \;\; \omega_4\right]\begin{bmatrix} 0 & 0 \\ 0 & 1 \\ 0 & 0 \\ 1 & 0 \end{bmatrix}\underline{u}(k) \qquad (55)$$

Using the procedure outlined in the appendix we can determine $g_1, g_2, \omega_1, \omega_2, \omega_3, \omega_4$. If we set $f=\frac{1}{2}$ this procedure yields

$$\omega=\left[1,-\frac{1}{2},\frac{3}{4},\frac{1}{4}\right], \qquad M=\begin{bmatrix} 2 \\ 1 \end{bmatrix}$$

$$g=\left[-\frac{7}{8},-\frac{5}{8}\right]$$

$$h=\begin{bmatrix} -\frac{5}{2} & -\frac{1}{2} \\ -\frac{3}{4} & -\frac{3}{4} \end{bmatrix}$$

Thus the observer becomes

$$z(k+1)=\frac{1}{2}z(k)+\left[-\frac{7}{8},-\frac{5}{8}\right]\underline{y}_1(k)+\left[1,-\frac{1}{2},\frac{3}{4},-\frac{1}{4}\right]\begin{bmatrix} 0 & 0 \\ 0 & 1 \\ 0 & 0 \\ 1 & 0 \end{bmatrix}\underline{u}(k)$$

or

$$z(k+1)=\frac{1}{2}z(k)+\left[-\frac{7}{8},-\frac{5}{8}\right]\underline{y}_1(k)+\left[1,-\frac{1}{2}\right]\underline{u}(k)$$

$$\underline{u}(k)=\begin{bmatrix}-5/2 & -1/2\\ -3/4 & -3/4\end{bmatrix}\underline{y}_1(k)+\begin{bmatrix}2\\ 1\end{bmatrix}\underline{Z}(k) \tag{56}$$

We observe that we are able to construct an observer of dimension *one* to implement the control $\underline{u}(k)$ whereas if we used a state estimating procedure to first estimate $x(k)$ and then implement the control, we would need an observer of dimension *two*. In large scale control realization problems this reduction could be quite beneficial.

5. CONCLUSION

The standard procedure for implementing a control for a discrete-time system which utilizes discrete time delayed measurements, in addition to the output, but with an overall limited-dimension-output (i.e. dimension of overall output less than the state dimension) is to use an observer to reconstruct the state and then implement the control. In this paper it is shown that implementation of the control directly from the output and the observer leads to an observer which under certain conditions has a dimension much lower than the dimension difference of the state and output. This approach could find important applications in large scale control system problems.

6. APPENDIX

Equations (41) and (42) can be written as follows:

$$GC=WA-FW \tag{A.1}$$

$$H_1C=K-MW \tag{A.2}$$

By multiplying equations (A.1) by the matrix A we obtain

$$GCA=WA^2-FWA \tag{A.3}$$

adding and subtracting F^2W on right hand side of (A.3) we obtain

$$\begin{aligned}&GCA+FWA-F^2W=WA^2-F^2W\\ &GCA+F(WA-FW)=WA^2-F^2W\end{aligned} \tag{A.4}$$

Substituting (A.1) into (A.4) we obtain

$$GCA+FGC=WA^2-F^2W \tag{A.5}$$

Multiplying (A.5) again by A and repeating the same process we obtain

$$GCA^2+FGCA=WA^3-F^2WA$$

or

$$GCA^2+FGCA+F^2WA-F^3W=WA^3-F^3W$$

or

$$GCA^2+FGCA+F^2GC=WA^3-F^3W \tag{A.6}$$

The objective of this formulation is to make use of the Cayley-Hamilton theorem by which F satisfies

$$\sum_{i=0}^{r} f_i F^{r-i}=0 \tag{A.7}$$

where f_i are the coefficients of the characteristic equation of F i.e.

$$\det(I-F)=\lambda^r+f_1\lambda^{r-1}+f_2\lambda^{r-2}+....+f_r \tag{A.8}$$

with $f_0=1$.

We shall repeat this proces described by equations (A.1) through (A.5) until we obtain the rth power of F. The rth step, therefore, gives

$$GCA^{r-1}+FGCA^{r-2}+r^2GCA^{r-3}+....+FGC=WA^r-F^rA \tag{A.9}$$

Multiplying each equation in the r-step-process by f_{r-i} and summing all the terms and making use of the expression (A.7) we obtain

$$GCA^{r-1}+(F+f_1I)GCA^{r-2}+(F^2+f_1F+f_2I)GCA^{r-3}+....+$$

$$+(F^{r-1}+f_1F^{r-2}+...+f_{r-1}I)GC=W(A^r+f_1A^{r-1}+...+f_rI) \tag{A.9}$$

If we multiply equation (A.9) by M and use (A.2) in (A.9) we obtain

$$(K-H_1C)(A^r+f_1A^{r-1}+...+f_rI)=MGCA^{r-1}+M(F+f_1I)GCA^{r-2}+....+$$

$$+M(F^{r-1}+f_1F^{r-2}+...+f_{r-1}I)GC \tag{A.10}$$

Equation (A.10) can also be written as

$$K(A^r+f_1A^{r-1}+\ldots+f_rI)=H_1C(A^r+f_1A^{r-1}+\ldots.+f_rI)$$
$$+MGCA^{r-1}+M(F+f_1I)GCA^{r-2}+\ldots+M(F^{r-1}+f_1F^{r-2}+\ldots+f_{r-1}I)GC \qquad (A.11)$$

The matrix equation (A.11) is a nonlinear matrix equation of the unknown matrices, H_1,M,F,G.

Let us denote the columns of G by $\underline{g}_i$, $i=1,\ldots,m$. If we put all these columns of G into a vector form we obtain a single vector $\underline{g}$ such that

$$\underline{g}=\begin{bmatrix}\underline{g}_1\\ \underline{g}_2\\ \vdots\\ \underline{g}_m\end{bmatrix}_{rq}$$

Equation (A.11) can then be written as

$$(A^{T(r-1)}C^T \oplus M)\underline{q}+(A^{T(r-1)}C^T \oplus M(F+f_1I)\underline{g}+\ldots+)C^T$$
$$\oplus M(F^{r-1}+f_1F^{r-2}+\ldots+f_{r-1}))\underline{g}+((A^{Tr}+fA^{T(r-1)}+\ldots+f_rI)C^T$$
$$\oplus I_p)\underline{h}=((A^{Tr}+f_1A^{T(r-1)}+\ldots+f_rI)\oplus I_p)\underline{k} \qquad (A.12)$$

where

$\underline{h}$ and $\underline{k}$ are vectors which contain the columns of H and K respectively, and

$\oplus$ denotes the Kronecker product.

If we put vectors $\underline{h}$ and $\underline{g}$ into one vector and form the vector $\begin{bmatrix}\underline{h}\\ \hline \underline{g}\end{bmatrix}$, equation (A.12) can be written as

$$\left[\begin{array}{cccc|cccc} \tilde{A}_1 & 0_{nxq} & \cdots & 0_{nxq} & \tilde{A}_2 & 0_{nxrq} & \cdots & 0_{nxrq} \\ 0_{nxq} & \tilde{A}_1 & \cdots & 0_{nxq} & 0_{nxrq} & \tilde{A}_2 & \cdots & 0_{nxrq} \\ \vdots & & \ddots & \vdots & \vdots & & \ddots & \vdots \\ \vdots & & \tilde{A}_1 & \vdots & \vdots & & \tilde{A}_2 & \vdots \end{array}\right] \left[\begin{array}{cccc} M_{11}I_q & M_{12}I_q & \cdots & M_{1r}I_q \\ M'_{11}I_q & M'_{12}I_q & \cdots & M'_{1r}I_q \\ M''_{11}I_q & M''_{12}I_q & \cdots & M''_{1r}I_q \\ \vdots & & & \\ M_{21}I_q & M_{22}I_q & \cdots & M_{2r}I_q \\ M'_{21}I_q & M'_{22}I_q & \cdots & M'_{2r}I_q \\ \vdots & & & \\ M_{p1}I_q & M_{p2}I_q & \cdots & M_{pr}I_q \\ M'_{p1}I_q & M'_{p2}I_q & \cdots & M'_{pr}I_q \\ M''_{p1}I_q & M''_{p2}I_q & \cdots & M''_{pr}I_q \end{array}\right]$$

$$\cdot \left[\frac{\underline{h}}{\underline{g}}\right] = \left[\begin{array}{cccc} \tilde{A}_3 & 0_{nxn} & \cdots & 0_{nxn} \\ 0_{nxn} & \tilde{A}_3 & \cdots & 0_{nxn} \\ \vdots & & \ddots & \vdots \\ \vdots & & & \tilde{A}_3 \end{array}\right] \underline{k}_{nq} \qquad (A.13)$$

where

$\tilde{A}_1 = (A^{T(r)} + f_1 A^{T(r-1)} + f_2 A^{T(r-2)} + \ldots + f_r I) C^T$,

$\tilde{A}_3 = (A^{T(r)} + f_1 A^{T(r-1)} + \ldots + f_r I)$

$\tilde{A}_2 = (A^{T(r-1)} C^T \vdots A^{T(r-2)} C^T \vdots \ldots C^T)$

M_{ij} are the elements of M

M'_{ij} are the elements of the matrix $M(F+f_1 I)$

M''_{ij} are the elements of the matrix $M(F^2 + f_1 F + f_2 I)$ etc.

Equations (A.13) can be written as

$$\left[\tilde{\tilde{A}}_1 \vdots \tilde{\tilde{A}}_2 M\right] \left[\frac{\underline{h}}{\underline{g}}\right] = \tilde{A}_3 \underline{k} \qquad (A.14)$$

where $\tilde{\tilde{A}}_1, \tilde{\tilde{A}}_2, M$ are the respective matrices on the left hand side of equation (A.13) and $\tilde{\tilde{A}}_3$ is the matrix on the right hand side of equation (A.13). If we rearrange the rows of the matrix M, we can easily see that the new matrix can be written as

$$\begin{bmatrix} M \\ M(F+f_1 I) \\ \vdots \\ M(F^{r-1}+f_1 F^{r-2}+\ldots+f_{r-1} I) \end{bmatrix} \oplus I_q \qquad \text{(A.15)}$$

If we choose M, F to form an observable pair, then the rank of the product $\tilde{\tilde{A}}_2 M$ will depend on the rank of the matrix $\tilde{\tilde{A}}_2$ which in turn depends on the rank of the matrix $\tilde{A}_2$. If we add to the product $\tilde{\tilde{A}}_2 M$ the columns of the matrix $\tilde{\tilde{A}}_1$ whose rank depends on the rank of the matrix $\tilde{A}_1$ then it is easily seen that the rank of the matrix $[\tilde{\tilde{A}}_1 \vdots \tilde{\tilde{A}}_2 M]$ depends on the rank of the matrix $[\tilde{A}_1 \vdots \tilde{A}_2]$. The columns of the matrix $[\tilde{A}_1 \vdots \tilde{A}_2]$ span the same space as the columns of the matrix.

$\tilde{A} = [A^{Tr} C^T \vdots A^{T(r-1)} C^T \vdots \ldots \vdots C^T]$. The equation given by (A.13) has a solution if $r \geqslant \alpha - 1$ where α is the observability index of A,C or the least integer such that the matrix $[A^{T(\alpha-1)} C^T \vdots A^{T(\alpha-2)} C^T \vdots \ldots \vdots C^T]$ has full rank n.

But if A_{nxn}, C_{qxq} is an observable pair, the observability index α could vary as follows

$$\frac{n}{q} \leqslant \alpha \leqslant n-q+1$$

Hence, $\frac{n-q}{q} \leqslant r \leqslant n-q$

REFERENCES

1. R.E. Kalman, "Contributions to the theory of optimal control", Bul. Soc. Mat. Mex., vol. 5, pp. 102-199, 1960.
2. M. Athans and P.L. Falb, Optimal Control. New York:McGraw-Hill, 1966.

3. H. Kwakernaak and R. Sivan, Linear Optimal Control Systems. New York:Wiley-Interscience, 1972.
4. A.P. Sage and C.C. White, III, Optimum Systems Control, Second Edition. Englewood Cliffs, New Jersey: Prentice-Hall, 1977.
5. R.E. Kalman and R.S. Bucy, "New results in linear filtering and prediction theory, "Trans. ASME, J. Basic Eng., ser. D, pp. 95-108, Mar. 1961.
6. D.G. Luenberger, "Observing the state of a linear system", IEEE Trans. Mil. Electron., vol. MIL-8, pp. 74-80, Apr. 1964.
7. D.G. Luenberger, "Observers for multivariable systems, "IEEE Trans. Automat. Contr., vol. AC-11, pp. 190-197, Apr. 1966.
8. D.G. Luenberger, "An introduction to observers", IEEE Trans. Automat. Contr.,vol. AC-16, pp.596-602, Dec. 1971.
9. B. Gopinath, "On the control of linear multiple input-output systems", Bell Syst.Tech.J., vol. 50, pp. 1063-1081, Mar. 1971.
10. W.A. Wolovich, "On state estimation of observable systems", in 1968 Joint Automatic Control Conf., Preprints, pp. 210-220.
11. J.J. Bongiorno, Jr. and D.C. Youla, "On observers in multivariable control systems", Int. J. Contr.,vol.8, pp. 221-243, Sept. 1968.
12. Y.O. Yuksel and J.J. Bongiorno, Jr., "Observers for linear multivariable systems with applications", IEEE Trans. Automat.Contr., vol.AC-16, pp. 603-613, Dec.1971.
13. M. Aoki and J.R. Huddle, "Estimation of the state vector of a linear stochastic system with a constrained estimator", IEEE Trans. Automat.Contr. , vol AC-12, pp. 432-433, Aug. 1967.
14. K.G. Brammer, "Lower order optimal linear filtering of nonstationary random sequences", IEEE Trans. Automat. Contr., vol. AC-13, pp. 198-199, Apr. 1968.
15. E. Tse and M. Athans, "Optimal minimal-order observer-estimators for discrete linear time-varying systems", IEEE Trans. Automat. Contr., vol. AC-15, pp. 416-426, Aug. 1970.
16. E. Tse and M. Athnas, "Observer theory for continuous-time linear systems", J. Inform. Contr., vol. 22,pp. 405-434, 1973.
17. C.T. Leondes and L.M. Novak, "Optimal minimal-order observers for discrete-time systems -a unified theory, "Automatica, vol. 8, pp. 379-387, 1972.
18. S.C. Iglehart and C.T. Leondes, "A design procedure for intermediate-order observer-estimators for linear discrete-time dynamical systems", Int. J. Contr., vol. 16, pp. 401-415, Mar. 1972.
19. J. O'Reilly and M.M. Newmann, "Minimum-order observer-estimators for continuous-time linear systems", Int.J.

Contr., vol. 22, pp. 573-590, Apr. 1975.

20. L.M. Novak, "Discrete-time optimal stochastic observers", in Control and Dynamic Systems; Advances in Theory and Applications, vol. 12, C.T. Leondes Ed. New York:Academic, 1976.
21. F.W. Fairman, "Hybrid estimators for discrete-time stochastic systems", IEEE Trans. Syst. Man. Cybern., vol. SMC-8, pp. 849-854, Dec. 1978.
22. C.D. Johnson, "Accomodation of external disturbances in linear regulator and servomechanism problems", IEEE Trans. Automat. Contr., vol. AC-16, pp. 635-644, Dec. 1971.
23. C.D. Johnson, "Accommodation of disturbances in optimal control problems", Int. J. Contr., vol. 15, pp. 209-231, Feb.1972.
24. C.D. Johnson , "Theory of disturbance-accommodating controller", in Control and Dynamic Systems; Advances in Theory and Applications, vol. 12, C.T. Leondes Ed. York: Academic, 1976.
25. N.K. Loh, T.C. Hutchings, and R.E. Kasten, "State space design for helicopter pointing and tracking systems", Proc. of the Second U.S. Army Symposium on Gun Dynamics, pp. II-52-II-68, Sept. 1978.
26. Special Issue on Disturbance-Accommodating Control Theory, J. Interdis. Model& Simul., vol. 3, N.K. Loh and C.D. Johnson Co-Ed., Jan. 1980.
27. N.K. Loh and D.H. Chyung, "State reconstruction from delayed observations", Proc. of the Fifth Annual Pittsburgh Conference on Modeling and Simulation,vol. 5, pp. 961-963, Apr. 1974.
28. D.H. Chyung and N.K. Loh, "State reconstruction from delayed partial measurements in discrete linear systems", Proc. of the Sixth Annual Pittsburgh Conference on Modeling and Simulation, vol. 6, pp. 159-161, Apr. 1975.
29. D.H. Chung, "On a method for reconstructing inaccessible state variables using time delays", Division of Information Engineering, University of Iowa, Iowa City, Iowa, 1979.
30. K.C. Cheok and N.K. Loh, "A microprocessor-based state estimator and optimal controller", Presented at the 23rd Midwest Symposium on Circuits and Systems, University of Toledo, Toledo, Ohio, Aug. 4-5, 1980.
31. N.K. Loh, W.Z. Chen and R.R. Beck, Delayed-Measurement Observer for Discrete-Time Linear Systems. TR 80-07-100, School of Engineering, Oakland University, Rochester MI 48063.
32. A Jameson and D. Rothchild, A Direct Approach to the Design of Asymptotically Optimal Controllers, Int.J.Control, Vol.13, No.6, 1971.
33. J.E. Fortmann and D. Williamson, Design of Low-Observers for Linear Feedback Control Laws, IEEE Trans. on Control,

vol.AC-17, 1972.
34. J.L. Casti, Dynamical Systems and their Applications: Linear Theory, New York:Academic Press, 1977.
35. T. Kailath, Linear Systems, Englewood Cliffs, New Jersey, Prentice Hall, 1980.

CHAPTER 18

SINGULAR PERTURBATION METHOD AND RECIPROCAL TRANSFORMATION ON TWO-TIME SCALE SYSTEMS

G. Dauphin-Tanguy, O. Moreigne, P. Borne
Institut Industriel du Nord
Laboratoire d'Automatique et d'Informatique Industrielle
B.P. 48
59651 Villeneuve d'Ascq Cédex
France

The singular perturbation method applied to two-time scale systems allows to reduce dimensionnality and to obtain easily the slow disconnected part of the global system. When the study concerns the fast part, this method does not generally involve the same precision. So we propose here a new transformation, called "reciprocal transformation" which inverses dynamic and frequencial behaviour. The fast initial part, become the slow reciprocal part, is then decoupled by singular perturbation technique, and an other application of reciprocal transformation gives the fast decoupled subsystem which is a better approach of the initial fast part and in addition enables to take into account the influence of the evolution of the slow part of the process.

The initial values for the fast variables are conserved, which suppresses the difficulties to connect slow and fast decoupled trajectories.

As an application of reciprocal transformation, a singular optimal problem, impossible to solve for the global process, is considered. We obtain a composite control which is a quasi-optimal approximation of optimal control.

1. INTRODUCTION

Many physical systems are composed by parts with different dynamic behaviours. As examples, we can propose robots with slow mechanical parts and fast electronic parts, slow system with fast actuators and sensors, processes controlled through high gain ... [1] [2]. It is often usefull to disconnect dynamics in order to reduce dimensionnality, to simplify calculation of optimal control, to suppress imprecisions due to very different magnitude order terms.

The singular perturbation method allows to obtain the reduced slow system which takes care of the influence of the fast neglected part. But when the study concerns the fast part (parasistics, fast transients, high frequency evolution), this technique of reduction may involve insufficient precision. So we propose a new transformation, called "reci-

S. G. Tzafestas (ed.), Multivariable Control, 327–342.

procal transformation", which inverses dynamic and frequential behaviour. By simultaneous application of singular perturbation method (SP) and of reciprocal transformation (R), we obtain for the fast decoupled part a very good approximation of the fast components of the initial global system.

2. PRESENTATION OF THE SINGULAR PERTURBATIONS METHOD [3]

Let us consider the global system characterized by :

$$\begin{bmatrix} \overset{\circ}{x} \\ \varepsilon\, \overset{\circ}{z} \end{bmatrix} = \begin{bmatrix} A_{11} & A_{12} \\ A_{21} & A_{22} \end{bmatrix} \begin{bmatrix} x \\ z \end{bmatrix} + \begin{bmatrix} B_1 \\ B_2 \end{bmatrix} u$$

$$y = \begin{bmatrix} C_1 & C_2 \end{bmatrix} \begin{bmatrix} x \\ z \end{bmatrix} \tag{1}$$

$x(0)$, $z(0)$ initial values

where $x \in \mathbb{R}^{n_1}$ is the slow state vector and $z \in \mathbb{R}^{n_2}$ is the fast state vector.

The parameter ε allows to compare the time scales and is defined by :

$$\varepsilon = \frac{t - t_o}{\tau} \qquad t_o \text{ : initial moment, } \varepsilon \in \left]0\, ,\, 1\right]$$

where t and τ are the time scales associated respectively to the slow and fast variables.

The system (1) is said to be modellized as a singularly perturbed system.

By setting ε equal to zero (if the fast part is stable and A_{22} non singular), we obtain the reduced slow system as :

$$\left\{\begin{array}{l} \overset{\circ}{x}_s = A_s\, x_s + B_s\, u \\ y_s = C_s\, x_s + D_s\, u \\ z_s = -A_{22}^{-1}\, (A_{21}\, x_s + B_2\, u) \\ x_s(0) = x(0) \\ \text{with} \begin{array}{l} \left\{\begin{array}{l} A_s = A_{11} - A_{12}\, A_{22}^{-1}\, A_{21} \\ B_s = B_1 - A_{12}\, A_{22}^{-1}\, B_2 \end{array}\right. \\ \left\{\begin{array}{l} C_s = C_1 - C_2\, A_{22}^{-1}\, A_{21} \\ D_s = -C_2\, A_{22}^{-1}\, B_2 \end{array}\right. \end{array} \end{array}\right. \tag{2}$$

The fast part of the system is then obtained by considering the slow part as static ($\mathring{x}_s = 0$) during the evolution of z(t) with the equation :

$$z_{\delta}(t) = z(t) - z_s(t) \tag{3}$$

and according to the equations (1) and (2), we obtained the state equations of the fast decoupled part called boundary layer equations.

$$\begin{cases} \varepsilon\, \mathring{z}_{\delta_{sp}} = A_{22}\, z_{\delta_{sp}} + B_2\, u \\ y_{\delta_{sp}} = C_2\, z_{\delta_{sp}} \\ z_{\delta_{sp}}(0) = z(0) - A_{22}^{-1} A_{21}\, x(0) \end{cases} \tag{4}$$

The mutual influence between the two dynamical parts appears for the slow disconnected part in the matrical quadruplet (A_s, B_s, C_s, D_s), which takes into account the motion of the fast part.

For the fast decoupled subsystem, the slow variables appear only in the initial values, which are different from the initial conditions of the global system.

When the precise study concerns the fast part, the singular perturbation technique appears sometimes not sufficient. So we propose a new tool called Reciprocal Transformation, which, when added to previous method, enables to improve the modelling of the fast part in the boundary layer domain.

3. DEFINITIONS AND PROPERTIES OF THE RECIPROCAL TRANSFORMATION

3.1. Definitions

The reciprocal transformation notion, as defined here, has been introduced by Hutton-Friedland, 1975 [4] for order reduction by Routh method, generalized and enlarged to multi time scale systems by Dauphin-Tanguy and al [5] [6] [7] [8].

Many definitions are proposed here, according to the initial description of the system, in linear monovariable case.

Definition 1 :

When the process (Σ) is described by a linear differential equation as :

$$\sum_{i=0}^{n} a_i\, y^{(i)} = \sum_{i=0}^{m} b_i\, u^{(i)} \tag{5}$$

where $u^{(i)}$ and $y^{(i)}$ denote respectively the $i^{\underline{th}}$ time-derivatives of the input u and output y then the reciprocal system $(\tilde{\Sigma})$ is defined by :

$$\left\{ \begin{array}{l} \sum_{i=0}^{n} a_i \, y^{(n-i)} = \sum_{i=0}^{m} b_i \, u^{(n-1-i)} \qquad (6) \\ \text{with the notation (when } m = n) \\ u^{(-1)} = \int_{t_o}^{t} u(t) \, dt \qquad (7) \end{array} \right.$$

Definition 2 :

When the process (Σ) is described by a transfert function, W(s), then the reciprocal system $(\tilde{\Sigma})$ is defined by :

$$\tilde{W}(s) = \frac{1}{s} W(\frac{1}{s}) \qquad (8)$$

Definition 3 :

When the process (Σ) admits a state space description corresponding to :

$$\left\{ \begin{array}{l} \overset{\circ}{x} = A\,x + B\,u \\ y = C\,x + D\,u \end{array} \right. \qquad (9)$$

where $x \in \mathbb{R}^n$ is the state vector, then the reciprocal state equation associated to the reciprocal system $(\tilde{\Sigma})$ is :

$$\left\{ \begin{array}{l} \overset{\circ}{\tilde{x}} = \tilde{A}\,\tilde{x} + \tilde{B}\,u \\ \tilde{y} = \tilde{C}\,\tilde{x} + \int_{t_o}^{t} D\,u\,dt \\ \text{with} \left\{ \begin{array}{ll} \tilde{A} = A^{-1} & \text{(condition : A non singular)} \\ \tilde{B} = B & \\ \tilde{C} = -\,C\,A^{-1} & \end{array} \right. \end{array} \right. \qquad (10)$$

The integral term can be suppressed by introducing a new state variable v, corresponding to a mode equal to zero,

$$v = \int_{t_o}^{t} u \, dt \tag{11}$$

which gives a canonical form for equation (10) :

$$\begin{cases} \overset{\circ}{\tilde{x}} = \tilde{A}\,\tilde{x} + \tilde{B}\,u \\ \overset{\circ}{v} = u \\ \tilde{y} = \tilde{C}\,\tilde{x} + D\,v \end{cases} \tag{12}$$

3.2. Properties

Property 1 :

The initial values are invariant by reciprocal transformation :

$$\tilde{x}(t_o) = x(t_o) \tag{13}$$

Proof : By use of Laplace transformation, the initial output is :

$$y(s) = \left[\, C\,(sI - A)^{-1} B + D \,\right] u(s) + C\,(sI - A)^{-1}\, x(t_o) \tag{14}$$

which becomes by reciprocal transformation (Definition 2) :

$$\begin{aligned} \tilde{y}(s) = {} & \left[\, -C A^{-1} (sI - A^{-1})^{-1} B + \frac{D}{s} \,\right] u(s) - \\ & - C A^{-1} (sI - A^{-1})^{-1}\, x(t_o) \end{aligned} \tag{15}$$

From Definition 3, we obtain :

$$\tilde{y}(s) = \left[\, \tilde{C}\,(sI - \tilde{A})^{-1} \tilde{B} + \frac{D}{s} \,\right] u(s) + \tilde{C}\,(sI - \tilde{A})^{-1}\, \tilde{x}(t_o) \tag{16}$$

Then, the proposed result by identification of the two expressions (15) and (16).

Property 2 :

The reciprocal transformation inverses dynamics and frequencial behaviour of the process, due to the change $s \to \frac{1}{s}$ (or $A \to A^{-1}$).

4. APPLICATION OF RECIPROCAL TRANSFORMATION TO SEPARATION OF DYNAMICS

The proposed approach is the following :

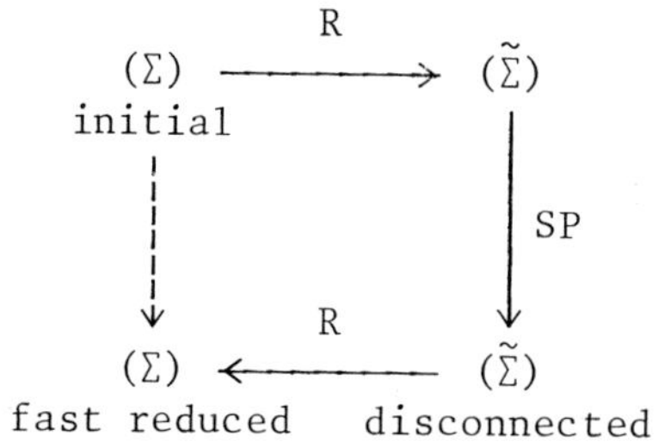

The initial system (Σ) is represented by equation (1) which can also be written as :

$$\left\{\begin{aligned} &\begin{bmatrix} \dot{x} \\ \dot{z} \end{bmatrix} = \underbrace{\begin{bmatrix} A_{11} & A_{12} \\ A_{21}^{*} & A_{22}^{*} \end{bmatrix}}_{A} \begin{bmatrix} x \\ z \end{bmatrix} + \underbrace{\begin{bmatrix} B_{1} \\ B_{2}^{*} \end{bmatrix}}_{B} u \\ &y = \underbrace{\begin{bmatrix} C_{1} & C_{2} \end{bmatrix}}_{C} \begin{bmatrix} x \\ z \end{bmatrix} \end{aligned}\right. \tag{17}$$

with $A_{21}^{*} = A_{21} / \varepsilon$; $A_{22}^{*} = A_{22} / \varepsilon$; $B_{2}^{*} = B_{2} / \varepsilon$.

The reciprocal system $(\tilde{\Sigma})$ is :

$$\left\{\begin{aligned} &\begin{bmatrix} \dot{\tilde{x}} \\ \dot{\tilde{z}} \end{bmatrix} = \begin{bmatrix} \tilde{A}_{11} & \tilde{A}_{12} \\ \tilde{A}_{21} & \tilde{A}_{22} \end{bmatrix} \begin{bmatrix} \tilde{x} \\ \tilde{z} \end{bmatrix} + \begin{bmatrix} \tilde{B}_{1} \\ \tilde{B}_{2} \end{bmatrix} u \\ &\tilde{y} = \begin{bmatrix} \tilde{C}_{1} & \tilde{C}_{2} \end{bmatrix} \begin{bmatrix} \tilde{x} \\ \tilde{z} \end{bmatrix} \end{aligned}\right. \tag{18}$$

The fast part of $(\tilde{\Sigma})$ is then composed by state variables $\tilde{x}$ and the slow part by $\tilde{z}$.

$(\tilde{\Sigma})$ can be modellized as a singularly perturbed system by :

$$\begin{cases} \begin{bmatrix} \varepsilon \,\mathring{\tilde{x}} \\ \mathring{\tilde{z}} \end{bmatrix} = \begin{bmatrix} \tilde{A}^*_{11} & \tilde{A}^*_{12} \\ \tilde{A}_{21} & \tilde{A}_{22} \end{bmatrix} \begin{bmatrix} \tilde{x} \\ \tilde{z} \end{bmatrix} + \begin{bmatrix} \tilde{B}^*_1 \\ \tilde{B}_2 \end{bmatrix} u \\ \tilde{y} = [\, \tilde{C}_1 \quad \tilde{C}_2 \,] \begin{bmatrix} \tilde{x} \\ \tilde{z} \end{bmatrix} \end{cases} \tag{19}$$

with $\tilde{A}^*_{11} = \tilde{\varepsilon}\,\tilde{A}_{11}$; $\tilde{A}^*_{12} = \tilde{\varepsilon}\,\tilde{A}_{12}$; $\tilde{B}^*_1 = \tilde{\varepsilon}\,\tilde{B}_1$.

By setting $\tilde{\varepsilon} = 0$ (SP method) and by application of reciprocal transformation (R) on the reciprocal slow reduced system, we obtain the following decoupled fast part $z_{\delta(SP+R)}$ defined by :

$$\begin{cases} \varepsilon\, \mathring{z}_{\delta(SP+R)} = A_{22}\, z_{\delta(SP+R)} + (B_2 + \varepsilon A_{22}^{-1} A_{21} B_1)\, u \\ y_{\delta(SP+R)} = C_2\, z_{\delta(SP+R)} + \int_{t_o}^{t} (C_1 - C_2 A_{22}^{-1} A_{21}) B_1\, u\, dt \\ z_{\delta(SP+R)}(0) = z(0) \end{cases} \tag{20}$$

Remarks :

- The differences between equations (4) and (20) appear in the integral term and in the command matrix which depends here of parameter ε. If the dynamics are not well separated, ε is not small (near 1) and the term $\varepsilon A_{22}^{-1} A_{21} B_1$ is not neglectable.
- The initial values are conserved by this method, which gives in boundary layer domain a suitable approximation of the system.

5. RESOLUTION OF A SINGULAR PROBLEM IN OPTIMAL CONTROL BY RECIPROCAL SYSTEM

Let us consider the global system represented by equation (1).

The problem consists in the determination of the optimal control minimizing the criterion :

$$J = \frac{1}{2} \int_0^{+\infty} y^T y \, dt \tag{21}$$

without constraints.

This is a singular problem in optimal control [9], and the maximum principle does not allow to determine the optimal trajectory of the global system.

The separation of dynamics by SP method for the slow part and by (SP+R) method for the fast part suppresses this singularity.

5.1. Control of the slow disconnected part

The optimal control problem is reduced here to the slow part and is presented as :

$$\begin{cases} \mathring{x}_s = A_s\, x_s + B_s\, u \\ y_s = C_s\, x_s + D_s\, u \\ \text{minimization of } J_s = \frac{1}{2}\displaystyle\int_0^\infty y_s^T\, y_s\, dt \end{cases} \tag{22}$$

which gives a new expression of J_s :

$$J_s = \frac{1}{2}\int_0^\infty (x_s^T\, C_s^T\, C_s\, x_s + 2\, x_s^T\, C_s^T\, D_s\, u + u^T\, D_s^T\, D_s\, u)\, dt \tag{23}$$

and with the variable change

$$\mathcal{U}_s = u + D_s^{-1}\, C_s\, x_s \qquad (D_s \text{ supposed non singular}) \tag{24}$$

the problem becomes a classical linear quadratic problem characterized by :

$$\begin{cases} \mathring{x}_s = (A_s - B_s\, D_s^{-1}\, C_s)\, x_s + B_s\, \mathcal{U}_s \\ y_s = D_s\, \mathcal{U}_s \\ J_s = \frac{1}{2}\displaystyle\int_0^{+\infty} \mathcal{U}_s^T\, D_s^T\, D_s\, \mathcal{U}_s\, dt \end{cases} \tag{25}$$

If $(A_s - B_s\, D_s^{-1}\, C_s)$ is stable, the quasi-optimal control for the slow part is $\mathcal{U}_s = 0$.

So :

$$\begin{cases} u_s = -\, D_s^{-1}\, C_s\, x_s \\ \text{with} \begin{cases} C_s = C_1 - C_2\, A_{22}^{-1}\, A_{21} \\ D_s = -\, C_2\, A_{22}^{-1}\, B_2 \end{cases} \end{cases} \tag{26}$$

If $(A_s - B_s D_s^{-1} C_s)$ is not stable, we have to research a control which stabilizes the system in closed loop.

5.2. Control of the fast disconnected part

The fast decoupled part, obtained by SP method is presented in (4). We see that the control u does not appear in the expression of the output $y_{\delta SP}$. So the optimal problem remains singular.

With the (SP+R) method, the reduced fast part is characterized by (20), which gives, by introducing a new fast state variable :

$$v = \frac{1}{\varepsilon} \int_{t_o}^{t} u \, dt \tag{27}$$

and with the variable change :

$$Z_{\delta(SP+R)} = z_{\delta(SP+R)} - (B_2 + \varepsilon A_{22}^{-1} A_{21} B_1) v \tag{28}$$

a new fast subsystem :

$$\begin{cases} \varepsilon \mathring{Z}_{\delta(SP+R)} = A_{22} Z_{\delta(SP+R)} + A_{22} (B_2 + \varepsilon A_{22}^{-1} A_{21} B_1) v \\ y_{\delta(SP+R)} = C_2 Z_{\delta(SP+R)} + (C_2 B_2 + \varepsilon C_1 B_1) v \end{cases} \tag{29}$$

The criterion J expressed by (21) can be decomposed into two parts :

$$J = \underbrace{\frac{1}{2} \int_{t_o}^{t_1} y^T y \, dt}_{J_\delta} + \underbrace{\frac{1}{2} \int_{t_1}^{+\infty} y^T y \, dt}_{J_s} \tag{10}$$

J_δ is associated to the fast variable, with t_1 infinitively small and J_s associated to slow variables. J_δ can be written, according to (29), as :

$$\begin{aligned} J_{\delta(SP+R)} = \frac{1}{2} \int_{t_o}^{t_1} \Big[& Z_{\delta(SP+R)}^T C_2^T C_2 Z_{\delta(SP+R)} + \\ & + 2 Z_{\delta(SP+R)}^T C_2^T (C_2 B_2 + \varepsilon C_1 B_1) v + \\ & + v^T (C_2 B_2 + \varepsilon C_1 B_1)^T (C_2 B_2 + \varepsilon C_1 B_1) v \Big] dt \end{aligned} \tag{30}$$

The variable change, in the same form as (24) :

$$u_{\delta(SP+R)} = v + (C_2 B_2 + \varepsilon C_1 B_1)^{-1} C_2 Z_{\delta(SP+R)} \tag{30}$$

associated to time scale change $\tau = t / \varepsilon$ gives a new formulation of the optimal problem :

$$\begin{cases} \dfrac{dZ_{\delta(SP+R)}}{d\tau} = \left[A_{22} - A_{22} (B_2 + \varepsilon A_{22}^{-1} A_{21} B_1)(C_2 B_2 + \varepsilon C_1 B_1)^{-1} C_2\right] \times \\ \qquad \times Z_{\delta(SP+R)} + A_{22} (B_2 + \varepsilon A_{22}^{-1} A_{21} B_1) u_{\delta(SP+R)} \\ y_{\delta(SP+R)} = (C_2 B_2 + \varepsilon C_1 B_1) u_{\delta(SP+R)} \\ J = \frac{1}{2} \displaystyle\int_0^{+\infty} u^T_{\delta(SP+R)} (C_2 B_2 + \varepsilon C_1 B_1)^T (C_2 B_2 + \varepsilon C_1 B_1) \times \\ \qquad \times u_{\delta(SP+R)} \, d\tau \end{cases} \tag{31}$$

The optimal control, in the case of stability of system (31) is then :

$$u_{\delta(SP+R)} = 0$$

So,

$$v_{\delta(SP+R)} = - (C_2 B_2 + \varepsilon C_1 B_1)^{-1} C_2 Z_{\delta(SP+R)} \tag{32}$$

With the relation :

$$\frac{dv_{\delta(SP+R)}}{d\tau} = u_{\delta(SP+R)}$$

and according to the equations (28) and (29), we obtain the fast quasi-optimal control :

$$u_{\delta(SP+R)} = - (C_2 B_2 + \varepsilon C_1 B_1)^{-1} C_2 A_{22} z_{\delta(SP+R)} \tag{33}$$

The composite control, which is an approximation of the global optimal control is then obtained by addition of (26) and (33) :

$$u_c^* = (C_2\, A_{22}^{-1}\, B_2)^{-1}\ (C_1 - C_2\, A_{22}^{-1}\, A_{21})\ x_s(t) - - (C_2\, B_2 + \varepsilon\, C_1\, B_1)^{-1}\ C_2\, A_{22}\ z_{f(SP+R)}(\tau) \tag{34}$$

6. COMPARISON ON A MECHANICAL EXAMPLE OF REDUCTION RESULTS OBTAINED BY SP AND (SP+R) METHODS

Let us consider the following mechanical system :

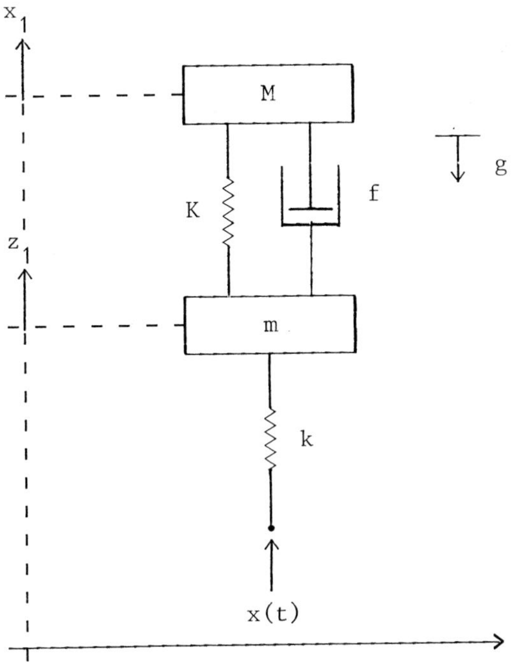

Figure 1. Physical model

x(t) and g (gravity) are inputs of the system supposed to be linear.

In phase variables (x_1, $\mathring{x}_1 = x_2$, z_1, $\mathring{z}_1 = z_2$), the state equation is then :

$$\left\{\begin{array}{l}
\begin{bmatrix} \mathring{x}_1 \\ \mathring{x}_2 \\ \mathring{z}_1 \\ \mathring{z}_2 \end{bmatrix} = \underbrace{\left[\begin{array}{cc|cc} 0 & 1 & 0 & 0 \\ -K/M & -f/M & K/M & f/M \\ \hline 0 & 0 & 0 & 1 \\ K/m & f/m & \frac{-(K+k)}{m} & -f/m \end{array}\right]}_{A} \begin{bmatrix} x_1 \\ x_2 \\ z_1 \\ z_2 \end{bmatrix} + \underbrace{\begin{bmatrix} 0 & 0 \\ 0 & -1 \\ 0 & 0 \\ \frac{k}{m} & -1 \end{bmatrix}}_{B} \begin{bmatrix} x \\ g \end{bmatrix} \\
y = \begin{bmatrix} 1 & 0 & 0 & 0 \\ 0 & 0 & 1 & 0 \end{bmatrix} \begin{bmatrix} x_1 \\ x_2 \\ z_1 \\ z_2 \end{bmatrix}
\end{array}\right. \qquad (35)$$

which gives in numerical values :

$$m = 1 \qquad M = 10 \qquad \frac{K}{M} = 60\ s^{-2} \qquad \frac{f}{m} = 4\ s^{-1} \qquad \frac{K+k}{M} = 1000\ s^{-2}$$

$$A = \left[\begin{array}{cc|cc} 0 & 1 & 0 & 0 \\ -60 & -4 & 60 & 4 \\ \hline 0 & 0 & 0 & 1 \\ 600 & 40 & -10000 & -40 \end{array}\right] \qquad B = \begin{bmatrix} 0 & 0 \\ 0 & -1 \\ 0 & 0 \\ 9400 & -1 \end{bmatrix} \qquad (36)$$

The eigenvalues of A can be organized into two disconnected sets, which allows to conclude to the two-time scale property of (35), (36), with a slow part associated to mass M and a fast part associated to mass m.

The slow parameter ε can be choosen in different ways. We shall take here ε equal to $m / M = 0.1$.

The system is then modellized as a singularly perturbed system :

$$\begin{bmatrix} \mathring{x}_1 \\ \mathring{x}_2 \\ \varepsilon\, \mathring{z}_1 \\ \varepsilon\, \mathring{z}_2 \end{bmatrix} = \begin{bmatrix} A_{11} & A_{12} \\ A_{21}^* & A_{22}^* \end{bmatrix} \begin{bmatrix} x_1 \\ x_2 \\ z_1 \\ z_2 \end{bmatrix} + \begin{bmatrix} B_{11} & B_{12} \\ B_{11}^* & B_{12}^* \end{bmatrix} \begin{bmatrix} x \\ g \end{bmatrix} \qquad (37)$$

with :

$$A^*_{21} = \varepsilon\, A_{21} = \begin{bmatrix} 0 & 0 \\ 60 & 4 \end{bmatrix} \qquad A^*_{22} = \varepsilon\, A_{22} = \begin{bmatrix} 0 & 0.1 \\ -1000 & -4 \end{bmatrix}$$

$$B^*_{21} = \varepsilon\, B_{21} = \begin{bmatrix} 0 \\ 940 \end{bmatrix} \qquad B^*_{22} = \varepsilon\, B_{22} = \begin{bmatrix} 0 \\ -0.1 \end{bmatrix}$$

The figure shows the step-responses of the slow part $x_1(t)$ and of the fast part $z_1(t)$ for inputs $x(t) = 0.1$ and $g(t) = 10$.

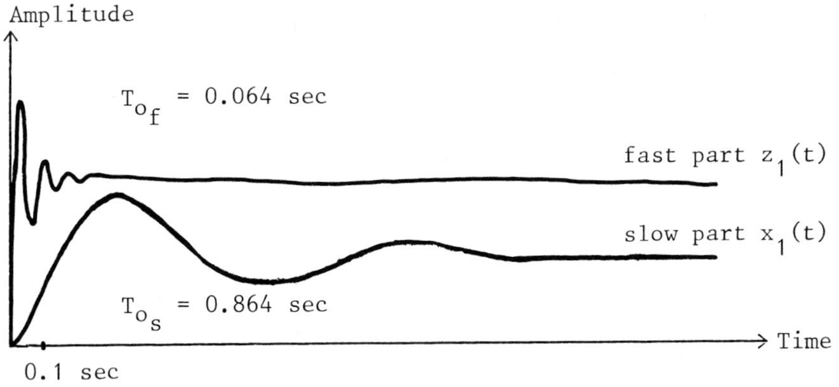

Figure 2. Slow and fast motions

We want here to compare the evolution of the fast part of the global system and of the two reduced subsystem obtained by SP or (SP+R) method.

Equations (4) give here :

$$\begin{bmatrix} \mathring{z}_1 \\ \mathring{z}_2 \end{bmatrix}_{\delta SP} = \begin{bmatrix} 0 & 1 \\ -10000 & -40 \end{bmatrix} \begin{bmatrix} z_1 \\ z_2 \end{bmatrix}_{\delta SP} + \begin{bmatrix} 0 & 0 \\ 9400 & -1 \end{bmatrix} \begin{bmatrix} x \\ g \end{bmatrix} \tag{38}$$

$$y_{\delta SP} = \begin{bmatrix} 1 & 0 \end{bmatrix} \begin{bmatrix} z_1 \\ z_2 \end{bmatrix}_{\delta SP}$$

Equations (20) give then :

$$\left\{\begin{array}{l} \begin{bmatrix} \mathring{z}_1 \\ \mathring{z}_2 \end{bmatrix}_{\delta(SP+R)} = \begin{bmatrix} 0 & 1 \\ -10000 & -40 \end{bmatrix} \begin{bmatrix} z_1 \\ z_2 \end{bmatrix}_{\delta(SP+R)} + \\ \qquad + \begin{bmatrix} 0 & 4/10000 \\ 9400 & -1 \end{bmatrix} \begin{bmatrix} x \\ g \end{bmatrix} \\ y_{\delta(SP+R)} = \begin{bmatrix} 1 & 0 \end{bmatrix} \begin{bmatrix} z_1 \\ z_2 \end{bmatrix}_{\delta(SP+R)} + \int_{t_o}^{t} \begin{bmatrix} 0 & -4.10^{-3} \end{bmatrix} \begin{bmatrix} x \\ g \end{bmatrix} dt \end{array}\right. \tag{39}$$

The different responses are presented Figure 3.

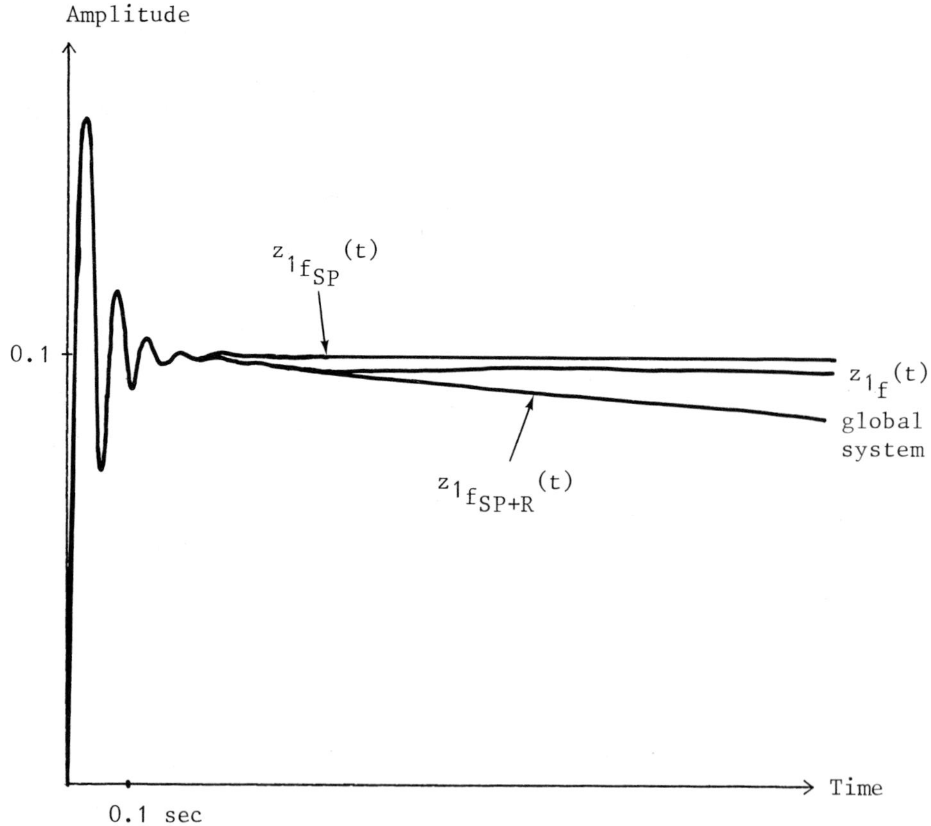

Figure 3. Fast global and decoupled outputs

Some fundamental remarks :

The curves presented Figure 3 show the great advantage of the (SP+R) method on SP method.

It appears that the global output $z_1(t)$ is the result of superposition of two effects. To the input $x(t)$ is associated a very fast oscillatory output $z_{1_\delta}(t)$ and to gravity correspond oscillatories of weak amplitude and low frequency associated to the slow behaviour $z_{1_s}(t)$ of the fast variable $z_1(t)$.

The SP method gives a very good approximation of $z_{1_\delta}(t)$, the two curves are not distinguishable during 80 % of damping time of $z_{1_\delta}(t)$.

The (SP+R) method gives a more exact value of $z_1(t)$, by taking into account the fast and the slow motion of the fast variable, and during a time twice longer than in SP method. The integral term, in (39) represents the slow component $z_{1_s}(t)$.

7. CONCLUSION

The (SP+R) method presented here brings more precision for the study of the fast part of a two-time scale system. To have the best disconnected process, it is possible to keep for the slow part the SP reduced slow subsystem and for the fast part the (SP+R) reduced fast subsystem. This method can be implemented very easily, and gives interesting results for simulation and optimal control. It can be extended to non-linear systems admitting a state space representation in the form $\mathring{x} = A(.)\,x + B(.)\,u$.

REFERENCES

(1) KOKOTOVIC P.V., O'MALLEY R.E., SANNUTI P.
"Singular perturbations and order reduction on control theory. An overview"
Automatica 1976, **12**, 123-132.

(2) SAKSENA V.R., O'REILLY J., KOKOTOVIC P.V.
"Singular perturbations in control theory" Survey 1976 - 1982
IFAC Workshop on Singular Perturbations Juillet 1982, Ohrid, Yougoslavie.

(3) FOSSARD A.J., MAGNI J.F.
"Modélisation, commande et applications des systèmes à échelles de temps multiples"
RAIRO Automatique / Syst. Anal. and Cont. 1982, **16**, n° 1, 5-23.

(4) HUTTON-FRIEDLAND
"Routh approximations for reducing order of linear time invariant systems"
IEEE Trans. on Automatic Control, **AC-20**, n° 3, 329-337.

[5] DAUPHIN-TANGUY G., EL MOUDNI A., BORNE P.
"Sur une estimation des régimes transitoires des processus à plusieurs échelles de temps"
Congrès MECO 1982, Tunis, Tunisie, 151-155.

[6] DAUPHIN-TANGUY G., BORNE P., MEIZEL D.
"On order reduction of multi-time scale systems by singular perturbation and frequency-like methods"
IEEE LSS Symposium 1982, Virginia Beach, U. S. A., 190-196.

[7] DAUPHIN-TANGUY G., LEBRUN M., BORNE P.
"Order reduction of multi-time scale systems using bond-graphs, reciprocal system and singular perturbation method"
To appear in *Journal of Franklin Institute*.

[8] DAUPHIN-TANGUY G.
"Sur une représentation multi-modèle des systèmes singulièrement perturbés"
D. Sc. Dissertation Octobre 1983, Lille, France.

[9] BELL D.J., JACOBSON D.H.
"Singular optimal control problems"
Mathematics in Science and Engineering, **117**, Academic Press, 1975.

[10] MAGNI J.F., FOSSARD A.J.
"Commande en deux étapes des systèmes linéaires à deux dynamiques"
RAIRO Automatique / Syst. Anal. and Cont. 1982, **16**, n° 1, 25-38.

CHAPTER 19

COORDINATED DECENTRALIZED CONTROL (CODECO) WITH MULTI-MODEL REPRESENTATION

Z. Binder, A.N. Hagras[+] and R. Perret
Laboratoire d'Automatique de Grenoble
E.N.S.I.E.G. - I.N.P.G.
38402 - Saint-Martin-d'Hères
France

ABSTRACT. The chapter describes a method for the control of large-scale system composed of a number of interconnected subsystems and whose characteristics have certain time evolution. The control structure proposed is essentially a two level structure with multi-model representation in each level. The tracking approach is employed both for the location of models and for the control. The proposed structure is well adapted to progressive implementation on a distributed microcomputer network.

1. INTRODUCTION

The development in the domain of distributed control systems based on microcomputers networks, increased the interest in the hierarchical and decentralized control methods for complex systems. These systems may represent electrical distribution network, industrial process or economical systems. The control of complex systems or their composing subsystems, that may have non linear characteristics or time evolution, is extremely difficult and there is no general methodology for treating this problem. The control methods usually make use of the linear representation of systems that gives sometimes the only solutions acceptable for real application.

A multi-model representation of systems was proposed [7] to overcome the limits of linear models. The multiple description allows representing the system in a wide range of functions where each linear model represents the system in a particular working conditions. Since then the multi-model technique has developed and its on-line application proved satisfactory [8].

Thus it was natural to think about introducing the multi-model technique to the hierarchical and decentralized control of large-scale systems. In this chapter a multi-model technique based on a tracking approach is proposed and introduced to the *Coordinated Decentralized Control methods* (*CODECO*) [1] [10].

+ died on January 6th, 1984

S. G. Tzafestas (ed.), Multivariable Control, 343–358.

The decomposition of the large-scale system may respect the geographical and functional composition of the system [9] and thus the proposed method lends itself to implementation on a distributed microcomputer network where the internal structure of the decision centers may be multiple, function of the number of models representing the system. Therefore, the implementation of the control structure may be progressive in the time function of the requirements of the system control.

In this chapter a method for the decentralized coordinated control of Large Scale Systems with multi-model representation is described. The large-scale system is supposed composed of or decomposed into a number of interconnected subsystems. The tracking approach is the basis of the two level control structure proposed. In each level linear multiple models represent the system around different operating points. The methods can be formulated as the simultaneous application of the tracking approach for the coordinated decentralized control and the selection of a best model and control law among the multiple model representation.

2. TRACKING APPROACH TO COORDINATED DECENTRALIZED CONTROL (CODECO)

The controlled system is supposed decomposed of or decomposable into a number of interconnected subsystems.

We can distinguish two control levels :

2.1. Local control level.

The local control of each subsystem is elaborated according to its model, the reference variables supplied by the second level and a tracking cost function.

The subsystems models are of the form

$$x_i(t+1) = A_i x_i(t) + B_{wi} w_i(t) + B_i u_i(t) \tag{1}$$

$$y_i(t) = C_i x_i(t) \tag{2}$$

$$v_i(t) = C_{vi} x_i(t) \tag{3}$$

$$w_i(t) = W_i v(t) \tag{4}$$

$$v = [v_1^T, \ldots, v_N^T]^T$$

where

x_i is the state vector of dimension n_i

w_i is the interaction input vector of dimension r_{wi}

u_i is the control vector of dimension r_i

v_i is the interaction output vector of dimension m_v and is defined such that w_i may be given by equation (4) which defines the interactions constraints.

A_i, B_{wi}, B_i, C_i, C_{vi} and W_i are matrices of appropriate dimensions.

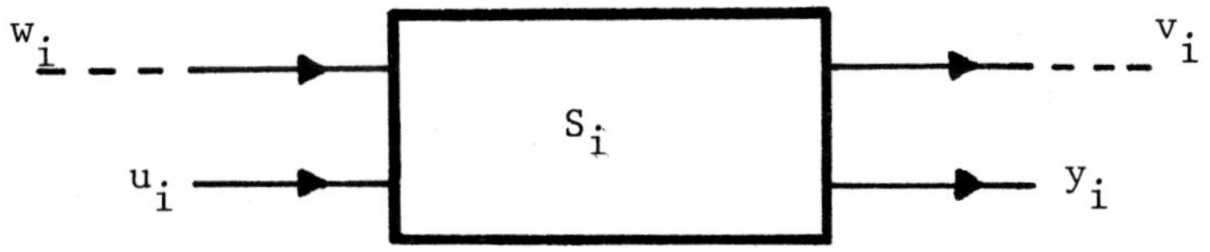

Fig. 1. Subsystem representation.

The tracking cost function will be of the form :

$$J_i = \sum_{t=o}^{T-1} \left[||zy_{ai}(t) - y_{ai}(t)||^2_{Q_i} + ||zu_{ai}(t) - u_{ai}(t)||^2_{R_i} \right] \tag{5}$$

where

$$y_{ai} = \begin{bmatrix} v_i \\ \\ y_i \end{bmatrix} \qquad u_{ai} = \begin{bmatrix} w_i \\ \\ u_i \end{bmatrix}$$

$$zy_{ai} = \begin{bmatrix} zv_i \\ \\ zy_i \end{bmatrix} \qquad zu_{ai} = \begin{bmatrix} zw_i \\ \\ zu_i \end{bmatrix}$$

The cost function is minimized w-r-t the input vector u_{ai} to ensure the optimal tracking of the reference variables.

zy_{ai} and zu_{ai} are the reference variables supplied by the second level. To simplify the solution and to have an algorithm suitable for on-line application, the horizon T in the cost function is taken as unitary [4]. The control law will be

$$u_{ai}(t) = L_i\, x_i(t) + M_{1i} zu_{ai}(t) - M_{2i} zy_{ai}(t) \tag{6}$$

2.2. Coordination level.

A simplified model for the overall system, coordinator model, is obtained either by a less precise identification of the system or by aggregation of the subsystems models. This coordinator model respects the system structure. Therefore, it contains explicitly the input and the output of the overall system, in addition to the interaction variables between the subsystems.

Out of this coordinator model and the external objectives describing the desired behaviour or the reference variables of the overall system, the coordinator will elaborate its control vector that assures the best tracking of its objectives according to a defined cost function. The evolution of the coordinator model variables, that reflect the inputs and the outputs of the local models, will represent the reference variables to be tracked by the corresponding variables of each local model. The tracking of these reference variables given by the coordinator represent its guiding and coordinating effect because they are elaborated out of knowledge, even simplified, of the overall system. Figure 2 represents the described control structure in the case of a system composed of two interconnected subsystems and shows the reference variables supplied by the coordinator MC to the local models M_1 and M_2.

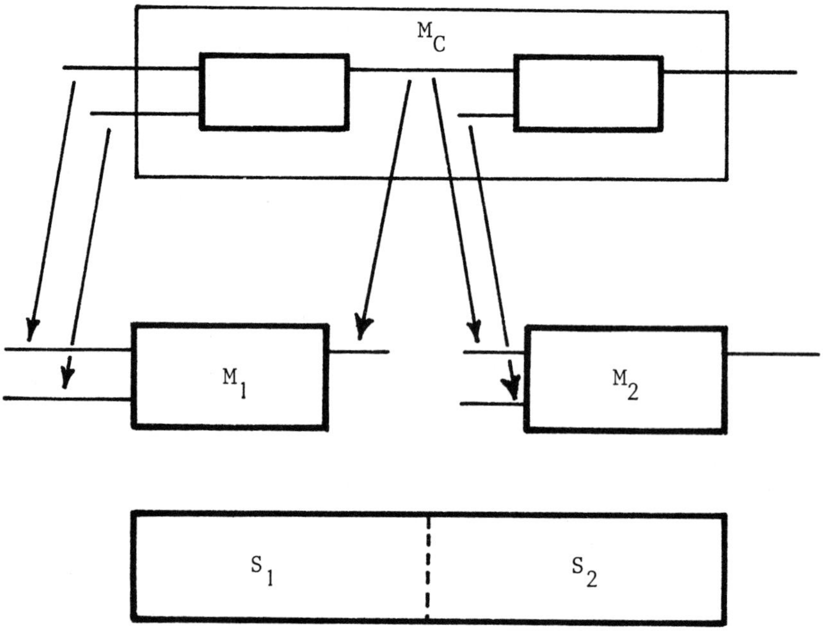

Fig. 2. Coordinated decentralized control.

The coordinator model is of the form :

$$x_M(t+1) = A_M\, x_M(t) + B_M\, u_M(t) \tag{7}$$

$$y_M(t) = C_M\, x_M(t) \tag{8}$$

where

x_M is the state vector of dimension $n < \sum_{i=1}^{N} n_i$

N is the number of subsystems and n_i the dimension of each one.

The coordinator model respects the structure of the complex system, hence

$$u_M = [w_1^T, \ldots, w_N^T, u_1^T, \ldots, u_N^T]^T \tag{9}$$

$$y_M = [y_1^T, \ldots, y_M^T]^T \tag{10}$$

The control u_M will minimize the tracking cost function :

$$J_M = \sum_{t=o}^{T-1} \left[\|\, zu - u_M(t) \|^2\, R_M + \| zy - y_M(t+1) \|^2 Q_M \right] \quad \text{for } T \to \infty \tag{11}$$

where :

$$\|\, z\, \|^2\, R = z^T R\, z$$

zu and zy are the reference variables defined either by static optimization or out of economical basis. In the next, they are considered constant.

The control u_m will be given by [10] :

$$\begin{aligned} u_M(t) &= -L_M\, x_M(t) + M_1 zu - M_2 zy \\ &= -L_M\, x_M(t) + [M_1, -M_2]\, \bar{u} \end{aligned} \tag{12}$$

$$\bar{u} = [zu^T, zy^T]^T$$

L_M, M_1 and M_2 are matrices function of the coordinator model parameters A_M, B_M and C_M and the weighting matrices Q_M and R_M.

The reference variables to be tracked by the local models variables can be put in the form [10] :

$$zw_i = CW_i\, x_M + DW_i\, \bar{u} \tag{13}$$

$$zy_i = CL_i\, x_M + DL_i\, \bar{u} \tag{14}$$

$$zy_i = CS_i\, x_M \tag{15}$$

The matrices CW_i, CL_i, DW_i and DL_i are matrices of appropriate dimensions defined as :

$$L_M = \begin{bmatrix} CW_i \\ CL_i \\ \vdots \\ CW_N \\ CL_N \end{bmatrix} \qquad [M_1, - M_2] = \begin{bmatrix} DW_i \\ DL_i \\ \vdots \\ DW_N \\ DL_N \end{bmatrix}$$

$$C_M = [CS_1, \dots, CS_N]$$

3. MULTIMODEL WITH TRACKING APPROACH

The technique of multi-model control lies within two main parts : location and control.

3.1. Location.

When a physical system is described by a finite set of models, how could one quantify at each instant, the validity of each model, or in other words, how should one locate the system w.r.t. its models ?

In order to answer these questions, it is necessary to ensure that the models are in continuous competition in their approach to the system. Each candidate model is governed by the same conditions of behaviour and tries to position itself as closely as possible (in the sense of a mathematical criterion) to the system that is, it tracks the system.

Once the models have taken positions in the neighbourhood of the system, one has to choose a criterion for classifying the models based on a proximity index. In general, the criterion for positioning and the proximity index are identical or at least correlated.

3.1.1. Positioning the Models. For dynamic system the positioning can be ensured by tracking the system with models which are in competition to approximate the system behaviour. Because of the tracking approach adopted [2], the models approach the system and their position depends principally on the system evolution, on the form of the criterion adopted and the control action applied for modifying the evolution of the models.

3.1.2. Classification of the models. The ultimate aim of the location of the system with respect to the set of its models is to classify the models according to their quality, i.e. according to their possibility of approximating the system during a given period or at the instant of classification. The quality of models could be judged by comparing with the system, its evolution or its variables. In practical applications, only the external variables of the system are accessible and can be compared with the corresponding variables of the models.

One can envisage the different forms of criteria called quality index according to which one may classify the models.

3.2. Control Synthesis.

The utilization of several models leads on to a synthesis of control in the following two stages :

1) the unitary control law for each model is independently computed,
2) the global control law for the system is computed as a function of the above and the results of location steps.

3.2.1. Unitary control (single model). The basic control laws are established mainly by the aims of the system control : either regulation or tracking along the transition between two set points. The algorithms depend as well on the representation one chooses for the system : deterministic or stochastic.

Generally the unitary control is elaborated from each model independently, one considers each model as an exact representation of the system to be controlled.

3.2.2. Global control law (multi-model). The synthesis of the control action in its final form, is a function of the different strategies concerning the number of models considered for the synthesis.

Here we notice two cases :

1) The single model control used at each moment for the control.
2) Several models approximating simultaneously the system ; in this case, the control applied, is obtained as the weighted combination of basic controls.

4. ALGORITHMS DEVELOPMENT

The development of algorithms is presented for the two levels described above : coordinator level and local level (Fig. 3).

4.1. Coordinator Models Location.

In general N_C simplified coordinator models are elaborated for the system around different operating points. The location of the coordinator model will be such as to select the coordinator model, that in the same time best approximates the system and test tracks the reference variables defined for the overall system.

4.1.1. Location w.r.t. the system. A quadratic criterion JCS_m is defined to express the deviation between the input and the output variables of the system (u,y) and the corresponding variables of the coordinator model MC_m (u_m, y_m). This criterion express the tracking of the system by the coordinator models :

$$JCS_m = \sum_{t=o}^{T-1} \left[|| u(t) - u_m(t) ||^2_R + ||y(t) - y_m(t) ||^2_Q \right] \qquad (16)$$

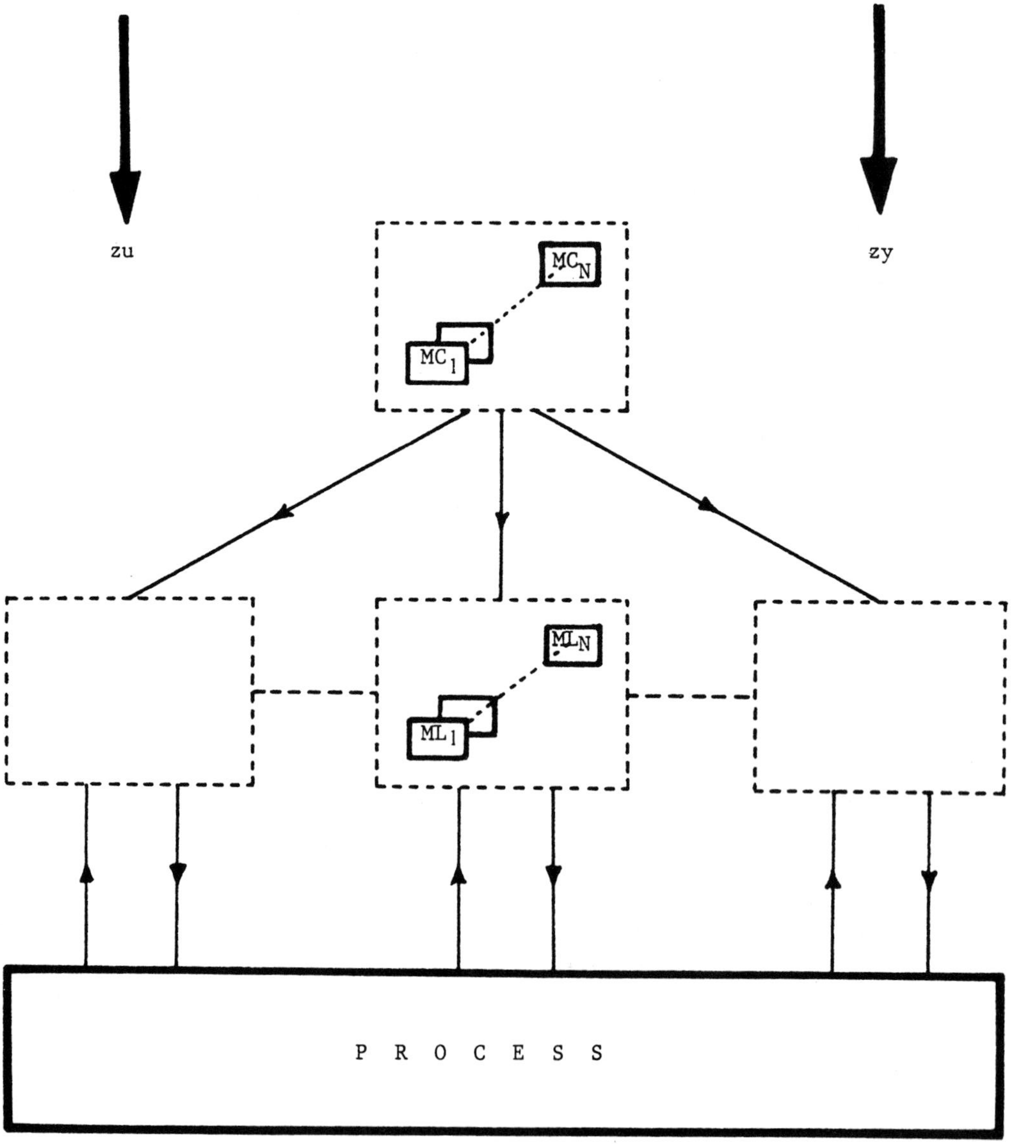

Fig.3. Multi-model coordinated decentralized control.

It is fairly difficult to obtain this solution for the optimisation problem. To determine the input u_m for the model which locates the model w.r.t. the system in an optimal fashion, a knowledge of the evolution of u and y (input and output respectively of the system) over the period (0,T) is indispensable. This is impossible here as the objectives are the real trajectories of the system, which themselves depend on the decisions taken in real time. Therefore one has to make simplifying hypotheses in order to have a solution to our problem. In this case, the computations would have to be done in reverse time starting from the final time T [1]. This is acceptable in simulation but poses problems for real life systems.

To obtain an useful solution, we have made the following assumptions :

- The model MC_m is a valid representation of the system over the period (0,T).
- The input $u(t)$ of the system is generated by a linear system over the period (0,T) which is presumed sufficiently long.

An explanation has been given in the first hypothesis. If we suppose that it is verified, then model MC_m is able to track variable objectives, in a perfect manner. On the contrary, deviations during tracking will occur, showing a difference of behaviour between the system and the model.

In the case where the input $u(t)$ takes a step at the initial instant, the input $u_m(t)$ of the model which minimises the quadratic cost function (16) could be expressed as follows :

$$u_m(t) = - L_m x_m(t) + M_{1m} u(t) - M_{2m} x(t) - M_{3m} y(t) \tag{17}$$

with $L_m = S B_m^T K_m A_m$

$$M_{1m} = S_m B_m^T [G_m^T - I]^{-1} [L_m^T R - Mo_m B_m] + S_m R$$

$$M_{2m} = S_m B_m^T [G_m^T]^{-1} Mo_m$$

$$M_{3m} = S_m B_m^T [G_m^T]^{-1} C_m^T Q$$

$$S_m = [R + B_m^T K_m B_m]^{-1}$$

$$G_m = A_m - B_m L_m$$

Mo_m = solution of the equation $G_m^T Mo_m A_m - Mo_m = C_m^T Q C_m$

K_m = steadystate solution of the Riccati equation :

$$K_m(t) = A_m^T K_m(t+1) A_m - A_m^T K_m(t+1) B_m [R + B_m^T K_m(t+1) B_m]^{-1}$$

$$B_m^T K_m(t+1) A_m + C_m^T Q C_m$$

We note that the knowledge of the state vector x(t) of the system is necessary in the absence of which x(t) would have to be constructed through an observer based on the mth model or in certain cases replaced by the states $x_m(t)$ of model MC_m.

4.1.2. Location w.r.t. the reference variables. A quadratic criterion JCR_M is defined to represent the tracking of the reference variables, defined for the system, by each coordinator model.

$$JCR_M = \sum_{t=o}^{T-1} \left[||zu - u_M(t)||^2 R + ||zy - y_M(t)||^2 Q \right]$$

The input vector u_M that minimises the criterion will be as given by eq. (12) $u_M(t) = - L_M x(t) + M_1 \ zu - M_2 \ zy$

4.1.3. Classification of the Coordinator Models. The coordinator models are classified according to their possibility to jointly approximate the system and track the reference variables. A quality index can be defined for each coordinator model as :

$$\begin{aligned} JC_m(t) &= K_S \ JCS_m(t) + K_R \ JCR_m(t) \\ &= K_S \ [\ ||u(t) - u_m(t)||^2 R + ||y(t) - y_m(t)||^2 Q] \qquad (18) \\ &+ K_R \ [\ ||zu - u_M(t)||^2 R + ||zy - y_M(t)||^2 Q] \end{aligned}$$

K_S and K_R are weighting scalers that determine the relative importance given to approximating the system by the coordinating models or tracking the reference variables defined for the system.

$$K_S + K_R = 1$$

The best model will be that one which gives the weaker value of the quality index

$$JC_m(t) = \inf \{JC_1(t), JC_2(t), \ldots, JC_{NC}(t)\} \qquad (19)$$

4.2. Reference Variables for Local Models.

As described in the problem formulation the coordinator model function is to supply the local models with the reference variables zu_i and zy_i.

The utilization of several models leads on to the following two stages :

1) The unitary control law for each coordinator model is independently computed as given by (12). The reference variables zu_{ij} and zy_{ij} allocated by the coordinator model MC_j to the subsystems S_i will be as given by eqs. (14) and (15).

2) The reference variables are allocated for the local level function of the above stage and the results of the location steps. We can notice two cases.

a) Only one coordinator model is used at each instant. In this case the local models will follow or track this model through the reference variables as given by eqs. (14) and (15).

b) Several coordinator models are used simultaneously. In this case the local models tracks simultaneously all the coordinator models but of course with weights corresponding to the quality of each one. The tracking cost function for each local (subsystem) model will be of the form:

$$J_i = \sum_{t=o}^{T-1} [\ ||zu_{i1}(t) - u_i(t)\ ||^2\ R_{i1} + ||zy_{i1}(t) - y_i(t)||^2\ Q_{i1} + \qquad (20)$$

$$+ \quad ||zu_{iN_C}(t) - u_i(t)||^2 R_{iN_C} + ||zy_{iN_C}(t) + y_i(t)||^2 Q_{iN_C}]$$

where

$$R_{ij}(t) = \alpha_{ij}(t)\ R_i \qquad\qquad Q_{ij} = \alpha_{ij}(t)\ Q_i$$

α_{ij} are weighting factors function of the real time quality index of the coordinator models.

$$\alpha_{ij}(t) = \frac{\dfrac{1}{JC_j(t)}}{\displaystyle\sum_{k=1}^{N_C} \frac{1}{JC_k(t)}}$$

$$\sum_{j=1}^{N_C} \alpha_{ij} = 1$$

The minimisation of the cost function (20) is equivalent to defining new reference variables zu_i and zy_i elaborated by the coordinator models together with the control vector.

$u_i(t)$ that minimises the cost function $\hat{J}_i$ [6] :

$$zu_i(t) = \alpha_{ij}(t)\ zu_{i1}(t) + \ldots\ldots\ldots + \alpha_{iN_C}(t)\ zu_{iN_C}(t)$$

$$zy_i(t) = \alpha_{ij}(t)\ zy_{i1}(t) + \ldots\ldots\ldots + \alpha_{iN_C}(t)\ zy_{iN_C}(t)$$

$$\hat{J}_i = \sum_{t=o}^{T-1} \left[||zu_i(t) - u_i(t)||^2\ R_i + ||zy_i(t) - y_i(t)||^2\ Q_i \right] \qquad (21)$$

the control vector u_i will be [6].

$$u_i(t) = -\ S_i(t+1)\ B_i^T\ K(t+1)\ A_i\ x_i(t) + S_i(t+1)\ R\ zu_i(t)$$
$$+\ S_i(i+1)\ B_i^T g_i(t+1)$$

where

$$S(t+1) = [R_i + B_i^T\ K_i(t+1)]^{-1}$$

$$g_i(t) = [A_i - B_i S_i(t+1)\ B_i^T\ K_i(t+1)\ A_i]^T g_i(t+1)$$
$$+\ [S_i(t+1)\ B_i^T\ K_i(t+1)\ A]^T\ R_i\ zu_i(t) + C_i^T\ Q_i zy_i(t)$$

To have an algorithm adaptable to real-time applications, the optimization period is reduced to one step. In this case the control law will be as given by eq. (6)

$$u_i(t) = -\ L_i\ x_i(t) + M_{1i}\ zu_i(t) - M_{2i}\ zy_i(t)$$

4.3. Subsystems Model Location.

The tracking approach is also used in the local level to select the subsystems models that best approximate it and best follow the coordinator model. Each subsystem S_i is represented by NL_i models elaborated around different operating points.

4.3.1. <u>Location with respect to the system</u>. A quadratic criterion J_{ij} is defined for the model j of the subsystem S_i (M_{ij})

$$JS_{ij} = \sum_{t=o}^{T-1} \left[||\ u_i(t) - u_{ij}(t)\ ||^2\ R_i + ||y_i(t) - y_{ij}||^2\ Q_i \right]$$

u_{ij} and y_{ij} represent the input and the output of the model M_{ij}. u_i and y_i represent the corresponding variables of the system.

The input vector u_{ij} that minimize the above criterion will be as given by eq. (16) and under the same assumptions

$$us_{ij}(t) = -\ L_{ij}x_{ij}(t) + M_{3ij}\ u_i(t) + M_{4ij}x_i(t) - M_{5ij}\ y_i(t) \qquad (22)$$

4.3.2. Location with respect to the coordinator model. A quadratic criterion is defined to position the subsystem models w.r.t. coordinator model. This criterion can be of the same form as given by eq.(5) to determine the subsystem control that ensures the best tracking of the reference variables defined by the coordinator model.

$$ur_{ij} = - L_{ij}\, x_{ij} + M_{1ij}\, zu_i(t) - M_{2ij}\, zy_{ij}(t)$$

4.3.3. Classification of the subsystems models. The subsystems models are classified according to their possibility to jointly approximate their subsystems and best tracking of the coordinator model.

A quality index is defined for each step and for each model as

$$J_{ij}(t) = L_S\, JS_{ij}(t) + L_C\, JC_{ij}(t)$$

L_S and L_C are weighting scaler numbers

$$L_S + L_C = 1$$

4.4. Control of Subsystems.

The multi-model representation leads on to a synthesis of control in the following two stages [2] [8].

1. An unitary control law for each subsystem model is independently computed as given by eq. (6).
2. The control applied to the subsystem is computed as a function of the above stage and the results of location and classification of the models.

5. CONCLUSION

The chapter treats the problem of control of large-scale systems whose characteristics have certain time evolution that can be represented by several models. A two level control method is proposed to solve this problem. The large-scale system is supposed decomposed of or decomposable into a number of interconnected subsystems. The first level represents the local control of subsystems. The coordination in the second level is represented by a simplified coordinator model of the overall system that supplies the first level with the reference variables to be tracked by the subsystems variables. The coordinator in the second level and the subsystems in the first level have multi-models representation. Fig. (4) represents the organigram of the multi-model coordinated decentralized control structure.

The proposed control structure represents several practical and conceptional advantages.

First, the extension from two levels to more levels is possible. The transfer of information is effected between adjacent levels. The higher level takes into account the external information represented by

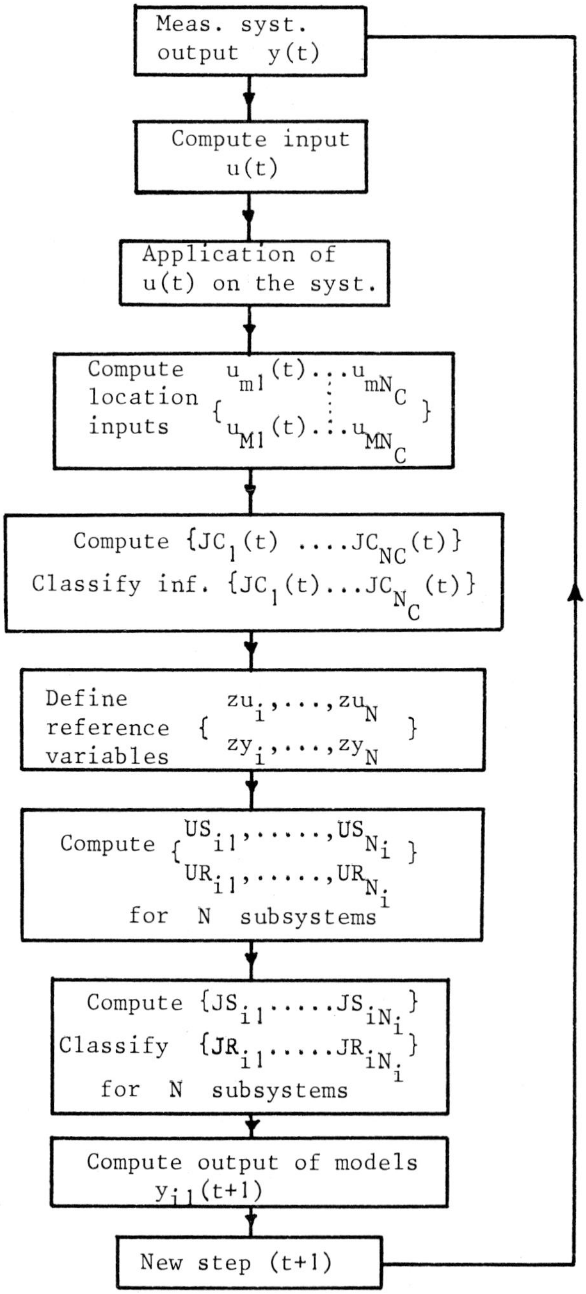

Fig. 4. Organigram of the control structure.

by the reference variables defined for the overall system and supplies the lower level controlers with the necessary reference variables.

A practical advantage of the proposed method is the possibility of progressive construction of the control structure starting by the local control of each subsystem. The multi-model representation can be progressively introduced function of the practical situation. The implementation of one or more higher levels does not introduce any modifications in the local level.

The realization of the control structure is well adapted to the distributed control systems with parallel structure for each decision center for the multi-model computation. A microcomputer network will be suitable for the progressive application and in the same time for the security and the reliability of the control structure.

6. REFERENCES.

[1] Binder, Z. (1977). 'Sur l'organisation et la Conduite des Systèmes Complexes' Thèse de Docteur-es-Sciences, Grenoble.

[2] Binder, Z., D. Baudois, H. Fontaine and M.F. Magalhaes (1981) 'About a Multi-Model Control Methodology, Algorithms, Microprocessors implementation - application' *In Control Science and Technology for the Progress of Society*, Proceeding of IFAC VII Congress, KYOTO.

[3] Coudurier, J.F. (1982). 'Robustesse de la Commande Décentralisée Coordonnée (CODECO)'. Thèse de Doctorat 3°Cycle, I.N.P. GRENOBLE.

[4] Hagras, A.N. (1979). 'Commande Décentralisée Coopérative des Systèmes Dynamiques Interconnectés'. Thèse de Docteur-Ingénieur, I.N.P. GRENOBLE.

[5] Hagras, A.N. and Z. Binder (1983). 'A Decentralized Cooperative Control Method' *Large Scale Systems, Theory and Application.* 4 (1983), pp. 263-277.

[6] Janex, A. (1979). 'Changement de Modèles dans la Commande à Plusieurs Niveaux' Thèse de Doctorat 3°Cycle - I.N.P. GRENOBLE.

[7] Lainiotis, D.G. (1969). 'Optimal Adaptive Estimation, Structure and Parameter Adaptation' *Proceedings of the 1969 I.E.E.E. Adapt. Process. Symp.*

[8] Ferreira Magalhaes, M. (1983). 'Sur une représentation multi-modèle - Etude d'une Technique de Localisation et Application à la Commande d'une Colonne à Distiller" - Thèse de Docteur-Ingénieur I.N.P. GRENOBLE.

[9] Perret, R. (1983). 'Sur une Méthode de Commande Hiérarchisée' Bulletin ASSPA/SGA - ZURICH, 3 (1983), pp. 3-10.

[10] Rey, D. (1978). 'Sur la Commande Décentralisée Coordonnée - Application à un Procédé Pilote de Distillation' Thèse de Docteur-Ingénieur, I.N.P. GRENOBLE.

CHAPTER 20

DESIGN OF TWO-LEVEL OPTIMAL REGULATORS WITH CONSTRAINED STRUCTURES

M. MARITON, M. DROUIN, H. ABOUKANDIL

Laboratoire des Signaux et Systèmes
CNRS - ESE
Plateau du Moulon - 91190 GIF SUR YVETTE - FRANCE.

ABSTRACT

In the time domain, a basic approach for multivariable problems is the now classical linear quadratic regulator. In this chapter the case of large scale systems is considered where one adds to the multivariable nature of the problem, its geographical distribution. Structural constraints such as partial decentralization are then prescribed for the control laws. Using a decomposition coordination method, a solution is sought within a two level framework. The first level consists of a feedback with constrained structures. Techniques are presented to adjust the feedback gain to derive a control law that drives the process to an almost optimal performance. This step of the design corresponds to a multivariable control problem with structural constraints, typically decentralization. The characteristic feature of that level is the use of a local criterion to obtain an initial feedback structure that is afterwards optimized by the tuning of a limited set of parameters. In this presentation, the upper level of the hierarchy is specialized to monitor the static behaviour of the system. Through a slow feedback, it is shown how it can counterback unmeasured disturbances. Finally a microprocessor implementation is presented.

I - INTRODUCTION

The increasing amount of research effort in the development of decomposition-coordination methods can be traced back to two main reasons :

- The complexity of the controlled process has increased and the classical tools of automatic control were often proved inefficient to solve these new challenging problems. Complexity comes either from the large dimensionnality of the process, either from nonlinearities or delays.
- Microcomputers prodigious emergence allows the use of cheap but powerfull distributed control structures. Single Board Computers appear as the evident unit controller for interconnected large scale systems.

S. G. Tzafestas (ed.), Multivariable Control, 359–379.

First introduced by Dantzig and Wolfe (Dantzig 61) to solve large linear programming problems, the basic idea of decomposition-coordination techniques is intuitively simple : the initial optimization problem is splitted into a set of smaller subproblems, easier to solve, coordinated by a supremal problem that ensures convergence towards the global optimal solution. This approach was further developped for large scale optimal control problem by pionnering workers (Mesarovic 70, Lasdon 65, Pearson 71, Bauman 68). However, due to the geographical expansion of systems and the lack of robustness of centralized control schemes, stress was laid on decentralized implementation of these structures (Singh 77, Titli 79, Sandell 78, Findeisen 78, Cohen 78).

In the second section of this paper, an original classification of decomposition-coordination methods is presented. While referring the reader to quoted papers for detailed presentations, it is believed that this classification provides some insight in this topical field. Then, in the next two sections, we proceed to the principle of a new method (Drouin 82) and its application to dynamical systems. In the fifth section procedures are proposed for the design of the regulation first level, while the next section gives special emphasis to a special version of the second level algorithm. Finally these ideas are illustrated on a realistic experimentation and conclusions proposed.

II - A CLASSIFICATION OF DECOMPOSITION-COORDINATION METHODS

The classification proposed here is based on the analysis of the "concepts" or "techniques" used by the different solutions. This analysis highlights two fundamental stages tied with the concepts of decomposition and coordination :

(i) Problem manipulation,
(ii) Solution strategy.

The manipulation transforms the initial problem into another one whose algorithmic solution is easier. The solution strategy indicates how the solution of the modified problem is sought. A classification of techniques for these two steps was first proposed by Geoffrion (Geoffrion 71) in the context of mathematical programming,(Wilson 79) recently extended this analysis to the optimal control field.

The following incomplete list is proposed for manipulations and strategies

Problem manipulation :
- parametric decomposition,
- structural decomposition,
- Lagrange transformation,
- Quasilinearization.

Research strategy :
- piecewise,
- restriction,
- relaxation,

- feasible direction.

The existing techniques use several problem manipulations and a research strategy. Depending on the strategy chosen, a method belongs to one of the two main families, that is sequential algorithms or parallel algorithms. This analysis is interesting from two point of views :

- It quickly features the advantages and the drawbacks of a given method from its strategical choices. A survey of existing techniques is given below in this spirit.
- New methods would arise from different combinations of the mentionned recipes. The new method proposed in the next section is such an example.

2.1 Hierarchical control methods

This approach, most often encountered in the litterature, is based on three problem manipulations :

- A variable transformation : new intermediate variable are introduced that might, or might not, have a physical meaning. They are called coupling or interaction variables

$$Z_i = \gamma_i(X,U)$$

- A Lagrange transformation : it replaces the contrained problem by an unconstrained one

$$\mathcal{L} = \sum_i J_i + \sum_i \psi_i^T f_i + \sum_i \mu_i^T (Z_i - \gamma_i)$$

- A parametric decomposition : the lagrangian is decomposed into independent sublagrangians

$$\mathcal{L} = \sum \mathcal{L}_i$$

The research strategy is either a feasible direction or a relaxation depending on the variable that are parameterized. These methods are very well suited for off-line problems but they become less efficient for on-line applications since the actual control depends only on the coordination parameters, that is the process is left uncontrolled if a communication link breaks down.

2.2 Relaxation control methods

This approach was originally developped in the field of mathematical programming (Auslender 70, Glowinski 71) and extended to control problems by Lhote and Miellou (Miellou 75, Lhote 79). For these methods the techniques used are

- A parametric decomposition : the interaction variables are parameterized
- A sequential resolution : the result of a subproblem is

immediately used by the next subproblem.
The upper level task is not present here so that local tasks often become heavy burdens for existing microprocessors.

2.3 Gradients control methods

Cohen (Cohen 78) proposed this approach as a generalization of most existing algorithms. The mark of this method is an initial manipulation of the problem :

- Problem transformation : the original problem $\underset{u}{\text{Min}}\ J(u)$ is transformed in an auxiliary problem where one seeks the solution iteratively

$$u^{\ell+1} = \text{Arg}\ \underset{V}{\text{Min}}\ \mathcal{G}^{\ell} = K(u) + [\varepsilon^{\ell} J'(u^{\ell}) - K'(u^{\ell})]^{T}\ u$$

where the kernel K(u) is additive and convex but otherwise arbitrary.
- Lagrange transformation,
- Parametrization.

Many of the existing techniques then simply reduce to a proper choice of the characteristic kernel K(u).

As a conclusion of this brief survey it is noticeable that decomposition coordination methods remained so far a powerfull tool for the solution of optimization problems off-line and have not fullfilled initial hopes for on-line control of complex processes.

III - PRINCIPLE OF A NEW METHOD

The basic idea of the method comes from the remark that approaches that yield closed-loop solutions either do not introduce the adjoint vector (dynamic programming) or get rid of it (Kalman's method). This path is followed here together with the fundamental notions of decomposition and coordination. Among the "concepts" mentionned above, the new method proposed here consists in :

- A problem transformation (step 1 and 2 in the following)
- A parametric decomposition (step 3)
- A relaxation (step 4)

The principle of the method is recalled for a static, unconstrained, optimization problem.

Let a static system described by

$$f(x,u) = 0 \tag{1}$$

and a performance index J(u,x). The problem is to minimize J :

$$\underset{u}{\text{Min}}\ J(u,x) \tag{2}$$

$x \in R^n$ is the state vector, and $u \in R^r$ the control vector. The following assumptions hold

- J is strictly convex
- There exists a matrix $G = \dfrac{\partial f}{\partial u}$
- The matrix $\dfrac{\partial f}{\partial x}$ is continuous and non-singular.

From these assumptions, it comes

i) There exist a function g such that

$- \forall\, u \in R^r,\ \exists x = g(u) \quad / \quad f(x,u) = 0$

$-\dfrac{dx}{du} = \dfrac{\partial f}{\partial x}^{-1} \cdot \left(\dfrac{\partial f}{\partial u}\right) = -\dfrac{dg}{du}$

ii) The performance index is in fact function solely of u

$I(u) = J(u,g(u)).$

The optimality of a control u* is then stated as

$$\left.\frac{dI(u)}{du}\right|_{u=u^*} = 0 \qquad (3)$$

Remark : The function g(u) was only supposed to exist. It will appear in the following that the only required knowledge is the first derivative $\frac{dg}{du}$, called the "influence matrix", that can be computed from test responses, as explained below.
The solution of the problem (1), (3) becomes difficult when the system is large or control structure imposed. The proposed method is characterized by the following steps :

Step 1 :

The control space U is decomposed in N subspaces

$U = U_1 \times U_2 \times \ldots \times U_N$

each local controller computes its control u_i to solve the ith part of the stationnarity conditions (3)

$$\left.\frac{dI(u)}{du_i}\right|_{u_i=u_i^*} = 0 \qquad i=1,N \qquad (4)$$

Step 2 :

The performance index J (or I) is decomposed in N different manners on the same pattern

$$I(u) = I_i(u_i) + \bar{I}_i(u_i) \qquad i = 1,N \qquad (5)$$

The condition (4) then becomes :

$$\left.\frac{dI_i(u_i)}{du_i}\right|_{u_i=u_i^*} + \left.\frac{d\bar{I}_i(u_i)}{du_i}\right|_{u_i=u_i^*} = 0 \qquad i=1,N \tag{6}$$

Step 3 :

In order to simplify the task of local controllers, a part of (6) is parameterized

$$\rho_i = \frac{d\bar{I}_i(u_i)}{du_i} \tag{7}$$

In the following ρ_i will be referred to as the coordination vector.
The task of the ith local controller is then to satisfy

$$\frac{dI_i(u_i)}{du_i} + \rho_i = 0 \tag{8}$$

It will be shown in the next section how a proper selection of the subcriterion I_i gives an analytical solution to (8), so that the control law is a function of x_i and ρ_i

$$u_i = u_i(x_i,\rho_i) \tag{9}$$

The upper level task is to compute the coordination from

$$\rho_i^* = \frac{d\bar{I}_i(u_i^*)}{du_i} \tag{10}$$

Step 4 :

Equation (10) explicits the dependence of ρ_i^* on u_i^*. As u_i^* is not known so far, it is clear that the control law (9),(10) together with the system equation (1) form some sort of two point boundary value problem that has to be solved by repetitive iterations. For a given estimate ρ_i^c of (10), local controllers compute a control u^c that gives through (1) a state x^c of the process. From the knowledge of (x^c,u^c) the upper level finds a new estimate $\rho_i^{c+1/2}$ of (10). To improve convergence, it is usefull to filter the estimates ρ_i through a relaxation scheme

$$\rho^{c+1} = \Omega\rho^{c+1/2} + (I-\Omega)\rho^c \tag{11}$$

When the upper level fails to compute ρ the uncoordinated control law (9), $u_i = u_i(x_i,o)$, corresponds to the optimization of I_i. However this local control law depends on the element $\frac{\partial x_i}{\partial u_i}$ of the influence matrix and in this sense it takes into account the system as a whole. This feature, not present in other methods, makes the control always feasible and most often stabilizing. The coordination vector simply informs the controller of its effect on the subcriterion $\bar{I}_i$.

IV - APPLICATION TO DYNAMIC PROCESSES

This section proposes the application of the method to the case of discrete time dynamic processes. These systems will be described by a state space recurrence equation since this representation permits an easy calculation of the influence matrix $\frac{dx}{du}$. The models will be linear for simplicity of exposition, but non linear problems have been successfully tackled in a similar way.
Depending on the decomposition, the control scheme will be said coordinated, when the tasks are assumed by one controller in a centralized manner, or hierarchical, when several distributed computers act simultaneously.

4.1 Coordinated control/temporal decomposition

Let us consider the linear quadratic dynamical problem

$$\underset{u_k}{\text{Min}}\ J = \frac{1}{2}\sum_{k=o}^{K} x_{k+1}^T Qx_{k+1} + u_k^T Ru_k \tag{12}$$

$$\text{s.t.} \quad x_{k+1} = Ax_k + Bu_k \tag{13}$$

with $Q \geqq 0$, $R > 0$

(A,B) commandable

(K+1) fictitious agents are supposed to act at each control instant k. The criterion J is decomposed into (K+1) different ways so that the part assigned to the kth agent is precisely the instantaneous value of the criterion

$$J_k = \frac{1}{2} x_{k+1}^T Qx_{k+1} + \frac{1}{2} u_k^T Ru_k \tag{14}$$

The complement of this subcriterion is affected to the upper coordinating level

$$\bar{J}_k = \frac{1}{2}\sum_{\ell\neq k} x_{\ell+1}^T Qx_{\ell+1} + u_\ell^T Ru_\ell \tag{15}$$

The problem to be solved by each agent is then

$$\underset{u_k}{\text{Min}}\ C_k = \frac{1}{2} x_{k+1}^T Qx_{k+1} + \frac{1}{2}u_k^T Ru_k + \rho_k^T u_k \tag{16}$$

$$\text{s.t.} \quad x_{k+1} = Ax_k = Bu_k$$

An obvious computation yields the following solution

$$u_k = \Gamma x_k + \theta_k \tag{17}$$

with Γ the feedback matrix $\Gamma = -(R+B^TQB)^{-1}B^TQA$ (18)

and $\theta_k = -(R+B^TQB)^{-1}\rho_k$ (19)

The coordinating vector ρ_k is evaluated at the second level as explained earlier

$$\rho_k = \frac{d\bar{I}_k}{du_k}$$

That is

$$\rho_k = \sum_{\ell \neq k} \frac{dx_{\ell+1}^T}{du_k} Qx_{\ell+1} + \frac{dx_\ell^T}{du_\ell} \Gamma^T Ru_\ell \tag{20}$$

The influence matrices, derivatives of x with respect to u, are given by

$$\frac{dx_{\ell+1}}{du_\ell} = \begin{cases} \tilde{A}^{\ell-k}B & \text{if } \ell \geq k \\ 0 & \text{if } \ell < K \end{cases} \tag{21}$$

with $\tilde{A} = A+B\Gamma$

The expression of ρ_k is obtained in a limited sum

$$\rho_k = \sum_{\ell \geq k+1}^{K} B^T\tilde{A}^{T(\ell-k)}Qx_{\ell+1} + B^T\tilde{A}^{T(\ell-k-1)}\Gamma^TRu_\ell \tag{22}$$

with $\rho_K=0$

It can be checked that ρ_k in (20) can be computed recursively by means of the equation

$$\begin{cases} \rho_k = B^T W_k \\ \text{with } W_{k-1} = \tilde{A}^T Qx_{k+1} + \tilde{A}^T W_k + \Gamma^T Ru_k \\ W_K = 0 \end{cases} \tag{23}$$

As in the static case, the coordination is relaxed at the upper level

$$\rho_k^{c+1/2} = \Omega_k\rho_k^{c+1/2} + (I - \Omega_k)\,\rho_k^c \tag{24}$$

It is possible to adjust off-line the relaxation matrices Ω_k, so that on-line the coordination algorithm converges quickly to the optimal solution

$$\rho_k^* = \lim_{c\to\infty} \rho_k^c \tag{25}$$

4.2 Hierarchical control / temporal and spatial decomposition

The process itself is supposed to be decomposed into subsystems ssi, i=1 N, each characterized by a state x_i, controlled by u_i and evolving under the interaction z_i. To each subsystem ssi is associated a controller S_i which may be a mini or a micro-computer. Another computer, Ssup, will assume the coordinating or supervising task. The state equation of the system is then

$$\text{ssi} \quad \begin{cases} x_{ik+1} = A_{ii}\, x_{ik} + B_{ii}\, u_{ik} + z_{ik} \\ \text{with } z_{ik} = \sum_{j \neq i} A_{ij}\, x_{jk} \end{cases} \tag{26}$$

The criterion (12) remains unchanged but the matrices Q and R are block-diagonal. To each local agent the local instantaneous criterion is assigned, that is

$$J_{ik} = \frac{1}{2} x_{ik+1}^T Q_i\, x_{ik+1} + \frac{1}{2} u_{ik}^T R_i u_{ik} \tag{27}$$

so that the local optimization problem becomes

$$\min_{u_{ik}} C_{ik} = J_{ik} + \rho_{ik}^T u_{ik} \tag{28}$$

The solution of (28) is straightforward and the control is

$$u_{ik} = -(R_i + B_{ii}^T Q_i B_{ii})^{-1} (B_{ii}^T Q_i A_{ii} x_{ik} + B_{ii}^T Q_i z_{ik} + \rho_{ik}) \tag{29}$$

That is, with obvious notations

$$u_{ik} = \Gamma_i\, x_{ik} + \Lambda_i\, z_{ik} + \theta_{ik} \tag{30}$$

The coordination vectors satisfy

$$\rho_{ik} = \frac{d\bar{I}_{ik}}{du_{ik}} \qquad \forall k \in (0,K),\ \forall i \in (1,N) \tag{31}$$

One can then easily show that (31) reduces to a recurrent equation

$$\rho_{ik} = B_i^T w_k \tag{32}$$

$$\text{with } w_k = \tilde{A}^T Q x_{k+2} + \tilde{\Gamma}^T R u_{k+1} + \tilde{A}^T w_{k+1}$$

$$\text{where } B_i^T = [0,0,\ldots,B_i^T,\ldots 0]$$

$$\tilde{A}_{ii} = A_{ii} + B_{ii}\Gamma_i$$

$$\tilde{A}_{ij} = (I + B_{ii}\Lambda_i)A_{ij}$$

$$\tilde{\Gamma}^T = [\Gamma_1^T,\ldots,\Gamma_N^T]$$

$$\tilde{\Gamma}_i = [\Lambda_i A_{i1}, \ldots, \Gamma_i, \ldots, \Lambda_i A_{iN}]$$

The decentralized control structure thus obtained is then computed and the control law (30) to be applied to each subsystem is composed of two parts : the first one is dependent only on the local state vector and the interactions, while the second is a function of the coordination parameters. Thus a control structure is obtained with local feedbacks ; The absence of coordination corresponds to solutions where only local optimization have been performed so that this operating mode may still be acceptable. As in the temporal decomposition approach (§ 4.1) the coordination parameters may be estimated off-line and then fixed at local levels in on-line applications. In that case the control structure is entirely decentralized.

V - DESIGN OF THE FIRST LEVEL - CONSTRAINED STRUCTURES

We now turn in detail to the design of the two levels of the hierarchical or coordinated scheme. This section considers the first level. In most cases the entire state vector is not measurable. A solution could be the use of a Kalman filter or a Luenberger observer to reconstruct the missing components of the state vector. This solution is not practical for the complex processes considered here, since it requires the synthesis of a structure of complexity equal to that of the process itself. On the other hand various methods have been proposed that impose the structure of the feedback and then try to tune the parameters to their best values. Levine (Levine 70) derived an algorithm that yields the optimal output feedback with respect to a quadratic criterion. The initialization of the procedure has been improved by Kosut (Kosut 70) in a suboptimal solution. Bingulac (Bingulac 74) introduced a sequential algorithm with three level to compute the parameters of a given structure. It is clear however that these methods have important computation requirements and they are less efficient for large scale systems.

Another approach is proposed here through a modification of the method presented in the previous sections. As opposed to other methods, the desired structure is obtained from partial optimization of a quadratic criterion and the parametric improvement of this structure is conducted on a limited number of variables.

5.1 Criterion modification

The flexibility of the proposed approach is used here to obtain desired structures for the local feedbacks. It is pointed out at this step that the control proposed above, equations (29)-(32), is not fully decentralized. The local control u_i is a function of the interaction variables z_i. Either these variables are measurable, either one has to compute them through A_{ij} coupling matrices and other substate vector x_j. This dependence originates in the form of the local instantaneous criterion (27) that is recalled here for convenience

$$J_{ik} = \frac{1}{2} x_{ik+1}^{T} Q_i x_{ik+1} + \frac{1}{2} u_{ik}^{T} R_i u_{ik} \quad (33)$$

The vector z_i appears in u_i when $Q_i x_{ik+1}$ is substituted following the expression of x_{ik+1} in (26). To remove this short coming a first modification is introduced

$$J_{ik} = x_{ik+1}^{T} Q_i x_{ik} + \frac{1}{2} u_{ik}^{T} R_i u_{ik} \quad (34)$$

With this subcriterion the local feedback is expressed by

$$u_i = \Gamma_i x_{ik} \quad (35)$$

$$\text{with } \Gamma_i = - R_i^{-1} B_{ii}^{T} Q_i \quad (36)$$

If optimality is needed, coordinating actions θ_i have to be added, they are computed at the upper level through an algorithm much similar to (32) (Drouin 82). Our concern here is with output feedbacks (the system outputs y_i are given by $y_i = C_{ii} x_i$) so that Q_i will be selected on the pattern of $C_{ii}^{T} C_{ii}$.
Equation (35) then turns to be an output feedback

$$u_i = -R_i^{-1} B_{ii}^{T} C_{ii}^{T} y_i \; (= K_i y_i) \quad (37)$$

However in most cases the product $C_{ii} \times B_{ii}$ is zero and the feedback gain K_i vanishes. This simply indicates that the input does not influence the output "in one step". To take into account this dynamic, a second modification is proposed

$$J_{ik} = x_{ik+2}^{T} Q_i x_{ik} + \frac{1}{2} u_{ik}^{T} R_i u_{ik} \quad (38)$$

or, more generally, with p_i a fixed integer

$$J_{ik} = x_{ik+p_i}^{T} Q_i x_{ik} + \frac{1}{2} u_{ik}^{T} R_i u_{ik} \quad (39)$$

The choice of the integer p_i will be explained below. In a final modification Q_i is substituted with S_i to introduce new free parameters that will be used in § 5.3 to tune the feedback gain. The structure of S_i fixes the structure of the local control, the most common choice will be to fit S_i to $C_{ii}^{T} C_{ii}$ to obtain decentralized output feedbacks. The modified criterion is now

$$J_{ik} = x_{ik+p_i}^{T} S_i x_{ik} + \frac{1}{2} u_{ik}^{T} R_i u_{ik} \quad (40)$$

For this criterion the control law is expressed by

$$u_{ik} = \Gamma_i x_{ik} + \theta_{ik} \quad (41)$$

$$\text{with } \Gamma_i = - R_i^{-1} B_{ii}^{T} A_{ii}^{T p_i - 1} S_i \quad (42)$$

$$\text{and } \theta_{ik} = B_{ii}^{T} W_k - \Gamma_i x_{ik} \quad (43)$$

where W_k is the same as in (32)

(S_i will be chosen proportionnal to $C_{ii}{}^T C_{ii}$ to obtain output feedbacks).

5.2 Choice of p_i

The modification that introduced p_i was an attempt to incorporate the system dynamics into the design. The integer p_i stands as a free parameter to achieve this incorporation with best income. The idea is to maximize the influence of the input u_i at instant k on the output y_i at instant $k+p_i$. To identify the instant of this maximum effect the appropriate tool is to visualize the output trajectories in response to test excitations on the inputs. For example if the system is excited with step-inputs the instant of the greatest slopes is a reasonnable choice for p_i.

For a multivariable system it can be expected to find several values of p depending on the excited input. To solve this case it is necessary to become a little more formal. A criterion is built that weights the different input-output couples (for simplicity of exposition it is assumed that dim y = dim u, other cases result in a slightly modified criterion). The discrete slope function of the ith step response is noted $\{\dot{y}_{ik}\}$ and n_i designs the corresponding weighting coefficient. A criterion to guide the choice of p is then defined

$$H(p) = \sum_{i=1}^{m} n_i |\dot{y}_{ip}| \tag{44}$$

with m = dim y

A natural choice for the n_i is $n_i = s_{ii}$ with $S = (s_{ij})_{i,j \in (1,m)}$, that is the different input-output couples are given the same importance in the choice criterion H (44) and in the performance index J.
The integer p is chosen at the maximum of H

$$p = \operatorname{Arg}\,\underset{n}{\operatorname{Max}}\, H(n) \tag{45}$$

For a SISO system this p clearly corresponds to the greatest slope on the step response, when for a MIMO system it is a weighted average value on the m outputs.

A value being given to $p(p_i)$, the feedback expression is obtained from (42) as a function of p

$$\Gamma_i(p_i) = - R_i^{-1} B_{ii}^T A_{ii}^{T\, p_i-1} S_i \tag{46}$$

Thus resulting in a closed loop dynamic matrix $\tilde{A}$ that depends on p

$$A(p) = \tilde{A} + B\Gamma(p) \tag{47}$$

where $\Gamma(p) = \operatorname{diag} \Gamma_i(p_i)$

and p is a notation for the set of $p_1, \ldots, p_i, \ldots p_N$

Experiments on various examples suggested the following attractive conjecture :

- The value of p given by (45) corresponds to the minimum of the spectral radius of the resulting closed loop dynamic matrix $\tilde{A}(p)$. This fact is illustrated on Figure I for the system studied in section VII. The trial of this conjecture was successfull on various other examples, and its theoretical grounds are presently investigated.

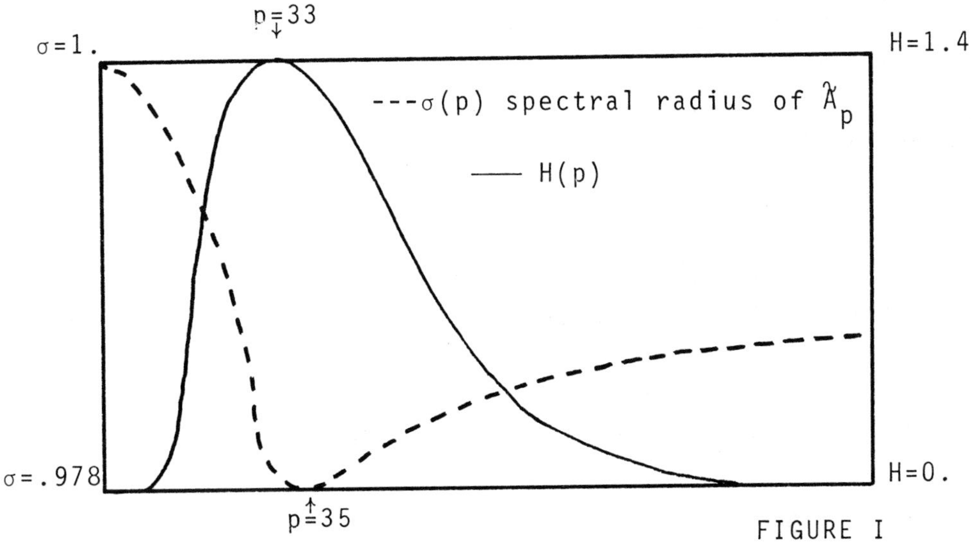

FIGURE I

5.3 Optimization of the diagonal matrix S

Equation (46) indicates a possible closed-loop control that can be gi-given, through the choice of non-zero coefficients in S, a desired structure. Most often $S = C^TC$ will be selected to obtain a decentralized output feedback. However optimality is achieved with respect to J only if the coordinating terme θ_k is added. The elimination of this term, with an acceptable cost increase, must be considered. The introduction of S provides the designer with a set of free parameters that can be used to improve the initial feedback within its constrained structure.

The control law (30. with $\theta \equiv 0$) is now parameterized by S ($\Gamma = \Gamma(S)$). As an initial guess for S a natural choice is $S^o = C^TQC$ where Q is the weighting matrix in the full state criterion (27). This means that the first feedback $\Gamma(S^o)$ corresponds to a step by step local solution, since J_{ik} is then the local instantaneous criterion, and is consequently very close to the optimal solution. From this good initialization, a minimization algorithm is carried out to adjust S

$$S^* = \text{Arg Min } J = \frac{1}{2}\sum_{k=0}^{K} y_{k+1}^T S\, y_{k+1} + u_k^T R\, u_k \qquad (48)$$

S diagonal

with $u = \Gamma(S)y$

As compared to other methods, a natural initialization is proposed that sets up the algorithm close to its final point and the optimization is conducted on a reduced number of parameters (S contains far less non-zero coefficients than Γ). When the second level is omitted a better criterion can be used for parameters tuning, for example an ITAE criterion (Duc 83). The importance of the upper level is thus reduced. A scalar variable, called the "coordination gain" measures the loss of performance for a non-coordinated law : with J_o the optimal value of the criterion and J_ℓ its value for a local feedback after optimization of S, g is defined by

$$g = \frac{J_\ell - J_o}{J_o}$$

The coordination gain is usually foundless than 10 % so that J_ℓ remains almost optimal.

VI - A SPECIAL CASE FOR THE SECOND LEVEL : DISTURBANCE REJECTION

The task of the upper-level is usually to compute a coordination vector from (23) or (32) to ensure optimality. Though, after the procedure described in section V, the lower level already performs a suboptimal control. Since weighting matrices have not very strict definition in optimal control theory, a cost increase of less than 10 % will be considered acceptable and the upper level freed from its dynamic optimization task.

The idea developped here is consequently to take advantage of the supervisor computing capabilities to reject perturbations. As opposed to the geometric approach, the aim is not to realize a perfect rejection but, less ambitiously, to counterback the effect of disturbances on the static regime of the process. Around its operating point the system (26) controlled by a law of the form (41) is described by

$$\Delta x_{k+1} = \tilde{A}\Delta x_k + B\theta_k + D_k \qquad (49)$$

$$\Delta y_k = C\Delta x_k$$

where Δ denotes the deviation from the nominal values, $\tilde{A} = A+B\Gamma$ the dynamic closed loop matrix and D_k represents the disturbance entering the system. This disturbance is supposed unknown and unmeasurable. The concern being only with the final regime, the static system associated with (49) is considered

$$\Delta X = - (\tilde{A} - I)^{-1} (B\theta + D) \tag{50}$$

Only constant or slow varying perturbations are considered so that $D_k = D$. The deviation between the outputs and their desired values is then expressed by

$$\Delta Y = Y - Y_d = - C(\tilde{A} - I)^{-1} (B\theta + D) \tag{51}$$

The coordination term θ, computed at the upperlevel, would minimize ΔY. An analytical solution of (51) is not possible since neither D is measurable, nor dimensionnality conditions on B, C and D usually satisfied.

To find a closed loop additive control θ, the performance of the static process (50) is evaluated by a quadratic criterion in ΔY and θ

$$J' = \frac{1}{2} \Delta Y^T Q' \Delta Y + \frac{1}{2} \theta^T R' \theta \tag{52}$$

The stationnarity of J' with respect to θ gives

$$\theta = - R'^{-1} \left(\frac{\partial \Delta Y^T}{\partial \theta}\right) Q' \Delta Y \tag{53}$$

where ΔY is expressed from (51) :

$$\frac{\partial \Delta Y^T}{\partial \theta} = -[C(\tilde{A} - I)^{-1} B]^T$$

The coordination term is then obtained in a closed loop fashion

$$\theta = M \Delta Y \tag{54}$$

with $M = R'^{-1} [C(\tilde{A} - I)^{-1} B]^T Q'$

From a practical point of view, the perturbation is detected once the error $y-y_d$ is measured over a given level ε. This detection starts the computation with

$$\theta^{c+1/2} = M \Delta Y^c \tag{55}$$

where c indicates the supervisor measuring interval, generally equal to five to ten times the control interval. This additive control $\theta^{c+1/2}$ is filtered through a first order relaxation to assure a good dynamic behaviour. The actual next control, derived from $\theta^{c+1/2}$ and θ^c is θ^{c+1}

$$\theta^{c+1} = \Omega \theta^{c+1/2} + (I - \Omega) \theta^c \tag{56}$$

If it is not sufficient to take the outputs back to their desired values, the computation is iterated with the new measurement y^{c+1} until the error $y-y_d$ is no longer significant.

VII - EXPERIMENT RESULT

These ideas have been applied on a light-hearted example consisting of two coupled third order subsystems (Fig. 2)

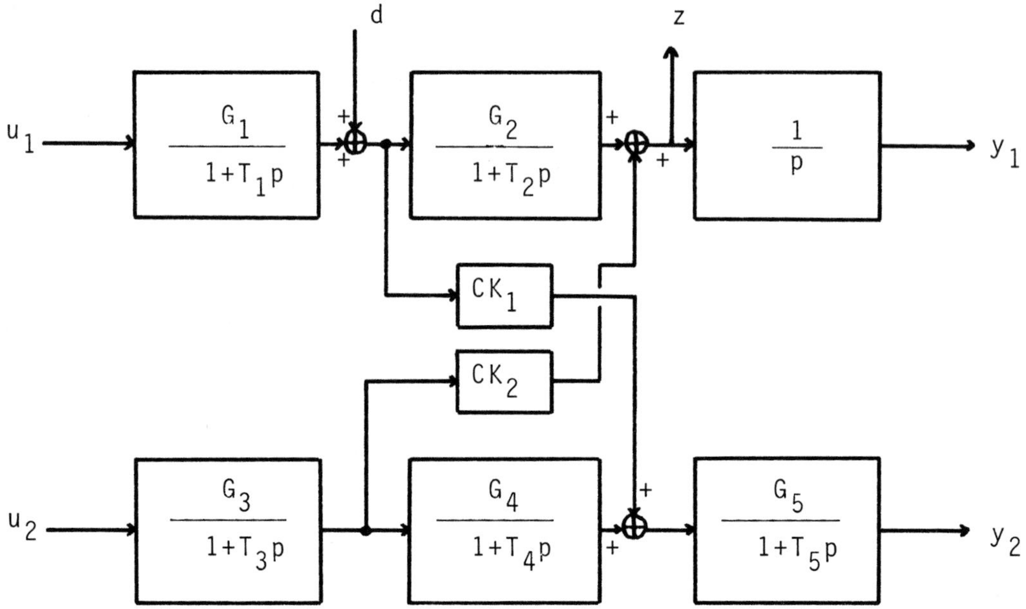

FIGURE II

The control laws derived in the previous sections were implemented on a physically distributed computing structure. The two local controllers consisted in microprocessors iSBC 80/10B (built on Intel processor 8080A) associated with a fast arithmetic unit. A microcomputer (MDS 800) assumed the upper level tasks (Fig. 3).

7.1 Regulation problem

In this experiment the system was left from a given initial condition (y=0) towards desired setpoints (y_{1d} = 5V , y_{2d} = 0V). The influence of the optimization of the weighting matrix S is illustrated on Figure 4. Figure 5 shows the corresponding control trajectories.

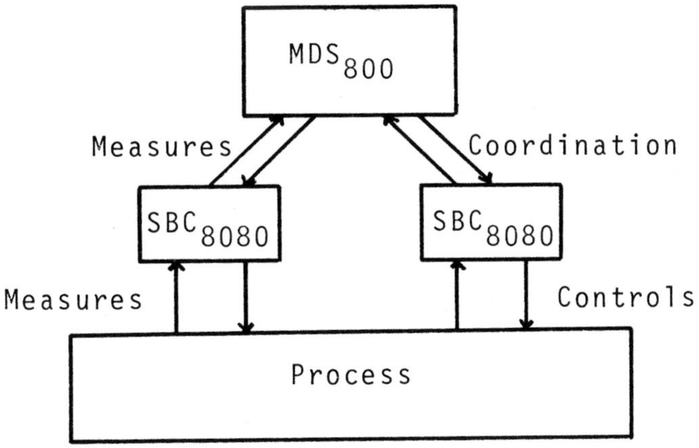

FIGURE III

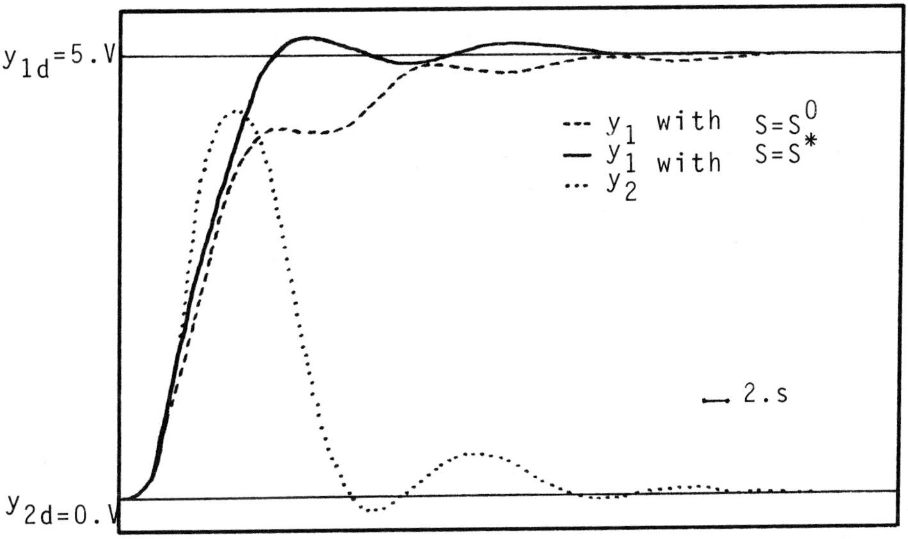

FIGURE IV

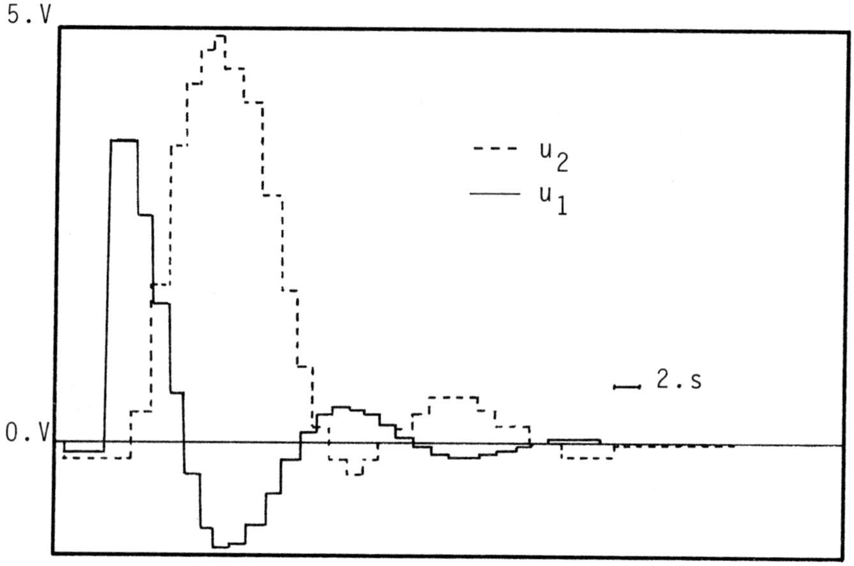

FIGURE V

7.2 Disturbance rejection

When a persisting disturbance occurs the upper level computes a correcting term. Through repetitive actions of this coordination the system is driven back to its set-points.

Figure 6 shows the output trajectories for a step disturbance applied on the perturbation input d. The correcting effect of the upperlevel is illustrated.

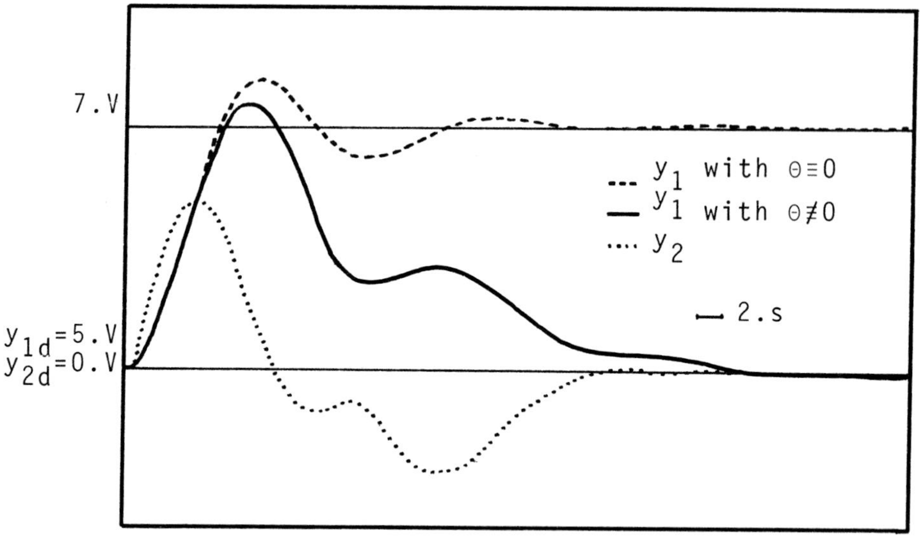

FIGURE VI

VIII - CONCLUSION

Decomposition-coordination techniques are by now a well established tool to deal with large scale systems. A new classification was proposed that discriminates methods on the basis of the techniques they use at two main anchor-points : problem manipulation and research strategy. This provides an easy diagnosis on a given method since advantages and drawbacks are directly apparent from the choice made for manipulations and strategies. Moreover it becomes clear that a shift on these two points generates new solutions.This is illustrated by the principle of the new method to which the rest of the paper is devoted. The ideas of this solution come close to Kalman's method or dynamic programming. The main characteristic of the obtained control law is its mixed constitution:the lower level computes a linear instantaneous feedback with constrained structures, the upper level adds a coordinating term that ensures optimality. An important field is that of linear processes associated with quadratic criterions. For this problem two decompositions, namely temporal and spatial-temporal, were proposed. Interest was then turned to the design of specific procedures for each level. At the lower level it resulted in the tuning of the local feedback gains as to obtain an almost optimal performance. The upper level consequently gave up its dynamic task to consider static optimization of the permanent regime. It was shown how this can improve the behaviour of the system when slow varying disturbances occur. The proposed solutions were illustrated on an example and the implementation described in some details.

Next steps will consider both theoretical and practical aspects. The connection of the method with the standard optimal control theory will be investigated, and it can be foreseen that new properties will result for the method. In the described implementation communications were asynchronous series point to point. Their monitoring represented a heavy burden for local units and they are replaced now by a Local Area Network (L.A.N.). This new technological tool allows enlarged communications and it will be possible to use measurements from other subsystems in the control law of a given subsystem. It is expected to further improve the first level and to allow partial decoupling of process units.

REFERENCES

AUSLENDER A.,70, "Une méthode générale pour la décomposition et la minimisation de fonctions non différentiables"
CRAS tome 271 Paris.

BAUMAN E., 68,"Multilevel optimization techniques with application to trajectory decomposition"
Advances in Control Systems, 6, pp 160-222.

BINGULAC S.P., N.M. CUK, M.S. CALOVIC, 75, "Calculation of optimum feedback gains for output constrained regulators"
IEEE Trans. AC-20, pp 164-166.

COHEN G., 78, "Optimization by decomposition and coordination : a unified approach"
IEEE Trans. AC-23, pp 222-232

DANTZIG G.B., P. WOLFE, 61, "The decomposition algorithm for linear program"
Econometrica, vol. 29, pp 767-778

DROUIN M., P. BERTRAND, 82, "A new coordination structure for on-line control of complex processes"
Large Scale Systems 3, pp 147-157

DUC G., M. DROUIN, P. BERTRAND 83, "On the design of simple structure control laws for complex processes"
Pro. IEEE CDC San Antonio.

FINDEISEN W. and co-workers, 78,"On-line hierarchical control for steady state systems"
IEEE Trans. AC-23, pp 189-204.

GEOFFRION A.M., 71, "Large scale linear and non linear programming"
in Wismer 71.

KOSUT R.L., 70, "Suboptimal control of linear time invariant systems subject to control structure constraints"
IEEE Trans. AC-15, pp 557-563

GLOWINSKI R., 71, "La méthode de relaxation : applications à la minimisation avec et sans contraintes de fonctionnels convexes"
Quaderni dei Rendiconti Matematica, n°14, Instituto Guido Castel nuovo, Roma.

LASDON L.S., J.D. SCHOEFFER, 65, "A multilevel technique for optimization"
Proc. JACC, Troy N.Y.

LEVINE W.S., M. ATHANS, 70, "On the determination of the optimal constant output feedback gains for linear multivariable systems" IEEE Trans. AC-15, pp 44-48.

LHOTE F., J.C. MIELLOU, 79, "Algorithmes de décentralisation et de coordination par relaxation en commande optimale" in Titli 79.

MESAROVIC M.D., D. MACKO, Y. TAKAHARA, 70, Theory of hierarchical, multilevel systems Academic Press, New York.

MIELLOU J.C., 75, "Algorithmes de relaxation chaotique à retards" RAIRO Analyse Numérique, 9ème année, pp 55-82.

PEARSON J.D., 71, "Dynamic decomposition techniques" in Wismer 71.

SANDELL N. and co-workers, 78, "Survey of decentralized control methods for large scale systems" IEEE Trans. AC-23, pp 108-128.

SINGH M.G., 77, Dynamical hierarchical control, North Holland, Amsterdam.

TITLI A. Editor, 79, Analyse et commande des systèmes complexes Cepadues Editions, Toulouse.

WILLSON I.D., 79, "Foundations of hierarchical control" Int. J. Control, vol. 29, n°6, pp 899-933.

WISMER D.A. Editor, 71, Optimization methods for large scale systems Mac Grawhill, New York.

PART IV MULTIDIMENSIONAL SYSTEMS

CHAPTER 21

A CANONICAL STATE-SPACE MODEL FOR m-DIMENSIONAL DISCRETE SYSTEMS

N. J. Theodorou
Airforce Research and
Technology Center (KETA)
Delta Falirou, Paleo Faliro
Athens, Greece

S. G. Tzafestas
Control Systems Laboratory
Eletrical Engineering Department
University of Patras
Patras, Greece

This chapter is concerned with SISO linear multidimensional systems,which may be represented by a transfer function that is a proper ratio of two multinomials with real constant coefficients. Assuming that a canonical state-space representation of any m-D (m-Dimensional) system, characterized by its m-D transfer function, is known, a procedure is presented for deriving a canonical state-space model for any (m+1)-D system, characterized by an (m+1)-D transfer function. As state-space models for up to 3-D systems exist, this procedure permits to construct state-space models of any dimension. Special state-space models for pure recursive and nonrecursive systems are also derived and the dimensionality of these models is investigated. A nontrivial example illustrates the validity of the method.

1. INTRODUCTION

Multidimensional systems and signals, i.e. systems and signals depending on more than one independent variable, has attracted much attention over the recent years [1-47]. The majority of the derived results refer to analysis, synthesis, stability, filtering, observer design and feedback control of 2-D and 3-D systems.

In this chapter an effort towards state-space realization of multidimensional single-input single-output (SISO) discrete linear systems, characterized by transfer functions with real constant coefficients, is attempted. Assumming that for any m-D transfer function an equivalent canonical state-space model is known, then a procedure is presented for a canonical state-space representation of any given (m+1)-D transfer function. Following this procedure and starting from a canonical 3-D state-space model, a 4-D model of this type is constructed and tested on the computer for its validity. Similarly, further state-space models of higher dimension may be realized at each step, going from m-D to (m+1)-D systems. Pure re-

S. G. Tzafestas (ed.), Multivariable Control, 383–407.

cursive and nonrecursive systems are also treated and their state-space models are investigated, as special cases. The dimensionality of these models, i.e. the dimension of their state-space matrix A, is calculated.

The chapter is organized as follows: In Sec. 2 the state-space model for m-D systems is given. In Sec. 3 the procedure for constructing an (m+1)-D state-space model is presented. In Sec. 4 the special cases of pure recursive and nonrecursive systems are examined, and in Sec. 5 an example is provided.

II. m-D STATE-SPACE MODEL

An m-D SISO linear discrete system can be described by the following transfer function with real space-invariant coefficients:

$$G(z_1,\ldots,z_m)=\frac{Y(z_1,\ldots,z_m)}{U(z_1,\ldots,z_m)}=\frac{\sum\limits_{i_1=0}^{K_1}\cdots\sum\limits_{i_m=0}^{K_m} b_{i1,\ldots,i_m} z_1^{-i_1}\cdots z_m^{-i_m}}{\sum\limits_{i_1=0}^{N_1}\cdots\sum\limits_{i_m=0}^{N_m} a_{i1,\ldots,im} z_1^{-i_1}\cdots z_m^{-i_m}} \tag{1}$$

where

$$(0,\ldots,0)\leq(K_1,\ldots,K_m)\leq(N_1,\ldots,N_m) \tag{1a}$$

$$a_{o,\ldots,o}=1 \tag{1b}$$

and $Y(z_1,\ldots,z_m)$ and $U(z_1,\ldots,z_m)$ are the m-D Z-transforms of the output $y(n_1,\ldots,n_m)$ and the input $u(n_1,\ldots,n_m)$ of the system respectively. Note that here $m\geq 3$ is assumed.
A canonical state-space representation of this m-D system is:

$$\begin{bmatrix} x^{(1)}(n_1+1,\ldots,n_m) \\ \vdots \\ x^{(m)}(n_1,\ldots,n_m+1) \end{bmatrix} = \begin{bmatrix} A_{1,1} & \cdots & A_{1,m} \\ \vdots & & \vdots \\ A_{m,1} & \cdots & A_{m,m} \end{bmatrix} \begin{bmatrix} x^{(1)}(n_1,\ldots,n_m) \\ \vdots \\ x^{(m)}(n_1,\ldots,n_m) \end{bmatrix} + \begin{bmatrix} B_1 \\ \vdots \\ B_m \end{bmatrix} u(n_1,\ldots,n_m) \tag{2a}$$

$$y(n_1,\ldots,n_m)=[C_1\ldots C_m]\begin{bmatrix}x^{(1)}(n_1,\ldots,n_m)\\ \vdots\\ x^{(m)}(n_1,\ldots,n_m)\end{bmatrix}+Du(n_1,\ldots,n_m) \tag{2b}$$

where $x=[x^{(1)}\ldots x^{(m)}]$ is the state-vector, and the dimensions of the above matrices are:

$$\dim(x^{(i)})\stackrel{\Delta}{=}N_i^! x1, \qquad \dim(x)=\sum_{i=1}^{m} N_i^! x1,$$

$$\dim(A_{i,j})=N_i^! x N_j^!,\quad \dim(B_i)=N_i^! x1, \dim(C_j)=1xN_j^!,\quad \dim(D)=1x1,$$

for $(1,1)\leqslant(i,j)\leqslant(m,m)$.

For reasons that will be justified in Sec.3,these vectors and matrices are further subdivided in the following form:

$$x^{(i)}=\begin{bmatrix}r^{(i)}\\ s^{(i)}\end{bmatrix} \quad \begin{matrix}\dim(r^{(i)})\stackrel{\Delta}{=}{}^rN_i x1\\ \dim(s^{(i)})\stackrel{\Delta}{=}{}^sN_i x1\end{matrix} \quad 1\leqslant i<m$$

$$x^{(m)}=t^{(m)} \quad \dim(t^{(m)}=N_m x1$$

$$A_{i,j}=\begin{bmatrix}{}^{rr}A_{ij} & {}^{rs}A_{i,j}\\ {}^{sr}A_{i,j} & {}^{ss}A_{i,j}\end{bmatrix}, \dim({}^{\alpha\beta}A_{i,j})={}^{\alpha}N_i x^{\beta}N_j$$

$$A_{i,m}=\begin{bmatrix}{}^{rt}A_{i,m}\\ {}^{st}A_{i,m}\end{bmatrix}, \qquad \dim({}^{\alpha t}A_{i,m})={}^{\alpha}N_i xN_m$$

$$A_{m,j}=[{}^{tr}A_{m,j} \quad {}^{ts}A_{m,j}], \dim({}^{t\beta}A_{m,j})=N_m x^{\beta}N_j$$

$$A_{m,m}={}^{tt}A_{m,m}, \qquad \dim({}^{tt}(A_{m,m})=N_m xN_m$$

$$B_i=\begin{bmatrix}{}^{r}B_i\\ {}^{s}B_i\end{bmatrix}, \quad \dim({}^{\alpha}B_i)={}^{\alpha}N_i x1$$

$$B_m={}^{t}B_m, \qquad \dim({}^{t}B_m)=N_m x1$$

$$C_j=[{}^{r}C_j \quad {}^{s}C_j], \quad \dim({}^{\beta}C_j)=1x^{\beta}N_j$$

$C_m = {}^tC_m$, $\dim({}^tC_m) = 1 \times N_m$

$D = d$, $\dim(d) = 1 \times 1$

where $\alpha = r,s$, $\beta = r,s$ and $(1,1) \leq (i,j) \leq (m-1,m-1)$.

Hence, according to the previous definitions, the dimensions of the partial vectors $x(i)$ are:

$$N_i' = {}^rN_i + {}^sN_i \quad \text{for } 1 \leq i \leq m-1, \quad N_m' = N_m \tag{3}$$

and the dimension of the total vector x which specifies the dimensionality of the model is:

$$N' = \sum_{i=1}^{m} N_i' = \sum_{i=1}^{m-1} N_i' + N_m = \sum_{i=1}^{m-1} {}^rN_i + \sum_{i=1}^{m-1} {}^sN_i + N_m \tag{4}$$

2.1. Pure recursive m-D systems

Such systems are characterized by transfer functions of the following form:

$$G(z_1,\ldots,z_m) = \frac{b^*_{o,\ldots,o}}{\sum_{i_1=0}^{N_1} \cdots \sum_{i_m=0}^{N_m} a_{i1,\ldots,im} z_1^{-i_1} \cdots z_m^{-i_m}}, \quad a_{0,\ldots,0} = 1 \tag{5}$$

and have the following state-space representation:

$$\begin{bmatrix} s^{(1)}(n_1+1,\ldots,n_{m-1},n_m) \\ \vdots \\ s^{(m-1)}(n_1,\ldots,n_{m-1}+1,n_m) \\ t^{(m)}(n_1,\ldots,n_{m-1},n_m+1) \end{bmatrix} = \begin{bmatrix} {}^{ss}A_{1,1} & \cdots & {}^{ss}A_{1,m-1} & {}^{st}A_{1,m} \\ & & \vdots & \\ {}^{ss}A_{m-1,1} & \cdots & {}^{ss}A_{m-1,m-1} & {}^{st}A_{m-1,m} \\ {}^{ts}A_{m,1} & \cdots & {}^{ts}A_{m,m-1} & {}^{tt}A_{m,m} \end{bmatrix}$$

$$\begin{bmatrix} s^{(1)}(n_1,\ldots,n_m) \\ \vdots \\ s^{(m-1)}(n_1,\ldots,n_m) \\ t^{(m)}(n_1,\ldots,n_m) \end{bmatrix} + \begin{bmatrix} {}^sB_1 \\ \vdots \\ {}^sB_{m-1} \\ {}^tB_m \end{bmatrix} u(n_1,\ldots,n_m) \tag{6a}$$

$$y(n_1,\ldots,n_m) = \begin{bmatrix} {}^sC_1 & \cdots & {}^sC_{m-1} & {}^tC_m \end{bmatrix} \begin{bmatrix} s^{(1)}(n_1,\ldots,n_m) \\ \vdots \\ s^{(m-1)}(n_1,\ldots,n_m) \\ t^{(m)}(n_1,\ldots,n_m) \end{bmatrix} + Du(n_1,\ldots,n_m) \tag{6b}$$

where the dimensions of the above matrices have been defined before, and

$$ {}^{s}D_m \overset{\Delta}{=} \sum_{i=1}^{m-1} {}^{s}N_i, \quad {}^{t}D_m = N_m \tag{7}$$

Hence the dimensionality of this model is ${}^{s}D_m + {}^{t}D_m$.

2.2. Pure nonrecursive m-D systems

Suche systems are characterized by transfer functions of the following form:

$$ G(z_1,\dots,z_m) = \sum_{i_1=0}^{K_1} \dots \sum_{i_m=0}^{K_m} b_{i1,\dots,im} z_1^{-i_1} \dots z_m^{-i_m} \tag{8}$$

and have the following state-space representation:

$$\begin{bmatrix} r^{(1)}(n_1+1,\dots,n_{m-1},n_m) \\ \vdots \\ r^{(m-1)}(n_1,\dots,n_{m-1}+1,n_m) \\ t^{(m)}(n_1,\dots,n_{m-1},n_m,n_m+1) \end{bmatrix} = \begin{bmatrix} {}^{rr}A_{1,1} & \dots & {}^{rr}A_{1,m-1} & {}^{rt}A_{1,m} \\ \vdots & & \vdots & \vdots \\ {}^{rr}A_{m-1,1} & \dots & {}^{rr}A_{m-1,m-1} & {}^{rt}A_{m-1,m} \\ {}^{tr}A_{m,1} & \dots & {}^{tr}A_{m,m-1} & {}^{tt}A_{m,m} \end{bmatrix}$$

$$\begin{bmatrix} r^{(1)}(n_1,\dots,n_m \\ \vdots \\ r^{(m-1)}(n_1,\dots,n_m) \\ t^{(m)} 9n_1,\dots,n_m) \end{bmatrix} + \begin{bmatrix} {}^{r}B_1 \\ \vdots \\ {}^{r}B_{m-1} \\ {}^{t}B_m \end{bmatrix} u(n_1,\dots,n_m) \tag{9a}$$

$$ y(n_1,\dots,n_m) = \begin{bmatrix} {}^{r}C_1 \dots {}^{r}C_{m-1} & {}^{t}C_m \end{bmatrix} \begin{bmatrix} r^{(1)}(n_1,\dots,n_m) \\ \vdots \\ r^{(m-1)}(n_1,\dots,n_m) \\ t^{(m)}(n_1,\dots,n_m) \end{bmatrix} + Du(n_1,\dots,n_m) \tag{9b}$$

$$ {}^{r}D_m = \sum_{i=1}^{m-1} {}^{r}N_i, \quad {}^{t}D_m = N_m \tag{10}$$

Hence the dimensiona lity of this model is ${}^{r}D_m + {}^{t}D_m$.

2. 3. Dimensionality of general m-D state-space models

According to (4), (7) and (10), the dimensionality of the state-space model (2), representing a general m-D system, is:

$$D_m = {}^rD_m + {}^sD_m = \sum_{i=1}^{m-1} ({}^rN_i + {}^sN_i) + N_m = \sum_{i=1}^{m-1} N'_i + N'_m = N' \tag{11}$$

Hence the state-space model realizing an m-D transfer function is assumed to have the canonical form given in this section. The elements of the various state-space matrices and submatrices are unknown. Only for 1-D, 2-D and 3-D systems these elements are known and can be written down simply by inspection of the coefficients of the transfer function.

3. (m+1)-D STATE-SPACE MODEL

Assume that a state-sapce realization of an m-D transfer function has the canonical form given in the previous section. Also assume that all the elements of the state-space matrices are known functions of the coefficients of the transfer function. In this section a procedure for constructing a state-space model for any given (m+1)-D transfer function is briefly presented. This (m+1)-D state-space model: i) is based on the known m-D state-space model, ii) its state-space submatrices are state-space matrices of the respective m-D model, and iii) has the same general structure as the m-D model. Following this procedure and starting from a 3-D state-space model, a 4-D model can be obtained. Then from this 4-D model, a 5-D model etc. Hence, state-space models realizing transfer functions of any dimension may be found, according to the above mentioned procedure. A summary of this procedure is now given.

An (m+1)-D system has the following transfer function:

$$G(z_1,\dots,z_m) = \frac{Y}{U} = \frac{\sum_{i_1=0}^{K_1} \cdots \sum_{i_m+1=0}^{K_{m+1}} b_{i1,\dots,i(m+1)} z_1^{-i_1} \cdots z_{m+1}^{-i(m+1)}}{\sum_{i_1=0}^{N_1} \cdots \sum_{i_{m+1}=0}^{N_{m+1}} a_{i1,\dots,i(m+1)} z_1^{-i_1} \cdots z_{m+1}^{-i(m+1)}}$$

$$= \frac{\sum_{i_1=0}^{K_1} \cdots \sum_{i_m=0}^{K_m} \left[\sum^{K_{m+1}} b_{i1,\dots,i(m+1)} z_{m+1}^{-i(m+1)} \right] z_1^{-i_1} \cdots z_m^{-i_m}}{\sum_{i_1=0}^{N_1} \cdots \sum_{i_m=0}^{N_m} \left[\sum_{i_{m+1}=0}^{N_{m+1}} a_{i1,\dots,i(m+1)} z_{m+1}^{-i(m+1)} \right] z_1^{-i_1} \cdots z_m^{-i_m}}$$

$$= \left[\sum_{\substack{i_1=0 \\ (i_1,\ldots i_m)\neq(0,\ldots,0)}}^{K_1} \cdots \sum_{i_m=0}^{K_m} \frac{\sum_{i_{m+1}=0}^{K_{m+1}} b_{i1,\ldots,im,i(m+1)} z_{m+1}^{-i(m+1)}}{\sum_{i_{m+1}=0}^{N_{m+1}} a_{0,\ldots,0,i(m+1)} z_{m+1}^{-i(m+1)}} z_1^{-i_1} \cdots z_m^{-i_m} \right.$$

$$\left. + \frac{\sum_{i_{m+1}=0}^{K_{m+1}} b_{0,\ldots,0,i(m+1)} z_{m+1}^{-i(m+1)}}{\sum_{i_{m+1}=0}^{N_{m+1}} a_{0,\ldots,0,i(m+1)} z_{m+1}^{-i(m+1)}} \right]$$

$$\left[\sum_{\substack{i_1=0 \\ (i_1,\ldots,i_m)\neq(0,\ldots,0)}}^{N_1} \cdots \sum_{i_m=0}^{N_m} \frac{\sum_{i_{m+1}=0}^{N_{m+1}} a_{i1,\ldots,im,i(m+1)} z^{-i(m+1)}}{\sum_{i_{m+1}=0}^{N_{m+1}} a_{0,\ldots,0,i(m+1)} z_{m+1}^{-i(m+1)}} z_1^{-i_1} \cdots z_m^{-i_m} + 1 \right]^{-1}$$

$$= \left\{ \left[\sum_{\substack{i_1=0 \\ (i_1,\ldots,i_m)\neq(0,\ldots,0)}}^{K_1} \sum_{i_m=0}^{K_m} \frac{\sum_{i_{m+1}=1}^{N_{m+1}} \tilde{b}_{i1,\ldots,im,i(m+1)} z^{-i(m+1)}}{\sum_{i_{m+1}=0}^{N_{m+1}} a_{0,\ldots,0,i(m+1)} z_{m+1}^{-i(m+1)}} z_1^{-i_1} \cdots z_m^{-i_m} \right. \right.$$

$$\left. + \sum_{\substack{i_1=0 \\ (i_1,\ldots,i_m)\neq(0,\ldots,0)}}^{K_1} \cdots \sum_{i_m=0}^{K_m} b_{i1,\ldots,im,0} z_1^{-i_1} z_m^{-i_m} \right] + b_{0,\ldots,0}$$

$$\left. + \left[\frac{\sum_{i_{m+1}=0}^{N_{m+1}} \tilde{b}_{0,\ldots,0,i(m+1)} z_{m+1}^{-i(m+1)}}{\sum_{i_m+1=0}^{N_{m+1}} a_{0,\ldots,0,i(m+1)} z_{m+1}^{-i(m+1)}} \right] \right\}$$

$$\left[\sum_{i_m=0}^{N_1} \cdots \sum_{\substack{i_m=0 \\ (i_1,\dots,i_m)\neq(0,\dots,0)}}^{N_m} \frac{\sum_{i_{m+1}=1}^{N_{m+1}} \tilde{a}_{i1,\dots,im,i(m+1)} z^{-i(m+1)}}{\sum_{i_{m+1}=0}^{N_{m+1}} a_{0,\dots,0,i(m+1)} z_{m+1}^{-i(m+1)}} z_1^{-i_1} \dots z_m^{-i_m}\right.$$

$$\left. + \sum_{i_1=0}^{N_1} \cdots \sum_{\substack{i_m=0 \\ (i_1,\dots,i_m)\neq(0,\dots,0)}}^{N_m} a_{i1,\dots,im,0} z_1^{-i_1} \dots z_m^{-i_m+1}\right]^{-1} \tag{12}$$

where

$(0,\dots,0) \leqslant (K_1,\dots,K_{m+1}) \leqslant (N_1,\dots,N_{m+1})$, $a_{0,\dots,0}=1$

$\tilde{a}_{i1,\dots,im,i(m+1)} = a_{i1,\dots,im,i(m+1)} - a_{i1,\dots,im,0}\, a_{0,\dots,0,i(m+1)}$

for $(0,\dots,0,1) \leqslant (i_1,\dots,i_m,i_{m+1}) \leqslant (N_1,\dots,N_m,N_{m+1})$,

with $(i_1,\dots,i_m) \neq (0,\dots,0)$, and (13)

$\tilde{b}_{i1,\dots,im,i(m+1)} = b_{i1,\dots,im,i(m+1)} - b_{i1,\dots,im,0}\, a_{0,\dots,0,i(m+1)}$

for $(0,\dots,0) \leqslant (i_1,\dots,i_m,i_{m+1}) \leqslant (K_1,\dots,K_m,K_{m+1})$ (14)

Now introducing the definitions:

$$\frac{Y_r}{Y_s} \triangleq \sum_{i_1=0}^{K_1} \cdots \sum_{\substack{i_m=0 \\ (i_1,\dots,i_m)\neq(0,\dots,0)}}^{K_2} \frac{\sum_{i_{m+1}=1}^{N_{m+1}} \tilde{b}_{i1,\dots,im,i(m+1)} z_{m+1}^{-i(m+1)}}{\sum_{i_{m+1}=0}^{N_{m+1}} a_{0,\dots,0,i(m+1)} z_{m+1}^{-i(m+1)}} z_1^{-i_1} \dots z_m^{-i_m}$$

$$+ \sum_{i_1=0}^{K_1} \cdots \sum_{\substack{i_m=0 \\ (i_1,\dots,i_m)\neq(0,\dots,0)}}^{K_m} b_{i1,\dots,im,0} z_1^{-i_1} \dots z_m^{-i_m} \tag{15}$$

$$\frac{Y_s}{U} \triangleq b_{0,\dots,0}\left[\sum_{i_1=0} \cdots \sum_{\substack{i_m=0 \\ (i_1,\dots,i_m)\neq(0,\dots,0)}} \frac{\sum_{i_{m+1}=1}^{N_{m+1}} \tilde{a}_{i1,\dots,im,i(m+1)} z^{-i(m+1)}}{\sum_{i_m+1=0}^{N_{m+1}} a_{0,\dots,0,i(m+1)} z_{m+1}^{-i(m+1)}} z^{-i_1} \dots z_m^{-i_m}\right.$$

$$+ \sum_{\substack{i_1=0 \\ (i_1,\dots,i_m)\neq(0,\dots,0)}}^{N_1} \cdots \sum_{i_m=0}^{N_m} a_{i1,\dots,im,0} z_1^{-i_1} \cdots z_m^{-i_{m+1}} \Bigg]^{-1} \tag{16}$$

and

$$\frac{Y_t}{Y_s} = \frac{\sum_{i_{m+1}=1}^{N_{m+1}} \tilde{b}_{0,\dots,0,i(m+1)} z_{m+1}^{-i(m+1)}}{\sum_{i_{m+1}=0}^{N_{m+1}} a_{0,\dots,0,i(m+1)} z_{m+1}^{-i(m+1)}} \tag{17}$$

the expression (12) is written as

$$G(z_1,\dots,z_{m+1}) = \frac{Y(z_1,\dots,z_{m+1})}{U(z_1,\dots,z_{m+1})} =$$

$$\frac{Y_r(z_1,\dots,z_{m+1})+Y_s(z_1,\dots,z_{m+1})+Y_t(z_1,\dots,z_{m+1})}{U(z_1,\dots,z_{m+1})}$$

which, by taking the inverse (m+1)-D Z transforms yields

$$y(n_1,\dots,n_{m+1})=y_r(n_1,\dots,n_{m+1})+y_s(n_1,\dots,n_{m+1})+y_t(n_1,\dots,n_{m+1}) \tag{18}$$

or $\quad y=y_r+y_s+y_t \quad$ (18a)

Now the realization of the three m-D transfer functions (15), (16) and (17) can be achieved separately, according to Section 2. During this procedure, the state-space matrices ${}^{\alpha\beta}A_{i,j}$, ${}^{\alpha}B_i$, ${}^{\beta}C_j$ and vectors r,s,t, where α or β=r,s,t i or j=1,...,m appear in the realizations.

Denoting these by ${}^{\alpha\beta}_{\;p}A_{\gamma,\delta,i,j}$, ${}^{\alpha}_{\;p}B_{\gamma,i}$, ${}^{\beta}_{\;p}C_{\gamma,j}$, ${}_{p}r^{i}_{\gamma}$, ${}_{p}s^{i}_{\gamma}$, ${}_{p}t^{i}_{\gamma}$, where γ or δ is r, s and t for (15), (16) and (17) respectively, $p=1,\dots,N_{m+1}$ and avoiding the tedious details of the procedure, the following state-space model representing the general (m+1)-D transfer function (12) is found:

$$\begin{bmatrix} r^{(1)}(n_1+1,\ldots,n_m,n_{m+1}) \\ s^{(1)}(n_1+1,\ldots,n_m,n_{m+1}) \\ \vdots \\ r^{(m)}(n_1,\ldots,n_m+1,n_{m+1}) \\ s^{(m)}(n_1,\ldots,n_m+1,n_{m+1}) \\ t\ (n_1,\ldots,n_m,n_{m+1}+1) \end{bmatrix}$$

$$= \begin{bmatrix} A_{rr1,1} & A_{rs1,1} & \ldots A_{rr1,m} & A_{rs1,m} & A_{rt1} \\ 0 & A_{ss1,1} & \ldots 0 & A_{ss1,m} & A_{st1} \\ \cdot & \cdot & \cdots\cdot & \cdot & \cdot \\ \cdot & \cdot & \cdots\cdot & \cdot & \cdot \\ \cdot & \cdot & \cdots\cdot & \cdot & \cdot \\ A_{rrm,1} & A_{rsm,1} & \ldots A_{rrm,m} & A_{rsm,m} & A_{rtm} \\ 0 & A_{ssm,1} & \ldots 0 & A_{ssm,m} & A_{stm} \\ 0 & A_{ts1} & \ldots 0 & A_{tsm} & A_{tt} \end{bmatrix} \begin{bmatrix} r^{(1)}(n_1,\ldots,n_{m+1}) \\ s^{(1)}(n_1,\ldots,n_{m+1}) \\ \cdot \\ \cdot \\ \cdot \\ r^{(m)}(n_1,\ldots,n_{m+1}) \\ s^{(m)}(n_1,\ldots,n_{m+1}) \\ t\ (n_1,\ldots,n_{m+1}) \end{bmatrix}$$

$$+ \begin{bmatrix} B_{r1} \\ B_{s1} \\ \vdots \\ B_{rm} \\ B_{sm} \\ B_t \end{bmatrix} u(n_1,\ldots,n_m) \qquad (19a)$$

$$y(n_1,\ldots,n_{m+1}) = \begin{bmatrix} C_{r1} & C_{s1} \ldots C_{rm} & C_t \end{bmatrix} \begin{bmatrix} r^{(1)}(n_1,\ldots,n_{m+1}) \\ s^{(1)}(n_1,\ldots,n_{m+1}) \\ \vdots \\ r^{(m)}(n_1,\ldots,n_{m+1}) \\ s^{(m)}(n_1,\ldots,n_{m+1}) \\ t\ (n_1,\ldots,n_{m+1}) \end{bmatrix} + Du(n_1,\ldots,n_{m+1}) \qquad (19b)$$

where

$$r^{(i)} = \begin{bmatrix} {}_1 r_r^{(i)} \\ \vdots \\ {}_{N(m+1)} r_r^{(i)} \\ {}_o r_r^{(i)} \end{bmatrix}, \quad s^{(i)} = \begin{bmatrix} {}_1 r_s^{(i)} \\ {}_1 s_s^{(i)} \\ \vdots \\ {}_{N(m+1)} r_s^{(i)} \\ {}_{N(m+1)} s_s^{(i)} \\ {}_o s_s^{(i)} \end{bmatrix} \qquad 1 \leqslant i \leqslant m-1$$

$$r^{(m)} = \begin{bmatrix} {}_1 t_r^{(m)} \\ \vdots \\ {}_{N(m+1)} t_r^{(m)} \\ {}_o t_r^{(m)} \end{bmatrix}, \quad s^{(m)} = \begin{bmatrix} {}_1 t_s^{(m)} \\ \vdots \\ {}_{N(m+1)} t_s^{(m)} \\ {}_o t_s^{(m)} \end{bmatrix}, \quad t = \begin{bmatrix} t_1 \\ \vdots \\ t_{N_{m+1}} \end{bmatrix}$$

$$A_{rri,j} = \begin{bmatrix} {}_1^{rr}A_{ri,j} & & & \\ & \ddots & & 0 \\ 0 & & {}_{N(m+1)}^{rr}A_{ri,j} & \\ & & & {}_o^{rr}A_{ri,j} \end{bmatrix}$$

$$A_{rri,m} = \begin{bmatrix} {}_1^{rt}A_{ri,m} & & & \\ & \ddots & & 0 \\ 0 & & {}_{N(m+1)}^{rt}A_{ri,m} & \\ & & & {}_o^{rt}A_{ri,m} \end{bmatrix}$$

$$A_{rrm,j} = \begin{bmatrix} {}_1^{tr}A_{rm,j} & & & \\ & \ddots & & 0 \\ 0 & & {}_{N(m+1)}^{tr}A_{rm,j} & \\ & & & {}_o^{tr}A_{rm,j} \end{bmatrix}$$

$$A_{rrm,m} = \begin{bmatrix} {}_1^{tt}A_{ri,m} & & & \\ & \ddots & & 0 \\ 0 & & {}_{N(m+1)}^{tt}A_{rm,m} & \\ & & & {}_o^{tt}A_{rm,m} \end{bmatrix}$$

$$A_{rsi,j} = \begin{bmatrix} & & 0 & & \\ {}^{rr}_{1}A_{rsi,j} & {}^{rs}A_{rsi,j} \cdots & {}^{rr}_{N(m+1)}A_{rsi,j} & {}^{rs}A_{rsi,j} & {}^{rs}A_{rsi,j} \end{bmatrix}$$

$$A_{rsi,m} = \begin{bmatrix} & 0 & \\ {}^{rt}A_{rsi,m} \cdots & {}^{rt}A_{rsi,m} & {}^{rt}A_{rsi,m} \end{bmatrix}$$

$$A_{rsm,j} = \begin{bmatrix} & & 0 & & \\ {}^{tr}_{1}A_{rsm,j} & {}^{ts}A_{rsm,j} \cdots & {}^{tr}_{N(m+1)}A_{rsm,j} & {}^{ts}A_{rsm,j} & {}^{ts}A_{rsm,j} \end{bmatrix}$$

$$A_{rsm,m} = \begin{bmatrix} {}^{tt}A_{rsm,m} \cdots & {}^{tt}A_{rsm,m} & {}^{tt}A_{rsm,m} \end{bmatrix}$$

$$A_{rtj} = \begin{bmatrix} {}^{r}_{1}B_{rj} & & 0 \\ & \ddots & \\ 0 & & {}^{r}_{N(m+1)}B_{rj} \\ & 0 & \end{bmatrix}, \quad A_{rtm} = \begin{bmatrix} {}^{t}_{1}B_{rm} & & 0 \\ & \ddots & \\ 0 & & {}^{t}_{N(m+1)}B_{rm} \\ & 0 & \end{bmatrix}$$

$$A_{ssi,j} = \begin{bmatrix} {}^{rr}_{1}A_{si,j} & {}^{rs}_{1}A_{si,j} & & & \\ {}^{sr}_{1}A_{si,j} & {}^{ss}A_{si,j} & & 0 & \\ & & \ddots & & \\ 0 & & {}^{rr}_{N(m+1)}A_{si,j} & {}^{rs}_{N(m+1)}A_{si,j} & \\ & & & & {}^{ss}A_{si,j} \end{bmatrix}$$

$$A_{ssi,m} = \begin{bmatrix} {}^{rt}_{1}A_{si,m} & & & \\ {}^{st}A_{si,m} & & 0 & \\ & \ddots & & \\ 0 & {}^{rt}_{N(m+1)}A_{si,m} & & \\ & {}^{st}A_{si,m} & & \\ & & {}^{st}A_{si,m} & \end{bmatrix}$$

$$A_{ssm,j}=\begin{bmatrix} {}^{tr}_{1}A_{sm,j} & {}^{ts}_{1}A_{sm,j} & & & 0 \\ & & \ddots & & \\ 0 & & & {}^{tr}_{N(m+1)}A_{sm,j} & {}^{ts}A_{sm,j} \\ & & & & {}^{ts}A_{sm,j} \end{bmatrix}$$

$$A_{ssm,m}=\begin{bmatrix} {}^{tt}A_{sm,m} & & 0 \\ & \ddots & \\ 0 & {}^{tt}A_{sm,m} & \\ & & {}^{tt}A_{sm,m} \end{bmatrix},\ A_{stj}=\begin{bmatrix} {}^{r}_{1}B_{sj} & & 0 \\ {}^{s}B_{sj} & \ddots & \\ & \ddots & {}^{r}_{N(m+1)}B_{sj} \\ 0 & & {}^{s}B_{sj} \\ & & 0 \end{bmatrix}$$

$$A_{stm}=\begin{bmatrix} {}^{t}B_{sm} & & 0 \\ & \ddots & {}^{t}B_{sm} \\ 0 & & \\ & & 0 \end{bmatrix},\ A_{tt}=\begin{bmatrix} -a_{0,\ldots,0,1} & \cdots & -a_{0,\ldots,N(m+1)} \\ 1 & \ddots & \\ & \ddots & \\ & 1 & 0 \end{bmatrix}$$

$$A_{tsi}=\begin{bmatrix} {}^{r}_{1}A_{tsi} & {}^{s}A_{tsi} & \cdots & {}^{r}_{N(m+1)}A_{tsi} & {}^{s}A_{tsi} & {}^{s}A_{tsi} \end{bmatrix}$$

$$A_{tsm}=\begin{bmatrix} {}^{t}A_{tsm} & \cdots & {}^{t}A_{tsm} & {}^{t}A_{tsm} \end{bmatrix}$$

$$B_{rj}=\begin{bmatrix} 0 \\ \\ {}^{r}B_{rj} \end{bmatrix},\ B_{rm}=\begin{bmatrix} 0 \\ \\ {}^{t}B_{rm} \end{bmatrix},\ B_{sj}=\begin{bmatrix} 0 \\ \\ {}^{s}B_{sj} \end{bmatrix},\ B_{sm}=\begin{bmatrix} 0 \\ \\ {}^{t}B_{sm} \end{bmatrix},\ B_{t}=\begin{bmatrix} 1 \\ 0 \\ \vdots \\ 0 \end{bmatrix}$$

$$C_{rj}=\begin{bmatrix} {}^{r}_{1}C_{rj} & \cdots & {}^{r}_{N(m+1)}C_{rj} & {}^{r}_{o}C_{rj} \end{bmatrix},\ C_{rm}=\begin{bmatrix} {}^{t}_{1}C_{rm} & \cdots & {}^{t}_{N(m+1)}C_{rm} & {}^{t}_{o}C_{rm} \end{bmatrix}$$

$$C_{sj}=\begin{bmatrix} {}^{r}_{1}C_{sj} & {}^{s}C_{sj} & \cdots & {}^{r}_{N(m+1)}C_{sj} & {}^{s}C_{sj} & {}^{s}C_{sj} \end{bmatrix} D$$

$$C_{sm}=\begin{bmatrix} {}^{t}C_{sm} & \cdots & {}^{t}C_{sm} & {}^{t}C_{sm} \end{bmatrix} D$$

$$C_{t}=\begin{bmatrix} b_{0,\ldots,0,1} & \cdots & b_{0,\ldots,0,N(m+1)} \end{bmatrix},\quad D=b_{0,\ldots,0}$$

where

$$\dim[r^{(i)}]=[{}^{r}N_i(N_{m+1}+1)]\times 1,\quad \dim[s^{(i)}]=[({}^{r}N_i+{}^{s}N_i)N_{m+1}+{}^{s}N_i]\times 1$$

$$\dim[r^{(m)}]=[N_m(N_{m+1}+1)]\times 1,\quad \dim[s^{(m)}]=[N_m(N_{m+1}+1)]\times 1$$

$$\dim(t)=N_{m+1} \tag{20}$$

$$\dim(A_{rri,j})=[{}^{r}N_i(N_{m+1}+1)]\times[{}^{s}N_j(N_{m+1}+1)]$$

$$\dim(A_{rri,m})=[{}^{r}N_i(N_{m+1}+1)]\times[N_m(N_{m+1}+1)]$$

$$\dim(A_{rrm,j})=[N_m(N_{m+1}+1)]\times[{}^{r}N_j(N_{m+1}+1)]$$

$$\dim(A_{rrm,m})=[N_m(N_{m+1}+1)]\times[N_m(N_{m+1}+1)]$$

$$\dim(A_{rsi,j})=[{}^{r}N_i(N_{m+1}+1)]\times[({}^{r}N_j+{}^{s}N_j)(N_{m+1}+1)]$$

$$\dim(A_{rsi,m})=[{}^{r}N_i(N_{m+1}+1)]\times[N_m(N_{m+1}+1)]$$

$$\dim(A_{rsm,j})=[N_m(N_{m+1}+1)]\times[({}^{r}N_j+{}^{s}N_j)(N_{m+1}+1)]$$

$$\dim(A_{rsm,m})=[N_m(N_{m+1}+1)]\times[N_m(N_{m+1}+1)]$$

$$\dim(A_{rtj})=[{}^{r}N_j(N_{m+1}+1)]\times N_{m+1},\quad \dim(A_{rtm})=[N_m(N_{m+1}+1)]\times N_{m+1}$$

$$\dim(A_{ssi,j})=[({}^{r}N_i+{}^{s}N_i)N_{m+1}+{}^{s}N_i]\times[({}^{r}N_j+{}^{s}N_j)N_{m+1}+{}^{s}N_j]$$

$$\dim(A_{ssi,m})=[({}^{r}N_i+{}^{s}N_i)N_{m+1}+{}^{s}N_i]\times[N_m(N_{m+1}+1)]$$

$$\dim(A_{ssm,j})=[N_m(N_{m+1}+1)]\times[({}^{r}N_j+{}^{s}N_j)N_{m+1}+{}^{s}N_j]$$

$$\dim(A_{ssm,m})=[N_m(N_{m+1}+1)]\times[N_m(N_{m+1}+1)]$$

$$\dim(A_{stj})=[({}^{r}N_j+{}^{s}N_j)N_{m+1}+{}^{s}N_j]\times N_{m+1},\quad \dim(A_{stm})=[N_m(N_{m+1}+1)]\times N_{m+1}$$

$$\dim(A_{tsi})=N_{m+1}\times[({}^{r}N_i+{}^{s}N_i)N_{m+1}+{}^{s}N_i],\quad \dim(A_{tsm})=[N_{m+1}\times]N_m(N_{m+1}+1)$$

$$\dim(A_{tt})=N_{m+1}\times N_{m+1}$$

$$\dim(B_{rj})=[{}^{r}N_j(N_{m+1}+1)]\times 1,\quad \dim(B_{rm})=[N_m(N_{m+1}+1)]\times 1$$

$$\dim(B_{sj})=[({}^{r}N_j+{}^{s}N_j)N_{m+1}+{}^{s}N_j]\times 1,\quad \dim(B_{sm})=[N_m(N_{m+1}+1)]\times 1$$

$$\dim(B_t)=N_{m+1}\times 1$$

$$\dim(C_{rj})=1\times[{}^{r}N_j(N_{m+1}+1)],\quad \dim(C_{rm})=1\times[N_m(N_{m+1}+1)]$$

$$\dim(C_{sj})=1\times[({}^{r}N_j+{}^{s}N_j)N_{m+1}+{}^{s}N_j],\quad \dim(C_{sm})=1\times[N_m(N_{m+1}+1)]$$

$$\dim(C_t)=1\times N_{m+1},\quad \dim(D)=1\times 1$$

where $(1,1)\leqslant(i,j)\leqslant(m-1,m-1)$.

3.1. Dimensionality of the (m+1)-D state-space model

From (7) and (10) for the m-D state-space model one finds that:

$$ {}^{r}D_m = \sum_{i=1}^{m-1} {}^{r}N_i, \quad {}^{s}D_m = \sum_{i=1}^{m-1} {}^{s}N_i, \quad {}^{t}D_m = N_m \tag{21} $$

and the dimensionality of the model, i.e. the dimension of the total state-vector is:

$$ D_m = {}^{r}D_m + {}^{s}D_m + {}^{t}D_m = \sum_{i=1}^{m} ({}^{r}N_i + {}^{s}N_i) + N_m \tag{22} $$

For the (m+1)-D state-space model, according to the dimensions of the partial states given in (20), it is found that:

$$ {}^{r}D_{m+1} = \sum_{i=1}^{m} \dim\left[r^{(i)}\right] \overset{(21)}{=} (N_{m+1}+1)\,{}^{r}D_m + (N_{m+1}+1)\,{}^{t}D_m \tag{23a} $$

$$ {}^{s}D_{m+1} = \sum_{i=1}^{m} \dim\left[s^{(i)}\right] \overset{(21)}{=} N_{m+1}\,{}^{r}D_m + (N_{m+1}+1)\,{}^{s}D_m + (N_{m+1}+1)\,{}^{t}D_m \tag{23b} $$

$$ {}^{t}D_{m+1} = N_{m+1} \tag{23c} $$

$$ D_{m+1} = {}^{r}D_{m+1} + {}^{s}D_{m+1} + {}^{t}D_{m+1} \tag{23d} $$

Equations (23) can be written in the more compact form:

$$ \begin{bmatrix} {}^{r}D_{m+1} \\ {}^{s}D_{m+1} \\ {}^{t}D_{m+1} \end{bmatrix} = \begin{bmatrix} N_{m+1}+1 & 0 & N_{m+1}+1 \\ N_{m+1} & N_{m+1}+1 & N_{m+1}+1 \\ 0 & 0 & 0 \end{bmatrix} \begin{bmatrix} {}^{r}D_m \\ {}^{s}D_m \\ {}^{t}D_m \end{bmatrix} + \begin{bmatrix} 0 \\ 0 \\ N_{m+1} \end{bmatrix} 1 \tag{24a} $$

$$ D_m = \begin{bmatrix} 1 & 1 & 1 \end{bmatrix} \begin{bmatrix} {}^{r}D_m \\ {}^{s}D_m \\ {}^{t}D_m \end{bmatrix} \tag{24b} $$

It is easily seen that (24) represent the state-space equations of a 1-D, third order SISO discrete linear system with time-varying coefficients. The initial conditions of this system are the dimensions of the partial states of the 3-D state-space model, given in [39]:

$$ {}^{r}D_3 = K_1N_3+K_1+K_2, \quad {}^{s}D_3 = N_1N_3+N_1+N_2, \quad {}^{t}D_3 = N_3 \tag{25} $$

Omitting the details, the solution of the recursive equations (24), with the initial conditons (25), is found to be:

$$ {}^{r}D_m = (N_m+1)(N_{m-1}+1)\ldots(N_4+1)(K_1N_3+K_1+K_2+N_3) \quad \text{for } m\geqslant 4 \tag{26a} $$

$$ {}^{s}D_m = (N_m+1)(N_{m-1}+1)\ldots(N_4+1)(N_1N_3+N_1+N_2)+\sum_{i=3}^{m-1}(N_m+1)(N_{m-1}+1)\ldots $$
$$ \ldots(N_{i+2}+1)\left[N_{i+1}(N_i+1)(N_{i-1}+1)\ldots(N_4+1)(K_1N_3+K_1+K_2+N_3)\right. $$
$$ \left.+(N_{i+1}+1)N_i\right] \quad \text{for } i\geqslant 4,\ m\geqslant 5 \tag{26b} $$

$$ {}^{s}D_4 = (N_4+1)(N_1N_3+N_1+N_2)+N_4(K_1N_3+K_1+K_2)+(N_4+1)N_3 \tag{26b'} $$

$$ {}^{t}D_m = N_m \quad \text{for } m\geqslant 1 \tag{26c} $$

$$ D_m = {}^{r}D_m + {}^{s}D_m + {}^{t}D_m \quad \text{for } m\geqslant 1 \tag{26d} $$

For the special case of equal order systems, i.e. $K_i=N_i=N$ for $1\leqslant i\leqslant m+1$, the above relations reduce to:

$$ {}^{r}D_m = (N+1)^{m-2}+N(N+2)(N+1)^{m-3}-(N+1) \quad \text{for } m\geqslant 4 \tag{27a} $$

$$ {}^{s}D_m = (N+1)^{m-4}N\left[(m-2)N^2+3(m-2)N+(m-3)\right] \quad \text{for } m\geqslant 5 \tag{27b} $$

$$ {}^{s}D_4 = N(2N^2+6N+3) \tag{27c} $$

$$ {}^{t}D_m = N_m \quad \text{for } m\geqslant 1 \tag{27c'} $$

$$ D_m = {}^{r}D_m + {}^{s}D_m + {}^{t}D_m \quad \text{for } m\geqslant 1 \tag{27d} $$

For the special case of first order systems, i.e. N=1, the above relations reduce to:

$$ {}^{r}D_m = 2^{m-2}+3.2^{m-3}-2 \quad \text{for } m\geqslant 4, \quad {}^{r}D_4 = 8 \tag{28a} $$

$$ {}^{s}D_m = (5m-11)2^{m-4} \quad \text{for } m\geqslant 5, \quad {}^{s}D_4 = 11 \tag{28b} $$

$$ {}^{t}D_m = 1 \quad \text{for } m\geqslant 1, \quad {}^{t}D_4 = 1 \tag{28c} $$

$$D_m = 2^{m-2} + 3.2^{m-3} + (5m-11)2^{m-4} - 1 \quad \text{for } m>5, \quad {}^4D = 20 \tag{28d}$$

For the above special case of first order systems the dimensionality D_m of the state-space model presented may be compared with the dimensionality $D'_m = 2^m - 1$ of Galkowski's model [38], i.e.

$$\left.\begin{array}{ll} D_m = D'_m & \text{for } m \leqslant 3 \quad [39] \\ D_m > D'_m & \text{for } m \geqslant 4 \end{array}\right\} \text{ implies } D_m \geqslant D'_m \tag{29}$$

Hence the model presented in this paper has greater dimensionality than Galkowski's model. The advantage of the present model is that it is more general, as it refers to systems of any order, and not only to first order systems as Galkowski's model.

4. SPECIAL CASES

4.1. Pure recursive systems

Given the m-D state-space model (2) and according to the analysis in section 3. an (m+1)-D system desribed by the transfer function:

$$G = \frac{b_{0,\dots,0}}{\sum_{i_1=0}^{N_1} \cdots \sum_{i_{m+1}=0}^{N_{m+1}} a_{i1,\dots,i(m+1)} z_1^{-i_1} \cdots z_{m+1}^{-i(m+1)}}, \quad a_{0,\dots,0} = 1 \tag{30}$$

has a state-space model realization of the type:

$$\begin{bmatrix} s^{(1)}(n_1+1,\dots,n_m,n_{m+1}) \\ \cdot \\ \cdot \\ \cdot \\ s^{(m)}(n_1,\dots,n_{m+1},n_{m+1}) \\ t\ \ (n_1,\dots,n_m,n_{m+1}+1) \end{bmatrix} = \begin{bmatrix} A_{ss1,1} & \dots & A_{ss1,m} & A_{st1} \\ \cdot & & \cdot & \cdot \\ \cdot & & \cdot & \cdot \\ \cdot & & \cdot & \cdot \\ A_{ssm,1} & \dots & A_{ssm,m} & A_{stm} \\ A_{ts1} & \dots & A_{tsm} & A_{tt} \end{bmatrix} \begin{bmatrix} s^{(1)}(n_1,\dots,n_{m+1}) \\ \cdot \\ \cdot \\ \cdot \\ s^{(m)}(n_1,\dots,n_{m+1}) \\ t\ \ (n_1,\dots,n_{m+1}) \end{bmatrix}$$

$$+ \begin{bmatrix} B_{s1} \\ \cdot \\ \cdot \\ \cdot \\ B_{sm} \\ B_t \end{bmatrix} u(n_1,\dots,n_{m+1}) \tag{31a}$$

$$y(n_1,\ldots,n_{m+1})=\begin{bmatrix}C_{s1}\ldots C_{sm} & C_t\end{bmatrix}\begin{bmatrix}s^{(1)}(n_1,\ldots,n_{m+1})\\ s^{(m)}(n_1,\ldots,n_{m+1})\\ t\ (n_1,\ldots,n_{m+1})\end{bmatrix}+Du(n_1,\ldots,n_{m+1}) \quad (31b)$$

where the various state-space vector and matrices have been defined in section 3.

The dimensionality of this model is

$$D_m={}^sD_m+{}^tD_m \quad (32)$$

where ${}^sD_m$ and ${}^tD_m$ are defined in (26) or (27) or (28). In the special case of first order systems, the dimensionality D_m is

$$D_m=(5m-11)2^{m-4}+1 \text{ for } m>5,\ D_4=12 \quad (33)$$

Comparing this dimensionality with the dimensionality $D'_m=2^m-1$ of Galkowski's model [38], it is found that:

$$D_m<D'_m \text{ for } m\leqslant 3 \text{ and } D_m>D'_m \text{ for } m\geqslant 4 \quad (34)$$

4.2. Nonrecursive systems

Given the m-D state-space model in (2) and according to the analysis in section 3,an (m+1)-D system described by the transfer function:

$$G=\sum_{i_1=0}^{K_1}\ldots\sum_{i_{m+1}=0}^{K_{m+1}} b_{i1,\ldots,i(m+1)}\, z_1^{-i_1}\ldots z_{m+1}^{-i(m+1)} \quad (35)$$

has a state-space model realizing G of the type:

$$\begin{bmatrix}r^{(1)}(n_1+1,\ldots,n_m,n_{m+1})\\ \cdot\\ \cdot\\ \cdot\\ r^{(m)}(n_1,\ldots,n_m+1,n_{m+1})\\ t\ (n_1,\ldots,n_m,n_{m+1}+1)\end{bmatrix}=\begin{bmatrix}A_{rr1,1}\ldots A_{rr1,m} & A_{rt1}\\ \cdot\ \ \cdots\cdot & \cdot\\ \cdot\ \ \cdots\cdot & \cdot\\ \cdot\ \ \cdots\cdot & \cdot\\ A_{rrm,1}\ldots A_{rrm,m} & A_{rtm}\\ A_{rt1}\ \ldots A_{rtm} & A_{tt}\end{bmatrix}\begin{bmatrix}r^{(1)}(n_1,\ldots,n_{m+1})\\ \cdot\\ \cdot\\ \cdot\\ r^{(m)}(n_1,\ldots,n_{m+1})\\ t\ (n_1,\ldots,n_{m+1})\end{bmatrix}$$

$$+\begin{bmatrix} B_{r1} \\ \vdots \\ B_{rm} \\ B_t \end{bmatrix} u(n_1,\ldots,n_{m+1}) \tag{36a}$$

$$y(n_1,\ldots,n_{m+1}) = \begin{bmatrix} C_{r1} \cdots C_{rm} & C_t \end{bmatrix} \begin{bmatrix} r^{(1)}(n_1,\ldots,n_{m+1}) \\ \vdots \\ r^{(m)}(n_1,\ldots,n_{m+1}) \end{bmatrix} + Du(n_1,\ldots,n_{m+1}) \tag{36b}$$

where the various state-space vectors and matrices have been defined in section 3.

The dimensionality of this model is

$$D_m = {}^{r}D_m + {}^{t}D_m \tag{37}$$

where ${}^{r}D_m$ and ${}^{t}D_m$ are defined in (26) or (27) or (28). In the special case of first order systems, the dimensionality D_m is

$$D_m = 2^{m-2} + 3.2^{m-3} - 1 \quad \text{for } m \geq 4 \tag{38}$$

Comparing this dimensionality with the dimensionality $D'_m = 2^m - 1$ of Galkowski's model [38], it is found that:

$$D_m < D'_m \quad \text{for } m \leq 4, \quad D_m > D'_m \text{ for } \quad m \geq 4 \tag{39}$$

5. EXAMPLE

Given the state-space model of a first order 3-D system in [39] and following the previous procedure, a canonical state-space model representing the pure recursive 4-D transfer function:

$$G = \frac{b_{0,0,0,0}}{\sum_{i_1=0}^{1} \sum_{i_2=0}^{1} \sum_{i_3=0}^{1} \sum_{i_4=0}^{1} a_{i1,i2,i3,i4} z_1^{-i_1} z_2^{-i_2} z_3^{-i_3} z_4^{-i_4}}, \quad a_{0,0,0,0} = 1 \tag{40}$$

is the following:

$$\begin{bmatrix} x_1(n_1+1,n_2,n_3,n_4) \\ x_2(n_1,n_2+1,n_3,n_4) \\ x_3(n_1,n_2,n_3+1,n_4) \\ x_4(n_1,n_2,n_3,n_4+1) \end{bmatrix} = \begin{bmatrix} A_{11} & A_{12} & A_{13} & A_{14} \\ A_{21} & A_{22} & A_{23} & A_{24} \\ A_{31} & A_{32} & A_{33} & A_{34} \\ A_{41} & A_{42} & A_{43} & A_{44} \end{bmatrix} \begin{bmatrix} x_1(n_1,n_2,n_3,n_4) \\ x_2(n_1,n_2,n_3,n_4) \\ x_3(n_1,n_2,n_3,n_4) \\ x_4(n_1,n_2,n_3,n_4) \end{bmatrix} + \begin{bmatrix} B_1 \\ B_2 \\ B_3 \\ B_4 \end{bmatrix} u(n_1,n_2,n_3,n_4) \tag{41a}$$

$$y(n_1,n_2,n_3,n_4) = \begin{bmatrix} C_1 & C_2 & C_3 & C_4 \end{bmatrix} \begin{bmatrix} x_1(n_1,n_2,n_3,n_4) \\ x_2(n_1,n_2,n_3,n_4) \\ x_3(n_1,n_2,n_3,n_4) \\ x_4(n_1,n_2,n_3,n_4) \end{bmatrix} + Du(n_1,n_2,n_3,n_4) \tag{41b}$$

where

$$x_1 = \begin{bmatrix} {}_1r_1^{(1)} \\ {}_1r_2^{(1)} \\ {}_1s_1^{(1)} \\ {}_1s_2^{(1)} \\ {}_os_1^{(1)} \\ {}_os_2^{(1)} \end{bmatrix}, \quad x_2 = \begin{bmatrix} {}_1r^{(2)} \\ {}_1s^{(2)} \\ {}_os^{(2)} \end{bmatrix}, \quad x_3 = \begin{bmatrix} {}_1t \\ {}_ot \end{bmatrix}, \quad x_4 = t$$

$$A_{11} = \begin{bmatrix} 0 & 0 & 0 & 0 & 0 & 0 \\ 0 & 0 & -\tilde{a}'_{1,0,1,0} & -a_{1,0,0,0} & 0 & 0 \\ 0 & 0 & 0 & 0 & 0 & 0 \\ 0 & 0 & -\tilde{a}'_{1,0,1,0} & -a_{1,0,0,0} & 0 & 0 \\ 0 & 0 & 0 & 0 & 0 & 0 \\ 0 & 0 & 0 & 0 & 0 & 0 \end{bmatrix}$$

$$A_{12} = \begin{bmatrix} 0 & 0 & 0 \\ 0 & 1 & 0 \\ 0 & 0 & 0 \\ 0 & 1 & 0 \\ 0 & 0 & 0 \\ 0 & 0 & 1 \end{bmatrix}, A_{13} = \begin{bmatrix} 1 & 0 \\ 0 & 0 \\ 1 & 0 \\ 0 & 0 \\ 0 & 1 \\ 0 & 0 \end{bmatrix}, A_{14} = \begin{bmatrix} 0 \\ 1 \\ 0 \\ 1 \\ 0 \\ 0 \end{bmatrix}$$

$$A_{21}=\begin{bmatrix} -\tilde{\tilde{a}}_{1,1,1,1} & -\tilde{a}_{1,1,0,1} & a'_{1,0,0,0} & a'_{1,0,0,0} & 0 & 0 \\ 0 & 0 & -a'_{1,1,1,1} & -\tilde{a}'_{1,1,0,0} & 0 & 0 \\ 0 & 0 & 0 & 0 & -a'_{1,1,1,1} & -\tilde{a}'_{1,1,0,0} \end{bmatrix}$$

$$A_{22}=\begin{bmatrix} 0 & -\tilde{a}_{0,1,0,1} & 0 \\ 0 & -a_{0,1,0,0} & 0 \\ 0 & 0 & -a_{0,1,0,0} \end{bmatrix}, A_{23}=\begin{bmatrix} -\tilde{\tilde{a}}_{0,1,1,1} & 0 \\ -\tilde{a}'_{0,1,1,0} & 0 \\ 0 & -a'_{0,1,1,0} \end{bmatrix}, A_{24}=\begin{bmatrix} -\tilde{a}_{0,1,0,1} \\ -a_{0,1,0,0} \\ 0 \end{bmatrix}$$

$$A_{31}=\begin{bmatrix} 0 & 0 & -\tilde{a}'_{1,0,1,0} & -a_{1,0,0,0} & 0 & 0 \\ 0 & 0 & 0 & 0 & -\tilde{a}'_{1,0,1,0} & -a_{1,0,0,0} \end{bmatrix}$$

$$A_{32}=\begin{bmatrix} 0 & 1 & 0 \\ 0 & 0 & 1 \end{bmatrix}, A_{33}=\begin{bmatrix} -a_{0,0,1,0} & 0 \\ 0 & -a_{0,0,1,0} \end{bmatrix}, A_{34}=\begin{bmatrix} 1 \\ 0 \end{bmatrix}$$

$$A_{41}=\begin{bmatrix} -\tilde{\tilde{a}}_{1,0,1,1} & -\tilde{a}_{1,0,0,1} & 0 & 0 & -\tilde{a}'_{1,0,1,0} & -a_{1,0,0,0} \end{bmatrix}$$

$$A_{42}=\begin{bmatrix} 1 & 0 & 1 \end{bmatrix}, \quad A_{43}=\begin{bmatrix} -\tilde{a}_{0,0,1,1} & -a_{0,0,1,0} \end{bmatrix}, A_{44}=\begin{bmatrix} -a_{0,0,0,1} \end{bmatrix}$$

$$B_1=\begin{bmatrix} 0 \\ 0 \\ 0 \\ 0 \\ 0 \\ 1 \end{bmatrix}, B_2=\begin{bmatrix} 0 \\ 0 \\ -a_{0,1,0,0} \end{bmatrix}, B_3=\begin{bmatrix} 0 \\ 1 \end{bmatrix}, \quad B_4=\begin{bmatrix} 1 \end{bmatrix}$$

$$C_1=\begin{bmatrix} -\tilde{\tilde{a}}_{1,0,1,1} & -\tilde{a}_{1,0,0,1} & 0 & 0 & -\tilde{a}'_{1,0,1,0} & -a_{1,0,0,0} \end{bmatrix} b_{0,0,0,0}$$

$$C_2=\begin{bmatrix} 1 & 0 & 1 \end{bmatrix} b_{0,0,0,0}, C_3=\begin{bmatrix} -\tilde{a}_{0,0,1,1} & -a_{0,0,1,0} \end{bmatrix} b_{0,0,0,0}$$

$$C_4=\begin{bmatrix} -a_{0,0,0,1} \end{bmatrix} b_{0,0,0,0}, \quad D=\begin{bmatrix} b_{0,0,0,0} \end{bmatrix}$$

where

$$\tilde{a}_{i1,i2,i3,1}=a_{i1,i2,i3,1}-a_{i1,i2,i3,0}a_{0,0,01} \quad \text{for } (0,0,0)<(i_1,i_2,i_3)\leqslant(1,1,1)$$

$$\tilde{a}'_{0,1,1,0}=a_{0,1,1,0}-a_{0,1,0,0}a_{0,0,1,0}$$

$$\tilde{a}'_{1,0,1,0}=a_{1,0,1,0}-a_{1,0,0,0}a_{0,0,1,0}$$

$$\tilde{a}'_{1,1,0,0}=a_{1,1,0,0}-a_{1,0,0,0}a_{0,1,0,0},\ \tilde{\tilde{a}}_{0,1,1,1}=\tilde{a}_{0,1,1,1}-\tilde{a}_{0,1,0,1}a_{0,0,1,0}$$

$$\tilde{\tilde{a}}_{1,0,1,1}=\tilde{a}_{1,0,1,1}-\tilde{a}_{1,0,0,1}a_{0,0,1,0},\ \tilde{\tilde{a}}_{1,1,1,1}=\tilde{a}_{1,1,1,1}-\tilde{a}_{1,1,0,1}a_{0,0,1,0}$$

$$\tilde{a}_{1,1,1,0}=a_{1,1,1,0}-a_{1,1,0,0}a_{0,0,1,0},\ a'_{1,1,1,1}=\tilde{a}_{1,1,1,0}-\tilde{a}'_{1,0,1,0}a_{0,1,0,0}$$

$$a'_{1,0,1,0}=\tilde{a}'_{1,0,1,0}\tilde{a}_{0,1,0,1},\ a'_{1,0,0,0}=a_{1,0,0,0}\tilde{a}_{0,1,0,1}$$

The dimensionality of this model is $D_4=12$. Comparing this with the dimensionality $D'_4=2^4-1=15$ of Galkowski's model it is found that the present model has lower dimensionality than Galkowski's model.

A computer simulation program has shown the validity of the results obtained for the above example. In this program numerical values of the output of the system obtained from the transfer function and the state-space model were identical for a wide range of values of the coefficients and the variables z_1, z_2, z_3, z_4.

5. CONCLUSION

In this chapter an inductive way of constructing (m+1)-D state-space models from m-D ones was presented. It is found that the dimensionality of the state-space matrices, included in the realization, is immensely increasing with increasing m. One possible direction of research is to find ways for decreasing of the model's dimensionality. This has partially been achieved for 3-D systems in [39]. Work needs to be done for m-D systems too for $m \geq 4$.

Other procedures for obtaining m-D state-space models of the form presented in this paper may be found. Also totally different m-D state-space models may arise in the future. Then mutual comparisons of the various state-space models may be done, as it was done for 2-D and 3-D systems in [44].

REFERENCES

1. D. D. Givone and R.P. Roesser, "Multidimensional Linear Iterative Circuits-General Properties", IEEE Trans. Comp., Vol. C-21, No.10, pp.1067-1073, October 1972.
2. M.K.Hu and A. Iosupovicz, "Analysis of the Terminal Behavior of Some Classes of Iterative Arrays of Linear Machines" IEEE Trans. Comp., Vol. C-22, No.12, pp 1394-1399, December 1972.

3. D.D .Givone and R.P. Roesser, "Minimization of Multidimensional Linear Iterative Circuits", IEEE Trans. Comp., Vol. C-22, No.7, pp. 673-678, July 1973.
4. B. Vilfan, "Another Proof of the Two-Dimensional Cayley-Hamilton Theorem", IEEE Trans. Comp., Vol. C-22, No.12, pp. 1140, December 1973.
5. R.P. Roesser, "A Discrete State-Space Model for Linear Image Processing", IEEE Trans. Aut. Control, Vol.AC-20, No.1, pp. 1-10, February 1975.
6. T. Ciftcibasi and O. Yuksel, "On the Cayley Hamilton Theorem for Two-Dimensional Systems", IEEE Trans. Aut. Control,Vol. AC-27, No.1, pp. 193-194, February 1982.
7. N.J. Theodorou and R.A. King, "State-Space Discrete Systems:Computation of the Transition Matrix", IEEE Trans. Aut. Control, Vol. AC-25, No.2, pp. 296-297, April 1980.
8. B.G. Mertzios and P.N. Paraskevopoulos, "Transfer Function Matrix of 2-D Systems", IEEE Trans. Aut. Control, Vol. 26, No.3, pp. 722-725, June 1981.
9. P.N. Paraskevopoulos and B.G. Mertzios, "Transfer Function Factorization of 2-D Systems Using State Feedback", Int. J. Systems Sci., Vol. 12, No.9, pp. 1135-1147, 1981.
10. P.N. Paraskevopoulos, "Eigenvalue Assignment of Linear 2-Dimensional Systems", Proc. IEE,Vol. 126, pp. 1204-1208, 1979.
11. P.N. Paraskevopoulos, "Characteristic Polynomial Assignement and Determination of the Residual Polynomial in 2-D Systems", IEEE Trans. Aut.Control,Vol. 26, NO.2, pp.541-543, April 1981.
12. P.N. Paraskevopoulos and O.I. Kosmidou, "Eigenvalue Assignement of Two-Dimensional Systems Using PID Controllers", Int. J. Systems Sci.,Vol.12, No.4, pp. 407-422, 1981.
13. P.N.Paraskevopoulos, "Transfer Function Matrix Synthesis of Two-Dimensional Systems", IEEE Trans. Aut. Control, Vol. AC-25, No.2,pp. 321-324, April 1980.
14. P.N. Paraskevopoulos, "Exact Model-Matching of 2-D Systems via State Feedback", J. Franklin Inst.,Vol. 308, No.5, pp. 475-486, November 1979.
15. P.N. Paraskevopoulos and O.I. Kosmidou, "Exact Model-matching of Two-Dimensional Systems Using PID Controllers", Proc. Second Int. Conf.Inf.Sci. and Systems, Univ. of Patras, Patras, Greece, Vol. II, pp. 210-217, July 1979.
16. P.N. Paraskevopoulos and O.I. Kosmidou, "Dynamic Compensation for Exact Model-Matching of Two-Dimensional Systems", Int. J.Systems Sci., Vol. 11, No.10, pp. 1163-1175, 1980.
17. P.N.Paraskevopoulos, "Feedback Design Techniques for Linear Multivariable 2-D Systems", Proc. fourth Int.Conf. Analysis and Optimization of Systems, Versailles, France, Vol. 28, pp. 763-780, 1980.
18. P. Stavroulakis and P.N.Paraskevopoulos, "Reduced-Order Feedback Law Implementation for 2D Digital Systems, Int. J.Systems Sci.,Vol.12, No.5, pp. 525-537, 1981.

19. P.N. Paraskevopoulos and B.G. Mertzios, "Sensitivity Analysis of 2-D Systems", IEEE Trans. Circ. and Systems, Vol. CAS-28, No.8, pp. 833-838, August 1981.
20. P. Stavroulakis and P.N. Paraskevopoulos, "Low Sensitivity Observer-Compensator Design for Two-Dimensional Digital Systems", Proc. IEE, Vol. 129, No.5, pp. 193-200, September 1982.
21. E. Fornasini and G. Marchesini, "Algebraic Realization of Two-Dimensional Filters", Variables Structure Systems, A. Ruberti and R. Mohler Eds. (Springer Lecture Notes in Economics and Mathematical Systems), pp. 64-82, 1975.
22. E. Fornasini and G. Marchesini, "State-Space Realization Theory of Two-Dimensional Filters", IEEE Trans.Aut.Control, Vol.AC-21, No.4, pp. 484-492, August 1976.
23. E. Fornasini and G. Marchesini, "Doubly Indexed Dynamical Systems:State-Space Models and Structural Properties", Math.Systems Theory, Vol.12, No.1, pp. 59-72, 1978.
24. E. Fornasini and G. Marchesini, "Global Properties and Duality in 2-D Systems", Systems and Control Letters, Vol. 2, No.1, pp. 30-38, July 1982.
25. E. Fornasini and G. Marchesini, "Some Aspects of the Duality Theory in 2-D Systems", Proc. First IASTED Symposium on Applied Informatics, Lille, France, Vol. I, pp. 197-200, March 1983.
26. S.Y.Kung, B.C.Levy, M.Morf and T.Kailath, "New Results in 2-D Systems Theory, Part II:2-D State-Space Models-Realization and the Notions of Controllability, Observability and Minimality", Proc. IEEE, Vol. 65, No.6, pp. 945-961, June 1977.
27. N.J. Theodorou, "State-Space Representation and Error Analysis of Two-Dimensional Discrete Systems", Ph.D.Thesis, Imperial College, Univ, London, 1978.
28. N.J. Theodorou and R.A. King, "A State-Space Realization for 2-D Discrete Systems and its Error Analysis", Proc. of the 1978 European Conf. on Circ. Theory and Design (ECCTD), Lausanne, pp. 350-354, 1978.
29. N.J. Theodorou and R.A. King, "Minimum Roundoff Noise in 2-D State-Space Digital Filters", Proc. Int.Conf.on Digital Signal Processing , Florence, Italy, pp. 271-278, 1978.
30. R. Eising, "Realization and Stabilization of 2-D Systems", IEEE Trans. Aut. Control,Vol. AC-23, No.5, pp. 793-799, October 1978.
31. R. Eising, "Controllability and Observability of 2-D Systems", IEEE Trans. Aut.Control,Vol. AC-24, No.1, pp. 132-133, February 1979.
32. R. Eising, "Low Order Realizations for 2-D Transfer Functions", Proc. IEEE, Vol. 67, No.5, pp. 856-858, May 1979.
33. R. Eising, "State-Space Realization and Invertion of 2-D Systems", IEEE Trans.Circ.Systems, Vol.CAS-27, No.7, pp.

612-619, July 1980.

34. S. G. Tzafestas and T.G.Pimenides, "Transfer Function Computation and Factorization of 3-D Systems in State-Space", Proc. IEE, in press.

35. S.G.Tzafestas and T.G. Pimenides, "Feedback Characteristic Polynomial Controller Design of 3-D Systems in State-Space", J. Franklin Inst.,Vol. 314, No.3, pp. 169-189, September 1982.

36. S. G. Tzafestas and T.G. Pimenides, "Exact Model-Matching Control of Three-Dimensional Systems Using State and Output Feedback", Int.J.Systems Sci.,Vol. 13, pp. 1982.

37. S. G. Tzafestas, P.N. Paraskevopoulos and T.G.Pimenides, "Modelling and Control of Multidimensional Systems", Proc. First IASTED Symposium on Applied Informatics, Lille, France,Vol. I, pp. 167-179, March 1983.

38. K. Galkoski, "The State-Space Realization of an n-Dimensional Transfer Function", Circ.Theory and Applications, Vol. 9, pp. 189-197, 1981.

39. N.J. Theodorou and S.G.Tzafestas, "A Canonical State-Space Model Representing 3-D Discrete Systems", Proc.First IASTED Symposium on Applied Informatics, Lille, France,Vol. I, pp. 201-207, March 1983.

40. N.J. Theodorou and S.G.Tzafestas, "A Canonical State-Space Model Representing 3-D Discrete Systems", submitted.

41. S.G. Tzafestas and N.J. Theodorou, "Feedback Factorizing Control of 3-D Transfer Functions via a Canonical State-Space Model", J. Franklin Inst., to be published.

42. N.J.Theodorou and S.G. Tzafestas, "Characteristic Polynomial Assignement of 3-D Discrete Systems Using a Canonical State-Space Model", Int.J.Syst. Sci., to be published.

43. S.G. Tzafestas and N.J.Theodorou, "Exact Model-Matching of 3-D Discrete Systems Using Feedback Control", to be be published.

44. S.G. Tzafestas and N.J.Theodorou, "A Comparative Overview of some Multidimensional State-Space Models", Proc.of the MECO 83 (Measurement and Control), Aug.-Sept.1983.

45. E. Fornasini, "On the Relevance of Noncommutative Power Series in Spatial Filters Realization", IEEE Trans. Circ. and Systems,Vol. CAS-25, No.5, pp. 290-299, May 1978.

46. S. Chakrabarti and S.K.Mitra, "Decision Methods and Realization of 2-D Digital Filters Using Minimum Number of Delay Elements", IEEE Trans. Circ. and Systems,Vol. CAS-27, No.8, pp. 657-666, August 1980.

47. S. Chakrabarti and S.K. Mitra, "Corrections to "Decision Methods and Realization of 2-D Digital Filters Using Minimum Number of Delay Elements"", IEEE Trans. Circ., and Systems, Vol.CAS-28, No.3, pp. 262-263, March 1981.

CHAPTER 22

EIGENVALUE ASSIGNMENT OF 3-D SYSTEMS

T. Kaczorek
Technical University of Warsaw
Electrical Engineering Faculty
ul. Koszykowa 75
00-662 Warszawa, Poland

Sufficient conditions are given for existence of a solution to the eigenvalues assignment problem for three-dimensional (3-D) linear systems with separable closed-loop characteristic polynomials. Three methods for finding the feedback gain matrix are presented. The method 3 is an extension for 3-D systems of the method presented in [3] for 2-D systems.

1. INTRODUCTION

The eigenvalue assignment problem and the characteristic function coefficients assignment problem using state feedback or output feedback for two-dimensional (2-D) linear systems have been receiving extensive attention in the last few years. First results were given by Eising in [1]. Next Paraskevopoulos has presented in [8,9] a general method for characteristic polynomial assignment and determination of the residual polynomial in 2-D systems. Eigenvalue assignment of 2-D systems using PID controllers was considered by Paraskevopoulos and Kosmidou in [10]. Sufficient conditions for existence of a solution to the eigenvalue assignment and algorithms for finding the feedback gain matrices for 2-D systems with separable characteristic polynomial were given by Kaczorek [3] and by Mertzios [7]. The same problem for 3-D systems was considered by Kaczorek in [4]. The problem of characteristic polynomial assignment by dynamic output feedback for 2-D systems was solved by Šebek [12] and for 3-D systems by Kaczorek [5]. The separability-assignment problem for n-D systems was considered by Kaczorek and Kurek in [6]. Tzafestas and Pimenides in [13] have extended the Roesser's model for 3-D systems. The purpose of this paper is to present three methods for solving the eigenvalue assignment problem using state feedback for 3-D systems with separable closed-loop

S. G. Tzafestas (ed.), Multivariable Control, 409–427.

characteristic polynomials.

2. PROBLEM STATEMENT

Consider a 3-D linear system described by the equations [13]

$$x' = Ax + Bu \tag{1}$$

where

$$x' = \begin{bmatrix} x^h(i+1,j,k) \\ x^v(i,j+1,k) \\ x^d(i,j,k+1) \end{bmatrix}, \quad x = \begin{bmatrix} x^h(i,j,k) \\ x^v(i,j,k) \\ x^d(i,j,k) \end{bmatrix}$$

$$A = \begin{bmatrix} A_1 & A_2 & A_3 \\ A_4 & A_5 & A_6 \\ A_7 & A_8 & A_9 \end{bmatrix}, \quad B = \begin{bmatrix} B_1 \\ B_2 \\ B_3 \end{bmatrix}$$

$x^h(i,j,k) \in R_{n_1}$ is the horizontal state vector, $x^v(i,j,k) \in R_{n_2}$ is the vertical state vector, $x^d(i,j,k) \in R_{n_3}$ is the depth state vector, $u(i,j,k) \in R_m$ is the input vector; A_i, B_i, are constant matrices of appropriate dimensions and $(i,j,k) \in ZxZxZ = Z^3$; Z is the set of nonnegative integer numbers.
The closed-loop system with the state feedback law

$$u = v - Fx$$

is described by the equation

$$x' = A_c x + Bv$$

where

$$A_c = A - BF \tag{2}$$

$F \in R_{mxn}$ $(n=n_1+n_2+n_3)$ is the feedback gain matrix and $v=v(i,j,k) \in R_m$ is the control input vector.
The 3-D characteristic polynomial $q(z_1,z_2,z_3)$ of the matrix A is defined by

$$q(z_1,z_2,z_3) = \det\left[Z-A\right] = \sum_{i=o}^{n_1} \sum_{j=o}^{n_2} \sum_{k=o}^{n_3} a_{ijk} z_1^i z_2^j z_3^k \tag{3}$$

where $a_{n_1 n_2 n_3} = 1$ and

$$Z = \begin{bmatrix} I_{n_1}z_1 & 0 & 0 \\ 0 & I_{n_2}z_2 & 0 \\ 0 & 0 & I_{n_3}z_3 \end{bmatrix}$$

I_k is the identity matrix of the order k.

The 3-D characteristic polynomial (3) will be called separable iff

$$q(z_1,z_2,z_3)=q_1(z_1)q_2(z_2)q_3(z_3) \tag{4}$$

where

$$q_i(z_i)=z_i^{n_i}+a_{in_1-1}z_i^{n_i-1}+\dots+a_{i1}z_i+a_{i0}$$

for i=1,2,3.

For the separable polynomial (4) the 3-D eigenvalues of A are the triples (z_1,z_2,z_3) which satisfy the equation $q_i(z_i)=0$ for i=1,2,3.

<u>The problem</u> can be stated as follows: given the matrix <u>A,B,</u> find a feedback gain matrix F such that the closed-loop system matrix (2) has a separable 3-D characteristic polynomial with the given set of 3-D eigenvalues:

$$E = \{(z_{1i},z_{2j},z_{3k})(i=1,\dots,n_1;j=1,\dots,n_2;k=1,\dots,n_3)\} \tag{5}$$

3. PROBLEM SOLUTION

3.1. <u>Method 1</u>

Let us denote

$$B_{ij}=\begin{bmatrix} B_i \\ B_j \end{bmatrix} \quad (i,j=1,2,3) \qquad \begin{matrix} A_{1i}=A_i \\ A_{2i}=A_{3+i} \\ A_{3i}=A_{6+i} \end{matrix} \quad (i=1,2,3)$$

<u>Theorem 1</u>

The problem has a solution if there exist i,j,k $(i\neq j\neq k)$ such that the matrix B_{ij} is of full row rank and

$$B_k B_{ij}^g \begin{bmatrix} A_{ik} \\ A_{jk} \end{bmatrix} = A_{kk}-A_{ckk} \tag{6}$$

where B^g_{ij} is the generalized inverse matrix for B_{ij}.

Proof

Without loss of generality we can assume $i=1, j=2, k=3$ and rank $B_{12}=n_1+n_2$. For $F=[F_1 F_2 F_3]$ we have

$$A_c=\begin{bmatrix} A_{c1} & A_{c2} & A_{c3} \\ A_{c4} & A_{c5} & A_{c6} \\ A_{c7} & A_{c8} & A_{c9}\end{bmatrix} = A-BF = \begin{bmatrix} A_1-B_1F_1 & A_2-B_1F_2 & A_3-B_1F_3 \\ A_4-B_2F_1 & A_5-B_2F_2 & A_6-B_2F_3 \\ A_7-B_3F_1 & A_8-B_3F_2 & A_9-B_3F_3\end{bmatrix} \quad (7a)$$

The closed-loop system has a separable 3-D characteristic polynomial with the desired eigenvalues if

$$A_{c2}=0\ ,\ A_{c3}=0\ ,\ \text{and } A_{c6}=0 \quad (7b)$$

and A_{c1}, A_{c5} and A_{c9} have the given eigenvalues $z_{11}, z_{12}, \ldots, z_{1n_1}$; $z_{21}, z_{22}, \ldots, z_{2n_2}$ and $z_{31}, z_{32}, \ldots, z_{3n_3}$, respectively. From (7) we obtain

$$\begin{bmatrix} A_{c1} \\ A_{c4}\end{bmatrix} = \begin{bmatrix} A_1 \\ A_4\end{bmatrix} - B_{12}F_1\ ,\quad \begin{bmatrix} 0 \\ A_{c5}\end{bmatrix} = \begin{bmatrix} A_2 \\ A_5\end{bmatrix} - B_{12}F_2\ , B_{12}F_3=\begin{bmatrix} A_3 \\ A_6\end{bmatrix}$$

and

$$F_1=B^g_{12}\left(\begin{bmatrix} A_1 \\ A_4\end{bmatrix} - \begin{bmatrix} A_{c1} \\ A_{c4}\end{bmatrix}\right)\ ,\ F_2=B^g_{12}\left(\begin{bmatrix} A_2 \\ A_5\end{bmatrix} - \begin{bmatrix} 0 \\ A_{c5}\end{bmatrix}\right)\ ,$$

$$F_3=B^g_{12}\begin{bmatrix} A_3 \\ A_6\end{bmatrix} \quad (8)$$

where

$$B^g_{12}=B^T_{12}\left[B_{12}B^T_{12}\right]^{-1}$$

If (6) is satisfied we have

$$A_9-B_3B^g_{12}\begin{bmatrix} A_3 \\ A_6\end{bmatrix} = A_{c9}$$

Algorithm 1

If the conditions of the theorem 1 are satisfied the matrix F can be found by the use of the following algorithm.

Step 1

Choose i,j $i\neq j$ such that B_{ij} is of full row rank and find

$$B^g_{ij}=B^T_{ij}\left[B_{ij}B^T_{ij}\right]^{-1} \tag{9}$$

Step 2

Choose A_{c1}, A_{c5}, A_{c9} so that they have the given eigenvalues $z_{11}, z_{12}, \ldots, z_{1n_1}$; $z_{21}, z_{22}, \ldots, z_{2n_2}$ and $z_{31}, z_{32}, \ldots, z_{3n_3}$ respectively;
for example in the form

$$A_{c1}=\begin{bmatrix} 0 & 1 & 0 & \ldots & 0 \\ 0 & 0 & 1 & \ldots & 0 \\ \ldots & \ldots & \ldots & \ldots & \ldots \\ 0 & 0 & 0 & \ldots & 1 \\ -a_0 & -a_1 & -a_2 & \ldots & -a_{n_1-1} \end{bmatrix}$$

and

$$(z-z_{11})(z-z_{12})\ldots(z-z_{1n_1})=z^{n_1}+a_{n_1-1}z^{n_1-1}+\ldots+$$

$$+a_1z+a_0$$

In a similar way choose A_{c5} and A_{c9}. Then choose a form of the matrix A_c which provides a separable 3-D closed-loop characteristic polynomial;
for example

$$A_c=\begin{bmatrix} A_{c1} & 0 & 0 \\ A_{c4} & A_{c5} & 0 \\ A_{c7} & A_{c8} & A_{c9} \end{bmatrix}, \quad A_c=\begin{bmatrix} A_{c1} & A_{c2} & A_{c3} \\ 0 & A_{c5} & A_{c6} \\ 0 & 0 & A_{c9} \end{bmatrix} \quad \text{or}$$

$$A_c=\begin{bmatrix} A_{c1} & 0 & A_{c3} \\ A_{c4} & A_{c5} & A_{c6} \\ 0 & 0 & A_{c9} \end{bmatrix}$$

Step 3

Using (8) find F_1, F_2, F_3 and $F=\left[F_1, F_2, F_3\right]$

Example 1

Given the matrices

$$A = \begin{bmatrix} A_1 & A_2 & A_3 \\ A_4 & A_5 & A_6 \\ A_7 & A_8 & A_9 \end{bmatrix} = \left[\begin{array}{c:cc:c} -1 & 1 & -3 & 1 \\ \hdashline 1 & 0 & -1 & 1 \\ 2 & -3 & -5 & -1 \\ \hdashline -1 & 0 & -2 & -2 \end{array}\right], B = \begin{bmatrix} B_1 \\ B_2 \\ B_3 \end{bmatrix} = \left[\begin{array}{cc} 1 & 0 \\ \hdashline 0 & 1 \\ 1 & -1 \\ \hdashline 0 & 1 \end{array}\right]$$

and the set of 3-D eigenvalues

$$E = \{z_{11}=1,\ z_{21}=z_{22}=-2,\ z_{31}=-3\} \tag{10}$$

find F such that the closed-loop system matrix A_c has the given set of eigenvalues (10).
In this case $n_1=n_3=1$ and $n_2=2$.

Step 1 We choose i=1, j=3 and

$$B^g_{13}=B_{13}=\begin{bmatrix} 1 & 0 \\ 0 & 1 \end{bmatrix}$$

Step 2 From (10) we have

$$A_{c1}=1,\ A_{c5}=\begin{bmatrix} 0 & 1 \\ -4 & -4 \end{bmatrix} \quad \text{and } A_{c9}=-3$$

and we assume

$$A_c = \left[\begin{array}{c:cc:c} 1 & 0 & 0 & a \\ \hdashline b & 0 & 1 & d \\ c & -4 & -4 & f \\ \hdashline 0 & 0 & 0 & -3 \end{array}\right]$$

where a,b,c,d,f, - arbitrary numbers.

Step 3

$$F_1=B^g_{13}\left(\begin{bmatrix} A_1 \\ A_7 \end{bmatrix} - \begin{bmatrix} A_{c1} \\ A_{c7} \end{bmatrix}\right) = \begin{bmatrix} -2 \\ -1 \end{bmatrix},\ F_2=B^g_{13}\begin{bmatrix} A_2 \\ A_8 \end{bmatrix} = \begin{bmatrix} 1 & -3 \\ 0 & -2 \end{bmatrix}$$

$$F_3=B^g_{13}\left(\begin{bmatrix} A_3 \\ A_9 \end{bmatrix} - \begin{bmatrix} A_{c3} \\ A_{c9} \end{bmatrix}\right) = \begin{bmatrix} 1-a \\ 1 \end{bmatrix}$$

and

$$F = [F_1 F_2 F_3] = \begin{bmatrix} -2, & 1, & -3, & 1-a \\ -1, & 0, & -2, & 1 \end{bmatrix}$$

It is easy to check that the matrix

$$A_c = A-BF = \left[\begin{array}{c:cc:c} 1 & 0 & 0 & a \\ \hdashline 2 & 0 & 1 & 0 \\ 3 & -4 & -4 & a-1 \\ \hdashline 0 & 0 & 0 & -3 \end{array}\right]$$

has the set of 3-D eigenvalues (10).

3.2. Method 2

Theorem 2

The problem has a solution if

i) $(I-B_1B_1^+)A_2=0$ and $(I-B_{12}B_{12}^+)A_{36}=0$

ii) the pairs (A_1,B_1), $(\overline{A}_5,\overline{B}_2)$ and $(\overline{A}_9,\overline{B}_3)$ are controllable

where $\overline{A}_5=A_5-B_2B_1^+A_2$, $\overline{B}_2=B_2(I-B_1^+B_1)$
$\overline{A}_9=A_9-B_3B_{12}^+A_{36}$, $\overline{B}_3=B_3(I-B_{12}^+B_{12})$ (11)

B_1^+ denotes any one-condition generalized inverse matrix of B_1 satisfying the condition $B_1B_1^+B_1=B_1$ [11].

Proof

From (6) it follows that the problem has a solution if (for example)

$$B_1F_2 = A_2 \quad , \quad B_{12}F_3 = A_{36} \tag{12}$$

and the matrices

$$A_{c1}=A_1-B_1F_1 \quad , \quad A_{c5}=A_5-B_2F_2 \quad , \quad A_{c9}=A_9-B_3F_3 \tag{13}$$

have the preassigned characteristic polynomials $\overline{q}_1(z_1)$, $\overline{q}_2(z_2)$ and $\overline{q}_3(z_3)$, respectively.

The equations (12) have solutions F_2 and F_3 iff the conditions i) are satisfied [11].
General solutions to (12) have the forms

$$\begin{aligned} F_2&=B_1^+A_2+(I-B_1^+B_1)F_{20} \\ F_3&=B_{12}^+A_{36}+(I-B_{12}^+B_{12})F_{30} \end{aligned} \tag{14}$$

where F_{20} and F_{30} are arbitrary matrices of appropriate dimensions. F_{20} and F_{30} can be chosen so that the matrices

$$A_{c1}=A_1-B_1F_1\ ,\quad A_{c5}=\bar{A}_5-\bar{B}_2F_{20}\ ,\quad A_{c9}=\bar{A}_9-\bar{B}_3F_{30}$$

have the preassigned characteristic polynomials iff the conditions ii) are satisfied. □

Remark 1 It can be easily shown [6] that the conditions i) can be written in the equivalent form

$$\text{rank } B_1 = \text{rank}\left[B_1 \vdots\ A_2\right] \text{ and rank } B_{12}=\text{rank}\left[B_{12} \vdots A_{36}\right]$$

Remark 2 Note that the pairs $(\bar{A}_5,\bar{B}_2),(\bar{A}_9,\bar{B}_3)$ are controllable only if $\bar{B}_2 \neq 0$ and $\bar{B}_3 \neq 0$. Therefore, from Remark 1 it follows that the pairs $(\bar{A}_5,\bar{B}_2),(\bar{A}_9,\bar{B}_3)$ are controllable only if

$$\text{rank } B_1 < \text{rank } B_{12} \quad \text{and} \quad \text{rank } B_{12} < \text{rank } B$$

Let rank $B_1=r \leqslant \min(n_1,m)$. It is possible to find full rank matrices $\bar{B}_1 \in R_{n_1 \times r}$, $\hat{B}_1 \in R_{r\times m}$ such that $B_1=\bar{B}_1\hat{B}_1$ and

$$B_1^+=\hat{B}_1^T\left[\hat{B}_1\hat{B}_1^T\right]^{-1}\left[\bar{B}_1^T\bar{B}_1\right]^{-1}\bar{B}_1^T \tag{15}$$

Algorithm 2

If the conditions i), ii) are satisfied the matrix F can be found by the use of the following algorithm

Step 1 Choose F_1 so that $A_{c1}=A_1-B_1F_1$ has the given eigenvalues $z_{11},z_{12},\dots,z_{1n_1}$.

Step 2 Using (15) and (11) find $B_1^+,\bar{B}_2,\bar{B}_3,\bar{A}_5$ and $\bar{A}_9$.

Step 3 Choose F_{20} and F_{30} so that $A_{c5}=\bar{A}_5-\bar{B}_2F_{20}$ and $A_{c9}=A_9-\bar{B}_3F_{30}$ have the given eigenvalues $z_{21},z_{22},\dots,z_{2n_2}$ and $z_{31},z_{32},\dots,z_{3n_3}$, respectively.

Step 4 Using (14) find F_2,F_3 and F.

Example 2

Given the matrices

$$A=\begin{bmatrix} A_1 & A_2 & A_3 \\ A_4 & A_5 & A_6 \\ A_7 & A_8 & A_9 \end{bmatrix}=\left[\begin{array}{c:cc:c} 1 & 0 & 3 & 3 \\ \hdashline 3 & 1 & -1 & 0 \\ 2 & -4 & -7 & -3 \\ \hdashline 1 & -1 & 2 & 2 \end{array}\right], \quad B=\begin{bmatrix} B_1 \\ B_2 \\ B_3 \end{bmatrix}=\left[\begin{array}{ccc} 0 & 1 & 0 \\ \hdashline 1 & 0 & 0 \\ 0 & -1 & 0 \\ \hdashline 0 & 0 & 1 \end{array}\right]$$

and the set of 3-D eigenvalues

$$E=\{z_{11}=-1,\ z_{21}=z_{22}=-2,\ z_{31}=3\} \quad , \tag{16}$$

find F such that the closed-loop system matrix A_c has the given set of 3-D eigenvalues (16).

In this case

$$B_1=[0\ 1\ 0]\ ,\quad B_{12}=\begin{bmatrix} 0 & 1 & 0 \\ 1 & 0 & 0 \\ 0 & -1 & 0 \end{bmatrix}=\begin{bmatrix} 0 & 1 \\ 1 & 0 \\ 0 & -1 \end{bmatrix}\begin{bmatrix} 1 & 0 & 0 \\ 0 & 1 & 0 \end{bmatrix}=\bar{B}_{12}\hat{B}_{12}$$

$$A_2=[0\ 3]\ ,\quad A_{36}=\begin{bmatrix} 3 \\ 0 \\ -3 \end{bmatrix}$$

and

$$B_1^+=B_1^T\left[B_1B_1^T\right]^{-1}=\begin{bmatrix} 0 \\ 1 \\ 0 \end{bmatrix}$$

$$B_{12}^+=\hat{B}_{12}^T\left[\hat{B}_{12}\hat{B}_{12}^T\right]^{-1}\left[\bar{B}_{12}^T\bar{B}_{12}\right]^{-1}\bar{B}_{12}^T=\begin{bmatrix} 0 & 1 & 0 \\ \frac{1}{2} & 0 & -\frac{1}{2} \\ 0 & 0 & 0 \end{bmatrix}$$

Hence

$$(I-B_1B_1^+)A_2=(1-1)\,[0\ 3]=[0\ 0]$$

and

$$(I-B_{12}B_{12}^+)A_{36}=\begin{bmatrix} \frac{1}{2} & 0 & \frac{1}{2} \\ 0 & 0 & 0 \\ \frac{1}{2} & 0 & \frac{1}{2} \end{bmatrix}\begin{bmatrix} 3 \\ 0 \\ -3 \end{bmatrix}=\begin{bmatrix} 0 \\ 0 \\ 0 \end{bmatrix}$$

Thus, the conditions i) of the theorem 2 are satisfied.

<u>Step 1</u> The matrix $F_1=\begin{bmatrix} 0 \\ 2 \\ 0 \end{bmatrix}$ satisfies the equation

$$A_{c1}=A_1-B_1F_1=1-\begin{bmatrix}0 & 1 & 0\end{bmatrix}\begin{bmatrix}0\\2\\0\end{bmatrix}=-1$$

Step 2

$$\overline{A}_5=A_5-B_2B_1^+A_2=\begin{bmatrix}1 & -1\\-4 & -7\end{bmatrix}-\begin{bmatrix}1 & 0 & 0\\0 & -1 & 0\end{bmatrix}\begin{bmatrix}0\\1\\0\end{bmatrix}\begin{bmatrix}0 & 3\end{bmatrix}=\begin{bmatrix}1 & -1\\-4 & -4\end{bmatrix}$$

$$\overline{B}_2=B_2(I-B_1^+B_1)=\begin{bmatrix}1 & 0 & 0\\0 & -1 & 0\end{bmatrix}\left(\begin{bmatrix}1 & 0 & 0\\0 & 1 & 0\\0 & 0 & 1\end{bmatrix}-\begin{bmatrix}0\\1\\0\end{bmatrix}\begin{bmatrix}0 & 1 & 0\end{bmatrix}\right)=\begin{bmatrix}1 & 0 & 0\\0 & 0 & 0\end{bmatrix} \tag{17}$$

$$\overline{A}_9=A_9-B_3B_{12}^+A_{36}=[2]-\begin{bmatrix}0 & 0 & 1\end{bmatrix}\begin{bmatrix}0 & 1 & 0\\\frac{1}{2} & 0 & -\frac{1}{2}\\0 & 0 & 0\end{bmatrix}\begin{bmatrix}3\\0\\-3\end{bmatrix}=[2]$$

$$\overline{B}_3=B_3(I-B_{12}^+B_{12})=\begin{bmatrix}0 & 0 & 1\end{bmatrix}\left(\begin{bmatrix}1 & 0 & 0\\0 & 1 & 0\\0 & 0 & 1\end{bmatrix}-\begin{bmatrix}0 & 1 & 0\\\frac{1}{2} & 0 & -\frac{1}{2}\\0 & 0 & 0\end{bmatrix}\begin{bmatrix}0 & 1 & 0\\1 & 0 & 0\\0 & -1 & 0\end{bmatrix}\right)=\begin{bmatrix}0 & 0 & 1\end{bmatrix} \tag{18}$$

It is easy to check that the pairs (17) and (18) are controllable. Thus, the conditions ii) of the theorem 2 are satisfied.

Step 3

The matrices

$$F_{20}=\begin{bmatrix}1 & -2\\0 & 0\\0 & 0\end{bmatrix},\quad F_{30}=\begin{bmatrix}0\\0\\-1\end{bmatrix}$$

satisfy the equations

$$A_{c5}=\overline{A}_5-\overline{B}_2F_{20}=\begin{bmatrix}1 & -1\\-4 & -4\end{bmatrix}-\begin{bmatrix}1 & 0 & 0\\0 & 0 & 0\end{bmatrix}\begin{bmatrix}1 & -2\\0 & 0\\0 & 0\end{bmatrix}=\begin{bmatrix}0 & 1\\-4 & -4\end{bmatrix}$$

$$A_{c9}=\overline{A}_9-\overline{B}_3F_{30}=[2]-\begin{bmatrix}0 & 0 & 1\end{bmatrix}\begin{bmatrix}0\\0\\-1\end{bmatrix}=[3]$$

Step 4

$$F_2=B_1^+A_2+(I-B_1^+B_1)F_{20}=\begin{bmatrix}0\\1\\0\end{bmatrix}\begin{bmatrix}0 & 3\end{bmatrix}+$$

$$+\left(\begin{bmatrix}1&0&0\\0&1&0\\0&0&1\end{bmatrix}-\begin{bmatrix}0\\1\\0\end{bmatrix}\begin{bmatrix}0&1&0\end{bmatrix}\right)\begin{bmatrix}1&-2\\0&0\\0&0\end{bmatrix}=\begin{bmatrix}1&-2\\0&3\\0&0\end{bmatrix}$$

$$F_3=B_{12}^{+}A_{36}+(I-B_{12}^{+}B_{12})F_{30}=\begin{bmatrix}0&1&0\\\frac{1}{2}&0&-\frac{1}{2}\\0&0&0\end{bmatrix}\begin{bmatrix}3\\0\\-3\end{bmatrix}+$$

$$+\left(\begin{bmatrix}1&0&0\\0&1&0\\0&0&1\end{bmatrix}-\begin{bmatrix}0&1&0\\\frac{1}{2}&0&-\frac{1}{2}\\0&0&0\end{bmatrix}\begin{bmatrix}0&1&0\\1&0&0\\0&-1&0\end{bmatrix}\right)\begin{bmatrix}0\\0\\-1\end{bmatrix}=\begin{bmatrix}0\\3\\-1\end{bmatrix}$$

and $F=\left[F_1F_2F_3\right]=\begin{bmatrix}0&1&-2&0\\2&0&3&3\\0&0&0&-1\end{bmatrix}$

It is easy to check that the matrix

$$A_c=A-BF=\left[\begin{array}{c:cc:c}-1&0&0&0\\\hdashline 3&0&1&0\\4&-4&-4&0\\\hdashline 1&-1&2&3\end{array}\right]$$

has the given set of 3-D eigenvalues (16).

3.3. Method 3

Let us assume that

$$F=F_t+F_p \tag{19}$$

where $F_t=\left[F_{t1}F_{t2}F_{t3}\right]$ (19a)

$$F_p=\left[F_{p1}F_{p2}F_{p3}\right] \tag{19b}$$

Matrix (19a) is chosen so that

$$\bar{A}=A-BF_t=\begin{bmatrix}\bar{A}_1&0&0\\\bar{A}_4&\bar{A}_5&0\\\bar{A}_7&\bar{A}_8&\bar{A}_9\end{bmatrix} \tag{20}$$

where

$$\bar{A}_1=A_1-B_1F_{t1},\ \bar{A}_4=A_4-B_2F_{t1},\ \bar{A}_5=A_5-B_2F_{t2},$$
$$\bar{A}_7=A_7-B_3F_{t1},\ \bar{A}_8=A_8-B_3F_{t2},\ \bar{A}_9=A_9-B_3F_{t2}$$

Lemma 1

The matrix (19a) can be chosen so that the matrix $\bar{A}$ has the triangular form (20) iff

$$\operatorname{rank} B_1 = \operatorname{rank}\left[B_1 \vdots A_2\right],\quad \operatorname{rank} B_{12}=\operatorname{rank}\left[B_{12} \vdots A_{36}\right] \tag{21}$$

Proof

The matrix $\bar{A}$ has the form (20) iff

$$B_1F_{t2}=A_2 \quad \text{and} \quad B_{12}F_{t3}=A_{36} \tag{22}$$

From Kronecker-Capelli's theorem it follows that (22) have solutions iff the conditions (21) hold.□
Note that if B_1 and B_{12} are non-singular matrices then (22) have solutions for any A_2 and A_{36}. F_{t1} can be chosen arbitrary, for example $F_{t1}=0$.

Lemma 2

Let B_{ij} be the j-th column of the matrix B_i ($i=1,2,3$; $j=1,2,\ldots,m$). If there exist $i\neq j_1\neq j_2$ such that

i) $B_{1k}=0$ for $k=j_1,j_2$ and the pair $(\bar{A}_1,B_{1i})$ is reachable.

ii) $B_{2j_2}=0$ and the pairs $(\bar{A}_5,B_{2j_1}),(\bar{A}_9,B_{3j_2})$ are reachable, then it is possible to find the matrix (19b) and the non-singular matrix

$$T=\begin{bmatrix} T_1 & 0 & 0 \\ 0 & T_2 & 0 \\ 0 & 0 & T_3 \end{bmatrix}$$

such that

$$\bar{A}_c=TA_cT^{-1}=T(\bar{A}-BF_p)T^{-1}=\begin{bmatrix} \bar{A}_{c1} & 0 & 0 \\ \bar{A}_{c4} & \bar{A}_{c5} & 0 \\ \bar{A}_{c7} & \bar{A}_{c8} & \bar{A}_{c9} \end{bmatrix} \tag{23}$$

where

$$\bar{A}_{c1}=T_1(\bar{A}_1-B_1F_{p1})T_1^{-1}=\begin{bmatrix} 0 & \vdots & I_{n_1-1} \\ \cdots & \cdots & \cdots \\ & -c_1 & \end{bmatrix} \tag{24a}$$

$$\bar{A}_{c5}=T_2(\bar{A}_5-B_2F_{p2})T_2^{-1}=\begin{bmatrix} 0 & \vdots & I_{n_2-1} \\ \cdots & \cdots & \cdots \\ & -c_2 & \end{bmatrix} \tag{24b}$$

$$\bar{A}_{c9}=T_3(\bar{A}_9-B_3F_{p3})T_3^{-1}=\begin{bmatrix} 0 & \vdots & I_{n_3-1} \\ \cdots & \cdots & \cdots \\ & -c_3 & \end{bmatrix} \tag{24c}$$

$c_i=\left[c_{i0}c_{i1}\cdots c_{in_i-1}\right]$, i=1,2,3; and the matrices

$$\bar{A}_{c4}=T_2(\bar{A}_4-B_2F_{p1})T_1^{-1} \quad , \quad \bar{A}_{c7}=T_3(\bar{A}_7-B_3F_{p1})T_1^{-1} \quad ,$$
$$A_{c8}=T_3(A_8-B_3F_{p2})T_2^{-1} \tag{24d}$$

have no special forms.

Proof

Without loss of generality we can assume $i=1$, $j_1=2$ and $j_2=3$. Hence

$$B_1=\left[B_{11}\ 0\ 0\ B_{14}\ \cdots\ B_{1m}\right]$$

Reachability of the pair $(\bar{A}_1,B_{11})$ implies the non-singularity of the matrix

$$R_1=\left[B_{11},\bar{A}_1B_{11},\ldots,\bar{A}_1^{n_1-1}B_{11}\right] \tag{25}$$

Let t_1 be n_1-th row of R_1^{-1} and

$$T_1=\begin{bmatrix} t_1 \\ t_1\bar{A}_1 \\ \cdots\cdots \\ t_1\bar{A}_1^{n_1-1} \end{bmatrix} \tag{26}$$

It is well known that

$$T_1\bar{A}_1T_1^{-1}=\begin{bmatrix} 0 & \vdots & I_{n_1-1} \\ \cdots & \cdots & \cdots \\ & -a_1 & \end{bmatrix}$$

and

$$T_1B_1=\left[e_{n_1}\ 0\ 0\ \bar{B}_{14}\ \cdots\ \bar{B}_{1m}\right]$$

where

$$a_1=\left[a_{10}\ a_{11}\ \cdots\ a_{1n_1-1}\right]$$

and e_{n_1} is the n_1-th column of I_{n_1}.

It is elementary to verify that for

$$F_{p1}T_1^{-1}=\begin{bmatrix} c_1-a_1 \\ \cdots\cdots \\ 0 \end{bmatrix}\in R_{m\times n_1}$$

the matrix $\bar{A}_{c1}$ has the form (24a).
Reachability of the pair $(\bar{A}_5,B_{22})$ implies the non-singularity of the matrix

$$R_2=\left[B_{22},\bar{A}_5B_{22},\ldots,\bar{A}_5^{n_2-1}B_{22}\right] \tag{27}$$

Let t_2 be the n_2-th row of R_2^{-1} and

$$T_2=\begin{bmatrix} t_2 \\ t_2\bar{A}_5 \\ \cdots\cdots\cdots \\ t_2\bar{A}_5^{n_2-1} \end{bmatrix} \tag{28}$$

It is well known that

$$T_2\bar{A}_5T_2^{-1}=\begin{bmatrix} 0 & \vdots & I_{n_2-1} \\ \cdots & \vdots & \cdots\cdots \\ & -a_2 & \end{bmatrix}$$

and

$$T_2B_2=\left[\bar{B}_{21}e_{n_2}0\ \bar{B}_{24}\ \cdots\ \bar{B}_{2m}\right]$$

where

$$a_2=\left[a_{20}\,a_{21}\ \cdots\ a_{2n_2-1}\right]$$

and e_{n_2} is the n_2-th column of I_{n_2}.
It can be easily shown that for

$$F_{p2}T_2^{-1}=\begin{bmatrix} 0 \\ \cdots\cdots \\ c_2-a_2 \\ \cdots\cdots \\ 0 \end{bmatrix}\begin{matrix} \}1 \\ \\ \\ \\ \}m-2 \end{matrix}\in R_{m\times n_2}$$

the matrix $\bar{A}_{c5}$ has the form (24b) and $T_1B_1F_{p2}T_2^{-1}=0$.

Reachability of the pair $(\bar{A}_9,B_{33})$ implies the non-singularity of the matrix

$$R_3=\left[B_{33},\bar{A}_9B_{33},\ldots,\bar{A}_9^{n_3-1}B_{33}\right] \tag{29}$$

Let t_3 be the n_3-th row of R_3^{-1} and

$$T_3=\begin{bmatrix} t_3 \\ t_3\bar{A}_9 \\ \cdots\cdots \\ t_3\bar{A}_9^{n_3-1} \end{bmatrix} \tag{30}$$

It is well known that

$$T_3\bar{A}_9T_3^{-1}=\begin{bmatrix} 0 & \vdots & I_{n_3-1} \\ \cdots & \cdots & \cdots \\ & -a_3 & \end{bmatrix}$$

and

$$T_3B_3=\left[\bar{B}_{31},\bar{B}_{32},e_{n_3}\bar{B}_{34}\cdots\bar{B}_{3m}\right]$$

where $a_3=\left[a_{30}\ a_{31}\ \cdots\ a_{3n_3-1}\right]$

and e_{n_3} is the n_3-th column of I_{n_3}.

It can be easily shown that for

$$F_{p3}T_3^{-1}=\begin{bmatrix} 0 \\ \cdots\cdots \\ c_3-a_3 \\ \cdots\cdots \\ 0 \end{bmatrix}\begin{matrix} \}\,2 \\ \\ \\ \\ \}\,m-3 \end{matrix} \in R_{mxn_3}$$

the matrix A_{c9} has the form (24c) and $T_1B_1F_{p3}T_3^{-1}=0$, $T_2B_1F_{p3}T_3^{-1}=0$. □

Theorem 3

The problem has a solution if the conditions (21) and i), ii) of lemma 2 are satisfied.

Proof

If the conditions (21) are satisfied F_t can be chosen so that $\bar{A}$ has the form (20) and if i), ii) of lemma 2 are satisfied it is possible to find F_p and T such that $\bar{A}_c$ has the form (23).

Note that

$$\det\left[Z-A_c\right]=\det\left[Z-\bar{A}_c\right]=\bar{q}_1(z_1)\bar{q}_2(z_2)\bar{q}_3(z_3)$$

where $\bar{q}_i(z_i)=z^{n_i}+c_{in_i-1}z_i^{n_i-1}+\dots+c_{i1}z_i+c_{i0}\quad(i=1,2,3)$

Thus, A_c has the separable characteristic polynomial with the given set of 3-D eigenvalues (5). □

Algorithm 3

If the conditions of theorem 3 are satisfied the matrix F can be found by the use of the following algorithm.

Step 1

Solving the equations

$$B_1F_{t2}=A_2 \quad , \quad B_{12}F_{t3}=A_{36}$$

find F_{t2} and F_{t3}.

Step 2

Assuming arbitrary F_{t1}, for example $F_{t1}=0$, find F_t and(20)

Step 3

For the given set (5) find c_1, c_2 and c_3.

Step 4

Using (25)-(30) find T_1, T_2, T_3 and the coefficient vectors a_1, a_2, a_3 of characteristic polynomials of $\bar{A}_1, \bar{A}_5$ and $\bar{A}_9$ respectively.

Step 5 Using

$$F_{p1}=\begin{bmatrix} c_1-a_1 \\ \cdots\cdots \\ 0 \end{bmatrix} T_1 \quad , \quad F_{p2}=\begin{bmatrix} 0 \\ \cdots\cdots \\ c_2-a_2 \\ \cdots\cdots \\ 0 \end{bmatrix} \begin{matrix} \}1 \\ \\ T_2 \\ \\ \}m-2 \end{matrix} \quad , \quad F_{p3}=\begin{bmatrix} 0 \\ \cdots\cdots \\ c_3-a_3 \\ \cdots\cdots \\ 0 \end{bmatrix} \begin{matrix} \}2 \\ \\ T_3 \\ \\ \}m-3 \end{matrix} \tag{31}$$

find F_p and $F=F_t+F_p$.

Remark 3

We obtain similar results if instead of (20) and (23) we assume $\bar{A}, \bar{A}_c$ in the forms

$$\bar{A}=\begin{bmatrix} \bar{A}_1 & \bar{A}_2 & \bar{A}_3 \\ 0 & \bar{A}_5 & \bar{A}_6 \\ 0 & 0 & \bar{A}_9 \end{bmatrix} \quad , \quad \bar{A}_c=\begin{bmatrix} \bar{A}_{c1} & \bar{A}_{c2} & \bar{A}_{c3} \\ 0 & \bar{A}_{c5} & \bar{A}_{c6} \\ 0 & 0 & \bar{A}_{c9} \end{bmatrix}$$

or

$$\bar{A}=\begin{bmatrix} \bar{A}_1 & \bar{A}_2 & 0 \\ 0 & \bar{A}_5 & 0 \\ \bar{A}_7 & \bar{A}_8 & \bar{A}_9 \end{bmatrix} \quad , \quad \bar{A}_c=\begin{bmatrix} \bar{A}_{c1} & \bar{A}_{c2} & 0 \\ 0 & \bar{A}_{c5} & 0 \\ \bar{A}_{c7} & \bar{A}_{c8} & \bar{A}_{c9} \end{bmatrix}$$

Example 3

Given the matrices

$$A=\begin{bmatrix} A_1 & A_2 & A_3 \\ A_4 & A_5 & A_6 \\ A_7 & A_8 & A_9 \end{bmatrix}=\left[\begin{array}{cc:c:c} 0 & 1 & 2 & 0 \\ -1 & 0 & 0 & 0 \\ \hdashline 0 & 1 & 0 & 1 \\ \hdashline -1 & 0 & 2 & 1 \end{array}\right], \quad B=\begin{bmatrix} B_1 \\ B_2 \\ B_3 \end{bmatrix}=\left[\begin{array}{ccc} 1 & 0 & 0 \\ 0 & 0 & 0 \\ \hdashline 0 & 1 & 0 \\ \hdashline -1 & 0 & 1 \end{array}\right]$$

and the set of 3-D eigenvalues

$$E=\{z_{11}=1, z_{12}=2, z_{21}=-1, z_{31}=-2\} \tag{32}$$

find F such that the closed-loop system matrix A_c has the set of 3-D eigenvalues equal to (32).
It is easy to check that the conditions of the theorem 3 are satisfied.

Step 1 The matrices

$$F_{t2}=\begin{bmatrix} 2 \\ 0 \\ 0 \end{bmatrix}, \quad F_{t3}=\begin{bmatrix} 0 \\ 1 \\ 0 \end{bmatrix}$$

satisfy the equations

$$\begin{bmatrix} 1 & 0 & 0 \\ 0 & 0 & 0 \end{bmatrix} F_{t2}=\begin{bmatrix} 2 \\ 0 \end{bmatrix}, \quad \begin{bmatrix} 1 & 0 & 0 \\ 0 & 0 & 0 \\ 0 & 1 & 0 \end{bmatrix} F_{t3}=\begin{bmatrix} 0 \\ 0 \\ 1 \end{bmatrix}$$

Step 2
Assuming $F_{t1}=0$ we obtain

$$F_t=\begin{bmatrix} F_{t1} F_{t2} F_{t3} \end{bmatrix}=\begin{bmatrix} 0 & 0 & 2 & 0 \\ 0 & 0 & 0 & 1 \\ 0 & 0 & 0 & 0 \end{bmatrix}$$

and

$$\bar{A}=A-BF_t=\left[\begin{array}{cc:c:c} 0 & 1 & 0 & 0 \\ -1 & 0 & 0 & 0 \\ \hdashline 0 & 1 & 0 & 0 \\ \hdashline -1 & 0 & 4 & 1 \end{array}\right]$$

Step 3
For (32) we have

$$(z_1-z_{11})(z_1-z_{12})=z_1^2-3z_1+2, \quad z_2-z_{21}=z_2+1,$$

$$z_3-z_{31}=z_3+2$$

and

$$c_1=[2 \;\; -3], \quad c_2=[1], \quad c_3=[2]$$

Step 4 Using (25)-(30) we obtain

$$R_1^{-1} = \left[B_{11}, \bar{A}_1 B_{11}\right]^{-1} = \begin{bmatrix} 1 & 0 \\ 0 & -1 \end{bmatrix}^{-1} = \begin{bmatrix} 1 & 0 \\ 0 & -1 \end{bmatrix}$$

and

$$T_1 = \begin{bmatrix} t_1 \\ t_1 \bar{A}_1 \end{bmatrix} = \begin{bmatrix} 0 & -1 \\ 1 & 0 \end{bmatrix}, \quad T_2 = [1], \quad T_3 = [1]$$

To find a_1 we calculate

$$\det\left[I_2 z - \bar{A}_1\right] = \begin{vmatrix} z & -1 \\ 1 & z \end{vmatrix} = z^2 + 1$$

Hence $a_1 = [1 \;\; 0]$ and $a_2 = [0]$, $a_3 = [-1]$.

Step 5 Using (31) we obtain

$$F_{p1} = \begin{bmatrix} c_1 - a_1 \\ \cdots\cdots \\ 0 \end{bmatrix} T_1 = -\begin{bmatrix} 3 & 1 \\ 0 & 0 \\ 0 & 0 \end{bmatrix}, \quad F_{p2} = \begin{bmatrix} 0 \\ \cdots\cdots \\ c_2 - a_2 \\ \cdots\cdots \\ 0 \end{bmatrix} T_2 = \begin{bmatrix} 0 \\ 1 \\ 0 \end{bmatrix}$$

$$F_{p3} = \begin{bmatrix} 0 \\ 0 \\ c_3 - a_3 \end{bmatrix} T_3 = \begin{bmatrix} 0 \\ 0 \\ 3 \end{bmatrix}, \quad F_p = \left[F_{p1} F_{p2} F_{p3}\right] = \begin{bmatrix} -3 & -1 & 0 & 0 \\ 0 & 0 & 1 & 0 \\ 0 & 0 & 0 & 3 \end{bmatrix}$$

and

$$F = F_t + F_p = \begin{bmatrix} -3 & -1 & 2 & 0 \\ 0 & 0 & 1 & 1 \\ 0 & 0 & 0 & 3 \end{bmatrix}$$

It is easy to check that the matrix

$$A_c = A - BF = \left[\begin{array}{cc:c:c} 3 & 2 & 0 & 0 \\ \hdashline -1 & 0 & 0 & 0 \\ 0 & 1 & -1 & 0 \\ \hdashline -4 & -1 & 4 & -2 \end{array}\right]$$

has the set of 3-D eigenvalues equal to (32).

References

1. R. Eising, 'Realization and stabilization of 2-D systems', IEEE Trans. Automat. Control, vol. AC-23, October 1978, pp. 793-799.
2. E. Emre, P.P. Khargonekar, 'Regulation of split linear systems over rings: coefficient-assignment and observers', IEEE Trans. Automat. Control, vol. AC-27, Febr. 1982, pp. 104-113.
3. T. Kaczorek, 'Pole assignment problem in two-dimensional linear systems', Int. J. Control, vol. 37, No 1, 1983, pp. 183-190.

4. T. Kaczorek, 'Pole assignment of 3-D linear systems with separable characteristic polynomials' Foundations of Control Engineering, vol. 8, 1983 (in press).
5. T. Kaczorek, 'Polynomial assignment via output dynamic feedback of 3-D systems' Bull. Acad. Polon. Sér. sci. techn. vol. 31, 1983 (in press)
6. T. Kaczorek, J. Kurek, 'Separability-assignment problem for q-dimensional linear discrete-time systems' Int. J. Control, 1983, vol. 37 (in press)
7. B.G. Mertzios, 'Pole assignment of 2-D systems for separable characteristic equations' Int. J. Control, 1983, vol. 37, (in press)
8. P.N. Paraskevopoulos, 'Characteristic polynomial assignment and determination of the residual polynomial in 2-D systems' IEEE Trans. Automat. Control, vol. AC-26, 1981, pp. 541-543.
9. P.N. Paraskevopoulos, 'Eigenvalue assignment of linear 2-dimensional systems' Proc. IEE., vol. 126, 1979, pp. 1204-1208.
10. P.N. Paraskevopoulos and O.I. Kosmidou, 'Eigenvalue assignment of two-dimensional systems using PID controllers' Int. J. Systems Sci., vol. 12, 1981, pp. 407-422.
11. R.M. Pringle, A.A. Rayner, 'Generalized inverse matrices with applications to statistics. London 1971 Griffin.
12. M. Šebek, 'On 2-D pole placement' IEEE Trans. Automat. Control, vol. AC-27, 1983 (in press).
13. S.G. Tzafestas and T.G. Pimenides, 'Exact model-matching control of three-dimensional systems using state and output feedback' Int. J. Systems Sci., vol. 13, No 11, 1982, pp. 1171-1187.

CHAPTER 23

FEEDBACK DEADBEAT CONTROL OF 2-DIMENSIONAL SYSTEMS

S. G. Tzafestas
University of Patras
Patras, Greece

N. J. Theodorou
Hellenic Airforce Tech. Center
Delta Falirou, Athens, Greece

Deadbeat control is among the most popular types of control for 1-dimensional discrete-time systems. Here the problem of designing deadbeat controllers is considered and solved for the case of 2-dimensional (2D) systems. Two important cases are examined. In the first case the system is described by its transfer function model and the objective is to design a dynamical series compensator such that to obtain a deadbeat closed-loop behaviour. In the second case the system is described by its Roesser state space model and the deadbeat controller is to be selected in static state feedback form. A simple example is worked out by using both the series compensator and the state feedback solutions.

1. INTRODUCTION

Much research effort has been devoted in the recent years to the study of signals and systems depending on many variables (time and space). The driving force behind this effort is that many practical systems and applications lead to multidimensional (MD) models i.e. to models involving state variables that depend on more than one independent variable. Among these applications we mention MD filtering, MD network synthesis, digital image processing, seismic data processing, x-ray enhancement, image restoration, the enhancement analysis of aerial photos for detection of forest fires on crop damage, the analysis and processing of satellite photos, sonar-radar and some array processing, object recognition, image deblurring, digital memory modelling, and distributed-parameter system analysis and design over multidimensional spatial domains. Recent surveys of the field are provided in [1-5].

The existing results may be grouped in four major areas as follows.

(i) *MD systems and filters* (transfer function representation, input-output properties, structural properties, recursibility, stability, etc.).

S. G. Tzafestas (ed.), Multivariable Control, 429–452.

(ii) *MD signal processing* (2D and 3D image processing, ranrom fields, space-time processing with application to x-ray enhancement, image deblurring, weather prediction, seismic analysis etc).

(iii) *MD state space modelling* (state space description, canonical models, structural properties, stability analysis, etc.).

(iv) *MD system control* (input-output decoupling, factorizing control, eigenvalue control, model-matching control, minimum energy control, adaptive control, deadbeat control, etc.).

The above results are significant contributions in the area of 2D and 3D signals and systems [6-15]. Our objective here is to provide two alternative solutions of the deadbeat controller design for 2D systems. Deadbeat controllers for 1D systems were provided by many authors e.g. [16-22]. A generalization of Kucera's [16] 1D deadbeat controller to 2D and MD systems was provided by Kaczorek [23-24].

The first solution derived here is based on the transfer function representation of the 2D system to be controlled, and the deadbeat controller is similarly designed in transfer function form. The second solution is based on the Roesser state space representation [3-5] of the system and the controller is derived in static linear state feedback form. The results are illustrated by a particular nontrivial example.

2. DEADBEAT CONTROLLER DESIGN USING THE TRANSFER FUNCTION MODEL.

A linear constant-coefficient 2-D single-input single-output (SISO) discrete system can be described by its difference input-output equation:

$$\sum_{i_1=0}^{N_1}\sum_{i_2=0}^{N_2} a_{i_1,i_2}\, y(n_1-i_1,n_2-i_2) = \sum_{i_1=0}^{N_1}\sum_{i_2=0}^{N_2} b_{i_1,i_2}\, u(n_1-i_1,n_2-i_2) \qquad (1)$$

with $a_{0,0}=1$, where u is the input and y the output of the system, or by its transfer function

$$G(z_1,z_2)=\frac{Y(z_1,z_2)}{U(z_1,z_2)}=\frac{\sum_{i_1=0}^{N_1}\sum_{i_2=0}^{N_2} b_{i_1,i_2} z_1^{-i_1} z_2^{-i_2}}{\sum_{i_1=0}^{N_1}\sum_{i_2=0}^{N_2} a_{i_1,i_2} z_1^{-i_1} z_2^{-i_2}} \qquad (2)$$

under the assumption of zero initial conditions.

Using a dynamical controller $G_R(z_1,z_2)$ in series with the system G and by feeding back the output to the input, as shown in fig. 1, the following closed loop transfer function $G_w(z_1,z_2)$ is obtained:

$$G_w(z_1,z_2)=\frac{Y(z_1,z_2)}{W(z_1,z_2)}=\frac{G_R(z_1,z_2)G(z_1,z_2)}{1+G_R(z_1,z_2)G(z_1,z_2)} \tag{3}$$

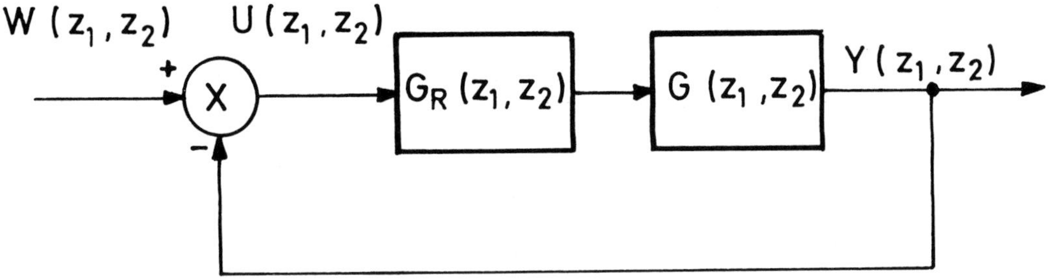

Fig.1.

Deadbeat behaviour of the closed loop system means that for a step change of the new input ("reference" variable) w, the input ("manipulated" variable) u and the output ("controlled" variable) y have to be in a steady state after a minimal settling 2D-space interval. In other words for

$$w(n_1,n_2)=1 \text{ for } n_1=0,1... \text{ and } n_2=0,1,...... \tag{4}$$

the input u and output y must assume constant values, after a minimal 2D-space interval (N_1,N_2), i.e.

$$u(n_1,n_2)=u(N_1,n_2) \text{ for } n_1\geqslant N_1 \text{ and } 0\leqslant n_2<N_2 \tag{5a}$$

$$u(n_1,n_2)=u(n_1,N_2) \text{ for } 0\leqslant n_1<N_1 \text{ and } n_2\geqslant N_2 \tag{5b}$$

$$u(n_1,n_2)=u(N_1,N_2)=\frac{1}{K} \text{ for } n_1\geqslant N_1 \text{ and } n_2\geqslant N_2 \tag{5c}$$

and

$$y(n_1,n_2)=y(N_1,n_2) \text{ for } n_1\geqslant N_1 \text{ and } 0\leqslant n_2<N_2 \tag{6a}$$

$$y(n_1,n_2)=y(n_1,N_2) \text{ for } 0\leqslant n_1<N_1 \text{ and } n_2\geqslant N_2 \tag{6b}$$

$$y(n_1,n_2)=y(N_1,N_2)=1 \text{ for } n_1\geqslant N_1 \text{ and } n_2\geqslant N_2 \tag{6c}$$

The justification of (5a,b) and (6a,b) is given in Appendix A. The important conclusion is that the deadbeat requirement is steady state input and output along parallel lines to the horizontal and vertical axes in the shaded areas I and II and along the whole double shaded area III of Fig. 2.

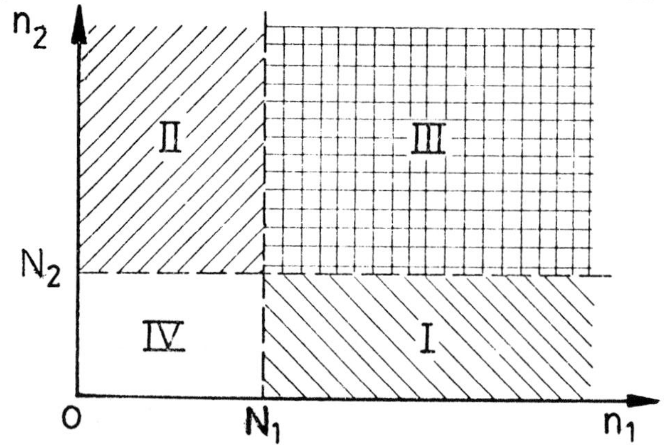

Fig.2.

In the following, the solution to the above deadbeat problem is presented. Consider the 2D Z-transforms of w,u, and y:

$$W(z_1,z_2)=\sum_{i_1=0}^{\infty}\sum_{i_2=0}^{\infty} w(i_1,i_2)z_1^{-i_1}z_2^{-i_2} \tag{7}$$

$$U(z_1,z_2)=\sum_{i_1=0}^{\infty}\sum_{i_2=0}^{\infty} u(i_1,i_2)z_1^{-i_1}z_2^{-i_2} \tag{8}$$

$$Y(z_1,z_2)=\sum_{i_1=0}^{\infty}\sum_{i_2=0}^{\infty} y(i_1,i_2)z_1^{-i_1}z_2^{-i_2} \tag{9}$$

Introducing (4) into (7), the Z-transform of the reference variable becomes:

$$W(z_1,z_2)=\sum_{i_1=0}^{\infty}\sum_{i_2=0}^{\infty} z_1^{-i_1}z_2^{-i_2}=\sum_{i_1=0}^{\infty} z_1^{-i_1}\sum_{i_2=0}^{\infty} z_2^{-i_2}$$

whence

$$W(z_1,z_2)=\frac{1}{(1-z_1^{-1})(1-z_2^{-1})} \tag{10}$$

From (8), (9) and (10) it is found that:

$$\frac{U(z_1,z_2)}{W(z_1,z_2)}=\sum_{i_1=0}^{\infty}\sum_{i_2=0}^{\infty} u(i_1,i_2)z_1^{-i_1}z_2^{-i_2}(1-z_1^{-1})(1-z_2^{-1}) \tag{11}$$

$$\frac{Y(z_1,z_2)}{W(z_1,z_2)}=\sum_{i_1=0}^{\infty}\sum_{i_2=0}^{\infty} y(i_1,i_2)z_1^{-i_1}z_2^{-i_2}(1-z_1^{-1})(1-z_2^{-1}) \tag{12}$$

Introducing (5) and (6) into (11) and (12) respectively, one finds

$$\frac{U(z_1,z_2)}{W(z_1,z_2)}=\sum_{i_1=0}^{N_1}\sum_{i_2=0}^{N_2} q_{i_1,i_2} z_1^{-i_1} z_2^{-i_2}=Q(z_1,z_2) \tag{13}$$

where

$$q_{0,0}=u(0,0) \tag{14a}$$

$$q_{0,i_2}=u(0,i_2)-u(0,i_2-1)\ ,\ q_{i_1,0}=u(i_1,0)-u(i_1-1,0) \tag{14b}$$

$$q_{i_1,0}=u(i_1,i_2)-u(i_1-1,i_2)-u(i_1,i_2-1)+u(i_1-1,i_2-1) \tag{14c}$$

for $(1,1)\leqslant(i_1,i_2)\leqslant(N_1,N_2)$, and

$$\frac{Y(z_1,z_2)}{W(z_1,z_2)}=\sum_{i_1=0}^{N_1}\sum_{i_2=0}^{N_2} p_{i_1,i_2} z_1^{-i_1} z_2^{-i_2}=P(z_1,z_2) \tag{15}$$

where

$$p_{0,0}=y(0,0) \tag{16a}$$

$$p_{0,i_2}=y(0,i_2)-y(0,i_2-1) \tag{16b}$$

$$p_{i_1,0}=y(i_1,0)-y(i_1-1,0) \tag{16c}$$

$$p_{i_1,i_2}=y(i_1,i_2)-y(i_1-1,i_2)-y(i_1,i_2-1)+y(i_1-1,i_2-1) \tag{16d}$$

for $(1,1)\leqslant(i_1,i_2)\leqslant(N_1,N_2)$.

The proof of the above is given in Appendix A. The following relations hold between p_{i_1,i_2} and q_{i_1,i_2}:

$$\sum_{i_1=0}^{N_1}\sum_{i_2=0}^{N_2} q_{i_1,i_2}=u(N_1,N_2)=\frac{1}{K} \tag{17}$$

$$\sum_{i_1=0}^{N_1}\sum_{i_2=0}^{N_2} p_{i_1,i_2}=y(N_1,N_2)=1 \tag{18}$$

the proof of which can be found in Appendix B.

From (13), (15) and (2) one gets:

$$G(z_1,z_2)=\frac{P(z_1,z_2)}{Q(z_1,z_2)}=\frac{\sum_{i_1=0}^{N_1}\sum_{i_2=0}^{N_2} b_{i_1,i_2} z_1^{-i_1} z_2^{-i_2}}{\sum_{i_1=0}^{N_1}\sum_{i_2=0}^{N_2} a_{i_1,i_2} z_1^{-i_1} z_2^{-i_2}} \tag{19}$$

Hence, by equating equal powers of z_1, z_2 one obtains the relations:

$$q_{i_1,i_2}=a_{i_1,i_2}q_{0,0} \tag{20}$$

$$p_{i_1,i_2}=b_{i_1,i_2}q_{0,0} \tag{21}$$

for $(0,0)\leqslant(i_1,i_2)\leqslant(N_1,N_2)$

From (17) and (20) it is found that:

$$q_{0,0}=1/\sum_{i_1=0}^{N_1}\sum_{i_2=0}^{N_2}b_{i_1,i_2} \tag{22}$$

From (3) and (15) one obtains:

$$G_W(z_1,z_2)=P(z_1,z_2) \tag{23}$$

Now the dynamical controller $G_R(z_1,z_2)$ may be found by solving (3) i.e.

$$G_R(z_1,z_2)=\frac{1}{G(z_1,z_2)}\cdot\frac{G_W(z_1,z_2)}{1-G_W(z_1,z_2)} \tag{24}$$

and substituting (19) and (23) into the above:

$$G_R(z_1,z_2)=\frac{Q(z_1,z_2)}{1-P(z_1,z_2)}=\frac{\sum_{i_1=0}^{N_1}\sum_{i_2=0}^{N_2}q_{i_1,i_2}z_1^{-i_1}z_2^{-i_2}}{1-\sum_{i_1=0}^{N_1}\sum_{i_2=0}^{N_2}p_{i_1,i_2}z_1^{-i_1}z_2^{-i_2}} \tag{25}$$

Finally, substituting the p_{i_1,i_2}'s and q_{i_1,i_2}'s from (20) and (21), considering (22), the controller $G_R(z_1,z_2)$ is completely specified:

$$G_R(z_1,z_2)=\frac{\sum_{i_1=0}^{N_1}\sum_{i_2=0}^{N_2}a_{i_1,i_2}z_1^{-i_1}z_2^{-i_2}}{\sum_{i_1=0}^{N_1}\sum_{i_2=0}^{N_2}b_{i_1,i_2}-\sum_{i_1=0}^{N_1}\sum_{i_2=0}^{N_2}b_{i_1,i_2}z_1^{-i_1}z_2^{-i_2}} \tag{26}$$

Next the input u and output y are calculated at several points in the 2D domain. To this end introducing (20) and (21) into (14) and (16), respectively, taking into account (22), two linear systems of $(N_1+1)(N_2+1)$ equations with respect to the same number of unknowns $u(n_1,n_2)$ and $y(n_1,n_2)$, for

$(0,0)\leqslant(n_1,n_2)\leqslant(N_1,N_2)$, are obtained. The solutions of these two systems are easily found to be:

$$u(n_1,n_2)=\frac{\sum_{i_1=0}^{n_1}\sum_{i_2=0}^{n_2}a_{i_1,i_2}}{\sum_{i_1=0}^{N_1}\sum_{i_2=0}^{N_2}b_{i_1,i_2}} \quad (27a), \quad y(n_1,n_2)=\frac{\sum_{i_1=0}^{n_1}\sum_{i_2=0}^{n_2}b_{i_1,i_2}}{\sum_{i_1=0}^{N_1}\sum_{i_2=0}^{N_2}b_{i_1,i_2}} \quad (28a)$$

for $(0,0)\leqslant(n_1,n_2)\leqslant(N_1,N_2)$

Considering (5) and (6) the manipulated input u and controlled output y are found at any point of the 2D space domain, namely

$$u(n_1,n_2)=u(N_1,n_2)=\Big(\sum_{i_1=0}^{N_1}\sum_{i_2=0}^{n_2}a_{i_1,i_2}\Big)/\Big(\sum_{i_1=0}^{N_1}\sum_{i_2=0}^{N_2}b_{i_1,i_2}\Big) \quad (27b)$$

$$y(n_1,n_2)=y(N_1,n_2)=\Big(\sum_{i_1=0}^{N_1}\sum_{i_2=0}^{n_2}b_{i_1,i_2}\Big)/\Big(\sum_{i_1=0}^{N_1}\sum_{i_2=0}^{N_2}b_{i_1,i_2}\Big) \quad (28b)$$

for $n_1\geqslant N_1$ and $0\leqslant n_2\leqslant N_2$,

$$u(n_1,n_2)=u(n_1,N_2)=\Big(\sum_{i_1=0}^{n_1}\sum_{i_2=0}^{N_2}a_{i_1,i_2}\Big)/\Big(\sum_{i_1=0}^{N_1}\sum_{i_2=0}^{N_2}b_{i_1,i_2}\Big) \quad (27c)$$

$$y(n_1,n_2)=y(n_1,N_2)=\Big(\sum_{i_1=0}^{n_1}\sum_{i_2=0}^{N_2}b_{i_1,i_2}\Big)/\Big(\sum_{i_1=0}^{N_1}\sum_{i_2=0}^{N_2}b_{i_1,i_2}\Big) \quad (28c)$$

for $0\leqslant n_1<N_1$ and $n_2\geqslant N_2$, and

$$u(n_1,n_2)=u(N_1,N_2)=\frac{\sum_{i_1=0}^{N_1}\sum_{i_2=0}^{N_2}a_{i_1,i_2}}{\sum_{i_1=0}^{N_1}\sum_{i_2=0}^{N_2}b_{i_1,i_2}}=\frac{1}{K}=\frac{1}{G(1,1)} \quad (27d)$$

$y(n_1,n_2)=y(N_1,N_2)=1$

for $n_1\geqslant N_1$ and $n_2\geqslant N_2$. (28d)

Introducing the excessive zero coefficients

$$a_{i_1,i_2} = b_{i_1,i_2} = 0 \text{ for } i_1 > N_1 \text{ or } i_2 > N_2 \tag{29}$$

eqs. (27) and (28) can be written in the form:

$$u(n_1,n_2) = \Big(\sum_{i_1=0}^{n_1} \sum_{i_2=0}^{n_2} a_{i_1,i_2}\Big) \Big/ \Big(\sum_{i_1=0}^{N_1} \sum_{i_2=0}^{N_2} b_{i_1,i_2}\Big) \tag{30}$$

$$y(n_1,n_2) = \Big(\sum_{i_1=0}^{n_1} \sum_{i_2=0}^{n_2} b_{i_1,i_2}\Big) \Big/ \Big(\sum_{i_1=0}^{N_1} \sum_{i_2=0}^{N_2} b_{i_1,i_2}\Big) \tag{31}$$

for $n_1 = 0,1,\ldots$ and $n_2 = 0,1,\ldots$

The ratio y/u is easily found to be:

$$h(n_1,n_2) = \frac{y(n_1,n_2)}{u(n_1,n_2)} = \frac{\sum_{i_1=0}^{n_1} \sum_{i_2=0}^{n_2} b_{i_1,i_2}}{\sum_{i_1=0}^{n_1} \sum_{i_2=0}^{n_2} a_{i_1,i_2}} \tag{32}$$

for all (n_1,n_2).

To summarize, a dynamical controller G_R has been specified in (26), which results in deadbeat behaviour of a given 2-D system, as it has been defined in equations (4), (5) and (6). For this deadbeat behaviour the input u and the output y were prespecified (at the beginning of the solution) to attain constant values along horizontal and vertical lines in the areas I and II and along the whole area III of fig.2. The value of y in area III was prespecified to be 1 and the rest remained unknown. Finally, the above constant values of u in the areas I,II and III, and y in I and II, as well as all the distinct values of u and y in the area IV were calculated

Remark 1: Characteristic polynomial

It was found above that:

$$G_W(z_1,z_2) = P(z_1,z_2) = \sum_{i_1=0}^{N_1} \sum_{i_2=0}^{N_2} p_{i_1,i_2} z_1^{-i_1} z_2^{-i_2} = \frac{Y(z_1,z_2)}{W(z_1,z_2)} \tag{33}$$

Hence the characteristic polynomial E_W of the closed loop transfer function is 1:

$$E_w(z_1,z_2)\ \sum_{i_1=0}^{N_1}\sum_{i_2=0}^{N_2} a_{w,i_1,i_2} z_1^{-i_1} z_2^{-i_2} = a_{w,0,0} = 1 \tag{34}$$

$$a_{w,i_1,i_2} = 0 \quad \text{for} \quad (0,0) < (i_1,i_2) \lesssim (N_1,N_2) \tag{35}$$

This means that the closed loop system is a nonrecursive one.

Remark 2. Verification of the method

$$y(n_1,n_2) = \sum_{i_1=0}^{N_1}\sum_{i_2=0}^{N_2} p_{i_1,i_2} w(n_1-i_1,n_2-i_2) \tag{36}$$

and from (21) and (22):

$$y(n_1,n_2) = \frac{\sum_{i_1=0}^{N_1}\sum_{i_2=0}^{N_2} b_{i_1,i_2} w(n_1-i_1,n_2-i_2)}{\sum_{i_1=0}^{N_1}\sum_{i_2=0}^{N_2} b_{i_1,i_2}} \tag{37}$$

Taking into account the fact that $w(n_1-i_1,n_2-i_2)$ equals one for $n_1 \geq i_1$ *and* $n_2 \geq i_2$, and zero for $n_1 < i_1$ *or* $n_2 < i_2$, the above reduces to (28) or (31). This is an alternative way to calculate y and results in the same values of y in the 2D space domain.

Remark 3. Generalization to m-D systems

The generalization of the above results from 2-D to m-D systems seems to be straight-forward.

3. DEADBEAT CONTROLLER DESIGN USING THE STATE SPACE MODEL

Referring to the same 2-D system described in Sec.2, an equivalent representation is by using its state space model, namely:

$$\begin{bmatrix} r(n_1+1,n_2) \\ s(n_1+1,n_2) \\ t(n_1,n_2+1) \end{bmatrix} = \begin{bmatrix} A_{11} & A_{12} & A_{13} \\ A_{21} & A_{22} & A_{23} \\ A_{31} & A_{32} & A_{33} \end{bmatrix} \begin{bmatrix} r(n_1,n_2) \\ s(n_1,n_2) \\ t(n_1,n_2) \end{bmatrix} + \begin{bmatrix} B_1 \\ B_2 \\ B_3 \end{bmatrix} u(n_1,n_2) \tag{38a}$$

$$y(n_1,n_2)=\begin{bmatrix}C_1 & C_2 & C_3\end{bmatrix}\begin{bmatrix}r(n_1,n_2)\\ s(n_1,n_2)\\ t(n_1,n_2)\end{bmatrix}+Du(n_1,n_2) \tag{38b}$$

where $x=[r,s,t]^T$ is the state vector and

$$\bar{x}(n_1,n_2)=\begin{bmatrix}r(n_1+1,n_2)\\ s(n_1+1,n_2)\\ t(n_1,n_2+1)\end{bmatrix},\quad A=\begin{bmatrix}A_{11} & A_{12} & A_{13}\\ A_{21} & A_{22} & A_{23}\\ A_{31} & A_{32} & A_{33}\end{bmatrix},\quad B=\begin{bmatrix}B_1\\ B_2\\ B_3\end{bmatrix}$$

$$C=\begin{bmatrix}C_1 & C_2 & C_3\end{bmatrix},\quad D=d \tag{39}$$

with

$$A_{11}=\begin{bmatrix}0 & 1 & & \\ & \ddots & \ddots & 0\\ & & \ddots & 1\\ 0 & & & 0\end{bmatrix}_{(N_1xN_1)},\quad A_{12}=\begin{bmatrix}b_{1,0} & & \\ \vdots & & 0\\ b_{N_1,0} & & \end{bmatrix}_{(N_1xN_1)},\quad A_{13}=\begin{bmatrix}b_{1,1} & \cdots & b_{1,N_2}\\ \vdots & & \vdots\\ b_{N_1,1} & \cdots & b_{N_1,N_2}\end{bmatrix}_{(N_1xN_2)}$$

$$A_{21}=\begin{bmatrix}0\end{bmatrix}_{(N_1xN_1)},\quad A_{22}=\begin{bmatrix}-a_{1,0} & 1 & & 0\\ \vdots & & \ddots & \\ \vdots & 0 & \ddots & 1\\ -a_{N_1,0} & & & 0\end{bmatrix}_{(N_1xN_1)},\quad A_{23}=\begin{bmatrix}-\tilde{a}_{1,1} & \cdots & -\tilde{a}_{1,N_2}\\ \vdots & & \vdots\\ -\tilde{a}_{N_1,1} & \cdots & -\tilde{a}_{N_1,N_2}\end{bmatrix}_{(N_1xN_2)}$$

$$A_{31}=\begin{bmatrix}0\end{bmatrix}_{(N_2xN_1)},\quad A_{32}=\begin{bmatrix}1 & \\ & 0\\ & \end{bmatrix}_{(N_2xN_1)},\quad A_{33}=\begin{bmatrix}-a_{0,1} & \cdots & -a_{0,N_2}\\ 1 & \ddots & 0\\ 0 & \ddots\ 1 & 0\end{bmatrix}_{(N_2xN_2)}$$

$$B_1=\begin{bmatrix}b_{1,0}\\ \vdots\\ b_{N_1,0}\end{bmatrix}_{(N_1x1)},\quad B_2=\begin{bmatrix}-a_{1,0}\\ \vdots\\ -a_{N_1,0}\end{bmatrix}_{(N_1x1)},\quad B_3=\begin{bmatrix}1\\ 0\\ \\ \end{bmatrix}_{(N_2x1)}$$

$$C_1=\begin{bmatrix}1 & 0\end{bmatrix}(1xN_1),\quad C_2=\begin{bmatrix}b_{0,0} & 0\end{bmatrix}(1xN_1),\quad C_3=\begin{bmatrix}\tilde{b}_{0,1}\cdots\tilde{b}_{0,N_2}\end{bmatrix}(1xN_2)$$

$$d=b_{0,0} \tag{40}$$

where

$$\tilde{a}_{i_1,i_2}=a_{i_1,i_2}-a_{i_1,0}a_{0,i_2} \quad \text{for } (1,1)\leqslant(i_1,i_2)\leqslant(N_1,N_2) \tag{41a}$$

$$\tilde{b}_{i_1,i_2}=b_{i_1,i_2}-b_{i_1,0}a_{0,i_2} \quad \text{for } (0,1)\leqslant(i_1,i_2)\leqslant(N_1,N_2) \tag{41b}$$

The feedback law adopted here, in order to achieve deadbeat control, has the form

$$u(n_1,n_2)=\lambda w(n_1,n_2)-F^T x(n_1,n_2) \tag{42}$$

where λ is a scaling factor and F the column vector feedback gain to be specified:

$$F^T=\left[f_1\ldots f_{N_1} \; g_1\ldots g_{N_1} \; h_1\ldots h_{N_2}\right]^T=\left[f,g,h\right]^T \tag{43}$$

Using this state feedback control, the closed loop system becomes:

$$\bar{x}=(A-BF^T)x+\lambda Bw=A_w x+B_w w \tag{44a}$$

$$y=(C-DF^T)x+\lambda Dw=C_w x+D_w w \tag{44b}$$

where

$$A_w=A-BF^T, \quad B_w=\lambda B \tag{45a}$$

$$C_w=C-BF^T, \quad D_w=\lambda D \tag{45b}$$

are the new closed-loop state-space matrices.

In order to obtain the same closed-loop deadbeat behaviour as in Sec.2, the closed-loop transfer function must take the form (see (21), (22), (23)):

$$G_w(z_1,z_2)=\sum_{i_1=0}^{N_1}\sum_{i_2=0}^{N_2} b_{i_1,i_2} q_{0,0} z_1^{-i_1} z_2^{-i_2} \tag{46}$$

Sufficient conditions for $G_w(z_1,z_2)$ to take the above form are the following:

i) A_w takes a nearly state space form, with zero the elements in the places of $a_{i_1,0}$, a_{0,i_2}, for $(1,1)\leqslant(i_1,i_2)\leqslant(N_1,N_2)$, and

ii) C_w takes the form of C but with $\tilde{b}_{0,i_2}=b_{0,i_2}$ for $1\leqslant i_2\leqslant N_2$ and a zero in the place of $b_{0,0}$

Proof: It is well known that

$$G_w = C_w (I - A_w Z_{1,2}^{-1})^{-1} Z_{1,2}^{-1} B_w + D_w \tag{47}$$

where

$$Z_{1,2} = \begin{bmatrix} Z_1 & 0 \\ 0 & Z_2 \end{bmatrix}, Z_1 = \begin{bmatrix} z_1 & & 0 \\ & \ddots & \\ 0 & & z_1 \end{bmatrix}_{(N_1 \times N_1)}, Z_2 = \begin{bmatrix} z_2 & & 0 \\ & \ddots & \\ 0 & & z_2 \end{bmatrix}_{(N_1 \times N_2)}$$

Using (39), (43) and (45) A_w and C_w take the form:

$$A_w = \begin{bmatrix} A_{w11} & A_{w12} & A_{w13} \\ A_{w21} & A_{w22} & A_{w23} \\ A_{w31} & A_{w32} & A_{w33} \end{bmatrix} = \begin{bmatrix} A_{11} - B_1 f^T & A_{12} - B_1 g^T & A_{13} - B_1 h^T \\ A_{21} - B_2 f^T & A_{22} - B_2 g^T & A_{23} - B_2 h^T \\ A_{31} - B_3 f^T & A_{32} - B_3 g^T & A_{33} - B_3 h^T \end{bmatrix} \tag{48a}$$

$$C_w = \begin{bmatrix} C_{w_1} & C_{w_2} & C_{w_3} \end{bmatrix} = \begin{bmatrix} C_1 - D f^T & C_2 - D g^T & C_3 - D h^T \end{bmatrix} \tag{48b}$$

The requirement i) means that $A_{wij}, i=1,2,3 \; j=1,2,3$, preserve their form with $a_{i_1,0}$, a_{0,i_2}, $(1,1) \leqslant (i_1, i_2) \leqslant (N_1, N_2)$, equal equal to zero, if possible. In order to preserve the form of A_{w11}, A_{w21}, A_{w31}, one must set:

$$f^T = \begin{bmatrix} f_1 \dots f_{N_1} \end{bmatrix}^T = 0 \tag{49}$$

For A_{w22} to keep its form, g it must have the form

$$g^T = \begin{bmatrix} g_1 \dots g_{N_1} \end{bmatrix}^T = \begin{bmatrix} 1 & 0 \dots 0 \end{bmatrix} \tag{50}$$

but then A_{w12} and A_{w32} become zero which is a discrepancy with $A_{12} \neq 0$ and $A_{32} \neq 0$.

For A_{w33} to keep its form, h must have the form

$$h^T = \begin{bmatrix} h_1 \dots h_{N_2} \end{bmatrix}^T = \begin{bmatrix} -a_{0,1} \dots -a_{0,N_2} \end{bmatrix}^T \tag{51}$$

Then A_{w13} and A_{w23} take the same form as A_{13} and A_{23} respectively, but the $\tilde{a}_{i_1,i_2}$'s and $\tilde{b}_{i_1,i_2}$'s are without hats i.e. a_{i_1,i_2} and b_{i_1,i_2}. As far as ii) is concerned, C_{w1} remains the same with C_1, C_{w2} is altered to become zero and C_{w3} keeps its form with the $\tilde{b}_{0,i_2}$'s without hats, i.e. b_{0,i_2}

Finally, it remains to be proved that the closed-loop system

(A_w, B_w, C_w, D_w) of the above form has a transfer function G_w, given by (47), equal to G_w, given by (46). This is a pure algebraic matter and the proof is given in Appendix C.

The conclusion which is drawn from the above results is that employing the static state feedback law (42), with the feedback gain vector F given by (49), (50), (51), the closed-loop system has the same transfer function with the closed loop system in Section 2 (where dynamic output feedback control has been used). Hence, both the dynamic output-and static state-feedback control of the same 2-D system, result in the same transfer function and to the same deadbeat behaviour.

Remark 1. Scaling factor λ

Comparing (46) with (47) and taking into account (45),it is easily seen that the scaling factor λ must take the value:

$$\lambda = q_{0,0} = 1 / \sum_{i_1=0}^{N_1} \sum_{i_2=0}^{N_2} b_{i_1,i_2} \tag{52}$$

so that $G_w(z_1, z_2)$ in (46) coincides with G_w in (47).

Remark 2. Characteristic polynomial:

According to the Cayley Hamilton theorem the matrix A_w satisfied its characteristic polynomial, i.e.

$$\sum_{i_1=0}^{N_1} \sum_{i_2=0}^{N_2} a_{w,i_1,i_2} A_w^{n_1-i_1, n_2-i_2} = 0 \quad \text{for } (n_1, n_2) \geq (N_1, N_2) \tag{53}$$

Because of (34), this gives:

$$A_w^{n_1,n_2} = - \sum_{\substack{i_1=0 \\ (i_1,i_2)\neq(0,0)}}^{N_1} \sum_{i_2=0}^{N_2} a_{w,i_1,i_2} A_w^{n_1-i_1, n_2-i_2} = 0 \tag{54}$$

for $(n_1, n_2) \geq (N_1, N_2)$. If $x(0,0)$ is the only different from zero initial state vector, then it is well known that:

$$x(n_1, n_2) = A^{n_1,n_2} x(0,0) \tag{55}$$

Then from (54)

$$x(n_1, n_2) = 0 \quad \text{for } (n_1, n_2) \geq (N_1, N_2) \tag{56}$$

Then according to (44b):

$$y(n_1,n_2)=D_w w(n_1,n_2) \quad \text{for} \quad (n_1,n_2)\geq(N_1,N_2) \tag{57}$$

i.e. the output reaches a value, equal to one, after a minimal 2D region space (N_1,N_2) and remains steady thereafter. This is also a verification that the procedure followed leads to a closed loop system with deadbeat behaviour. In the above approach use of the Cayley Hamilton theorem was made, whereas in the previous approach (Sec.2) other algebraic techniques were followed.

Remark 3. m-D systems

Work is in progress towards the end of extending the state-feedback deadbeat control to m-D systems.

4. EXAMPLE

Consider a 2D system with transfer function taken from [23]:

$$G(z_1,z_2)=\frac{z_2^{-1}}{1+z_2^{-1}+z_1^{-1}z_2^{-1}+z_2^{-2}+2z_1^{-1}z_2^{-2}+z_2^{-3}+z_1^{-1}z_2^{-3}} \tag{58}$$

According to (2) we have $N_1=1$, $N_2=3$, and

$$\begin{gathered} a_{0,0}=1,\ a_{1,0}=0,\ a_{0,1}=1, a_{1,1}=1,\ a_{0,2}=1, a_{1,2}=2, \\ a_{0,3}=1, a_{1,3}=1,\ b_{i_1,i_2}=0\ \forall(i_1,i_2)\ \text{except}\ b_{0,1}=1 \end{gathered} \tag{59}$$

According to (20), (21) and (22):

$$\begin{gathered} q_{0,0}=1, q_{1,0}=0, q_{0,1}=1, q_{1,1}=1, q_{0,2}=1, q_{1,2}=2, \\ q_{0,3}=1,\ q_{1,3}=1; p_{i_1,i_2}=0\ \forall(i_1,i_2)\ \text{except}\ p_{0,1}=1 \end{gathered} \tag{60}$$

From (25) the output deadbeat controller is found to be:

$$G_R(z_1,z_2)=\frac{1+z_2^{-1}+z_1^{-1}z_2^{-1}+z_2^{-2}+2z_1^{-1}z_2^{-2}+z_2^{-3}+z_1^{-1}z_2^{-3}}{1-z_2^{-1}} \tag{61}$$

From (27) and (28) or from (30) and (31) the values of u and y are found in the 2D-space domain:

$u(0,0)=1,\ u(1,0)=1,\ u(0,1)=2,\ u(1,1)=3,$

$u(0,2)=3,\ u(1,2)=6,\ u(0,3)=4,\ u(1,3)=8$

$u(n_1,0)=1\ ,\ u(n_1,1)=3,\ u(n_1,2)=6,\ u(n_1,3)=8$ for $n_1\geq 1,$

$u(0,n_2)=4,\ u(1,n_2)=8$ for $n_2\geq 3$

$y(0,0)=0,\ y(1,0)=0,\ y(0,1)=1,\ y(1,1)=1,$

$y(0,2)=1,\ y(1,2)=1,\ y(0,3)=1,\ y(1,3)=1,$

$y(n_1,0)=0,\ y(n_1,1)=1,\ y(n_1,2)=1,\ y(n_1,3)=1$ for $n_1\geq 1,$

$y(0,n_2)=1,\ y(1,n_2)=1$ for $n_2\geq 3$

In Fig. 3 the values of u and y are shown:

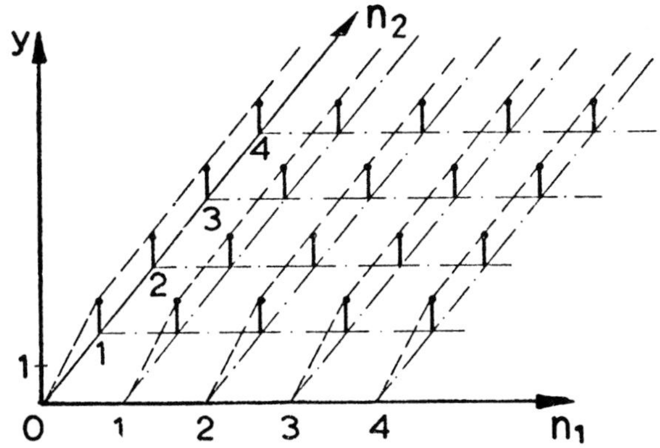

Fig. 3: 2D deadbeat performance

Finally, from (42) and (49)-(52) the state-feedback deadbeat controller is found to be:

$$\begin{aligned} u&=w-s+a_{0,1}t_1+a_{0,2}t_2+a_{0,3}t_3 \\ &=w-s+t_1+t_2+t_3 \end{aligned} \tag{62}$$

which, according to Sec.3, gives exactly the same deadbeat behaviour of the system as the controller (61).

5. APPENDICES

5.1. Appendix A: Proof of Eqs. (13) to (16)

From (11) it is found that:

$$\frac{U}{W} = \sum_{i_1=0}^{\infty} \sum_{i_2=0}^{\infty} u(i_1,i_2) z_1^{-i_1} z_2^{-i_2} (1-z_1^{-1})(1-z_2^{-1})$$

$$= \sum_{i_1=0}^{\infty} \sum_{i_2=0}^{\infty} \Big[u(i_1,i_2) z_1^{-i_1} z_2^{-i_2} - u(i_1,i_2) z_1^{-i_1-1} z_2^{-i_2} - u(i_1,i_2) z_1^{-i_1} z_2^{-i_2-1} + u(i_1,i_2) z_1^{-i_1-1} z_2^{-i_2-1} \Big]$$

$$= \sum_{i_1=0}^{\infty} \sum_{i_2=0}^{\infty} u(i_1,i_2) z_1^{-i_1} z_2^{-i_2} - \sum_{i_1=0}^{\infty} \sum_{i_2=0}^{\infty} u(i_1,i_2) z_1^{-i_1-1} z_2^{-i_2} -$$

$$- \sum_{i_1=0}^{\infty} \sum_{i_2=0}^{\infty} u(i_1,i_2) z_1^{-i_1} z^{-i_2-1} + \sum_{i_1=0}^{\infty} \sum_{i_2=0}^{\infty} u(i_1,i_2) z_1^{-i_1-1} z_2^{-i_2-1}$$

$$= \sum_{i_1=0}^{\infty} \sum_{i_2=0}^{\infty} u(i_1,i_2) z_1^{-i_1} z_2^{-i_2} - \sum_{i_1=1}^{\infty} \sum_{i_2=0}^{\infty} u(i_1-1,i_2) z_1^{-i_1} z_2^{-i_2} -$$

$$- \sum_{i_1=0}^{\infty} \sum_{i_2=0}^{\infty} u(i_1,i_2-1) z_1^{-i_1} z_2^{-i_2} + \sum_{i_1=1}^{\infty} \sum_{i_2=1}^{\infty} u(i_1-1,i_2-1) z_1^{-i_1} z_2^{-i_2} =$$

$$= \Big[u(0,0) + \sum_{i_2=1}^{\infty} u(0,i_2) z_2^{-i_2} + \sum_{i_1=0}^{\infty} u(i_1,0) z_1^{-i_1} + \sum_{i_1=1}^{\infty} \sum_{i_2=1}^{\infty} u(i_1,i_2) z_1^{-i_1} z_2^{-i_2} \Big]$$

$$- \Big[\sum_{i_1=0}^{\infty} u(i_1-1,0) z_1^{-i_1} + \sum_{i_1=1}^{\infty} \sum_{i_2=1}^{\infty} u(i_1-1,i_2) z_1^{-i_1} z_2^{-i_2} \Big]$$

$$-\left[\sum_{i_2=1}^{\infty} u(0,i_2-1)z_2^{-i_2}+\sum_{i_1=1}^{\infty}\sum_{i_2=1}^{\infty} u(i_1,i_2-1)z_1^{-i_1}z_2^{-i_2}\right]$$

$$+\left[\sum_{i_1=1}^{\infty}\sum_{i_2=1}^{\infty} u(i_1,i_2)z_1^{-i_1}z_2^{-i_2}\right]$$

$$=u(0,0)+\sum_{i_2=1}^{\infty}\left[u(0,i_2)-u(0,i_2-1)\right]z_2^{-i_2}+\sum_{i_1=1}^{\infty}\left[u(i_1,0)-u(i_1-1,0)\right]z_1^{-i_1}$$

$$+\sum_{i_1=1}^{\infty}\sum_{i_2=1}^{\infty}\left[u(i_1,i_2)-u(i_1-1,i_2)-u(i_1,i_2-1)+u(i_1-1,i_2-1)\right]z_1^{-i_1}z_2^{-i_2} \tag{A1}$$

Sufficiency of (5): Because of (5) the second and third terms vanish for $i_1>N_1$ or $i_2>N_2$. Especially in the fourth term the four summands are cancelled, the first with the second and the third with the fourth, i.e.

$$u(i_1,i_2)-u(i_1-1,i_2)=0 \text{ for } i_1>N_1 \text{ and for all } i_2 \tag{A2a}$$

$$u(i_1,i_2-1)-u(i_1-1,i_2-1)=0 \text{ for } i_1>N_1 \text{ and for all } i_2 \tag{A2b}$$

Hence in order to cancel the fourth term after N_1 steps, (2a) and (2b) must hold. For $i_2<N_2$, (2a) and (2b) give (5a) and for $i_2\geq N_2$, they give (5c). In a similar manner, taking into account the cancellations between the first and third, and between the second and fourth summands, (5b) and (5c) are found to be sufficient for the fourth term to vanish after N_1 *or* N_2 steps.

Necessity of (5): The necessiry of (5) for the second and third term to vanish after N_1 or N_2 steps is probable. The necessity of (5) for the fourth term to vanish, can be seen from the following. If (2a) does not hold, but rather the relation

$$u(i_1,i_2)-u(i_1-1,i_2)-u(i_1,i_2-1)+u(i_1-1,i_2-1)=0$$

for $i_1>N_1$ and for all i_2, then for $i_2=1$, it would hold for

$i_1 > N_1$, i.e.

$$u(i_1,1)-u(i_1-1,1)=u(i_1,0)-u(i_1-1,0).$$

But the right hand side of the above is zero (it is a rquirement that the third term of (1) vanishes after N_1 steps). Proceeding for $i_2=2$, it is found that

$u(i_1,2)-u(i_1-1,2)=0$ for $i_1>N_1$, and generally

$u(i_1,i_2)-u(i_1-1,i_2)=0$ for $i_1>N_1$ and all i_2.

In a similar way,

$u(i_1,i_2)-u(i_1,i_2-1)=0$ for all i_1 and $i_2>N_2$.

Hence the necessity of conditions (5) has also been proved.

Now, considering that the second, third and fourth terms vanish after N_1 or N_2 steps,(A1) is written

$$\frac{U}{W}=u(0,0)+\sum_{i_2=1}^{N_2}[u(0,i_2)-u(0,i_2-1)]z_2^{-i_2}+\sum_{i_1=1}^{N_1}[u(i_1,0)-u(i_1-1,0)]z_1^{-i_1}$$
$$+\sum_{i_1=0}^{N_1}\sum_{i_2=0}^{N_2}[u(i_1,i_2)-u(i_1-1,i_2)-u(i_1,i_2-1)+u(i_1-1,i_2-1)]z_1^{-i_1}z_2^{-i_2}$$

and the proof is complete. Equations (15) and (16) are proved in a similar manner.

5.2. Appendix B:Proof of (17), (18)

$$\sum_{i_1=0}^{N_1}\sum_{i_2=0}^{N_2}q_{i_1,i_2}=q_{0,0}+\sum_{i_2=1}^{N_2}q_{0,i_2}+\sum_{i_1=1}^{N_1}q_{i_1,0}+\sum_{i_1=1}^{N_1}\sum_{i_2=1}^{N_2}q_{i_1,i_2}$$

by (14) $=u(0,0)+\sum_{i_2=1}^{N_2}[u(0,i_2)-u(0,i_2-1)]+\sum_{i_1=1}^{N_1}[u(i_1,0)-u(i_1-1,0)]$

$$+\sum_{i_1=1}^{N_1}\sum_{i_2=1}^{N_2}\left[u(i_1,i_2)-u(i_1-1,i_2)-u(i_1,i_2-1)+u(i_1-1,i_2-1)\right]$$

$$=u(0,0)+\sum_{i_2=1}^{N_2}u(0,i_2)-\sum_{i_2=1}^{N_2}u(0,i_2-1)+\sum_{i_1=1}^{N_1}u(i_1,0)-\sum_{i_1=1}^{N_1}u(i_1-1,0)$$

$$+\sum_{i_1=1}^{N_1}\sum_{i_2=1}^{N_2}u(i_1,i_2)-\sum_{i_1=1}^{N_1}\sum_{i_2=1}^{N_2}u(i_1-1,i_2)$$

$$-\sum_{i_1=1}^{N_1}\sum_{i_2=1}^{N_2}u(i_1,i_2-1)+\sum_{i_1=1}^{N_1}\sum_{i_2=1}^{N_2}u(i_1-1,i_2-1)$$

$$=u(0,0)+\sum_{i_2=1}^{N_2}u(0,i_2)-\sum_{i_2=0}^{N_2}u(0,i_2)+\sum_{i_1=1}^{N_1}u(i_1,0)-\sum_{i_1=0}^{N_1}u(i_1,0)$$

$$+\sum_{i_1=1}^{N_1}\sum_{i_2=1}^{N_2}u(i_1,i_2)-\sum_{i_1=0}^{N_1-1}\sum_{i_2=0}^{N_2}u(i_1,i_2)$$

$$-\sum_{i_1=0}^{N_1}\sum_{i_2=0}^{N_2-1}u(i_1,i_2)+\sum_{i_1=0}^{N_1-1}\sum_{i_2=0}^{N_2-1}u(i_1,i_2)$$

$$=u(0,0)+\left[\sum_{i_2=0}^{N_2-1}u(0,i_2)-u(0,0)+u(0,N_2)\right]-\sum_{i_2=0}^{N_2-1}u(0,i_2)$$

$$+\left[\sum_{i_1=0}^{N_1-1}u(i_1,0)-u(0,0)+u(N_1,0)\right]-\sum_{i_1=0}^{N_1-1}u(i_1,0)$$

$$+\left[\sum_{i_2=1}^{N_2-1}u(N_1,i_2)+\sum_{i_1=1}^{N_1-1}u(i_1,N_2)+\sum_{i_1=1}^{N_1-1}\sum_{i_2=1}^{N_2-1}u(i_1,i_2)+u(N_1,N_2)\right]$$

$$-\left[\sum_{i_2=1}^{N_2-1}u(0,i_2)+u(0,N_2)+\sum_{i_1=1}^{N_1-1}\sum_{i_2=1}^{N_2-1}u(i_1,i_2)+\sum_{i_1=1}^{N_1-1}u(i_1,N_2)\right]$$

$$-\left[\sum_{i_1=1}^{N_1-1}u(i_1,0)+u(N_1,0)+\sum_{i_1=1}^{N_1-1}\sum_{i_2=1}^{N_2-1}u(i_1,i_2)+\sum_{i_2=1}^{N_2-1}u(N_1,i_2)\right]$$

$$+\left[u(0,0)+\sum_{i_2=1}^{N_2-1} u(0,i_2)+\sum_{i_1=1}^{N_1-1} u(i_1,0)+\sum_{i_1=1}^{N_1-1}\sum_{i_2=1}^{N_2-1} u(i_1,i_2)\right]$$

$=u(N_1,N_2)$ Q.E.D.

A graphical illustration of term cancellations is shown in Fig. 4, where the "+" and "-" denote that the corresponding term is added or subtracted.

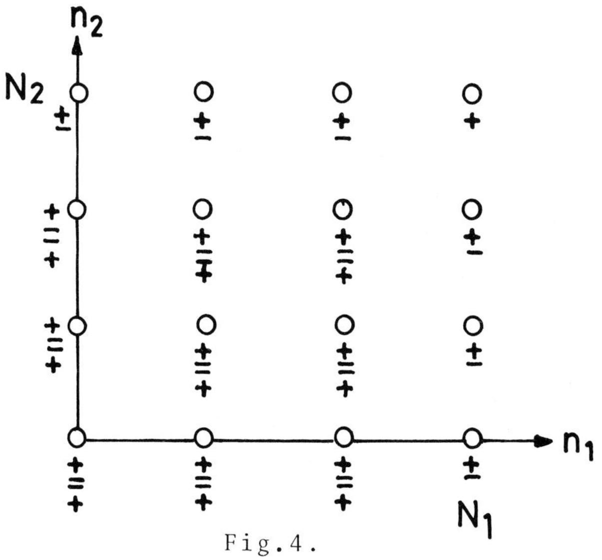

Fig.4.

One observes that all terms are twice added and twice subtracted, except of the last row and column, where they are added and subtracted only once.

5.3. Appendix C: Proof of (46)

The closed-loop state space matrices have the form

$$A_{w11}=\begin{bmatrix} 0 & 1 & & 0 \\ & \ddots & \ddots & \\ & & \ddots & 1 \\ 0 & & & 0 \end{bmatrix}_{(N_1 x N_1)}, A_{w12}=\begin{bmatrix} \\ 0 \\ \\ \end{bmatrix}_{(N_1 x N_1)}, A_{w13}=\begin{bmatrix} b_{1,1} \cdots b_{1,N_2} \\ \cdots\cdots\cdots \\ \cdots\cdots\cdots \\ \cdots\cdots\cdots \\ b_{N_1,1} \cdots b_{N_1 N_2} \end{bmatrix}_{(N_1 x N_2)}$$

$$A_{w21}=\begin{bmatrix} \\ 0 \\ \\ \end{bmatrix}_{(N_1xN_1)},A_{w22}=\begin{bmatrix} 0 & 1 & & 0 \\ & \ddots & \ddots & \\ 0 & & \ddots & 1 \\ & & & 0 \end{bmatrix}_{(N_1xN_1)},A_{w23}=\begin{bmatrix} -a_{1,1} & \cdots & -a_{1,N_2} \\ \vdots & & \vdots \\ -a_{N_1,1} & \cdots & -a_{N_1,N_2} \end{bmatrix}_{(N_1xN_2)}$$

$$A_{w31}=\begin{bmatrix} \\ 0 \\ \\ \end{bmatrix}_{(N_2xN_1)},A_{w32}=\begin{bmatrix} \\ 0 \\ \\ \end{bmatrix}_{(N_2xN_1)},A_{w33}=\begin{bmatrix} 0 & 1 & & 0 \\ & \ddots & \ddots & \\ 0 & & \ddots & 1 \\ & & & 0 \end{bmatrix}_{(N_2xN_2)}$$

$$B_{w1}=\begin{bmatrix} b_{1,0} \\ \vdots \\ b_{N_1,0} \end{bmatrix}_{(N_1x1)}\lambda ,B_{w2}=\begin{bmatrix} -a_{1,0} \\ \vdots \\ -a_{N_1,0} \end{bmatrix}_{(N_1x1)}\lambda ,B_{w3}=\begin{bmatrix} 1 \\ 0 \\ \\ \end{bmatrix}_{(N_2x1)}\lambda$$

$$C_{w1}=[1 \quad 0]_{(1xN_1)},C_{w2}=[0]_{(1xN_1)},\ C_{w3}=[b_{0,1}\cdots b_{0,N_2}]_{(1xN_2)}$$

$$D_w=b_{0,0}\cdot\lambda \tag{C1}$$

Equation (47) gives:

$$\frac{1}{\lambda}G_w=[C_{w1} \quad 0 \quad C_{w3}]\begin{bmatrix} I-A_{w11}z_1^{-1} & 0 & -A_{13}z_1^{-1} \\ 0 & I-A_{w22}z_1^{-1} & -A_{23}z_1^{-1} \\ 0 & 0 & I-A_{w33}z_2^{-1} \end{bmatrix}^{-1}\begin{bmatrix} B_{w1}z_1^{-1} \\ B_{w2}z_1^{-1} \\ B_{w3}z_2^{-1} \end{bmatrix}+D_w \tag{C2}$$

According to the matrix invertion property

$$\begin{bmatrix} A & B \\ 0 & D \end{bmatrix}^{-1}=\begin{bmatrix} A^{-1} & -A^{-1}B\ D^{-1} \\ 0 & D^{-1} \end{bmatrix}$$

where A,D are square invertible matrices, (C2) becomes

$$\frac{1}{\lambda}G_w=[C_{w1} \quad 0 \quad C_{w3}]$$

$$\begin{bmatrix} (I-A_{w11}z_1^{-1})^{-1} & 0 & (I-A_{w11}z_1^{-1})^{-1}A_{w13}z_1^{-1}(I-A_{w33}z_2^{-1})^{-1} \\ 0 & (I-A_{w22}z_1^{-1})^{-1} & (I-A_{w21}z_1^{-1})^{-1}A_{w23}z_1^{-1}(I-A_{w23}z_2^{-1})^{-1} \\ 0 & 0 & (I-A_{w33}z_1^{-1})^{-1} \end{bmatrix}\begin{bmatrix} B_{w1}z_1^{-1} \\ B_{w2}z_1^{-1} \\ B_{w3}z_2^{-1} \end{bmatrix}+b_{0,0}$$

Whence

$$\frac{1}{\lambda}G_w = C_{w1}(I-A_{w11}z_1^{-1})^{-1}B_{w1}z_1^{-1}$$
$$+C_{w1}(I-A_{w11}z_1^{-1})^{-1}A_{w13}z_1^{-1}(I-A_{w33}z_2^{-1})^{-1}B_{w3}z_2^{-1}$$
$$+C_{w3}(I-A_{w33}z_2^{-1})^{-1}B_{w3}z_2^{-1}+b_{0,0} \qquad (C3)$$

It is easily seen that

$$C_{w1}(I-A_{w11}z_1^{-1})^{-1}B_{w1}z_1^{-1} = \sum_{i_1=1}^{N_1} b_{i_1,0}z_1^{-i_1} \qquad (C4)$$

$$C_{w3}(I-A_{w33}z_2^{-1})^{-1}B_{w3}z_2^{-1} = \sum_{i_2=1}^{N_2} b_{0,i_2}z_2^{-i_2} \qquad (C5)$$

$$C_{w1}(I-A_{w11}z_1^{-1})^{-1}A_{w13}z_1^{-1}(I-A_{w33}z_2^{-1})^{-1}B_{w3}z_2^{-1}$$

$$=C_{w1}(I-A_{w11}z_1^{-1})^{-1}\begin{bmatrix} b_{1,1} & & b_{1,N_2} \\ \vdots & \cdots & \\ b_{N_1,1} & & b_{N_1N_2}\end{bmatrix} z_1^{-1}(I-A_{w33}z_2^{-1})^{-1}B_{w3}z_2^{-1}$$

$$\text{by (C4)} = \left[\sum_{i_1=1}^{N_1} b_{i_1,1}z_1^{-i_1}, \ldots, \sum_{i_1=1}^{N_1} b_{i_1,N_2}z_1^{-i_1}\right](I-A_{w33}z_2^{-1})^{-1}B_{w3}z_2^{-1}$$

$$\text{by (C5)} = \sum_{i_2=1}^{N_2}\sum_{i_1=1}^{N_1} b_{i_1,i_2}z_1^{-i_1}z_2^{-i_2}$$

Thus $G_{w1}(I-A_{w11}z_1^{-1})^{-1}A_{w13}z_1^{-1}(I-A_{w33}z_2^{-1})^{-1} = \sum_{i_1=1}^{N_1}\sum_{i_2=1}^{N_2} b_{i_1,i_2}z_1^{-i_1}z_2^{-i_2}$

Introducing (C4) to (C6) into (C3), Eq. (46) is proved.

It is remarked that the model (A_w, B_w, C_w, D_w) is a new state space model representing a nonrecursive 2-D system. An other state-space model for a nonrecursive 2-D system can be derived from the model of Kung-Levy and has the following matrices:

$$A=\left[\begin{array}{ccccc|c} 0 & 1 & & & & \\ & 0 & \ddots & 0 & & \\ 0 & & \ddots & 1 & & B_{12} \\ & & & 0 & & \\ \hline & & & 0 & & \\ & 0 & & 1 & 0 \ddots & 0 \\ & & & & \ddots & 0 \\ & & & 0 & & 1 \end{array}\right], \quad B=\begin{bmatrix} b_{1,0} \\ \vdots \\ b_{N_1,0} \\ \hline 0 \\ \vdots \\ 0 \end{bmatrix}$$

$$D=b_{0,0}$$

$$B_{12}=[b_{ij}]\ (N_1 x N_2)$$

$$C=[1,0\ldots,0\ b_{0,1}\ldots b_{0,N_2}]$$

In this model A in $(N_1+N_2)x(N_1xN_2)$-dimensional, where in the model of the paper A is $(2N_1+N_2)x(2N_1+N_2)$-dimensional with the elements a_{ij} involved in A_{w23} and B_{w2} not appearing in the transfer function (being cancelled). The flexibility due to the increased dimensionality can be used with profit in solving various control problems.

6. CONCLUSION

The control of 2D and 3D systems has received during the last years a growing interest and many important results have been derived regarding various types of control (eigenvalue control, model matching, decoupling, separating control, minimum energy control, deadbeat control, etc.).

In the present work we have provided two new deadbeat controllers; one in the form of a precompensator in a closed-loop configuration, and one in state-feedback form for a state-space 2D model of the Roesser type. Deadbeat controllers have found many applications due to their property of driving the controlled system to the desired conditions in minimum time. Work is in progress for deriving other versions of deadbeat controllers as well as self-tuning controllers for 2D and 3D systems.

REFERENCES

1. Special issue on digital filter, Proceedings of the IEEE, Vol. 63, No.4. 1975.
2. Special issue on multidimensional systems, Proc.of the IEEE, Vol. 65, No.6., 1977.
3. Paraskevopoulos, P.N.: 1980, "Feedback Design Techniques for Linear Multivariable 2-D Systems", in "Analysis and Optimization of Systems" (A. Bensoussan and J.L.Lions, Editors) pp. 763-780.
4. Tzafestas, S.G., Paraskevopoulos, P.N. and Pimenides,T.G.: 1983, "Modelling and Control of Multidimensional Systems", Proc. IASTED Applied Informatics Symp., (P.Borne, Editor), Vol. 1, pp. 167-179, Lille Univ., France.
5. Tzafestas, S.G.:1984, "State Feedback Control of 3-D Systems"in Multidimensional Systems:Techniques and Applications, (S.G.Tzafestas, Editor) Marcel Dekker.
6. Bose, N.K.:1979, "Multidimensional Systems:Theory and Applications", IEEE Press, New York.
7. Mersereau, R.M.and Dudgeon, D.E.:1975, "Two-Dimensional Digital Filtering", Proc.IEEE, Vol.63, pp. 610-623.
8. Shanks, J.L., Treitel, S. and Justice, J.H.:1979 "Stability and Synthesis of Two-Dimensional Recursive Filters", IEEE-Trans.Audio Electracoust., Vol.20,pp.115-128.

9. Dudgeon, D.E.:1977, "Fundamental of Digital Array Processing", Proc. IEEE, Vol.65, pp. 898-904.
10. Katayama, T.:1979, "Restoration of Noisy Images using a 2-Dimensional Linear Model", IEEE Trans. Syst.Man,Cybern., Vol.SMC-9, No.11, pp. 711-717.
11. Katayama, T.:1982, "Restoration of Images Degraded by Motion Blur and Noise, IEEE Trans. Auto.Control,Vol.AC-27, No.5, pp. 1024-1033.
12. Aboutabib, A.O., Murphy, M.S. and Silverman, L.M.:1977, "Digital Restoration of Images Degraded by General Motion Blurs", IEEE Trans. Auto.Control, Vol.AC-22, No.3, pp. 294-301.
13. Murphy, M.S. and Silverman, L.M.:1978, "Image Model Representation and Line-by-Line Recursive Restoration", IEEE Trans. Auto.Control, Vol.AC-23, No.5, pp. 809-816.
14. Jain, A.K.:1977, "Partial Differential Equations and Finite Difference Methods in Image Processing-Part I:Image Representation, J. Optimiz.Theory and Appl., Vol.23, pp. 65-91.
15. Jain, A.K. and Jain, J.R.:1978, "Partial Difference Methods in Image Processing-Part II:Image Representation, IEEE Trans. Auto Control, Vol.AC-23, No.5, pp.817-834.
16. Kucera, V.:1980, "A deadbeat Servo Problem", Int.J.Control, Vol.32, p. 107.
17. Kucera, V.:1979,"Discrete Linear Control-The Polynomial Equation Approach", J.Wiley, Chichester.
18. Poster,B and Bradshaw,A.:1975, "Design of Deadbeat Controller and Full-Order Observers for Linear Multivariable Discrete-Time Plants", Int.J.Control, Vol.22, pp. 149-155.
19. Sebek, M.:1980, "Multivariable Deadbeat Servo Problem", Kibernetica, Vol.16, pp. 442-453.
20. Seraji, H.:1975:"Deadbeat Control of Discrete-Time Systems Using Output Feedback", Int.J.Control, Vol.21, pp.213-223.
21. Iserman,:1980, Digital Control Systems", Springer Verlag, Berlin.
22. Tzafestas, S.G.:1983, "Microprocessors in Signal Processing, Measurement and Control (Ch.12)" D.Reidel, Dordrecht, Holland.
23. Kaczorek, T.:1983. "Dead-Beat Servo Problem for 2-Dimensional Linear Systems, Int.J.Control, Vol.37, pp. 1349-1353.
24. Kaczorek, T.:1983, "Dead-Beat Control of Multi-Input-Output n-D Linear Systems", Proc.3rd IFAC/IFORS Symp. LSSTA (A. Straszak, Ed.), Pergamon Press, Oxford.

CHAPTER 24

STATE OBSERVERS FOR 2-D AND 3-D SYSTEMS

S. Tzafestas, T. Pimenides
University of Patras
Patras, Greece

P. Stavroulakis
Mediterranean College
Athens, Greece

This chapter is devoted to the study of state observers for 2-dimensional and 3-dimensional systems. Firstly, the following structural properties of 3-D state-space models are considered: observability and controllability (local and separately local), and transformation of triangular systems to canonical form. Secondly, some 1-D results on adaptive state-observers and feedback implementation are extended to 2-D systems described by the Roesser model. Finally, the state observer design problem for 3-D systems is considered. The observer equations are derived and shown to provide an error that converges to zero independently of the choice of the initial conditions. Then a procedure is developed for constructing a 3-D state observer of triangular form.

1. INTRODUCTION

The theory of 1-Dimensional observers is by now very well developed and expanded [1-3]. The observer is actually a system that generates (reconstructs) the state of the original system using as inputs the measured (observed) output record. Some results regarding the design and utilization of 2-D and 3-D state observers were provided in [4-8].

Here we shall present some additional results. Specifically in the 2-D case the basic problem of identification and control of an unknown plant will be considered and treated using the model reference adaptive control approach [9-15]. The 2-D state observer will be used when the system state is not fully available but measured through the output. The procedure will be based on Lyapunov's direct method which leads to an asymptotically stable adaptive algorithm.

In the 3-D case the basic problem of state-observer design

S. G. Tzafestas (ed.), Multivariable Control, 453–477.

will be treated by extending the 2-D theory developed in |4|. First, the conditions, under which an observer exists, will be given, and then a design method will be developed for the important case in which the 3-D observer sought is of diagonal form in the sense that its matrix F is triangular.

The chapter starts by presenting in Section 2 two useful structural aspects of 3-D systems, namely the observability/contrallability, and transformation of triangular systems to canonical form. These aspects are of great value in solving analysis and design problems of 3-D systems, and obviously involve the corresponding results of 2-D systems as special case.

2. OBSERVABILITY AND CONTROLLABILITY OF 3-D SYSTEMS

It is well-known that for 1-D systems which are observable and controllable, there is a similarity transformation that brings the system into state space canonical form, and a minimum state space realization. Some results on observability and controllability (local) of 2-D systems were provided by Givone and Roesser [16]. These results are generalized here to the case of 3-D systems. Consider the 3-D system

$$\underline{x}'(i,j,k)=\underline{A}\ \underline{x}(i,j,k)+\underline{B}\ \underline{u}(i,j,k) \qquad i,j,k\geqslant 0 \qquad (1)$$
$$\underline{y}(i,j,k)=\underline{C}\ \underline{x}(i,jk)$$

where

$$\underline{x}=\begin{bmatrix}\underline{x}^h\\ \underline{x}^v\\ \underline{x}^d\end{bmatrix},\ \underline{x}'=\begin{bmatrix}\underline{x}^h(i+1,j,k)\\ \underline{x}^v(i,j+1,k)\\ \underline{x}^d(i,j,k+1)\end{bmatrix},\ \underline{B}=\begin{bmatrix}\underline{B}_1\\ \underline{B}_2\\ \underline{B}_3\end{bmatrix},\ \underline{A}=\begin{bmatrix}\underline{A}_1 & \underline{A}_2 & \underline{A}_3\\ \underline{A}_4 & \underline{A}_5 & \underline{A}_6\\ \underline{A}_7 & \underline{A}_8 & \underline{A}_9\end{bmatrix},\ \underline{C}=\left[\underline{C}_1,\underline{C}_2,\underline{C}_3\right] \qquad (2)$$

$x^h \varepsilon R^{n_1}$, $\underline{x}^v \varepsilon R^{n_2}$, $\underline{x}^d \varepsilon \underline{R}^{n_3}$, $\underline{u} \varepsilon R^r$, $y \varepsilon R^P$, and $\underline{A}_i$, $\underline{C}_i$ are constant matrices of appropriate dimensions.

DEFINITION 1: *The 3-D system (1)-(2) is said to be locally controllable if for* $\underline{x}^h(0,j,k)$, $j\geqslant 0, k\geqslant 0$; $\underline{x}^v(i,0,k)=0$, $i\geqslant 0$, $k\geqslant 0$; $\underline{x}^d(i,j,0)=0$, $i\geqslant 0, j\geqslant 0$ *and an arbitrary vector* $\underline{\xi} \varepsilon R^{n_1+n_2+n_3}$, *there exist* $N_1, N_2, N_3>0$ *and an input sequence* $u(i,j,k)$ *with* $(0,0,0) \leqslant (i,j,k)<(N_1,N_2,N_3)$, *such that* $\underline{x}(\underline{N}_1,N_2,N_3)=\underline{\xi}$.

This means that the system can reach an arbitrary state after a finite number of steps, under the assumption of zero initial conditions.

Analogously to the 1-D case we can show that: *A necessary condition for a 3-D system to be locally controllable is*

$$\text{ran } \underline{Q}=n_1+n_2+n_3 \tag{3a}$$

where $\underline{Q}$ *is the 3-D controllability matrix*

$$\underline{Q}=\left[\underline{M}(1,0,0),\ldots,\underline{M}(n_1,0,0);\underline{M}(0,1,0),\ldots,\underline{M}(n_1,n_2,n_3)\right] \tag{3b}$$

$$\underline{M}(i,j,k)=\underline{A}^{i-1,j,k}\underline{b}^{100}+\underline{A}^{i,j-1,k}\underline{b}^{010}+\underline{A}^{i,j,k-1}\underline{b}^{001} \tag{3c}$$

with $A^{i,j,k}=A^{i-1,j,k}A^{1,0,0}+A^{i,j-1,k}A^{0,1,0}+A^{i,j,k-1}A^{0,0,1}$

In fact, if the state $\underline{x}(i,j,k)$ cannot take the value $\bar{\xi}$ for some input $u(i,j,k)$ with $(i,j,k)\geqslant(0,0,0)$ then the system is not locally controllable. From the 3-D Cauley-Hamilton theorem it follows that each matrix $A^{\ell,m,n}$ is linearly dependent with $\underline{A}^{i,j,k}$ for $(0,0,0)\leqslant(i,j,k)<(n_1,n_2,n_3)$. Thus the condition of non local controllability is written as $\underline{Q}\ \underline{u}\neq\underline{\xi}$ for all $\underline{u}$ and some $\underline{\xi}$ where

$$\underline{u}=\text{col}\left[\underline{u}(1,0,0),\underline{u}(2,0,0),\ldots,\underline{u}(0,1,0),\ldots,\underline{u}(n_1,n_2,n_3)\right]$$

Clearly the maximum rank of $\underline{Q}$ is $n_1+n_2+n_3$, (i.e. equal to the dimensionality of the state vector). If rank $\underline{Q}=n_1+n_2+n_3$, then the transformation $\underline{Q}\ \underline{u}$ leads to the space of dimensionality $n_1+n_2+n_3$, i.e. to the state space, and it is not possible to exist a $\underline{\xi}$ such that $\underline{Q}\ \underline{u}\neq\xi$ for all $\underline{u}$. Thus the system is locally controllable. QED.

DEFINITION 2: *The 3-D system (1)-(2) is said to be locally observable, if for* $\underline{x}^h(0,j,k)=0$, $(j\geqslant1,\ k\geqslant1)$, $\underline{x}^v(i,0,k)=0$ $(i\geqslant1,\ k\geqslant1)$ *and* $\underline{x}^d(i,j,0)=0$ $(i\geqslant1, j\geqslant1)$ *there does not exist a nonzero initial state* $\underline{x}(0,0,0)$ *such that, for zero input* $\underline{u}(i,j,k)=0; i,j,\ k>0$, *the output* $\underline{y}(i,j,k)$ *is zero for all* $i,j,k\geqslant0$.

To check the local observability, we must compare the zero-initial-state output $\underline{y}_0(i,j,k)$ with the output $\underline{y}(i,j,k)$, $(i,j,k)>(0,0,0)$ obtained with any possible initial conditions, and any possible input. If $\underline{y}(i,j,k)=\underline{y}_0(i,j,k)$ the system is not locally observable. This means that

$\underline{C}\ \underline{A}^{i,j,k}\underline{x}(0,0,0)=0$ for all $(i,j,k)<(0,0,0)$ or due to the Cayley-Hamilton theorem for all $(0,0,0)\leq(i,j,k)<(n_1,n_2,n_3)$, in which case we have $\underline{K}\ \underline{x}=0$ where

$$\underline{K}=\left[(\underline{C}\ \underline{A}^{0,0,0})^T,\ (\underline{C}\ \underline{A}^{0,0,1})^T,\ldots,(\underline{C}\ \underline{A}^{n_1-1,n_2,n_3})^T\right]^T \tag{4}$$

is the observability matrix.

Clearly in order for the system to be locally observable we must have

$$\text{rank } \underline{K}=n_1+n_2+n_3 \tag{5}$$

Now as in the 1-D case we can determine a similarity tranformation which reveals the non controllable and non observable parts of the state vector $\underline{x}$. Since the state vector has three substates $\underline{x}^h, \underline{x}^v$ and $\underline{x}^d$ we need to introduce the separate local controllability and observability.

DEFINITION 3: *The 3-D system (1)-(2) is separately local controllable (correspondingly observable)*, *if*

$$\text{rank } \underline{Q}^h=n_1,\ \text{rank } \underline{Q}^v=n_2,\ \text{and } \underline{Q}^d=n_3, \tag{6}$$

(correspondingly

$$\text{rank } \underline{K}^h=n_1,\ \text{rank } \underline{K}^v=n_2,\ \text{rank } \underline{K}^d=n_3), \tag{7}$$

where

$$\underline{Q}=\begin{bmatrix}\underline{Q}^h\\ \underline{Q}^v\\ \underline{Q}^d\end{bmatrix}\begin{matrix}\}n_1\\ \}n_2\\ \}n_3\end{matrix}\qquad \underline{K}=[\underbrace{\underline{K}^h}_{n_1}\ \ \underbrace{\underline{K}^v}_{n_2}\ \ \underbrace{\underline{K}^d}_{n_3}] \tag{8}$$

Then the controllable state space is

$$\underline{X}^c=\underline{X}_h^c \oplus \underline{X}_v^c \oplus \underline{X}_d^c \tag{9a}$$

and the unobservable state space is

$$\underline{X}^u=\underline{X}_h^u \oplus \underline{X}_v^u \oplus \underline{X}_d^u \tag{9b}$$

where

$$\underline{X}_h^c=\{\underline{x}_1\varepsilon R^{n_1},\underline{x}_1\varepsilon \text{Range}[\underline{Q}^h]\},\ \text{etc} \tag{9c}$$

$$\underline{X}_h^u=\{\underline{x}_1\varepsilon R^{n_1},\ \underline{x}_1\varepsilon \text{Null}[\underline{K}^h]\},\text{etc.} \tag{9d}$$

If a 3-D system is of minimum state dimensionality then it is separately locally controllable and observable.

The similarity transformation is sought in the form $\underline{T}=\mathrm{diag}\left[\underline{T}_1,\underline{T}_5,\underline{T}_9\right]$, i.e.

$$\begin{bmatrix} \hat{\underline{x}}^h(i,j,k) \\ \hat{\underline{x}}^v(i,j,k) \\ \hat{\underline{x}}^d(i,j,k) \end{bmatrix} = \begin{bmatrix} \underline{T}_1 & 0 & 0 \\ 0 & \underline{T}_5 & 0 \\ 0 & 0 & \underline{T}_9 \end{bmatrix} \begin{bmatrix} \underline{x}^h(i,j,k) \\ \underline{x}^v(i,j,k) \\ \underline{x}^d(i,j,k) \end{bmatrix} \tag{10}$$

with $\underline{T}_1 \varepsilon R^{n_1 x n_1}, \underline{T}_5 \varepsilon R^{n_2 x n_2}, \underline{T}_9 \varepsilon R^{n_3 x n_3}$

Applying (10) the original system (1)-(2) is transformed to

$$\begin{aligned} \hat{\underline{x}}'(i,j,k) &= \underline{A}\hat{\underline{x}}(i,j,k)+\hat{\underline{B}}\ \underline{u}(i,j,k) \\ \underline{y}(i,j,k) &= \hat{\underline{C}}\ \hat{\underline{x}}(i,j,k) \end{aligned} \tag{11}$$

where

$$\begin{aligned} &\hat{\underline{A}}_1=\underline{T}_1\underline{A}_1\underline{T}_1^{-1},\ \underline{A}_2=\underline{T}_1\underline{A}_2\underline{T}_5^{-1},\ \hat{\underline{A}}_3=\underline{T}_1\underline{A}_3\underline{T}_9^{-1},\dots. \\ &\hat{\underline{B}}_1=\underline{T}_1\underline{B}_1,\ \hat{\underline{B}}_2=\underline{T}_5\underline{B}_2,\ \hat{\underline{B}}_3=\underline{T}_9\underline{B}_3;\hat{\underline{C}}_1=\underline{C}_1\underline{T}_1^{-1},\hat{\underline{C}}_2=\underline{C}_2\underline{T}_5^{-1},\hat{\underline{C}}_3=\underline{C}_3\underline{T}_9^{-1} \end{aligned} \tag{12}$$

It is easy to verify that the similar system (1)-(2) and (11) have the same transfer function.

The transformation (10) may not be able to reveal the non controllable (nonobservable) parts of the state of a locally noncontrollable (nonobservable) 3-D system, since we are not sure which form

$$\begin{bmatrix} \xi_1 \\ 0 \\ 0 \end{bmatrix} \text{ or } \begin{bmatrix} 0 \\ \xi_2 \\ 0 \end{bmatrix} \text{ or } \begin{bmatrix} 0 \\ 0 \\ \xi_3 \end{bmatrix}$$

has the vector $\underline{\xi}\varepsilon\mathrm{Ker}\underline{Q}$. However this is always possible when the system is separable locally noncontrollable (non observable).

3. TRANSFORMATION OF TRIANGULAR 3-D SYSTEMS TO CANONICAL FORM

A 3-D system (1)-(2) is said to be triangular if its matrix $\underline{A}$ has the form

$$\underline{A}=\begin{bmatrix}\underline{A}_1 & 0 & 0\\ \underline{A}_4 & \underline{A}_5 & 0\\ \underline{A}_9 & \underline{A}_8 & \underline{A}_9\end{bmatrix} \tag{13}$$

The value of triangular systems is that the denominator of their transfer function is expressed as the product of three 1-D polynomials. Thus their stability analysis and control is very easy [17].

THEOREM: For a triangular 3-D system which is separately local controllable and observable there exists a similarity transformation of the form (10) which reduces the system into the following canonical form:

$$\hat{\underline{A}}=\begin{bmatrix}\hat{\underline{A}}_1 & 0 & 0\\ \underline{A}_4 & \hat{\underline{A}}_5 & 0\\ \hat{\underline{A}}_7 & \hat{\underline{A}}_8 & \hat{\underline{A}}_9\end{bmatrix},\ \hat{\underline{B}}=\begin{bmatrix}\hat{\underline{B}}_1\\ \hat{\underline{B}}_2\\ \hat{\underline{B}}_3\end{bmatrix},\ \hat{\underline{C}}=\begin{bmatrix}\hat{\underline{C}}_1 & \hat{\underline{C}}_2 & \hat{\underline{C}}_3\end{bmatrix}$$

where

$$\hat{\underline{A}}_\ell=\begin{bmatrix}\underline{A}_{11}^{(\ell)} & \underline{A}_{12}^{(\ell)} & \underline{A}_{13}^{(\ell)} & \underline{A}_{14}^{(\ell)}\\ 0 & \underline{A}_{22}^{(\ell)} & \underline{A}_{23}^{(\ell)} & \underline{A}_{24}^{(\ell)}\\ 0 & 0 & \underline{A}_{33}^{(\ell)} & \underline{A}_{34}^{(\ell)}\\ 0 & 0 & 0 & \underline{A}_{44}^{(\ell)}\end{bmatrix},\ \hat{\underline{B}}_\ell=\begin{bmatrix}\underline{b}_1^{(\ell)}\\ \underline{b}_2^{(\ell)}\\ 0\\ 0\end{bmatrix},\ \hat{\underline{C}}_\ell=\begin{bmatrix}0 & \underline{c}^{(\ell)} & 0 & \underline{c}_4^{(\ell)}\end{bmatrix} \tag{14}$$

Since the complete proof is lengthy here we shall only give a sketch of it.

First we mention the following properties of the subspaces (9a-d).

P1: The horizontal controllable state space $\underline{X}_h^c$ and the horizontal nonobservable space $\underline{X}_h^u$ are invariant with respect to the matrix $\underline{A}_1$.

P2: The vertical controllable sate space $\underline{X}_v^c$ and the vertical nonobservable space $\underline{X}_v^u$ are invariant with respect to the matrix $\underline{A}_5$.

P3: The depth controllable space $\underline{X}_d^c$ are invariant with respect to the matrix $\underline{A}_9$.

<u>P4</u>: The following are true

$$A_4 X_h^c \subset X_v^c;\ A_4 X_h^u \subset X_v^u,\ A_7 X_h^c \subset X_d^c;\ A_7 X_v^u \subset X_d^u,\ A_8 X_v^c \subset X_d^c;\ A_8 X_v^u \subset X_d^u$$

Then using the separate local controllability and observability assumptions, and working as in the 1-D case we can obtain the following decompositions

$$\underline{X}_{h_1} = \underline{X}_h^c \cap \underline{X}_h^u;\ \underline{X}_h^c = \underline{X}_{h_1} \oplus \underline{X}_{x_2};\ \underline{X}_h^u = \underline{X}_{h_1} \oplus \underline{X}_{h_3};\ R^{n_1} = \underline{X}_{h_1} \oplus \underline{X}_{h_2} \oplus \underline{X}_{h_3} \oplus \underline{X}_{h_4}$$

$$\underline{X}_{v_1} = \underline{X}_v^c \cap \underline{X}_v^u;\ \underline{X}_v^c = \underline{X}_{v_1} \oplus \underline{X}_{v_2};\ \underline{X}_v^u = \underline{X}_{v_1} \oplus \underline{X}_{v_3};\ R^{n_2} = \underline{X}_{v_1} \oplus \underline{X}_{v_2} \oplus \underline{X}_{v_3} \oplus \underline{X}_{v_4}$$

$$\underline{X}_{d_1} = \underline{X}_d^c \cap \underline{X}_d^u;\ \underline{X}_d^c = \underline{X}_{d_1} \oplus \underline{X}_{d_2};\ \underline{X}_d^u = \underline{X}_{d_1} \oplus \underline{X}_{d_3};\ R^{n_3} = \underline{X}_{d_1} \oplus \underline{X}_{d_2} \oplus \underline{X}_{d_3} \oplus \underline{X}_{d_4}$$

with

$$\dim \underline{X}_{h_i} = s_{h_i}\ (i=1,2,3,4),\ \sum_i s_{h_i} = n_1$$

$$\dim \underline{X}_{v_i} = s_{v_i}\ (i=1,2,3,4),\ \sum_i s_{v_i} = n_2$$

$$\dim \underline{X}_{d_i} = s_{d_i}\ (i=1,2,3,4),\ \sum_i s_{d_i} = n_3$$

The dimensions of the submatrices in (14) are as follows:

$$\underline{A}_{ij}^{(1)}: s_{h_i} x s_{h_j},\quad \underline{A}_{ij}^{(4)}\ s_{v_i} x s_{h_j},\quad \underline{A}_{ij}^{(5)}: s_{v_i} x s_{v_j}$$

$$\underline{A}_{ij}^{(7)}: s_{d_i} x s_{h_j},\quad \underline{A}_{ij}^{(8)}: s_{d_i} x s_{v_j},\ \underline{A}_{ij}^{(9)}: s_{d_i} x s_{d_j}$$

$$\underline{b}_i^{(1)}: s_{h_i} x1,\ \underline{b}_i^{(2)}: s_{v_i} x1,\ \underline{b}_i^{(3)}: s_{d_i} x1 \qquad (i=1,2,3,4)$$

$$\underline{c}_i^{(1)}: 1xs_{h_i},\ \underline{c}_i^{(2)}: 1xs_{v_i},\ \underline{c}_i^{(3)}: 1xs_{d_i} \qquad (1=2,4)$$

The submatrices $\underline{T}_1$, $\underline{T}_5$ and $\underline{T}_9$ of the similarity transformation (10) are chosen as

$$\underline{T}_1 = \left[\underline{T}_{h_1}, \underline{T}_{h_2}, \underline{T}_{h_3}, \underline{T}_{h_4}\right]^{-1},\ \underline{T}_5 = \left[\underline{T}_{v_1}, \underline{T}_{v_2}, \underline{T}_{v_3}, \underline{T}_{v_4}\right]^{-1},$$

$$\underline{T}_9 = \left[\underline{T}_{d_1}, \underline{T}_{d_2}, \underline{T}_{d_3}, \underline{T}_{d_4}\right]^{-1}$$

where $T_{h_i}, T_{v_i}, T_{d_i}$, $i=1,2,3,4$ are $n_1 x s_{h_i}, n_2 x s_{v_i}, n_3 x s_{d_i}$ dimensional real-valued matrices with columns constituting bases of $X_{h_i}, X_{v_i}, X_{d_i}$ respectively. Then using P1 we have

$$\underline{A}_1 T^{-1} = \left[\underline{A}_1 T_{h_1}, \underline{A}_1 T_{h_2}, \underline{A}_1 \underline{T}_{h_3}, \underline{A}_1 \underline{T}_{h_4}\right]$$

$$= \left[\underline{T}_{h_1}, \underline{T}_{h_2}, \underline{T}_{h_3}, \underline{T}_{h_4}\right] \hat{\underline{A}}_1 = T_1^{-1} \hat{\underline{A}}_1$$

In the same way one can determine $\hat{\underline{A}}_4, \hat{\underline{A}}_5, \hat{\underline{A}}_7, \hat{\underline{A}}_8$, and $\hat{\underline{A}}_9$ using the properties P2-P4. Regarding $\hat{b}$ we observe that

$\underline{b}_1 \varepsilon X_h^c$, $\underline{b}_2 \varepsilon X_v^c$ and $\underline{b}_3 \varepsilon X_d^c$. Then

$$\underline{b}_1 = \underline{T}_1^{-1} \hat{\underline{b}}_1 ; \underline{b}_2 = \underline{T}_5^{-1} \hat{\underline{b}}_2 ; \underline{b}_3 = \underline{T}_9^{-1} \hat{\underline{b}}_3$$

where $\hat{b}_i$ have the form shown in (14). Similarly we find that

$$\hat{\underline{C}}_1 = \underline{C}_1 \underline{T}_1^{-1} = \left[\underline{C}_1 \underline{T}_{h_1}, \underline{C}_1 \underline{T}_{h_2}, \underline{C}_1 \underline{T}_{h_3}, \underline{C}_1 \underline{T}_{h_4}\right] = \left[0, \underline{c}_2^{(1)}, 0, \underline{c}_4^{(1)}\right]$$

$$\hat{\underline{C}}_2 = \underline{C}_2 \underline{T}_5^{-1} = \left[0, \underline{c}^{(2)}, 0, \underline{c}_4^{(2)}\right] ; \hat{\underline{C}}_3 = \underline{C}_3 \underline{T}_9^{-1} = \left[0, \underline{c}_2^{(3)}, 0, \underline{c}_4^{(3)}\right]$$

4. THE 2-D PARAMETER IDENTIFICATION PROBLEM

Consider the 2-D linear, time-invariant multivariable system described in state space as follows [5-7]:

$$\begin{bmatrix} \underline{x}_p^h(i+1,j) \\ \underline{x}_p^v(i,j+1) \end{bmatrix} = \begin{bmatrix} \underline{A}_1 & \underline{A}_2 \\ \underline{A}_3 & \underline{A}_4 \end{bmatrix} \begin{bmatrix} \underline{x}_p^h(i,j) \\ \underline{x}_p^v(i,j) \end{bmatrix} + \begin{bmatrix} \underline{B}_1 \\ \underline{B}_2 \end{bmatrix} \underline{u}(i,j)$$

$$\underline{y}(i,j) = \left[\underline{C}_1 \;\; \underline{C}_2\right] \begin{bmatrix} \underline{x}_p^h(i,j) \\ \underline{x}_p^v(i,j) \end{bmatrix}, i\ j>0 \qquad (15)$$

where $\underline{x}_p^h \varepsilon R^{n_1}$ is the *horizontal* state vector, $\underline{x}_p^v \varepsilon R^{n_2}$ is the vertical state vector, $\underline{u} \varepsilon R^n$ is the input vector, $\underline{y} \varepsilon R^{\ell}$ is the output vector and $\underline{A}_i, \underline{B}_i, \underline{C}_i$ are constant matrices of appropriate dimensions. System (15) is usually written more compactly as follows:

$$\tilde{\underline{x}}_p = \underline{A}\underline{x}_p + \underline{B}\ \underline{u}, \quad \underline{y} = \underline{C}\underline{x}_p \tag{16}$$

where

$$\underline{x}_p = \begin{bmatrix} \underline{x}_p^h(i,j) \\ \\ \underline{x}_p^v(i,j) \end{bmatrix}, \tilde{\underline{x}}_p = \begin{bmatrix} \underline{x}_p^h(i+1,j) \\ \\ \underline{x}_p^h(i,j+1) \end{bmatrix}, A = \begin{bmatrix} \underline{A}_1 & \underline{A}_2 \\ \\ \underline{A}_3 & \underline{A}_4 \end{bmatrix}, B = \begin{bmatrix} \underline{B}_1 \\ \\ \underline{B}_2 \end{bmatrix}, C = [\underline{C}_1 \ \ \underline{C}_2] \tag{17}$$

The initial conditions $\underline{x}_p^h(0,k)$, $0<k<\infty$ and $\underline{x}_p^v(k,0)$, $0<k<\infty$ required for the unique solution of (1) given that the elements of the matrices $\underline{A}$ and $\underline{B}$ are known. For the cases when the elements of the matrices are unknown constants, we generally seek to set up a model and suitably adjust its parameters so that the elements of these matrices can be estimated. The procedure to be described which is clearly a parameter identification procedure provides the motivation and framework for the schemes used for the observer which will discuss in the subsequent section for adaptive control problems.

5. IDENTIFICATION VIA MODEL REFERENCE APPROACH:ALL STATES ACCESSIBLE

Given the plant of equation (15), we shall define the model

$$\begin{bmatrix} \underline{x}_m^h(i+1,j) \\ \\ \underline{x}_m^v(i,j+1) \end{bmatrix} = \begin{bmatrix} \hat{\underline{A}}_1(i+1,j) & \hat{\underline{A}}_2(i+1,j) \\ \\ \hat{\underline{A}}_3(i,j+1) & \hat{\underline{A}}_4(i,j+1) \end{bmatrix} \begin{bmatrix} \underline{x}_p^h(i,j) \\ \\ \underline{x}_p^v(i,j) \end{bmatrix} + \begin{bmatrix} \hat{\underline{B}}_1(i+1,j) \\ \\ \hat{\underline{B}}_2(i,j+1) \end{bmatrix} u(i,j) \tag{18}$$

or

We shall determine an identification scheme for this 2-D digital system such that

$$\lim_{\substack{i\to\infty \\ j\to\infty}} \hat{\underline{A}}_1(i,j) = \underline{A}_1 \quad \lim_{\substack{i\to\infty \\ j\to\infty}} \hat{\underline{A}}_3(i,j) = \underline{A}_3 \quad \lim_{\substack{i\to\infty \\ j\to\infty}} \hat{\underline{B}}_1(i,j) = \underline{B}_1$$

$$\lim_{\substack{i\to\infty \\ j\to\infty}} \hat{\underline{A}}_2(i,j) = \underline{A}_2 \quad \lim_{\substack{i\to\infty \\ j\to\infty}} \hat{\underline{A}}_4(i,j) = \underline{A}_4 \quad \lim_{\substack{i\to\infty \\ j\to\infty}} \hat{\underline{B}}_2(i,j) = \underline{B}_2$$

and $\lim_{\substack{i\to\infty \\ j\to\infty}} (\underline{x}_m^h(i,j) - \underline{x}_p^h(i,j)) = \lim_{\substack{i\to\infty \\ j\to\infty}} \underline{e}^h(i,j) = 0$

and $\lim_{\substack{i\to\infty \\ j\to\infty}} (\underline{x}_m^v(i,j)-\underline{x}_p^v(i,j)=\lim_{\substack{i\to\infty \\ j\to\infty}} \underline{e}^v(i,j)=0$ (19)

Subtracting (15) from (18) we obtain

$$\begin{bmatrix} \underline{e}^h(i+1,j) \\ \underline{e}^v(i,j+1) \end{bmatrix} = \left[\begin{array}{c|c} \hat{\underline{A}}_1(i+1,j)-\underline{A}_1 & \hat{\underline{A}}_2(i+1,j)-\underline{A}_2 \\ \hline \hat{\underline{A}}_3(i,j+1)-\underline{A}_3 & \hat{\underline{A}}_4(i,j+1)-\underline{A}_4 \end{array}\right] \begin{bmatrix} \underline{x}_p^h(i,j) \\ \underline{x}_p^v(i,j) \end{bmatrix}$$

$$+\begin{bmatrix} \hat{\underline{B}}_1(i+1,j)-\underline{B}_1 \\ \hat{\underline{B}}_2(i+1,j)-\underline{B}_2 \end{bmatrix} \underline{u}(i,j) \tag{20}$$

or

$$\begin{bmatrix} \underline{e}^h(i+1,j) \\ \underline{e}^v(i,j+1) \end{bmatrix} = \begin{bmatrix} \tilde{\underline{A}}_1(i+1,j) & \tilde{\underline{A}}_2(i+1,j) \\ \tilde{\underline{A}}_3(i,j+1) & \tilde{\underline{A}}_4(i,j+1) \end{bmatrix} \begin{bmatrix} \underline{x}_p^h(i,j) \\ \underline{x}_p^v(i,j) \end{bmatrix} + \begin{bmatrix} \tilde{\underline{B}}_1(i+1,j) \\ \tilde{\underline{B}}_2(i,j+1) \end{bmatrix} u(i,j) \tag{21}$$

where

$\underline{e}^h(i+1,j)=\underline{x}_m^h(i+1,j)-\underline{x}_p^h(i+1,j)$; $\underline{e}^v(i,j+1)=\underline{x}_m^v(i,j+1)-\underline{x}_p^v(i,j+1)$

$\tilde{\underline{A}}_1(i+1,j)=\hat{\underline{A}}_1(i+1,j)-\underline{A}_1$; $\tilde{\underline{A}}_2(i+1,j)=\hat{\underline{A}}_2(i+1,j)-\underline{A}_2$

$\tilde{\underline{A}}_3(i,j+1)=\hat{\underline{A}}_3(i+1,j)-\underline{A}_3$; $\tilde{\underline{A}}_4(i,j+1)=\hat{\underline{A}}_4(i,j+1)-\underline{A}_4$

$\tilde{\underline{B}}_1(i+1,j)=\hat{\underline{B}}_1(i+1,j)-\underline{B}_1$; $\tilde{\underline{B}}_2(i,j+1)=\hat{\underline{B}}_2(i,j+1)-\underline{B}_2$

Equation (21) can be written as:

$$\begin{bmatrix} \underline{e}^h(i+1,j) \\ \underline{e}^v(i,j+1) \end{bmatrix} = \begin{bmatrix} \underline{\theta}_1(i+1,j) \\ \underline{\theta}_2(i,j+1) \end{bmatrix} \underline{x}(i,j) \tag{22}$$

where

$\underline{\theta}_1(i+1,j)=[\tilde{\underline{A}}_1(i+1,j)\ \ \tilde{\underline{A}}_2(i+1,j)\ \ \tilde{\underline{B}}_1(i+1,j)]$

$\underline{\theta}_2(i,j+1)=[\tilde{\underline{A}}_3(i,j+1)\ \ \tilde{\underline{A}}_4(i,j+1)\ \ \tilde{\underline{B}}_2(i,j+1)]$

and

$$\underline{x}(i,j)=\begin{bmatrix} \underline{x}_p^h(i,j) \\ \underline{x}_p^v(i,j) \\ \underline{u}(i,j) \end{bmatrix}$$

In what follows, we shall show that if we choose the rule

$$\underline{\theta}_1(i+1,j)=\underline{\theta}_1(i,j)-\underline{F}_1(i,j),\quad \underline{\theta}_2(i,j+1)=\underline{\theta}_2(i,j)-\underline{F}_2(i,j) \qquad (23)$$

where

$$\underline{F}_1(i,j)=a_1(i,j)\underline{P}_1\underline{e}^h(i,j)\underline{x}^T(i-1,j) \qquad (24)$$

$$\underline{F}_2(i,j)=a_2(i,j)P_2\underline{e}^v(i,j)\underline{x}^T(i,j-1) \qquad (24)$$

and the scalars

$a_1(i,j), a_2(i,j)>0$ and $\underline{P}_1$ and $\underline{P}_2$ are positive definite matrices, it leads to an identification law for the parameters of the matrices $\underline{A}$ and $\underline{B}$.

Consider the following candidates for the Lyapunov functions.

$$V_1(i,j)=\text{Trace }(\underline{\theta}_1^T(i,j)\underline{\theta}_1(i,j)) \qquad (25)$$

$$V_2(i,j)=\text{Trace }(\underline{\theta}_2^T(i,j)\underline{\theta}_2(i,j)) \qquad (25)$$

If we define

$$\Delta V_1(i,j)=V_1(i+1,j)-V_1(i,j) \qquad (26)$$

$$\Delta V_2(i,j)=V_2(i,j+1)-V_2(i,j) \qquad (26)$$

and substitute (25) into (26), we obtain

$$\Delta V_1(i,j)=\text{Trace}(\underline{\theta}_1^T(i+1,j)\underline{\theta}_1(i+1,j))-\text{Trace}(\underline{\theta}_1^T(i,j)\underline{\theta}_1(i,j)) \quad (27)$$

$$\Delta V_2(i,j)=\text{Trace}(\underline{\theta}_2^T(i,j+1)\underline{\theta}_2(i,j+1))-\text{Trace}(\underline{\theta}_2^T(i,j)\underline{\theta}_2(i,j)) \quad (27)$$

Using (24) into (27) and (24) into (27) we obtain:

$$\Delta V_1=-\underline{e}^{hT}(i,j)\left[2a_1(i,j)\underline{P}_1-a_1^2(i,j)\underline{\tilde{x}}^T(i-1,j)\underline{\tilde{x}}(i-1,j)\underline{P}_1^2\right]\underline{e}^v(i,j) \qquad (28)$$

$$\Delta V_2=-\underline{e}^{vT}(i,j)\left[2a_2(i,j)\underline{P}_2-a_2^2(i,j)\underline{\tilde{x}}^T(i,j-1)\underline{\tilde{x}}(i,j-1)\underline{P}_2^2\right]\underline{e}^v(i,j) \qquad (28)$$

This formulation is identical to the formulation presented in [11] for discrete linear multivariable systems.

Using lemma 1 in [11], it can be shown that if the scalars a_1 and a_2 are chosen such that

$$a_1(i,j)=\frac{a}{\lambda_{1max}\underline{\tilde{x}}^T(i-1,j)\underline{\tilde{x}}(i-1,j)} \tag{29}$$

$$a_2(i,j)=\frac{a}{\lambda_{2max}\underline{\tilde{x}}^T(i,j-1)\underline{\tilde{x}}(i,j-1)} \tag{29}$$

where $0<a<2$ and λ_{1max} and λ_{2max} are the largest eigenvalues of the matrices $\underline{P}_1$ and $\underline{P}_2$ respectively the Lyapunov functions ΔV_1 and ΔV_2 become negative semidefinite. Further, it is shown that under certain mild conditions on the input function $\underline{u}(i,j)$ the overall scheme is asymptotically stable.

6. ADAPTIVE CONTROLLER

In this particular case, it is desired to use state feedback so that the output behavior or the system asymptotically approaches that of a specified stable model. In other words, given the plant described by (15) for which the parameters of the matrices A_1,A_2,A_3,A_4 are unknown we shall assume a controller of the form

$$\underline{u}(i,j)=\underline{K}(i,j)\underline{x}_p+\underline{u}_m(i,j) \tag{30}$$

so that the plant (15) and the model

$$\begin{bmatrix}\underline{x}_m^h(i+1,j)\\ \underline{x}_m^v(i,j+1)\end{bmatrix}=\begin{bmatrix}\underline{A}_{1m} & \underline{A}_{2m}\\ A_{3m} & A_{4m}\end{bmatrix}\begin{bmatrix}\underline{x}_m^h(i,j)\\ \underline{x}_m^v(i,j)\end{bmatrix}+\begin{bmatrix}\underline{B}_1\\ B_2\end{bmatrix}\underline{u}_m(i,j) \tag{31}$$

achieve asymptotically the same output behaviour.

This model and the plant (15) yield a state error $x_m(i,j)-\underline{x}_p(i,j)=\underline{e}(i,j)$ which satisfied the set of equations

$$\begin{bmatrix}\underline{e}^h(i+1,j)\\ \underline{e}^v(i,j+1)\end{bmatrix}=\left[\begin{array}{c|c}\underline{A}_{1m} & \underline{A}_{2m}\\ \hline \underline{A}_{3m} & \underline{A}_{4m}\end{array}\right]\begin{bmatrix}\underline{e}^h(i,j)\\ \underline{e}^v(i,j)\end{bmatrix}+\left[\begin{array}{c|c}\underline{A}_{1m}-\underline{A}_1+\underline{B}_1\underline{K}_1(i,j) & \underline{A}_{2m}-\underline{A}_2+\underline{B}_1\underline{K}_2(i,j)\\ \hline \underline{A}_{3m}-\underline{A}_3+\underline{B}_2\underline{K}_1(i,j) & \underline{A}_{4m}-\underline{A}_4+\underline{B}_2\underline{K}_2(i,j)\end{array}\right]\begin{bmatrix}\underline{x}^h(i,j)\\ \underline{x}^v(i,j)\end{bmatrix} \tag{32}$$

If we define

$$\underline{\Phi}_{11}(i,j)=\underline{A}_{1m}-\underline{A}_1+\underline{B}_1\underline{K}_1(i,j),\ \underline{\Phi}_{21}(i,j)=\underline{A}_{3m}-\underline{A}_3+\underline{B}_2\underline{K}_1(i,j)$$

$$\underline{\Phi}_{12}(i,j)=\underline{A}_{2m}-\underline{A}_2+\underline{B}_1\underline{K}_2(i,j),\ \underline{\Phi}_{22}(i,j)=\underline{A}_{4m}-\underline{A}_4+\underline{B}_2\underline{K}_2(i,j)$$

Then equation (32) becomes:

$$\begin{bmatrix}\underline{e}^h(i+1,j)\\ \\ \underline{e}^v(i,j+1)\end{bmatrix}=\begin{bmatrix}\underline{A}_{1m} & \underline{A}_{2m}\\ \hline \underline{A}_{3m} & \underline{A}_{4m}\end{bmatrix}\begin{bmatrix}\underline{e}_h(i,j)\\ \\ \underline{e}^v(i,j)\end{bmatrix}+\left[\begin{array}{c|c}\underline{\Phi}_{11}(i,j) & \underline{\Phi}_{12}(i,j)\\ \hline \underline{\Phi}_{21}(i,j) & \underline{\Phi}_{22}(i,j)\end{array}\right]\begin{bmatrix}\underline{x}^h(i,j)\\ \\ \underline{x}^v(i,j)\end{bmatrix} \tag{33}$$

we can also write equation (33) as follows:

$$\begin{bmatrix}\underline{e}^h(i+1,j)\\ \\ \underline{e}^v(i,j+1)\end{bmatrix}=\begin{bmatrix}\underline{A}_1\\ \hline \underline{A}_2\end{bmatrix}\underline{e}(i,j)+\begin{bmatrix}\underline{z}^h(i,j)\\ \\ \underline{z}^v(i,j)\end{bmatrix} \tag{34}$$

where $\underline{e}(i,j)=\begin{bmatrix}\underline{e}^h(i,j)\\ \\ \underline{e}^v(i,j)\end{bmatrix}$ and

$$\underline{z}^h(i,j)=\underline{\Phi}_{11}(i,j)\underline{x}^h(i,j)+\underline{\Phi}_{12}(i,j)\underline{x}^v(i,j)$$

$$\underline{z}^v(i,j)=\underline{\Phi}_{21}(i,j)\underline{x}^h(i,j)+\underline{\Phi}_{22}(i,j)\underline{x}^v(i,j)$$

Using a methodology similar to that described in [15], we can apply the following adaptive model.

Let

$$\underline{e}_1^h(i,j)=\underline{C}^h\underline{e}(i,j)+\underline{D}^h\underline{z}^h(i,j)$$

$$\underline{e}_1^v(i,j)=\underline{C}^v\underline{e}(i,j)+\underline{D}^v\underline{z}^v(i,j)$$

$$\underline{z}^h(i,j)=\underline{\theta}^{hT}(i,j)\underline{x}^h(i,j)-a^h\underline{x}^{hT}(i,j)\underline{x}^h(i,j)\underline{e}_1^h(i,j)$$

$$\underline{z}^v(i,j)=\underline{\theta}^{vT}(i,j)\underline{x}^v(i,j)-a^v\underline{x}^{vT}(i,j)\underline{x}^v(i,j)\underline{e}_1^v(i,j)$$

$$\underline{\theta}^h(i+1,j)-\underline{\theta}^h(i,j)=-\underline{e}_1^h(i,j)\underline{x}^T(i,j)$$

$$\underline{\theta}^v(i,j+1)-\underline{\theta}^v(i,j)=-\underline{e}_1^v(i,j)\underline{x}^T(i,j) \tag{35}$$

Defining a Lyapunov function candidate for the set of difference equations (34), (35) as

$$V_1(\underline{e}_1^h(i,j),\underline{\theta}(i,j)) = 2\underline{e}^{hT}(i,j)\underline{P}^h\underline{e}^{hT}(i,j) + \mathrm{trace}\,\underline{\theta}^{hT}(i,j)\theta(i,j) \tag{36}$$

then

$$\begin{aligned}\Delta V_1 &= V_1(i+1,j) - V_1(i,j)\\ &= 2\underline{e}^{hT}(i+1,j)\underline{P}^h\underline{e}^h(i,j) + \mathrm{trace}\,\underline{\theta}^{hT}(i+1,j)\underline{\theta}^h(i+1,j)\\ &\quad -2\underline{e}^{hT}(i,j)\underline{P}^h\underline{e}^h(i,j) - \mathrm{trace}\,\underline{\theta}^{hT}(i,j)\underline{\theta}^h(i,j)\\ &= 2\underline{e}^T(i,j)\underline{A}_1^T\underline{P}^h\underline{A}_1\underline{e}(i,j) + 4\underline{e}^T(i,j)\underline{A}_1^T\underline{P}^h\underline{z}^h + 2\underline{z}^{hT}\underline{P}^h\underline{z}^h\\ &\quad +\mathrm{trace}\,\underline{x}^h(i,j)\underline{e}_1^{hT}(i,j)\underline{e}_1^h(i,j)\underline{x}^{hT}(i,j) -\\ &\quad -2\mathrm{trace}\,\underline{\theta}^{hT}(i,j)\underline{e}_1^h(i,j)\underline{x}^{hT}(i,j) - 2\underline{e}^{hT}(i,j)\underline{P}^h\underline{e}^h(i,j)\end{aligned} \tag{37}$$

Similarly we can define

$$\Delta V_2(\underline{e}_2^h(i,j),\underline{\theta}(i,j)) = \underline{e}^{vT}(i,j)\underline{P}^v\underline{e}^v(i,j) + \mathrm{trace}\,\underline{\theta}^{vT}(i,j)\underline{\theta}^v(i,j) \tag{38}$$

Hence

$$\begin{aligned}\Delta V_2 &= V_2(i,j+1) - V_2(i,j)\\ &= \underline{e}^{vT}(i,j+1)\underline{P}^v\underline{e}^{vT}(i,j+1) + \mathrm{trace}\,\underline{\theta}^{vT}(i,j+1)\underline{\theta}^v(i,j+1) -\\ &\quad -\mathrm{trace}\,\underline{\theta}^{vT}(i,j)\underline{\theta}(i,j) - \underline{e}^{vT}(i,j)\underline{P}^v\underline{e}^v - \mathrm{trace}\,\underline{\theta}^{vT}(i,j)\underline{\theta}^v(i,j)\\ &= \underline{e}^T(i,j)\underline{A}_2^T\underline{P}^v\underline{A}_2\underline{e}(i,j) + 2\underline{e}^T(i,j)\underline{A}_2^T\underline{P}^v\underline{z}^v(i,j) + \underline{z}^{vT}(i,j)\underline{P}^v\underline{z}^v(i,j)\\ &\quad +\mathrm{trace}\,\underline{\theta}\underline{x}^v(i,j)\underline{e}_1^{vT}(i,j)\underline{e}_1^v(i,j)\underline{x}^{vT}(i,j)\\ &\quad -2\mathrm{trace}\,\underline{\theta}^{vT}(i,j)\underline{e}_1^v(i,j)\underline{x}^{vT}(i,j) - \underline{e}^{vT}(i,j)\underline{P}^v\underline{e}^v(i,j)\end{aligned} \tag{39}$$

Equation (37) can be written as

$$\begin{aligned}&2\underline{e}^T(i,j)\left[\underline{A}_1^T\underline{P}^h\underline{A}_1 - \begin{bmatrix}\underline{P}^h & 0\\ 0 & 0\end{bmatrix}\right]\underline{e}(i,j) + 2\underline{e}^T(i,j)\underline{A}_1^T P^h\underline{z}^h + \underline{z}^{hT}\underline{P}^h\underline{z}^h\\ &\quad +\mathrm{trace}\,\underline{x}^h(i,j)\underline{e}^{hT}(i,j)\underline{e}_1^h(i,j)\underline{x}^{hT}(i,j) -\\ &\qquad -2\mathrm{trace}\,\underline{\theta}^{hT}(i,j)\underline{e}_1^h(i,j)\underline{x}^h(i,j)\end{aligned} \tag{40}$$

Assuming that $\underline{P}^h = \underline{P}^{hT} > 0$ and also

$$\left[\underline{A}_1^T\underline{P}^h\underline{A}_1 - \begin{bmatrix}\underline{P}^h & 0\\ 0 & 0\end{bmatrix}\right] = -\underline{q}\ \underline{q}^T - \varepsilon\underline{L}$$

$$\underline{A}_1^T \underline{P}^h = \tfrac{1}{2}\underline{C}^{hT} + \underline{v}\underline{q}^T, \quad \underline{D}^h - \underline{P}^h = \underline{v}\ \underline{v}^T \tag{41}$$

for some vectors $\underline{q}$, $\underline{v}$ and a matrix $\underline{L}=\underline{L}^T>0$, equation (40), then becomes

$$2\underline{e}^T(i,j)(-\underline{q}\ \underline{q}^T-\varepsilon\underline{L})\underline{e}(i,j)+4\underline{e}^T(i,j)(\tfrac{1}{2}\underline{C}^{hT}+\underline{v}\ \underline{q}^T)\underline{z}^h$$
$$+2\underline{z}^{hT}(\underline{D}^{hT}-\underline{v}\ \underline{v}^T)\underline{z}^h$$
$$+\underline{e}^{hT}(i,j)\underline{e}_1^h(i,j)\underline{x}^{hT}(i,j)\underline{x}^h(i,j)-2\,\mathrm{trace}\,\underline{\theta}^{hT}(i,j)\underline{e}_1^h(i,j)\underline{x}^{hT}(i,j) \tag{42}$$

Simplifying equation (42) we obtain

$$-2\left[\underline{e}^T(i,j)\underline{q}-\underline{v}^T\underline{z}^h(i,j)\right]^2-2\varepsilon\underline{e}^T(i,j)\underline{L}\underline{e}(i,j)+2\underline{e}^{hT}(i,j)\underline{z}^h$$
$$-2\underline{e}_1^{hT}\theta^h(i,j)\underline{x}^h(i,j)+\underline{e}_1^{hT}(i,j)\underline{e}_1^h(i,j)\underline{x}^{hT}(i,j)\underline{x}^h(i,j) \tag{43}$$

or

$$-2\left[\underline{e}^T(i,j)\underline{q}-\underline{v}^T\underline{z}^h(i,j)\right]^2-2\varepsilon\underline{e}^T(i,j)\underline{L}\underline{e}(i,j)$$
$$+(1-2\alpha^h)\underline{e}_1^{hT}(i,j)\underline{e}_1^h(i,j)\underline{x}^{hT}(i,j)\underline{x}^h(i,j) \tag{44}$$

we observe that if $\alpha^h>\tfrac{1}{2}$, then $\Delta V_1<0$

From equation (36) and (37), it follows that

$\underline{e}(i,j)\to 0$, $\underline{e}^{hT}(i,j)\underline{e}_1^h(i,j)\underline{x}^{hT}(i,j)\underline{x}^h(i,j)\to 0$ and $\underline{e}_1^h(i,j)\to 0$ as

$i\to\infty$.

Since $\Delta\underline{\theta}^h(i,j)=-\alpha^h\underline{e}_1^h(i,j)\underline{x}^{hT}(i,j)$, it follows that $\Delta\underline{\theta}^h(i,j)\to 0$ as $i\to\infty$, and if $\underline{x}^h(i,j)$ remains uniformly bounded it follows from (35) that $\underline{\Phi}_{ii}^h(i,j)$ will tend to a constant value as desired.

7. ADAPTIVE OBSERVER

For the case when the output $\underline{y}(i,j)$ is only available and the matrix $\underline{C}$ is known, we can easily see that the model (15) can be put into the the form

$$\begin{bmatrix}\underline{x}^h(i+1,j)\\ \\ \underline{x}^v(i,j+1)\end{bmatrix}=\left[\begin{array}{c|c}\underline{A}_1+\underline{K}_{p1}\underline{C}_1 & \underline{A}_2+\underline{K}_{p1}\underline{C}_2\\ \hline \underline{A}_3+\underline{K}_{p2}\underline{C}_1 & \underline{A}_4+\underline{K}_{p2}\underline{C}_2\end{array}\right]\begin{bmatrix}\underline{x}^h(i,j)\\ \\ \underline{x}^v(i,j)\end{bmatrix}+\begin{bmatrix}\underline{B}_{p1}\\ \\ \underline{B}_{p2}\end{bmatrix}\underline{u}(i,j) \tag{45}$$

where $\underline{K}_{p1}, \underline{K}_{p2}, \underline{B}_{p1}$ and $\underline{B}_{p2}$ are matrices of appropriate dimension the parameters of which need be identified whereas the matrices $\underline{A}_1, \underline{A}_2, \underline{A}_3, \underline{A}_4, \underline{C}_1, \underline{C}_2$ are known.

The model of equation (45) can be written in the form

$$\begin{bmatrix} \underline{x}^h(i+1,j) \\ \underline{x}^v(i,j+1) \end{bmatrix} = \left[\begin{array}{c|c} \underline{A}_1+\underline{K}_{01}\underline{C}_1 & \underline{A}_2+\underline{K}_{01}\underline{C}_1 \\ \hline \underline{A}_3+\underline{K}_{02}\underline{C}_1 & \underline{A}_4+\underline{K}_{02}\underline{C}_2 \end{array}\right] \begin{bmatrix} \underline{x}^h(i,j) \\ \underline{x}^v(i,j) \end{bmatrix}$$

$$+ \left[\begin{bmatrix} \underline{K}_{p1} \\ \underline{K}_{p2} \end{bmatrix} - \begin{bmatrix} \underline{K}_{01} \\ \underline{K}_{02} \end{bmatrix}\right] \underline{y}(i,j) + \begin{bmatrix} \underline{B}_{p1} \\ \underline{B}_{p2} \end{bmatrix} \underline{u}(i,j) \tag{46}$$

This model is in the form that points out the form of the adaptive model which can have the following form

$$\begin{bmatrix} \hat{\underline{x}}^h(i+1,j) \\ \hat{\underline{x}}^v(i,j+1) \end{bmatrix} =$$

$$\left[\begin{array}{c|c} \underline{A}_1+\underline{K}_{01}\underline{C}_1 & \underline{A}_2+\underline{K}_{01}\underline{C}_2 \\ \hline \underline{A}_3+\underline{K}_{02}\underline{C}_2 & \underline{A}_4+\underline{K}_{02}\underline{C}_2 \end{array}\right] \begin{bmatrix} \hat{\underline{x}}^h(i,j) \\ \hat{\underline{x}}^v(i,j) \end{bmatrix} + \begin{bmatrix} \hat{\underline{K}}_1(i,j) \\ \hat{\underline{K}}_2(i,j) \end{bmatrix} - \begin{bmatrix} \underline{K}_{01} \\ \underline{K}_{02} \end{bmatrix} \underline{y}(i,j) + \begin{bmatrix} \underline{B}_1(i,j) \\ \underline{B}_2(i,j) \end{bmatrix} \underline{u}(i,j) + \underline{w}(i,j)$$

$$\hat{\underline{y}}(i,j) = [\underline{C}_1, \underline{C}_2] \begin{bmatrix} \hat{\underline{x}}^h(i,j) \\ \hat{\underline{x}}^v(i,j) \end{bmatrix} \tag{47}$$

Substracting (47) from (46) yields

$$\begin{bmatrix} \underline{e}^h(i+1,j) \\ \underline{e}^v(i,j+1) \end{bmatrix} = \left[\begin{array}{c|c} \underline{A}_1+\underline{K}_{01}\underline{C}_1 & \underline{A}_2+\underline{K}_{01}\underline{C}_2 \\ \hline \underline{A}_3+\underline{K}_{02}\underline{C}_1 & \underline{A}_4+\underline{K}_{02}\underline{C}_2 \end{array}\right] \begin{bmatrix} \underline{e}^h(i,j) \\ \underline{e}^v(i,j) \end{bmatrix} + \left[\begin{bmatrix} \hat{\underline{K}}_1(i,j) \\ \hat{\underline{K}}_2(i,j) \end{bmatrix} - \begin{bmatrix} \underline{K}_{p1} \\ \underline{K}_{p2} \end{bmatrix}\right] \underline{y}(i,j) +$$

$$\left[\begin{bmatrix} \hat{\underline{B}}_1(i,j) \\ \hat{\underline{B}}_2(i,j) \end{bmatrix} - \begin{bmatrix} \underline{B}_{p1} \\ \underline{B}_{p2} \end{bmatrix}\right] \underline{u}(i,j) + \underline{w}(i,j) \tag{48}$$

and

$$\hat{\underline{y}}(i,j) - \underline{y}(i,j) = [\underline{C}_1 \;\; \underline{C}_2] \begin{bmatrix} \underline{e}^h(i,j) \\ \underline{e}^v(i,j) \end{bmatrix} \tag{49}$$

where $\underline{w}(i,j)$ are auxiliary signals which are required to stabilize the adaptive observer. In order to use the procedure described in the previous section, we can use the proposition proved in (38) by which equation (46) can be converted to a form similar to that described in section 6. After that the same procedure applies.

8. DESCRIPTION OF 3-D STATE OBSERVER

In this section we give the definition and equations of 3-D state observers. Consider a 3-D system in the Roesser state-space form (1)-(2). Let the vector $\underline{\omega}(i,j,k)\varepsilon R^q$ be defined as a linear function of the system state vector, namely

$$\underline{\omega}(i,j,k)=\left[\underline{K}_1,\underline{K}_2,\underline{K}_3\right]\underline{x}(i,j,k) \tag{50}$$

where $\underline{K}_1,\underline{K}_2,\underline{K}_3$ are constant matrices of dimensions qxn_1, qxn_2 and qxn_3 respectively.

Finally, consider the following 3-D systems which provides some estimate (approximation) $\hat{\omega}(i,j,k)$ of $\omega(i,j,k)$ [1-4]:

$$\underline{z}'(i,j,k)=\underline{F}\ \underline{z}(i,j,k)+\underline{G}\ \underline{u}(i,j,k)+\underline{H}\ \underline{y}(i,j,k) \tag{51a}$$

$$\underline{\hat{\omega}}(i,j,k)=\left[\underline{N}_1,\underline{N}_2,\underline{N}_3\right]z(i,j,k)+\underline{M}\ \underline{y}(i,j,k) \tag{51b}$$

where $\underline{F}$ has the same form as $\underline{A}$, for $i,j,k>0$ where $\underline{z}^h\varepsilon R^{\nu_1}$, $\underline{z}^v\varepsilon R^{\nu_2}$, $\underline{z}^d\varepsilon R^{\nu_3}$, $\hat{\omega}\varepsilon R^q$. Definition the 3-D system (51a,b) is said to be a 3-D functional observer of the vector $\underline{\omega}(i,j,k)$ if

$$\lim_{i,j,k\to\infty}\underline{\hat{\omega}}(i,j,k)=\lim_{i,j,k\to\infty}\underline{\omega}(i,j,k)$$

independently of the known input vector $\underline{u}(i,j,k)$ and of the (known or unknown) initial conditions.

The observer error vector $\underline{e}(i,j,k)$ is defined by

$$\underline{e}'(i,j,k)=\underline{z}(i,j,k)-\underline{T}\ \underline{x}(i,j,k) \tag{52}$$

where $\underline{T}=\text{diag}\left[\underline{T}_1,\underline{T}_5,\underline{T}_9\right]$ for the time being the constant matrices are somehow arbitrary with dimensions $\nu_1 x n_1$, $\nu_2 x n_2$ and $\nu_3 x n_3$ respectively, which will be determined in the sequel. From (1), (51a) and (52) it follows that

$$\underline{e}'(i,j,k)=\underline{F}\ \underline{e}(i,j,k)+\underline{\Phi}\ \underline{x}(i,j,k)+\underline{\Lambda}\ \underline{u}(i,j,k) \tag{53}$$

where

$$\underline{\Phi}_1=\underline{F}_1\underline{T}_1+\underline{H}_1\underline{C}_1-\underline{T}_1\underline{A}_1,\quad \underline{\Phi}_4=\underline{F}_4\underline{T}_1+\underline{H}_2\underline{C}_1-\underline{T}_5\underline{A}_4,\quad \underline{\Phi}_7=\underline{F}_7\underline{T}_1+\underline{H}_3\underline{C}_1-\underline{T}_9\underline{A}_7$$
$$\underline{\Phi}_2=\underline{F}_2\underline{T}_5+\underline{H}_1\underline{C}_2-\underline{T}_1\underline{A}_2,\quad \underline{\Phi}_5=\underline{F}_5\underline{T}_5+\underline{H}_2\underline{C}_2-\underline{T}_5\underline{A}_5,\quad \underline{\Phi}_8=\underline{F}_8\underline{T}_5+\underline{H}_3\underline{C}_2-\underline{T}_9\underline{A}_8$$
$$\underline{\Phi}_3=\underline{F}_3\underline{T}_9+\underline{H}_1\underline{C}_3-\underline{T}_1\underline{A}_3,\quad \underline{\Phi}_6=\underline{F}_6\underline{T}_9+\underline{H}_2\underline{C}_3-\underline{T}_5\underline{A}_6,\quad \underline{\Phi}_9=\underline{F}_9\underline{T}_9+\underline{H}_3\underline{C}_3-\underline{T}_9\underline{A}_9$$
$$\underline{\Lambda}_1=\underline{G}_1-\underline{T}_1\underline{B}_1,\ \underline{\Lambda}_2=\underline{G}_2-\underline{T}_5\underline{B}_2$$
$$\underline{\Lambda}_3=\underline{G}_3-\underline{T}_9\underline{B}_3 \tag{54}$$

From (53) it follows that actually the convergence of the error to zero depends on the matrix $\underline{F}$ (with submatrices F_i, $i=1,2,...,9$) and that (51,b) in an observer if the error converges to zero as $i,j,k\to\infty$.
The following theorem is true:

THEOREM 1: The 3-D system (51a,b) is a functional observer for the system (1) if the following conditions hold

$$\begin{array}{lll}
\underline{F}_1\underline{T}_1=\underline{T}_1\underline{A}_1-\underline{H}_1\underline{C}_1 & \underline{F}_6\underline{T}_9=\underline{T}_5\underline{A}_6-\underline{H}_2\underline{C}_3 & \underline{G}_1=\underline{T}_1\underline{B}_1,\ \underline{G}_2=\underline{T}_5\underline{B}_2,\ \underline{G}_3=\underline{T}_9\underline{B}_3 \\
\underline{F}_2\underline{T}_5=\underline{T}_1\underline{A}_2-\underline{H}_1\underline{C}_2 & \underline{F}_7\underline{T}_1=\underline{T}_9\underline{A}_7-\underline{H}_3\underline{C}_1 & \underline{K}_1=\underline{N}_1\underline{T}_1+\underline{M}\ \underline{C}_1 \\
\underline{F}_3\underline{T}_9=\underline{T}_1\underline{A}_3-\underline{H}_1\underline{C}_3 & \underline{F}_8\underline{T}_5=\underline{T}_9\underline{A}_8-\underline{H}_3\underline{C}_2 & \underline{K}_2=\underline{N}_2\underline{T}_5+\underline{M}\ \underline{C}_2 \\
\underline{F}_4\underline{T}_1=\underline{T}_5\underline{A}_4-\underline{H}_2\underline{C}_1 & \underline{F}_9\underline{T}_9=\underline{T}_9\underline{A}_9-\underline{H}_3\underline{C}_3 & \underline{K}_3=\underline{N}_3\underline{T}_9+\underline{M}\ \underline{C}_3 \\
\underline{F}_5\underline{T}_5=\underline{T}_5\underline{A}_5-\underline{H}_2\underline{C}_2 & &
\end{array} \tag{55}$$

If in addition the initial conditions are unknown the following conditions must also hold, where $\underline{F}^{ijk}$ is the fundamental (transition) matrix of the observer:

$$\lim_{i\to\infty}\underline{F}^{ijk}=0;\ \lim_{j\to\infty}\underline{F}^{ijk}=0;\ \lim_{k\to\infty}\underline{F}^{ijk}=0 \tag{56}$$

PROOF: From (50) and (51b) we find

$$\hat{\underline{\omega}}(i,j,k)-\underline{\omega}(i,j,k)=\left[\underline{N}_1,\underline{N}_2,\underline{N}_3\right]+\left[\underline{M}\ \underline{C}_1-\underline{K}_1,\underline{M}\ \underline{C}_2-\underline{K}_2,\underline{M}\ \underline{C}_3-\underline{K}_3\right]\underline{x} \tag{57}$$

and so if the conditions for $\underline{K}_1,\underline{K}_2,\underline{K}_3$ in (55) hold the error $\tilde{\underline{\omega}}=\hat{\underline{\omega}}-\underline{\omega}$ is written as

$$\tilde{\underline{\omega}}(i,j,k)=\left[\underline{N}_1,\underline{N}_2,\underline{N}_3\right]\underline{e}(i,j,k) \tag{58}$$

This shows that when the error $\underline{e}(i,j,k)$ is zero the vectors $\hat{\underline{\omega}}(i,j,k)$ and $\underline{\omega}(i,j,k)$ are identical. Now using the conditions (55) in (53) we find

$$\underline{e}'(i,j,k)=\underline{F}\ \underline{e}(i,j,k)$$

which has the solution [19,21,23]:

$$\underline{e}(i,j,k) = \sum_{m=0}^{j}\sum_{n=0}^{k} \underline{F}^{i,j-m,k-n}\begin{bmatrix}\underline{e}^{h}(0,m,n)\\0\\0\end{bmatrix} + \sum_{\ell=0}^{i}\sum_{n=0}^{k} \underline{F}^{i-\ell,j,k-n}\begin{bmatrix}0\\\underline{e}^{v}(\ell,0,n)\\0\end{bmatrix} + \sum_{\ell=0}^{i}\sum_{m=0}^{j} \underline{F}^{i-\ell,j-m,k}\begin{bmatrix}0\\0\\\underline{e}^{d}(\ell,m,0)\end{bmatrix} \tag{59}$$

From (59) it is obvious that when conditions (56) hold we have

$$\lim_{i,j,k\to\infty} e(i,j,k)=0$$

independently of the input u(i,j,k) and of the initial conditions of the error.

If the initial conditions of the system (1) are known, then we (52) to determine the initial conditions of the observer such that the initial value of the error is zero, which by (59) implies that it will remain zero for all i,j,k.

9. STATE OBSERVER DESIGN FOR TRIANGULAR 3-D SYSTEMS

In the design (construction) of an observer, which is a dynamic system, care must be taken to ensure its stability, by choosing the matrices $\underline{T}_1, \underline{T}_2, \underline{T}_3$ and all other parameters of the observer. Here we shall study the case of triangular systems ($\underline{A}$ is of triangular form), the transfer function of which is the product of three 1-D transfer functions. Therefore the stability requirement can be met relatively easily. This problem was solved for 2-D systems by Hinamoto et.al. [4]: For simplicity we consider the case where

$$\underline{K}_1 = \begin{bmatrix}\underline{I}_{n_1}\\0\\0\end{bmatrix}, \underline{K}_2 = \begin{bmatrix}0\\\underline{I}_{n_2}\\0\end{bmatrix}, \underline{K}_3 = \begin{bmatrix}0\\0\\\underline{I}_{n_3}\end{bmatrix} \tag{60}$$

In this case we see from (50) that the vector $\underline{\omega}(i,j,k)$ coincides with the state vector $\underline{x}(i,j,k)$ of the system.

In order to facilitate the determination of the eigenvalues of the matrix $\underline{F}$ when the system initial conditions are unknown, we impose the following conditions

$$\begin{gathered} \underline{F}_2=0,\ \underline{F}_3=0,\ \underline{F}_6=0 \\ (\underline{F}_5,\underline{C}_2) \text{ and } (\underline{F}_9,\underline{C}_3) \text{ observable} \end{gathered} \tag{61}$$

From the two inputs $\underline{u}$ and $\underline{y}$ of the observer (51a) information about the system state give only the independent components of $\underline{y}(i,j,k)$. Hence right from the beginning we reject the dependent components of $\underline{y}(i,j,k)$. This can be done by introducing the similarity transformation $\hat{\underline{x}}=\underline{V}\ \underline{x}$ or $\underline{x}=\underline{V}^{-1}\hat{\underline{x}}$, where $\underline{V}=\mathrm{diag}\left[\underline{V}_1,\underline{V}_5,\underline{V}_9\right]$, and $\underline{V}_5,\underline{V}_9$ are chosen such that

$$\underline{C}_2\underline{V}_5^{-1}=\left[\begin{array}{c|c} \underline{I}_{\nu_2} & 0 \\ \hline \tilde{\underline{C}}_2 & 0 \end{array}\right]\}\rho-\nu_2,\quad \underline{C}_3\underline{V}_9^{-1}=\left[\begin{array}{c|c} \underline{I}_{\nu_3} & 0 \\ \hline \tilde{\underline{C}}_3 & 0 \end{array}\right]\}\rho-\nu_3 \tag{62}$$

where $\rho=\mathrm{rank}\left[\underline{C}_1,\underline{C}_2,\underline{C}_3\right]$ rank $\underline{C}_2=\nu_2<\rho$, rank $\underline{C}_3=\nu_3<\rho$.

Now, from (55) and the second of (62) (again denoting for simplicity the matrix $\underline{C}_3\underline{V}_9^{-1}$ by $\underline{C}_3$) we find

$$\begin{aligned} \underline{K}_3 &= \underline{N}_3\underline{T}_9+\underline{M}\left[\begin{array}{c|c} \underline{I}_{\nu_3} & 0 \\ \hline \tilde{\underline{C}}_3 & 0 \end{array}\right] \\ &= \underline{N}_3\underline{T}_9+\tilde{\underline{M}}\left[\underline{I}_{\nu_3}\ \vdots\ 0\right],\ \tilde{\underline{M}}=\left[\begin{array}{c} \underline{I}_{\nu_3} \\ \hline \tilde{\underline{C}}_3 \end{array}\right] \end{aligned} \tag{63}$$

Suppose that $\underline{T}_9$ is selected in the form

$$\underline{T}_9=\left[-\underline{S}_9\ \vdots\ \underline{I}_{n_3-\nu_3}\right] \tag{64}$$

where $\underline{S}_9$ is a constant $(n_3-\nu_3)\mathrm{x}\nu_3$ matrix, with elements chosen so as all eigenvalues of $\underline{F}_9$ are zero [4].
Then using (64),(63) gives {viz.(60)}:

$$\underline{K}_3=\left[\begin{array}{c} 0 \\ \hline \underline{I}_{n_3} \end{array}\right]=\left[\tilde{\underline{M}}\ \vdots\ \underline{N}_3\right]\left[\begin{array}{c|c} \underline{I}_{\nu_3} & 0 \\ \hline -\underline{S}_9 & \underline{I}_{n_3-\nu_3} \end{array}\right]$$

whence

$$\underline{\tilde{M}}^T = \left[0 \mid \underline{I}_{\nu_3} \mid \underline{S}_9^T\right], \quad \underline{N}_3^T = \left[0 \mid 0 \mid \underline{I}_{n_3-\nu_3}\right] \tag{65}$$

or using (63):

$$\underline{\tilde{M}} = \underline{M}\begin{bmatrix}\underline{I}_{\nu_3}\\ ---\\ \underline{\tilde{C}}_3\end{bmatrix} = \begin{bmatrix}\underline{M}_1\\ ---\\ \underline{M}_2\\ ---\\ \underline{M}_3\end{bmatrix}\begin{bmatrix}\underline{I}_{\nu_3}\\ ---\\ \underline{\tilde{C}}_3\end{bmatrix}$$

whence

$$\underline{M}_1 = 0, \underline{M}_2 = I_{\nu_3}\begin{bmatrix}\underline{I}_{\nu_3}\\ ---\\ \underline{\tilde{C}}_3\end{bmatrix}^+, \underline{M}_3 = \underline{S}_9\begin{bmatrix}\underline{I}_{\nu_3}\\ ---\\ \underline{\tilde{C}}_3\end{bmatrix}^+ \tag{66a}$$

with

$$\begin{bmatrix}\underline{I}_{\nu_3}\\ ---\\ \underline{C}_3\end{bmatrix}^+ = \left[\underline{I}_{\nu_3} + \underline{\tilde{C}}_3^T\ \underline{\tilde{C}}_3\right]^{-1}\left[\underline{I}_{\nu_3} \mid \underline{\tilde{C}}_3^T\right] \tag{66b}$$

being the generalized matrix inverse [22].

Similarly from the condition

$\underline{K}_2 = \underline{N}_2\underline{T}_5 + \underline{M}\ \underline{C}_2$ {viz (55)} and the choice

$$\underline{T}_5 = \left[-\underline{S}_5 \mid \underline{I}_{n_2-\nu_2}\right] \tag{67}$$

where $\underline{S}_5$ is selected such that all eigenvalued of $\underline{F}_5$ are zero, we obtain $\underline{K}_2 = \underline{N}_2\left[-\underline{S}_5 \mid I_{n_2-\nu_2}\right] + \underline{M}\ \underline{C}_2$, whence

$$\underline{N}_2 = (\underline{K}_2 - \underline{M}\ \underline{C}_2)\underline{R}^+ + \underline{Q}(I_p - \underline{R}\ \underline{R}^+) \tag{68}$$

with $\underline{R} = \left[-\underline{S}_5 \mid I_{n_2-\nu_2}\right]$. Equation (68) is used to compute $\underline{N}_2$ on the basis write $\underline{K}_1 = \underline{N}_1\underline{T}_1 + \underline{M}\ \underline{C}_1$ {viz (55)} as

$$\begin{bmatrix}\underline{I}_{n_1}\\ 0\\ 0\end{bmatrix} = \begin{bmatrix}\underline{N}_{11}\\ \underline{N}_{21}\\ \underline{N}_{31}\end{bmatrix}\underline{T}_1 + \begin{bmatrix}\underline{M}_1\\ \underline{M}_2\\ \underline{M}_3\end{bmatrix}\underline{C}_1$$

and obtain (since $\underline{M}_1 = 0$):

$$\underline{T}_1=\underline{I}_{n_1};\underline{N}_{11}=\underline{I}_{n_1};\underline{N}_{21}=-\underline{M}_2\underline{C}_1;\underline{N}_{31}=-\underline{M}_3\underline{C}_1 \tag{69}$$

Now to determine $\underline{F}$ we use (55), taking into account that $\underline{F}_2=0$, $\underline{F}_3$ and $\underline{F}_6=0$ {viz (61)}, and obtain

$$\underline{T}_1\underline{A}_2-\underline{H}_1\underline{C}_2=0$$

$$\underline{T}_1\underline{A}_3-\underline{H}_1\underline{C}_3=0;\ \underline{T}_5\underline{A}_6-\underline{H}_2\underline{C}_2=0$$

Then using (67)

$$[\underline{A}_2\,|\,\underline{A}_3]=\underline{H}_1[\underline{C}_2\,|\,\underline{C}_3];\underline{T}_5\underline{A}_6-\underline{H}_2\underline{C}_2=0$$

which have the solution ($\underline{Q}_1,\underline{Q}_2$ are arbitrary of appropriate dimensions):

$$\underline{H}_1=[\underline{A}_2\,|\,\underline{A}_3][\underline{C}_2\,|\,\underline{C}_3]^+ +\underline{Q}_1\{I_p-[\underline{C}_2\,|\,\underline{C}_3][\underline{C}_2\,|\,\underline{C}_3]^T\} \tag{70a}$$

$$\underline{H}_2=\underline{T}_5\underline{A}_6\underline{C}_3^+ +\underline{Q}_2(I_p-\underline{C}_2\underline{C}_3^+) \tag{70b}$$

under the condition

range$[\underline{A}_2\,|\,\underline{A}_3]^T\subseteq$ range$[\underline{C}_2\,|\,\underline{C}_3]^T$, range$[\underline{T}_5\ \underline{A}_6]^T\subseteq$ range $\underline{C}_3^T$.

To determine $\underline{F}_1$ we start from $\underline{F}_1=\underline{A}_1-\underline{H}_1\underline{C}_1$ {viz (55)} and choose $\underline{Q}_1$ such that $\underline{F}_1$ has eigenvalues inside the unit circle. This is possible if the pair

$$(\underline{A}_1-[\underline{A}_2\,|\,\underline{A}_3][\underline{C}_2\,|\,\underline{C}_3]^+\underline{C}_1,\{\underline{I}_p-[\underline{C}_2\,|\,\underline{C}_3][\underline{C}_2\,|\,\underline{C}_3]^T\ \underline{C}_1\})$$

is detectable [3]. Then

$$\underline{F}_1=\underline{A}_1-[\underline{A}_2\,|\,\underline{A}_3][\underline{C}_2\,|\,\underline{C}_3]^+\underline{C}_1+\underline{Q}_1\{\underline{I}_p-[\underline{C}_2\,|\,\underline{C}_3][\underline{C}_2\,|\,\underline{C}_3]^T\}\underline{C}_1 \tag{71}$$

To determine $\underline{F}_9$ we start from $\underline{F}_9\underline{T}_9=\underline{T}_9\underline{A}_9-\underline{H}_3\underline{C}_3$ which, by (62) and (64), is written as

$$\underline{F}_9[-\underline{S}_9\,|\,\underline{I}_{n_3-v_3}]=[-\underline{S}_9\,|\,\underline{I}_{n_3-v_3}]\underline{A}=-\underline{\tilde{H}}_3[\underline{I}_{v_3}\,|\,0],\quad \underline{\tilde{H}}_3=\underline{H}_3\begin{bmatrix}\underline{I}_{v_3}\\ \hline \underline{\tilde{C}}_3\end{bmatrix}$$

which gives

$$[\underline{\tilde{H}}_3\,|\,\underline{F}_9]=[-\underline{S}_9\,|\,\underline{I}_{n_3-v_3}]\begin{bmatrix}\underline{A}_{91}+\underline{S}_9 & \underline{A}_{92} & \underline{A}_{92}\\ \hline \underline{A}_{93}+\underline{S}_9 & \underline{A}_{94} & \underline{A}_{94}\end{bmatrix}$$

whence

$$\underline{F}_9 = -\underline{S}_9\underline{A}_{92} + \underline{A}_{94}, \quad \tilde{\underline{H}}_3 = -\underline{S}_9\underline{A}_{91} + \underline{A}_{93} + \underline{S}_9\underline{F}_9 \tag{72a}$$

$$\text{or } \underline{H}_3 = \left[-\underline{S}_9\underline{A}_{91} + \underline{A}_{93} + \underline{S}_9\underline{F}_9\right] \left[\begin{array}{c} \underline{I}\,\underline{V}_3 \\ \hline \tilde{\underline{C}}_3 \end{array}\right]^+ \tag{72b}$$

In order for $\underline{F}_9$ to be stable the pair $(\underline{A}_{94}, \underline{A}_{92})$ must be detectable or equivalently $(\underline{A}_9, \underline{C}_3)$ observable.

In the same way one can determine $\underline{F}_4, \underline{F}_5, \underline{F}_7$ and $\underline{F}_8$ starting from the corresponding conditions (55). The computation of $\underline{G}_1, \underline{G}_2$ and $\underline{G}_3$ is straight forward {see (55)}.

10. CONCLUDING REMARKS

This chapter was concerned with the study of the basic structural aspects (observability, controllability, transformation to canonical form) and state-observer design problem for 2-D Dimensional and 3-Dimensional systems.

It was seen that an adaptive algorithm can be defined for adjusting the parameters of a model for the identification of 2-D digital multivariable systems. A model reference approach was used and the emphasis was on the stability of the overall system. The algorithm was first developed for the case of systems with all the states accessible and then was applied to the case when the state vector is known only through the output. In this case an adaptive observer was developed for the determination of the unknown parameters.

The 3-D observer design procedure developed gives a set of values for the matrices $\underline{F}$, $\underline{G}$, $\underline{H}$, $\underline{N}$ and $\underline{M}$ which fully dtermine our 3-D observer (51a,b). Of course the solution is not unique. The matrices which play a primary role in the design are $\underline{S}_5$, $\underline{S}_9$ and $\underline{Q}_1$. The first two matrices are chosen such that $\underline{F}_5$ and $\underline{F}_9$ respectively, have all their eigenvalues inside the unit circle. In this way the stability of our observer is assured. To establish the stability of the designed observer one employs the properties of $\underline{F}$ ($\underline{F}_2=0$, $\underline{F}_3=0$, $\underline{F}_4=0$; $\underline{F}_5$ and $\underline{F}_9$ have all zero eigenvalues, $\underline{F}_1$ stable) in the error response expression (59). The full development will be given elsewhere.

The 3-D observer can be utilized in a state feedback controller which stabilizes the original system at hand in a way similar to the 2-D case.

Work is in progress to see the effect of using the 3-D observer for implementing the state feedback controllers provided in [18-21].

REFERENCES

1. Luenberger,D., 1974: "Observing the state of a Linear System" IEEE Trans. Milit. Electron, Vol. 8, pp. 74-80.
2. Luenberger,D. 1971: "An Introduction to Observers" IEEE Trans. Autom. Control, Vol. 16, pp. 596-602,
3. Kwakernaak,H. and Sivan R.,1972: "Linear Optimal Control systems", Wiley.
4. Hinamoto,T. et.al.,1982: "Stabilization of 2D Filters using 2D Observers", Int. J. Systems Sci., Vol.13, 177-191.
5. Stavroulakis,P. and Paraskevopoulos P.N.,1981: "Reduced Order Feedback Law Implementation for 2-D Digital Systems", Int.J. Systems Science, Vol.12, 525-537.
6. Stavroulakis, P.,1981: "Low sensitivity feedback law implementation for 2-D Digital Systems", J. Franklin Inst., Vol.312, 217-229.
7. Stavroulakis,P. and Paraskevopoulos, P., 1982: "Low Sensitivity Observer Design for 2-Dimensional Digital Systems", Proc. IEE:Pt.D., Vol.129, 193-200.
8. Stavroulakis,P.and Tzafestas,S., 1983: "State Reconstruction in Low-Sensitivity Design of 3-Dimensional Systems", Proc. IEE:Pt.D.,Vol.130, 333-340.
9. Carrol,R.L. and Lindorff, D.P.,1973: "An Adaptive Observer for Single-Input Single-Output Linear System", IEEE Transactions on Auto Control, Vol.AC-18, No.5.
10. Luders, G. and Narendra, K.S., 1973:"An Adaptive Observer and Identifier for a Linear System", IEEE Trans. on Auto Control, AC-18, No.5.
11. Kudva, P. and Narendra, K.S.,1974: "An Identification Procedure for Discrete Multivariable Systems", IEEE Transactions on Auto Control, AC-19, No.5.
12. Narendra, K.S. and Valavani, L.S., 1976:"Stable Adaptive Observers and Controllers", Proceedings of IEEE, Vol.64, No.8.
13. Kreisselmeier, G., 1980:"Adaptive Control via Adaptive Observation and Asymptotic Feedback Matric Synthesis", IEEE Trans. on Auto Control, AC-25, No.4.
14. Kreisselmeier, G., 1982: "On Adaptive State Regulation", IEEE Trans, on Auto Control, AC-27, No.1.
15. Lin, H-Y and Narendra, K.S., 1980: "A New Error Model for Adaptive Systems", IEEE Transactions on Automatic Control, Vol. AC-25, No.3.
16. Givone, D.P. and Roesser, R.P., 1972:"Multidimensional Linear Iterative Circuits:General Properties", IEEE Trans., Vol.C-21, No.10.

17. Hinamoto, T., 1980:"Realizations of a State Space Model from Two-Dimensional Input-Output Map", IEEE Trans.,Vol. CS-27, 36-44.
18. Tzafestas, S., and Pimenides, T., 1982:"Feedback Characteristic Polynomial Controller Design of 3-D Systems in State Space", J. Franklin Inst., Vol. 314-169-189.
19. Pimenides, T. and Tzafestas, S., 1982:"Feedback Decoupling Controller Design of 3-D Filters in State Space", Math. Comp.Simul., Vol.24, 341-352.
20. Tzafestas, S. and Pimenides, T., 1982:"Exact Model Matching of 3-D Systems using State and Output Feedback", Int.J. Systems Sci., Vol.13, 1171-1187.
21. Tzafestas, S. and Pimenides, T., 1983: "Transfer Function Computation and Factorization of 3-D Systems in State Space", Proc. IEE:Pt.D., Vol.130, 231-242.
22. Rao, C., and Mitra, S., 1971:"Generalized Inverse of Matrices and its Applications", Wiley.
23. T.G.Pimenides:"Analysis and Design of 3-D Systems in State Space", Doctoral Thesis, Electr.Eng.Dept., Univ. of Patras, 1984.

CHAPTER 25

EIGENVALUE-GENERALIZED EIGENVECTOR ASSIGNMENT USING PID CONTROLLER

A. I. A. Salama
Research Scientist
Dept. of Energy, Mines and Resources, CANADA
P. O. Box 3294, Sherwood Park,
Alberta, CANADA, T8A 2A6

ABSTRACT. A time-domain design technique for eigenvalue-generalized eigenvector assignment using a proportional-plus-integral-plus-derivative controller is presented. The proposed design approach utilizes a null space formulation of the eigenstructure problem. The design techniques for continuous and discrete multivariable systems are investigated. In the continuous case the plant output is assumed to be available for measurement, while in the discrete case the plant state is assumed to be available for measurement. The resulting controller assigns a prescribed set of self-conjugate eigenvalues and a corresponding set of admissible self-conjugate generalized eigenvectors to the closed-loop system matrix. The proposed technique has some advantages over known techniques.

PART. 1 - CONTINUOUS-TIME SYSTEM

1. INTRODUCTION

Recently there has been much interest in the problem of designing proportional-plus-integral-plus-derivative (PID) controllers for linear continuous multivariable systems. This type of controller utilizes two extra actions (i.e. integral of error and derivative of output) compared to the proportional output feedback controller, to achieve acceptable performance in regard to static accuracy, disturbance rejection and transient response

Most of the PID controller design techniques assume both system state and output are available for measurement [1-5] or system output only is available for measurement [6,7]. The latter approach is superior compared to the former in that it is more practical since in many situations some of the system states are not available for measurement. The design techniques proposed assume the system under control (controllable) in state space formulation and subsequently determine the PID controller feedback gain matrices either in the time-domain [1-6] or in the frequency-domain [7]. The time-domain approach has some advantages

S. G. Tzafestas (ed.), Multivariable Control, 479–500.

over the frequency-domain approach in that it avoids the computation of the system transfer function matrix from its state variable formulation. This computation can be highly inaccurate [8]. Some authors [3,7,9] assume structural constraint on the feedback gain matrices of the PID controller in their design techniques, namely, that of unit rank. consequently the degrees of freedom in the controller parameters are greatly reduced. The Lypanuv technique is utilized in designing a state feedback PID controller [4,5]. A procedure for designing a PID controller for a multivariable system to track a polynomial type input is reported in [9].

The main feature of most of the PID controller design techniques reported [1-7,9], is that they are mainly addressed to the eigenvalue-assignment problem. However, the problem of eigenvalue-eigenvector assignment has practical significance (specifically the modal control), and consequently it will be the subject of the present paper.

In this paper, a time-domain approach for designing PID controller for linear continuous time-invariant system is adopted. The system output is assumed to be available for measurement. The eigenstructure formulation is utilized to determine the controller feedback gain matrices. The resulting matrices assign a prescribed set of self-conjugate eigenvalues and a corresponding set of admissible self-conjugate generalized eigenvectors to the closed-loop system matrix. The proposed PID controller design technique has the following advantages: i) it involves only output feedback thus covering most practical situations, ii) it is a time-domain approach, thus there is no computation of the transfer function matrix of the closed-loop system from its state variable formulation, iii) it imposes no structural constraint on the output feedback matrices, thus allowing greater freedom in the design, and iv) it assigns prescribed eigenvalues and admissible generalized eigenvectors to the closed-loop system matrix which are required for modal control. Thus the proposed technique has some advantages over those techniques reported elsewhere [1-7,9]. The problem investigated in this paper can be considered an extension to the problem of eigenvalue-generalized eigenvector assignment using proportional-plus-integral controller reported in [10].

2. PROBLEM STATEMENT

Consider the linear continuous time-invariant multivariable system described in state space form as follows

$$\begin{aligned} \dot{X} &= AX + BU \\ Y &= CX \end{aligned} \tag{1}$$

where X is the nx1 state vector, U is the mx1 input vector, Y is the qx1 output vector and A,B,C are constant matrices of appropriate dimensions. The output feedback law is assumed to be of the form (see Figure 1)

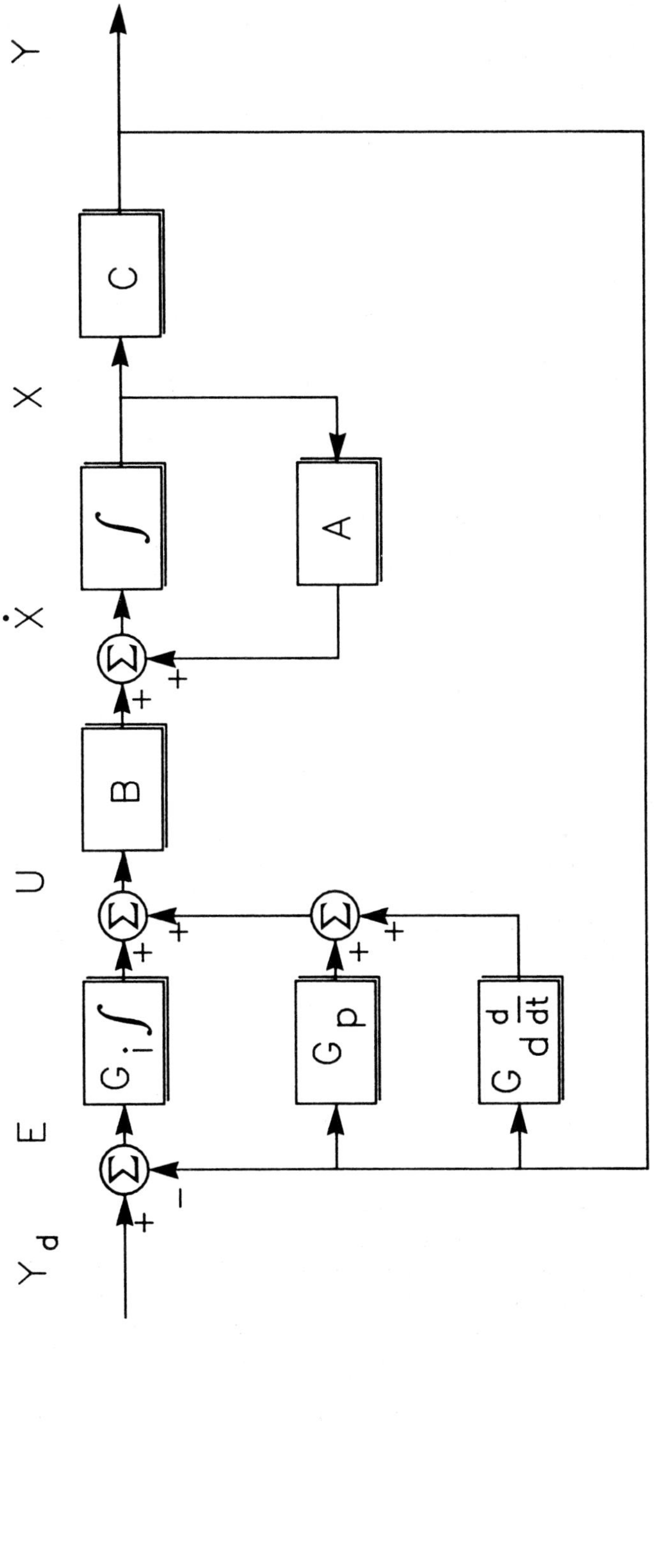

Figure 1 - A continuous multivariable system with PID controller

$$U = G_p Y + G_d \dot{Y} + G_i \int_0^t (Y - Y_d)\,dt, \tag{2}$$

where G_p, G_i and G_d are constant mxq proportional, integral and derivative output feedback gain matrices, respectively. Substitution of the output equation Y=CX of system (1) in (2) yields

$$\begin{aligned} U &= G_p CX + G_d C\dot{X} + G_i C\int_0^t X\,dt + \Theta \\ \Theta &\triangleq -G_i \int_0^t Y_d\,dt . \end{aligned} \tag{3}$$

Substitution of (3) in (1) yields the closed-loop system

$$\dot{X} = (I - BG_d C)^{-1}((A + BG_p C)X + BG_i C\int_0^t X\,dt + B\Theta), \tag{4}$$

providing that $(I-BG_dC)$ is invertible. Now let us introduce the variable Z as

$$Z = \int_0^t Y\,dt, \tag{5}$$

then (4) and (5) can be written in state variable form as

$$\begin{aligned} \dot{X}_c &= A_c X_c + B_c \Theta \\ Y &= C_c X_c \end{aligned} \tag{6}$$

where

$$A_c = \left[\begin{array}{c|c} (I-BG_dC)^{-1}(A+BG_pC) & (I-BG_dC)^{-1}BG_i \\ \hline C & 0 \end{array}\right], \tag{7}$$

$$X_c = \left[\begin{array}{c} X \\ \hline Z \end{array}\right], \tag{8}$$

$$B_c = \left[\begin{array}{c} (I-BG_dC)^{-1}B \\ \hline 0 \end{array}\right], \qquad C_c = [C \mid 0]. \tag{9}$$

The dimensions of X_c, A_c, B_c, C_c are (n+q), (n+q)x(n+q), (n+q)xm, qx(n+q).

The problem of eigenvalue-generalized eigenvector assignment of the system represented by (1) via a proportional, integral and derivative output feedback law of the form (3) , where the integrator state is governed by (5) is one of determining the real matrices G_p, G_i and G_d such that the closed-loop system matrix A_c has a prescribed set of self-conjugate eigenvalues and a corresponding set of admissible self-conjugate generalized eigenvectors. The case of distinct eigenvalues can be considered as a special case of the problem treated in this paper.

3. EIGENVALUE-GENERALIZED EIGENVECTOR STRUCTURE

3.1. Eigenvalues not Equal Zero

Consider a particular set of self-conjugate eigenvalues of A_c, $\lambda_i \neq 0$, $i=1,2,\ldots\ldots,l$, $l \leq (n+q)$. Let the eigenvalue λ_i has a geometric multiplicity k_i and algebraic multiplicity m_i. The generalized eigenvectors $U^{i,\mu,\rho}$'s associated with λ_i must satisfy

$$[A_c-\lambda_i I]U^{i,\mu,\rho} = U^{i,\mu,\rho-1} \qquad (10)$$
$$\rho = 1,2,\ldots\ldots,m_{i,\mu}, \qquad \mu = 1,2,\ldots\ldots,k_i,$$
$$i = 1,2,\ldots\ldots,l,$$

with $U^{i,\mu,0}=0$, also $m_{i,\mu}$ must satisfy

$$m_i = \sum_{\mu=1}^{k_i} m_{i,\mu} . \qquad (11)$$

The dimension of A_c, namely $(n+q)$, is related to m_i as

$$(n+q) = \sum_{i=1}^{l} m_i . \qquad (12)$$

Let the generalized eigenvector $U^{i,\mu,\rho}$ be partitioned as

$$U^{i,\mu,\rho} = ((U_1^{i,\mu,\rho})',(\mathcal{U}^{i,\mu,\rho})')', \qquad (13)$$

where the ' denotes vector transpose. The dimensions of the subvectors $U_1^{i,\mu,\rho}$ and $\mathcal{U}^{i,\mu,\rho}$ are nx1 and qx1 respectively. using (7) and (10) we get

$$((I-BG_dC)^{-1}(A+BG_pC)-\lambda_i I)U_1^{i,\mu,\rho}+(I-BG_dC)^{-1}BG_i\mathcal{U}^{i,\mu,\rho}=U_1^{i,\mu,\rho-1} \qquad (14)$$
$$CU_1^{i,\mu,\rho}- \qquad \lambda_i\mathcal{U}^{i,\mu,\rho}=\mathcal{U}^{i,\mu,\rho-1} \qquad (15)$$

Using (15) we get

$$\mathcal{U}^{i,\mu,\rho} = (CU_1^{i,\mu,\rho} - \mathcal{U}^{i,\mu,\rho-1})/\lambda_i . \qquad (16)$$

From (16) it is obvious that $\mathcal{U}^{i,\mu,\rho}$ can be expressed in terms of $U_1^{i,\mu,\rho}$. Substituting (16) into (14) yields

$$(A-\lambda_i I)U_1^{i,\mu,\rho} + B\Big[G_pCU_1^{i,\mu,\rho}+G_i(CU_1^{i,\mu,\rho}-\mathcal{U}^{i,\mu,\rho-1})/\lambda_i+$$
$$+G_d(\lambda_iCU_1^{i,\mu,\rho}+CU_1^{i,\mu,\rho-1})\Big] = U_1^{i,\mu,\rho-1} \qquad (17)$$

or in a matrix form as

$$\left[A-\lambda_i I, B\right]\begin{bmatrix} U_1^{i,\mu,\rho} \\ W^{i,\mu,\rho} \end{bmatrix} = U_1^{i,\mu,\rho-1} \tag{18}$$

$$\rho = 1,2,\ldots\ldots,m_{i,\mu}, \qquad \mu = 1,2,\ldots\ldots,k_i, \qquad i = 1,2,\ldots\ldots,l,$$

where

$$W^{i,\mu,\rho} = G_p CU_1^{i,\mu,\rho} + G_i(CU_1^{i,\mu,\rho} - U^{i,\mu,\rho-1})/\lambda_i + + G_d(\lambda_i CU_1^{i,\mu,\rho} + CU_1^{i,\mu,\rho-1}) \ . \tag{19}$$

Let the eigenvalues λ_i, $i=1,2,\ldots\ldots,l$, form a complete set of self-conjugate eigenvalues, and the integers $m_{i,\mu}$ and k_i are chosen so that the entire set of vectors $U^{i,\mu,\rho}$'s not only satisfies (10), but also linearly independent and self-conjugate, then the real output feedback gain matrices G_p, G_i and G_d can be expressed using (18) and (19) as

$$\left[G_p, G_i, G_d\right]\begin{bmatrix} \ldots\ldots, & CU_1^{i,\mu,\rho} & ,\ldots\ldots \\ \ldots\ldots, & (CU_1^{i,\mu,\rho} - U^{i,\mu,\rho-1})/\lambda_i & ,\ldots\ldots \\ \ldots\ldots, & (\lambda_i CU_1^{i,\mu,\rho} + CU_1^{i,\mu,\rho-1}) & ,\ldots\ldots \end{bmatrix} = \left[\ldots\ldots, W^{i,\mu,\rho}, \ldots\ldots\right] \tag{20}$$

$$\rho = 1,2,\ldots\ldots,m_{i,\mu}, \qquad \mu = 1,2,\ldots\ldots,k_i, \qquad i = 1,2,\ldots\ldots,l,$$

or in a compact form as

$$\left[G_p, G_i, G_d\right] U = W \ , \tag{21}$$

where the matrix U of dimension 3qx(n+q). If the matrix U is invertible then we can determine G_p, G_i and G_d as follows

$$\left[G_p, G_i, G_d\right] = WU^{-1} \ . \tag{22}$$

Three cases are to be investigated, namely $(n+q)=3q$, $(n+q)<3q$ and $(n+q)>3q$. In the case $(n+q)=3q$ or $n=2q$, (22) yields a unique solution for G_p, G_i and G_d, if it exists. However if $(n+q)<3q$ or $n<2q$, then we require to impose some constraints on the elements of G_p, G_i and G_d, such as setting some elements, and/or sum of some elements, and/or row sum of some rows equal to zero. This can be achieved by introducing suitable $(2q-n)$ columns in U such that it becomes square and invertible, and corresponding suitable zero columns in W. Then employing (22), we can determine G_p, G_i and G_d. The third case where $(n+q)>3q$ or $n>2q$, the eigenvalue-generalized eigenvector assignment is not possible. However adjustment of the number of system outputs may correct this situation, or the degrees of freedom in choosing the admissible generalized eigenvectors (18), could be utilized to assign a prescribed set of eigenvalues to the closed-loop system matrix A_c. To complete the presentation, the case where some eigenvalues are zeros will be treated next.

3.2. Eigenvalue $\lambda_i=0$

Let λ_i be zero eigenvalue , in this situation (14) and (15) become

$$(I-BG_dC)^{-1}(A+BG_pC)U_1^{i,\mu,\rho} + (I-BG_dC)^{-1}BG_i\mathcal{U}^{i,\mu,\rho} = U_1^{i,\mu,\rho-1}, \quad (23)$$

$$CU_1^{i,\mu,\rho} = \mathcal{U}^{i,\mu,\rho-1}. \quad (24)$$

Let us assume without loss of generality that the output matrix C be expressed as $[I_q,0]$. Examination of (24) reveals that q elements of $U_1^{i,\mu,\rho}$ are known in terms of $\mathcal{U}^{i,\mu,\rho-1}$. The remaining (n-q) elements of $U_1^{i,\mu,\rho}$ along with $\mathcal{U}^{i,\mu,\rho}$ are to be specified. Using (23) and (24) it can be shown that

$$[A,B]\begin{bmatrix} U_1^{i,\mu,\rho} \\ W^{i,\mu,\rho} \end{bmatrix} = U_1^{i,\mu,\rho-1} \quad (25)$$

$$W^{i,\mu,\rho} = G_p\mathcal{U}^{i,\mu,\rho-1} + G_i\mathcal{U}^{i,\mu,\rho} + G_d\mathcal{U}^{i,\mu,\rho-2} \quad (26)$$

$$\rho = 1,2,\ldots\ldots,m_{i,\mu}, \qquad \mu = 1,2,\ldots\ldots,m_i .$$

Using (25) and (26), we can express G_p, G_i and G_d in a matrix form as

$$[G_p,G_i,G_d]\begin{bmatrix} \ldots\ldots,\mathcal{U}^{i,\mu,\rho-1},\ldots\ldots \\ \ldots\ldots,\mathcal{U}^{i,\mu,\rho},\ldots\ldots \\ \ldots\ldots,\mathcal{U}^{i,\mu,\rho-2},\ldots\ldots \end{bmatrix} = [\ldots\ldots,\ W^{i,\mu,\rho}\ ,\ldots\ldots]. \quad (27)$$

Repeating this procedure for all the zero eigenvalues and following the procedure discussed previously for the non-zero eigenvalues, we can determine the output feedback gain matrices G_p, G_i and G_d via (22).

3.3. Single-Input Single-Output System with Distinct Eigenvalues ($\neq 0$)

Consider a linear continuous single-input single-output system having the state variable form (1) and expressed in the canonical form

$$A = \begin{bmatrix} 0 & 1 & 0 & \cdots & 0 \\ 0 & 0 & 1 & \cdots & 0 \\ \cdots & \cdots & \cdots & \cdots & \cdots \\ 0 & 0 & 0 & \cdots & 1 \\ -a_n & -a_{n-1} & -a_{n-2} & \cdots & -a_1 \end{bmatrix}, \quad B = \begin{bmatrix} 0 \\ 0 \\ . \\ 0 \\ 1 \end{bmatrix},$$

$$C = [1 \quad 0 \quad 0 \cdots\cdots 0],$$

where in this case m=q=1. Utilizing the results obtained in Section 3.1., it can be shown that the output feedback scalar gains g_p, g_i and g_d may be expressed as

$$[g_p, g_i, g_d]\begin{bmatrix} 1 & 1 & \cdots & 1 \\ 1/\lambda_1 & 1/\lambda_2 & \cdots & 1/\lambda_{n+1} \\ \lambda_1 & \lambda_2 & \cdots & \lambda_{n+1} \end{bmatrix} = \left[\sum_{j=0}^{n} a_j \lambda_1^{n-j}, \cdots\cdots, \sum_{j=0}^{n} a_j \lambda_{n+1}^{n-j}\right].$$

From the last expression it is obvious that for a second order system the output feedback scalar gains g_p, g_i and g_d are uniquely determined. These gains will assign an arbitrary set of distinct eigenvalues to the closed-loop matrix A_c. For higher order systems, n>2, the eigenvalue spectrum can not be completely assigned, which is noticed in the example reported in [6].

4. NUMERICAL EXAMPLES

Two numerical examples will be considered to illustrate the proposed technique.

4.1. Example (1)

Consider a third order system of the form (1), where

$$A = \begin{bmatrix} 0 & 1 & 0 \\ 0 & 0 & 1 \\ 1 & 2 & 1 \end{bmatrix}, \quad B = \begin{bmatrix} 1 & 0 & 0 \\ 0 & 1 & 0 \\ 1 & 0 & 1 \end{bmatrix}, \quad C = \begin{bmatrix} 1 & 0 & 0 \\ 0 & 1 & 0 \end{bmatrix},$$

n=m=3 and q=2. Let us assume that we are interested in having a closed-loop eigenvalue spectrum of the following description : $\lambda_1=-1$ with geometric multiplicity $k_1=1$ and algebraic multiplicity $m_1=3$ and $\lambda_{2,3}= -1\pm j$ with geometric multiplicity $k_2=k_3=1$ and algebraic multiplicity $m_2=m_3=1$. The design technique objective is to determine the output feedback gain matrices G_p, G_i, G_d such that the closed-loop system matrix A_c has the eigenvalue spectrum just described and the corresponding set of admissible self-conjugate generalized eigenvectors :

$$U^{1,1,1} = \begin{bmatrix} 1 \\ 1 \\ 0 \\ -1 \\ -1 \end{bmatrix}, \quad U^{1,1,2} = \begin{bmatrix} -1 \\ 0 \\ 1 \\ 0 \\ -1 \end{bmatrix}, \quad U^{1,1,3} = \begin{bmatrix} 1 \\ 1 \\ -1 \\ -1 \\ -2 \end{bmatrix},$$

$$U^{2,1,1} = \begin{bmatrix} 1 \\ j \\ -j \\ -1/2-j/2 \\ 1/2-j/2 \end{bmatrix}, \quad U^{3,1,1} = \begin{bmatrix} 1 \\ -j \\ j \\ -1/2+j/2 \\ 1/2+j/2 \end{bmatrix}.$$

The eigenvalue spectrum does not contain zero eigenvalues, then we can apply the results reported in Section 3.1. . Since in this example $n<2q$, then we impose some constraints on the elements of G_p, G_i and G_d. Three constraints are introduced and the results obtained are:

i) first column of G_d is set to zero and the corresponding G_p, G_i and G_d matrices are,

$$G_p = \begin{bmatrix} -3 & -1 \\ 0 & 0 \\ 1 & -4 \end{bmatrix}, \quad G_i = \begin{bmatrix} -3 & 1 \\ 3 & -1 \\ 8 & -4 \end{bmatrix}, \quad G_d = \begin{bmatrix} 0 & 0 \\ 0 & -1 \\ 0 & -6 \end{bmatrix},$$

ii) row sum of G_p is set to zero and the corresponding G_p, G_i and G_d matrices are,

$$G_p = \begin{bmatrix} 1 & -1 \\ 0 & 0 \\ 4 & -4 \end{bmatrix}, \quad G_i = \begin{bmatrix} 1 & -1/3 \\ 3 & -1 \\ 11 & -5 \end{bmatrix}, \quad G_d = \begin{bmatrix} 4/3 & 0 \\ 0 & -1 \\ 1 & -6 \end{bmatrix},$$

iii) row sum of G_d is set to zero and the corresponding G_p, G_i and G_d matrices are,

$$G_p = \begin{bmatrix} -3 & -1 \\ 3 & 0 \\ 19 & -4 \end{bmatrix}, \quad G_i = \begin{bmatrix} -3 & 1 \\ 6 & -2 \\ 26 & -10 \end{bmatrix}, \quad G_d = \begin{bmatrix} 0 & 0 \\ 1 & -1 \\ 6 & -6 \end{bmatrix},$$

Substituting the numerical values of G_p, G_i and G_d and the system A,B,C matrices into A_c, it can be shown for the three constraints that

$$A_c = \left[\begin{array}{ccc:cc} -3 & 0 & 0 & -3 & 1 \\ 0 & 0 & 1/2 & 3/2 & -1/2 \\ -1 & -3 & -2 & -4 & 0 \\ \hdashline 1 & 0 & 0 & 0 & 0 \\ 0 & 1 & 0 & 0 & 0 \end{array}\right].$$

It can be easily verified that A_c has the given set of eigenvalues and the corresponding set of self-conjugate generalized eigenvectors. This will complete the design technique.

4.2. Example (2)

Consider the same system as given in Section 4.1., except the eigenvalue spectrum is changed to: $\lambda_1=-1$ with geometric multiplicity $k_1=1$ and algebraic multiplicity $m_1=3$ and $\lambda_2=0$ with geometric multiplicity $k_2=1$ and algebraic multiplicity $m_2=2$. The set of admissible generalized eigenvectors to be assigned is given as:

$$U^{1,1,1}=\begin{bmatrix}1\\1\\0\\-1\\-1\end{bmatrix},\quad U^{1,1,2}=\begin{bmatrix}-1\\0\\1\\0\\-1\end{bmatrix},\quad U^{1,1,3}=\begin{bmatrix}1\\1\\-1\\-1\\-2\end{bmatrix},$$

$$U^{2,1,1}=\begin{bmatrix}0\\0\\1\\1\\-1\end{bmatrix},\quad U^{2,1,2}=\begin{bmatrix}1\\-1\\1\\2\\1\end{bmatrix}.$$

Since the eigenvalue spectrum contains zero eigenvalue, then the results reported in Section 3.2. are applicable. Note in this example $CU_1^{2,1,1}=U^{2,1,0}=0$ and $CU_1^{2,1,2}=U^{2,1,1}$. Once again $n<2q$, then we impose some constraints on the elements of G_p, G_i and G_d. Three constraints are introduced and the results obtained are:

i) first column of G_d is set to zero and the corresponding G_p, G_i and G_d matrices are

$$G_p=\begin{bmatrix}1 & 3\\-1 & -5\\-3 & -17\end{bmatrix},\quad G_i=\begin{bmatrix}1 & 1\\-2 & -1\\-5 & -4\end{bmatrix},\quad G_d=\begin{bmatrix}0 & 4\\0 & -2\\0 & -10\end{bmatrix},$$

ii) row sum of G_p is set to zero and the corresponding G_p, G_i and G_d matrices are

$$G_p=\begin{bmatrix}0.2 & -0.2\\0.2 & -0.2\\1 & -1\end{bmatrix},\quad G_i=\begin{bmatrix}0.2 & 0.2\\-0.8 & 0.2\\-1 & 0\end{bmatrix},\quad G_d=\begin{bmatrix}0.8 & 0.8\\-1.2 & 2.8\\-4 & 6\end{bmatrix},$$

iii) row sum of G_d is set to zero and the corresponding G_p, G_i and G_d matrices are

$$G_p=\begin{bmatrix}-1/3 & -7/3\\-1/3 & -7/3\\1/2 & -11/3\end{bmatrix},\quad G_i=\begin{bmatrix}-1/3 & -1/3\\-4/3 & -1/3\\-5/3 & -2/3\end{bmatrix},\quad G_d=\begin{bmatrix}4/3 & -4/3\\-2/3 & 2/3\\-10/3 & 10/3\end{bmatrix}.$$

Substituting the numerical values of G_p, G_i and G_d and the system A,B,C matrices into A_c, it can be shown for the three constraints that

$$A_c = \left[\begin{array}{ccc:cc} -1/3 & -8/3 & 4/3 & -5/3 & -1/3 \\ -1/3 & -5/3 & 1/3 & -2/3 & -1/3 \\ 1 & -2 & -1 & 0 & -1 \\ \hdashline 1 & 0 & 0 & 0 & 0 \\ 0 & 1 & 0 & 0 & 0 \end{array}\right].$$

It can be easily verified that A_c has the given set of eigenvalues and the corresponding set of generalized eigenvectors. This will complete the design procedure.

PART. 2 - DISCRETE-TIME SYSTEM

1. INTRODUCTION

The problem of designing proportional-plus-integral-plus-derivative (PID) controllers for linear continuous multivariable systems has received considerable attention recently [1-7,9]. However a modest interest was devoted for the problem of designing PID controllers for linear discrete-time multivariable systems [11]. While the technique proposed in [11] is useful it is of some importance to investigate the problem of designing PID controllers for eigenvalue-eigenvector assignment in discrete-time systems.

In this paper a time-domain technique for designing PID controller for linear discrete-time multivariable system is proposed. The state of the plant is assumed to be available for measurement. The eigenstructure formulation is utilized to determine the controller feedback gains. The resulting controller assigns a prescribed set of self-conjugate eigenvalues and a corresponding set of admissible self-conjugate eigenvectors to the closed-loop system matrix.

2. PROBLEM STATEMENT

Consider the linear discrete-time, time-invariant multivariable system described in state space form as follows

$$\begin{aligned} X(k+1) &= AX(k) + BU(k) \\ k &= 0,1,2,\ldots\ldots\ldots\ldots \end{aligned} \tag{28}$$

where $X(k)$ is the nx1 state vector at sample k, $U(k)$ is the mx1 input vector and A,B are constant matrices of appropriate dimensions. Let us introduce a q-dimensional integrator state vector $\mathcal{X}(k)$ which satisfies

$$\begin{aligned} \mathcal{X}(k+1) &= \mathcal{X}(k) + TE_i(X(k)-X_d) \\ k &= 0,1,2,\ldots\ldots\ldots\ldots \end{aligned} \tag{29}$$

where E_i is the matrix composed of some q rows of nxn identity matrix, E_iX_d is a desired vector for $E_iX(k)$ and T is the sampling interval. The state feedback law is assumed to be of the form (see Figure 2)

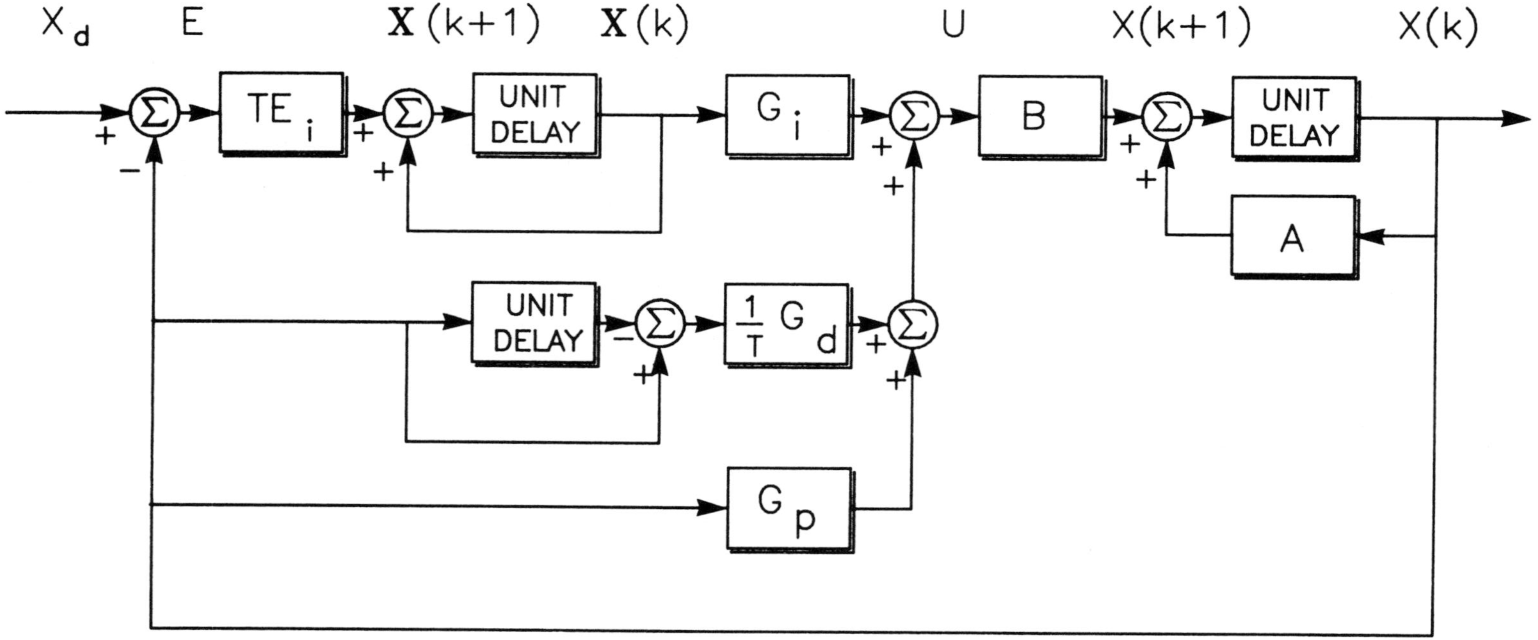

Figure 2 - A discrete-time multivariable system with PID controller

$$U(k) = G_p X(k) + G_i \mathcal{X}(k) + G_d (X(k)-X(k-1))/T, \tag{30}$$

where G_p, G_i and G_d are constant matrices of dimensions (mxn), (mxq) and (mxn) and representing the proportional, integral and derivative feedback gain matrices, respectively. Without loss of generality we can assume T is unity. Substitution of (30) in (28) yields the closed-loop system

$$X(k+1) = (A+BG_p+BG_d)X(k) - BG_d X(k-1) + BG_i \mathcal{X}(k). \tag{31}$$

Let us denote X(k-1) and X(k) by $Z_1(k)$ and $Z_2(k)$, respectively, then (29) and (31) can be written in state variable form as

$$Z_c(k+1) = A_c Z_c(k) + \Theta, \tag{32}$$

where

$$A_c = \begin{bmatrix} 0 & I & 0 \\ -BG_d & A+BG_p+BG_d & BG_i \\ 0 & E_i & I_q \end{bmatrix}, \qquad \Theta = \begin{bmatrix} 0 \\ 0 \\ -E_i X_d \end{bmatrix}, \tag{33}$$

$$Z_c = \begin{bmatrix} Z_1(k) \\ Z_2(k) \\ \mathcal{X}(k) \end{bmatrix}. \tag{34}$$

The closed-loop matrix A_c has a dimension (2n+q)x(2n+q).

The study of the eigenvalue assignment problem requires the establishment of state controllability. To this end the closed-loop matrix A_c in (33) can be rewritten as

$$A_c = \begin{bmatrix} 0 & I & 0 \\ 0 & A & 0 \\ 0 & E_i & I_q \end{bmatrix} + \begin{bmatrix} 0 \\ B \\ 0 \end{bmatrix} [-G_d, \; G_p+G_d, \; G_i] . \tag{35}$$

Using (35) and Cayley-Hamilton Theorem, the state controllability requires that the matrix

$$S = \begin{bmatrix} 0 & B & AB & \cdots\cdots & A^{n-1}B \\ B & AB & A^2B & \cdots\cdots & A^nB \\ 0 & E_iB & E_i(B+AB) & \cdots\cdots & E_i \sum_{i=0}^{n-2} A^iB \end{bmatrix}, \tag{36}$$

has a rank (2n+q). If this rank condition is not met, adjustment of the number of inputs and/or number of integrator inputs may correct this situation.

The problem of eigenvalue-generalized eigenvector assignment of the system represented by (28) via a proportional, integral and derivative feedback law of the form (30), where the integrator state is governed by (29) is one of determining the real matrices G_p, G_i and G_d such that the closed-loop system matrix A_c has a prescribed set of self-conjugate eigenvalues and a corresponding set of admissible self-conjugate generalized eigenvectors.

3. EIGENVALUE-GENERALIZED EIGENVECTOR ASSIGNMENT

3.1. Eigenvalue not Equal Zero or One

Consider a particular set of self-conjugate eigenvalues of A_c, $\lambda_i \neq 0$ or 1, $i=1,2,\ldots\ldots,l$, $l \leq (2n+q)$. Let the eigenvalue λ_i has a geometric multiplicity k_i and algebraic multiplicity m_i. The generalized eigenvectors $U^{i,\mu,\rho}$'s associated with λ_i must satisfy

$$[A_c - \lambda_i I]U^{i,\mu,\rho} = U^{i,\mu,\rho-1}$$
$$\rho = 1,2,\ldots\ldots,m_{i,\mu}, \qquad \mu = 1,2,\ldots\ldots,k_i, \tag{37}$$
$$i = 1,2,\ldots\ldots,l,$$

with $U^{i,\mu,0} = 0$, also $m_{i,\mu}$ must satisfy

$$m_i = \sum_{\mu=1}^{k_i} m_{i,\mu} . \tag{38}$$

The dimension of A_c, namely $(2n+q)$, is related to m_i as

$$(2n+q) = \sum_{i=1}^{l} m_i . \tag{39}$$

Let the generalized eigenvector $U^{i,\mu,\rho}$ be partitioned as

$$U^{i,\mu,\rho} = ((U_1^{i,\mu,\rho})',(U_2^{i,\mu,\rho})',(\mathcal{U}^{i,\mu,\rho})')', \tag{40}$$

where the ' denotes vector transpose. The dimensions of the subvectors $U_1^{i,\mu,\rho}$, $U_2^{i,\mu,\rho}$ and $\mathcal{U}^{i,\mu,\rho}$ are nx1, nx1 and qx1 respectively. Using (33) and (37) we get

$$-\lambda_i U_1^{i,\mu,\rho} + U_2^{i,\mu,\rho} = U_1^{i,\mu,\rho-1}, \tag{41}$$

$$-BG_d U_1^{i,\mu,\rho} + ((A-\lambda_i I)+B(G_p+G_d))U_2^{i,\mu,\rho} + BG_i \mathcal{U}^{i,\mu,\rho} = U_2^{i,\mu,\rho-1}, \tag{42}$$

$$E_i U_2^{i,\mu,\rho} + (1-\lambda_i)\mathcal{U}^{i,\mu,\rho} = \mathcal{U}^{i,\mu,\rho-1}. \tag{43}$$

From (41) and (43) we get , respectively,

$$U_1^{i,\mu,\rho} = (U_2^{i,\mu,\rho} - U_1^{i,\mu,\rho-1})/\lambda_i \ , \tag{44}$$

$$\mathcal{U}^{i,\mu,\rho} = (\mathcal{U}^{i,\mu,\rho-1} - E_i U_2^{i,\mu,\rho})/(1-\lambda_i) \ . \tag{45}$$

From (44) and (45) it is obvious that $U_1^{i,\mu,\rho}$ and $\mathcal{U}^{i,\mu,\rho}$ can be expressed in terms of $U_2^{i,\mu,\rho}$. Then (42) can be written in matrix form as

$$\begin{bmatrix} A-\lambda_i I, B \end{bmatrix} \begin{bmatrix} U_2^{i,\mu,\rho} \\ W^{i,\mu,\rho} \end{bmatrix} = U_2^{i,\mu,\rho-1} \ , \tag{46}$$

$$\rho = 1,2,\ldots\ldots,m_{i,\mu}, \qquad \mu = 1,2,\ldots\ldots,k_i,$$
$$i = 1,2,\ldots\ldots,l \ ,$$

where

$$W^{i,\mu,\rho} = G_p U_2^{i,\mu,\rho} + G_i \mathcal{U}^{i,\mu,\rho} + G_d (U_2^{i,\mu,\rho} - U_1^{i,\mu,\rho}) \ . \tag{47}$$

In the case where the eigenvalues λ_i, $i=1,2,\ldots\ldots,l$, form a complete set of self-conjugate eigenvalues and the integers $m_{i,\mu}$ and k_i are chosen so that the entire set of vectors $U^{i,\mu,\rho}$'s not only satisfies (41-43), but also linearly independent and self-conjugate, then the real feedback gain matrices can be expressed using (46) and (47) as

$$\begin{bmatrix} G_p, G_i, G_d \end{bmatrix} \begin{bmatrix} \ldots\ldots\ldots, & U_2^{i,\mu,\rho} & ,\ldots\ldots\ldots \\ \ldots\ldots\ldots, & \mathcal{U}^{i,\mu,\rho} & ,\ldots\ldots\ldots \\ \ldots\ldots\ldots, & (U_2^{i,\mu,\rho} - U_1^{i,\mu,\rho}) & ,\ldots\ldots\ldots \end{bmatrix}$$
$$= \begin{bmatrix} \ldots\ldots\ldots, W^{i,\mu,\rho}, \ldots\ldots\ldots \end{bmatrix} , \tag{48}$$

$$\rho = 1,2,\ldots\ldots,m_{i,\mu}, \qquad \mu = 1,2,\ldots\ldots,k_i,$$
$$i = 1,2,\ldots\ldots,l \ ,$$

or in a compact form as

$$\begin{bmatrix} G_p, G_i, G_d \end{bmatrix} U = W \ . \tag{49}$$

The matrix U is square with dimension (2n+q), if U is invertible then

$$\begin{bmatrix} G_p, G_i, G_d \end{bmatrix} = WU^{-1} \ . \tag{50}$$

The determination of the feedback gain matrices G_p, G_i and G_d of the PID controller using (50) will complete the design procedure. In the previous analysis, it was assumed that all the eigenvalues not equal zero, however the case where all the eigenvalues are identically zero has practical application (deadbeat response). Therefore the case where an eigenvalue is zero will be discussed in the next section.

3.2. Eigenvalue λ_i Equal Zero

Let λ_i be zero eigenvalue, then (41-43) reduce to

$$U_2^{i,\mu,\rho} = U_1^{i,\mu,\rho-1}, \tag{51}$$

$$-BG_d U_1^{i,\mu,\rho} + (A+B(G_p+G_d))U_2^{i,\mu,\rho} + BG_i \mathcal{U}^{i,\mu,\rho} = U_2^{i,\mu,\rho-1}, \tag{52}$$

$$E_i U_2^{i,\mu,\rho} + \mathcal{U}^{i,\mu,\rho} = \mathcal{U}^{i,\mu,\rho-1}. \tag{53}$$

From (51) and (53), we get

$$\mathcal{U}^{i,\mu,\rho} = \mathcal{U}^{i,\mu,\rho-1} - E_i U_1^{i,\mu,\rho-1}. \tag{54}$$

Examination of (51) and (54) indicates that $U_2^{i,\mu,\rho}$ and $\mathcal{U}^{i,\mu,\rho}$ can be expressed in terms of $U_1^{i,\mu,\rho-1}$. Using (51) and substituting (54) into (52) we obtain

$$AU_1^{i,\mu,\rho-1} + B\Big[G_p U_1^{i,\mu,\rho-1} + G_i(\mathcal{U}^{i,\mu,\rho-1} - E_i U_1^{i,\mu,\rho-1}) + G_d(U_1^{i,\mu,\rho-1} - U_1^{i,\mu,\rho})\Big] = U_1^{i,\mu,\rho-2}, \tag{55}$$

$$\rho = 1,2,\ldots\ldots,m_{i,\mu}, \qquad \mu = 1,2,\ldots\ldots,k_i,$$

or in a matrix form

$$[A,B]\begin{bmatrix} U_1^{i,\mu,\rho-1} \\ W^{i,\mu,\rho-1} \end{bmatrix} = U_1^{i,\mu,\rho-2}, \tag{56}$$

$$\rho = 2,3,\ldots\ldots,m_{i,\mu}+1, \qquad \mu = 1,2,\ldots\ldots,k_i,$$

where

$$W^{i,\mu,\rho-1} = G_p U_1^{i,\mu,\rho-1} + G_i(\mathcal{U}^{i,\mu,\rho-1} - E_i U_1^{i,\mu,\rho-1}) + G_d(U_1^{i,\mu,\rho-1} - U_1^{i,\mu,\rho}). \tag{57}$$

For $\rho=1$, (56) reduces to

$$-G_d U_1^{i,\mu,\rho} = 0. \tag{58}$$

Using (56) and (57), the real feedback gain matrices can be expressed as

$$[G_p,G_i,G_d]\begin{bmatrix} \ldots, & 0 & , & U_1^{i,\mu,1} & , & U_1^{i,\mu,2} & ,\ldots \\ \ldots, & 0 & , & \mathcal{U}^{i,\mu,1}-E_i U_1^{i,\mu,1} & , & \mathcal{U}^{i,\mu,2}-E_i U_1^{i,\mu,2} & ,\ldots \\ \ldots, & -U_1^{i,\mu,1} & , & U_1^{i,\mu,1}-U_1^{i,\mu,2} & , & U_1^{i,\mu,2}-U_1^{i,\mu,3} & ,\ldots \end{bmatrix}$$

$$= \left[\ldots, \quad 0 \quad , \quad W^{i,\mu,1} \quad , \quad W^{i,\mu,2} \quad , \ldots \right] . \qquad (59)$$

In the case where all the eigenvalues are identically zero, then from (58) and (59), it is clear that G_d is identically zero i.e. the resulting controller is a proportional-plus-integral (PI) controller.

3.3. Eigenvalue $\lambda_i=1$

Let λ_i be unity eigenvalue, then (41-43) reduce to

$$-U_1^{i,\mu,\rho} + U_2^{i,\mu,\rho} = U_1^{i,\mu,\rho-1} , \qquad (60)$$

$$-BG_d U_1^{i,\mu,\rho} + ((A-I)+B(G_p+G_d))U_2^{i,\mu,\rho} + BG_i \mathcal{U}^{i,\mu,\rho} = U_2^{i,\mu,\rho-1} , \qquad (61)$$

$$E_i U_2^{i,\mu,\rho} = \mathcal{U}^{i,\mu,\rho-1} . \qquad (62)$$

We can assume without loss of generality that the matrix E_i be expressed as $[I_q,0]$. Then from (62), it is obvious that q elements of $U_2^{i,\mu,\rho}$ are known in terms of $\mathcal{U}^{i,\mu,\rho-1}$. The remaining (n-q) elements of $U_2^{i,\mu,\rho}$ and $\mathcal{U}^{i,\mu,\rho}$ are to be assigned. From (60) we get

$$U_1^{i,\mu,\rho} = U_2^{i,\mu,\rho} - U_1^{i,\mu,\rho-1} , \qquad (63)$$

which indicates that $U_1^{i,\mu,\rho}$ can be expressed in terms of $U_2^{i,\mu,\rho}$. Equation (61) can be written in a matrix form as

$$[A-I,B]\begin{bmatrix} U_2^{i,\mu,\rho} \\ W^{i,\mu,\rho} \end{bmatrix} = U_2^{i,\mu,\rho-1} \qquad (64)$$

$$\rho = 1,2,\ldots\ldots,m_{i,\mu}, \qquad \mu = 1,2,\ldots\ldots,k_i,$$

where

$$W^{i,\mu,\rho} = G_p U_2^{i,\mu,\rho} + G_i \mathcal{U}^{i,\mu,\rho} + G_d U_1^{i,\mu,\rho-1} . \qquad (65)$$

Using (64) and (65), we can express the real feedback gain matrices as

$$[G_p,G_i,G_d]\begin{bmatrix} \ldots\ldots, & U_2^{i,\mu,\rho} & ,\ldots\ldots \\ \ldots\ldots, & \mathcal{U}^{i,\mu,\rho} & ,\ldots\ldots \\ \ldots\ldots, & U_1^{i,\mu,\rho-1} & ,\ldots\ldots \end{bmatrix} = [\ldots\ldots, W^{i,\mu,\rho}, \ldots\ldots] . \qquad (66)$$

For a general eigenvalue spectrum, we use the procedure discussed in Section 3.1. for the non-zero non-unity eigenvalues, the procedure discussed in Section 3.2. for the zero eigenvalues and the present procedure for the unity eigenvalues, then employing (50), we can determine the PID controller gain matrices G_p, G_i and G_d.

4. ILLUSTRATIVE EXAMPLES

Two numerical examples will be considered to illustrate the proposed technique.

4.1. Example (1)

Consider a third order system of the form (28) where

$$A = \begin{bmatrix} 1 & 3 & 2 \\ 0 & 1 & 2 \\ 0 & 0 & 1 \end{bmatrix}, \qquad B = \begin{bmatrix} 1 & 0 & 0 \\ 0 & 1 & 0 \\ 0 & 1 & 1 \end{bmatrix},$$

and three choices of E_i will be treated:

$$\text{(a) } E_i = \begin{bmatrix} 1 & 0 & 0 \end{bmatrix}, \quad \text{(b) } E_i = \begin{bmatrix} 1 & 0 & 0 \\ 0 & 0 & 1 \end{bmatrix}, \quad \text{(c) } E_i = I_3.$$

Using (36), it can be shown that the given system is state controllable.

The design objective is to determine the PID controller gain matrices such that the closed-loop system matrix A_c has the eigenvalue spectrum whose description :

$\lambda_1=-0.1$, $k_1=1$, $m_1=2$, $\lambda_2=0.1$, $k_2=1$, $m_2=2$,

$\lambda_3=-0.2$, $k_3=1$, $m_3=2$,

case (a) $\lambda_4=0.2$, $k_4=1$, $m_4=1$,

case (b) $\lambda_4=0.2$, $k4=1$, $m_4=2$,

case (c) $\lambda_4=0.2$, $k_4=1$, $m_4=2$,

$\lambda_5=0.5$, $k_5=1$, $m_5=1$,

and the corresponding set of admissible generalized eigenvector subvectors $U_2^{i,\mu,\rho}$'s from :

$$U_2^{1,1,1}=\begin{bmatrix} 1 \\ -1 \\ 1 \end{bmatrix}, \quad U_2^{1,1,2}=\begin{bmatrix} 0 \\ 1 \\ 1 \end{bmatrix}, \quad U_2^{2,1,1}=\begin{bmatrix} 1 \\ 0 \\ 1 \end{bmatrix}, \quad U_2^{2,1,2}=\begin{bmatrix} 1 \\ 1 \\ 0 \end{bmatrix},$$

$$U_2^{3,1,1}=\begin{bmatrix} -1 \\ 1 \\ 1 \end{bmatrix}, \quad U_2^{3,1,2}=\begin{bmatrix} 1 \\ 0 \\ 0 \end{bmatrix}, \quad U_2^{4,1,1}=\begin{bmatrix} 0 \\ 0 \\ 1 \end{bmatrix}, \quad U_2^{4,1,2}=\begin{bmatrix} 1 \\ 0 \\ 1 \end{bmatrix},$$

$$U_2^{5,1,1}=\begin{bmatrix} -1 \\ 1 \\ 0 \end{bmatrix}.$$

Note that

case (a) $q=1$, $i=1,2,3,4$, $m_{1,1}=m_{2,1}=m_{3,1}=2$, $m_{4,1}=1$,

case (b) $q=2$, $i=1,2,3,4$, $m_{1,1}=m_{2,1}=m_{3,1}=m_{4,1}=2$,

case (c) $q=3$, $i=1,2,3,4$, $m_{1,1}=m_{2,1}=m_{3,1}=m_{4,1}=2$, $m_{5,1}=1$.

Since the eigenvalue spectrum does not contain zero or unity eigenvalues, then the results obtained in Section 3.1. can be applied. Using (46) and (48), the resulting feedback gain matrices of the PID controllers for cases (a), (b) and (c) are :

$$[G_p, G_i, G_d] = \left[\begin{array}{ccc|c|ccc} -2.307 & -3.077 & -2.000 & -1.227 & -0.005 & -0.006 & 0.000 \\ 0.825 & -0.968 & -2.001 & 0.656 & 0.011 & 0.019 & 0.000 \\ 0.042 & 1.048 & 1.112 & -0.065 & 0.036 & 0.047 & -0.022 \end{array}\right],$$

$$[G_p, G_i, G_d] = \left[\begin{array}{ccc|cc|ccc} -2.300 & -3.080 & -1.976 & -1.223 & 0.021 & -0.004 & -0.005 & 0.000 \\ 0.818 & -0.966 & -2.026 & 0.653 & -0.021 & 0.010 & 0.018 & 0.000 \\ -0.215 & 1.123 & 0.294 & -0.180 & -0.710 & 0.001 & -0.007 & -0.005 \end{array}\right],$$

$$[G_p, G_i, G_d] = \left[\begin{array}{ccc|ccc|ccc} -2.802 & -4.029 & -1.962 & -1.614 & -0.974 & 0.032 & -0.011 & -0.019 & 0.000 \\ 0.827 & -0.948 & -2.026 & 0.660 & 0.019 & -0.021 & 0.010 & 0.018 & 0.000 \\ 0.121 & 1.759 & 0.285 & 0.082 & 0.653 & -0.717 & 0.006 & 0.003 & -0.005 \end{array}\right],$$

respectively. It can be shown that the resulting feedback gain matrices assign the desired eigenvalue spectrum and the set of admissible generalized eigenvectors to the closed-loop system matrix A_c. This will complete the design technique.

4.2. Example (2)

Consider the same system as given in Section 4.1., except the choices of E_i are:

$$\text{(a) } E_i = [0 \quad 0 \quad 1], \quad \text{(b) } E_i = \begin{bmatrix} 0 & 1 & 0 \\ 0 & 0 & 1 \end{bmatrix}, \quad \text{(c) } E_i = I_3 .$$

The objective is to design a PID controller such that the closed-loop system matrix A_c has the eigenvalue spectrum of the following description:

$$\begin{array}{lll} \lambda_1 = 0, & k_1 = 3, & \\ m_1 = 7 & & \text{case (a)}, \\ m_1 = 8 & & \text{case (b)}, \\ m_1 = 9 & & \text{case (c)}, \end{array}$$

note $k_1=3$ the same as the dimension of the null space in (56) with $\rho=2$.

The corresponding set of admissible generalized eigenvector subvectors $U_1^{i,\mu,\rho-1}$'s to be assigned is from :

$$U_1^{1,1,1} = \begin{bmatrix} 1 \\ 1 \\ -1 \end{bmatrix}, \quad U_1^{1,1,2} = \begin{bmatrix} 0 \\ 1 \\ 1 \end{bmatrix}, \quad U_1^{1,1,3} = \begin{bmatrix} 1 \\ 0 \\ 1 \end{bmatrix},$$

$$U_1^{1,2,1} = \begin{bmatrix} 0 \\ 1 \\ 1 \end{bmatrix}, \quad U_1^{1,2,2} = \begin{bmatrix} 1 \\ -1 \\ 0 \end{bmatrix}, \quad U_1^{1,2,3} = \begin{bmatrix} 1 \\ 0 \\ -1 \end{bmatrix},$$

$$U_1^{1,3,1} = \begin{bmatrix} 0 \\ 0 \\ 1 \end{bmatrix}, \quad U_1^{1,3,2} = \begin{bmatrix} 1 \\ 1 \\ 0 \end{bmatrix}, \quad U_1^{1,3,3} \begin{bmatrix} 1 \\ -1 \\ 1 \end{bmatrix}.$$

Note that

case (a) $q=1$, $i=1$, $\mu=1,2,3$, $m_{1,1}=m_{1,2}=3$ and $m_{1,3}=1$,
case (b) $q=2$, $i=1$, $\mu=1,2,3$, $m_{1,1}=m_{1,2}=3$ and $m_{1,3}=2$,
case (c) $q=3$, $i=1$, $\mu=1,2,3$, $m_{1,1}=m_{1,2}=m_{1,3}=3$.

Since the eigenvalues are zeros, then the results obtained in Section 3.2. can be applied. Using (56) and (59), the resulting feedback gain matrices of the PID controllers for cases (a), (b) and (c) are :

$$[G_p,G_i,G_d]=\left[\begin{array}{ccc|c|ccc} -1/3 & -10/3 & -2/3 & 1 & 0 & 0 & 0 \\ 4/3 & -5/3 & -1/3 & 1 & 0 & 0 & 0 \\ -1 & 5 & -1 & -3 & 0 & 0 & 0 \end{array}\right],$$

$$[G_p,G_i,G_d]=\left[\begin{array}{ccc|cc|ccc} -1 & -7/2 & -3/2 & -1/2 & 1/2 & 0 & 0 & 0 \\ 0 & -2 & -2 & -1 & 0 & 0 & 0 & 0 \\ 1 & 11/2 & 3/2 & 3/2 & -3/2 & 0 & 0 & 0 \end{array}\right],$$

$$[G_p,G_i,G_d]=\left[\begin{array}{ccc|ccc|ccc} -2 & -3 & -2 & -1 & 0 & 0 & 0 & 0 & 0 \\ 0 & -2 & -2 & 0 & -1 & 0 & 0 & 0 & 0 \\ 2 & 5 & 2 & 1 & 1 & -1 & 0 & 0 & 0 \end{array}\right],$$

respectively. Substituting the numerical values of G_p, G_i and G_d and the system A,B matrices into the expression of A_c, it can be shown that the closed-loop system matrix A_c has zero eigenvalues and the closed-loop response reaches the desired state in two samples. Note for deadbeat response (i.e. assigned eigenvalues are identically zero), the resulting controller is a PI controller. This will complete the design procedure.

CONCLUSIONS

In this paper, time-domain design techniques for PID controllers have been presented. The analysis for continuous-time and discrete-time multivariable systems are considered. In the continuous case the design technique utilizes system output, while in the discrete-time case the design technique utilizes system state. The resulting controller assigns a prescribed set of self-conjugate eigenvalues and a corresponding set of admissible self-conjugate eigenvectors to the closed-loop system matrix. The proposed techniques appear to be flexible, straightforward and computationally simpler over known techniques since they do not impose any structural constraints on the PID controller feedback gain matrices of unity rank [3,7,9], and the computation of the open-loop transfer function matrix is avoided [7]. It has been found in the case of a discrete-time system where the assigned eigenvalues are identically zero, the resulting controller is a PI controller. The flexibility in assigning the generalized eigenvectors could be utilized in minimizing eigenvalue sensitivities, quadratic performance index or any other criterion.

The eigenstructure appraoch adopted in this paper is similar to the eigenstructure approach utilized in the eigenvalue-generalized eigenvector assignment using state feedback [12,13], except that it is extended to the case of PID controller.

ACKNOWLEDGEMENT

The author would like to acknowledge the support of Canada Centre for Mineral and Energy Technology (CANMET), Department of Energy, Mines and Resources, Canada , and for the permission to publish this contribution.

REFERENCES

1. E. J. Davison and H. W. Smith, ' Pole assignment in linear time-invariant multivariable systems with constant disturbances ', Automatica, 7, pp. 489-498, 1971.

2. E. J. Davison, ' The output control of linear time-invariant multivariable systems with unmeasurable arbitrary disturbances ', IEEE Trans. Automat. Contr., 17, pp. 621-629, 1972.

3. P. C. Young and J. C. Willems, ' An approach to the linear multivariable servomechanism problem ', Int. J. Control, 15, pp. 961-979, 1972.

4. B. Porter, ' Synthesis of asymptotically stable linear time-invariant closed-loop systems incorporating multivariable 3-term controllers ', Electron. Lett., 5, pp. 557-558, 1969.

5. A. Simpson, ' Synthesis of asymptotically stable linear time-invariant closed-loop systems incorporating multivariable 3-term controllers ', Electron. Lett., 6, pp. 251-252, 1970.

6. P. N. Paraskevopoulos, ' On the design of PID output feedback controllers for linear multivariable systems ', IEEE Trans. Indust. Electr. and Contr., 27, pp. 16-18, 1980.

7. H. Seraji and M. Tarokh, ' Design of PID controllers for multivariable servomechanism problem ', Int. J. Control, 26, pp. 75-83, 1977.

8. M. J. Bosley, H. W. Kropholler, F. P. less and R. M. Neale, ' The determination of transfer functions from state variable models ', Automatica, 8, pp. 213-216, 1972.

9. V. Gourishankar and K. Ramar, ' Design of proportional-integral-derivative controllers for tracking polynomial-type inputs ', Proc. IEE, 119, pp. 911-914, 1972.

10. A. I. A. Salama and V. Gourishankar, ' Eigenvalue-eigenvector assignment using PI controller ', Proceedings of the MECO '81, Cairo-Egypt, 1981.

11. H. Seraji, ' Design of digital two-term and three-term controllers for discrete-time multivariable systems ', Int. J. Control, 38, pp. 843-865, 1983.

12. G. Klein and B. C. Moore, ' Eigenvalue-generalized eigenvector assignment with state feedback ', IEEE Trans. Automat. Contr., 22, pp. 140-141, 1977.

13. B. porter and J. J. D'Azzo, ' Closed-loop eigenstructure assignment by state feedback in multivariable linear systems ', Int. J. Control, 27, pp. 487-492, 1978.

SUBJECT INDEX